Vector Analysis for Engineers and Scientists

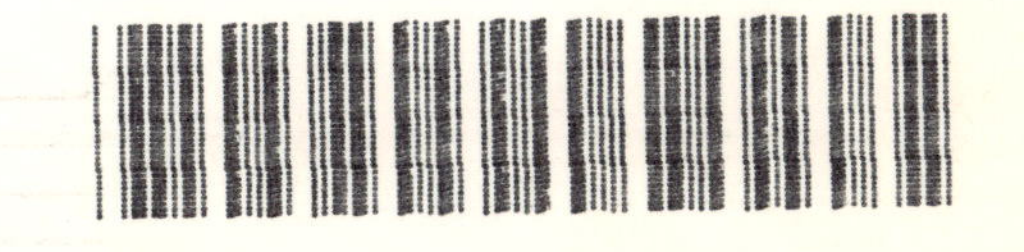

MODERN APPLICATIONS OF MATHEMATICS

Vector Analysis for Engineers and Scientists

P. E. Lewis and J. P. Ward
Loughborough University of Technology

ADDISON-WESLEY PUBLISHING COMPANY

Wokingham, England · Reading, Massachusetts · Menlo Park, California
New York · Don Mills, Ontario · Amsterdam · Bonn
Sydney · Singapore · Tokyo · Madrid · San Juan

Cover designed by Crayon Design of Henley-on-Thames and
printed by The Riverside Printing Co. (Reading) Ltd.
Artwork by Key Graphics, Aldermaston, Berks.
Filmset by Mid-County Press, London SW15.
Printed and bound in Great Britain by T.J. Press (Padstow), Cornwall.

First printed 1989.

British Library Cataloguing in Publication Data
Lewis, P. E. (Peter Edward)
Vector analysis for engineers and scientists.
1. Vector analysis
I. Title II. Ward, J. P. (Joseph Patrick)
515.'63
ISBN 0–201–17577–0

Library of Congress Cataloging in Publication Data
Lewis, P. E.
Vector analysis for engineers and scientists / P. E. Lewis and J. P. Ward.
p. cm.
Includes bibliographical references.
ISBN 0–201–17577–0
1. Vector analysis. I. Ward, J. P. II. Title.
QA433.L44 1989 89–17527
515′.63–dc20 CIP

To my mother and to the memory of my father
P. E. Lewis

To the memory of my parents Bridget Bridie and Michael Francis Ward
J. P. Ward

QUEEN MARY
UNIVERSITY OF LONDON
LIBRARY

Preface

The methods of vector algebra and of vector calculus are a most useful part of the mathematical knowledge of engineers and physicists, bearing in mind the applications of these methods to such areas as electromagnetism, solid mechanics and fluid mechanics. It is therefore our overall aim in writing this book to help undergraduates and others to acquire a sound understanding of vector methods and to be able to apply them in their particular areas.

The writers are both experienced teachers of mathematics to engineering and science students and are aware that the subject of vector analysis is considered difficult by many of them. Accordingly, although the text is mathematical in nature, we have not used an over-rigorous approach which we feel often confuses and even alienates students whose major area of study is not mathematics. Rather, our approach has been to adopt a fairly leisurely style of writing with detailed but informal explanations, plenty of worked examples, emphasis on applications and many more diagrams than are generally included in texts of this type. It is hoped that this style will also be appropriate to practising engineers and scientists whose mathematics may be rusty but who encounter vector field concepts.

Routine sets of exercises are interspersed at appropriate points in the text and the serious reader is urged to work through these before proceeding further in his or her reading. At the end of each chapter are to be found rather longer 'additional' exercises and full worked solutions to *all* of these are provided at the end of the book. Chapter summaries are also provided detailing major results. A number of chapters are followed by supplementary material for the more advanced (or the more interested) reader.

The order in which topics are treated is largely determined by the subject itself and no particular originality is claimed in this respect.

Chapter 1 introduces vectors and basic vector algebra including scalar and vector products and triple products. Simple applications in geometry and mechanics are covered and there is also an introduction to the vector analysis of stress which, although it could be omitted on a first reading, is not in fact over-taxing. The supplementary material to this chapter deals with the index (or tensor) notation for vectors and we present here some of the basic results of vector algebra in index form. The Kronecker delta and the permutation symbol are introduced and the topics of symmetry, skew-symmetry and contraction are discussed. The material in this supplement and in later chapters could form an introduction to a more advanced course in vector analysis or in areas of application such as solid and fluid mechanics and electromagnetism.

In Chapter 2 we consider a vector approach to the treatment of curves and surfaces. We define the derivative of a vector and apply the concept to velocity and acceleration with illustrative examples in ballistics and curvature problems. There are two supplements to this chapter, one on differential geometry and a more applicable section on impulsive forces.

Chapter 3 begins with a discussion of scalar fields and the important concept of the directional derivative. This leads to the idea of the gradient of a scalar field which is followed by an optional section on the vector analysis of strain which is used as an application of the gradient operator. This chapter also introduces vector fields and their divergence and curl in Cartesian coordinates. Properties of gradient, divergence and curl are examined and applications to gravitation and solid mechanics briefly mentioned. Index notation treatment of these concepts is covered in a supplement and there is a second supplement on the use of harmonic functions.

In Chapter 4 we cover integrals involving vectors, specifically line, surface and volume (or triple) integrals. Non-Cartesian coordinate systems, particularly cylindrical polar and spherical polar, are used where appropriate to evaluate surface and volume integrals, as is the projection method for surface integrals. This chapter contains numerous worked examples illustrating many different types of problem and the applicability of the subject to physics and engineering. There is also a supplementary section on the use of surface integrals in elasticity theory.

In Chapter 5 we study fundamental theorems linking the integrals, namely the divergence theorem and Stokes' theorem. These are used to give us more general definitions of divergence and curl, and applications of the theorems to continuity equations, heat transfer, strain energy and conservative vector fields are also covered.

Chapter 6 examines the extension of vector analysis to non-Cartesian coordinate systems and essentially completes the basic theory of the subject.

In Chapter 7 we use the full power of vector field theory in analysing electric and magnetic fields and waves. The approach should be useful as an illustration of vector field concepts even to readers who are not knowledgeable in this specific area.

Chapter 8, which is novel in a book of this kind, illustrates the use of vector methods in describing rotations (finite and infinitesimal) and the angular velocity concept. We also examine the mechanical effects expected when one observer is rotating with respect to another, as well as the Coriolis force.

There are three appendices. Two are brief background discussions on determinants and double integrals respectively. The third is a glimpse at the fascinating historical development of the subject of vector analysis.

P. E. Lewis
J. P. Ward
Loughborough, September, 1989

Contents

1

Elements of Vector Algebra

PREVIEW

In this chapter we introduce the idea of a vector quantity and develop the basic algebra of vectors. We show that vectors may be added and subtracted in a manner similar to that used for ordinary numbers (scalars). There is no concept of vector division, however, although there are two kinds of 'vector multiplication' – the scalar product and the vector product. The algebra of vectors is fully developed and applied to simple problems in geometry and mechanics to illustrate the basic concepts and to prepare the groundwork for later chapters.

1.1 Introduction

One of the main aims of physical scientists is to make sense of the world around them. There is little doubt that one of their greatest achievements (which has taken place over many centuries) has been to show that a large collection of apparently disparate entities, such as volume, mass, temperature and electric charge, are all related, in a mathematical sense, in that each may be characterized by a single entity – a **scalar** (a number). We have come to accept that this 'relationship' is obvious, but a child would have to *learn* that a quantity of water in a tall narrow glass has the same volume when poured into a short wide glass and that two volumes simply 'add up' as two numbers add up. Similarly, the mass of a lump of modelling clay is not dependent on its shape but can indeed be characterized by a single number.

It is now accepted that these so-called scalar quantities satisfy the basic rules of algebra (perhaps more properly called scalar algebra) that govern the way in which ordinary numbers may be combined. Indeed, it is this inherent characterization that allows us to replace the physical entity by an algebraic symbol.

A more recent achievement (in the nineteenth century) is the discovery that other apparently disparate entities, such as force, velocity and displacement, can be described mathematically using the single concept of a **vector**†. A vector is defined as a quantity which is characterized by both a **magnitude** and a **direction**, and is such that two vectors combine ('add up') according to a rule known as the **parallelogram law**.

Before formally developing the algebra of vectors, we shall briefly show that the claims made above about the vector nature of displacements, forces, etc. are indeed valid.

The directed line segment

A directed line segment (or displacement) from a point A to a point B (see Figure 1.1a) is represented by a straight line segment with an arrow to indicate the direction. We shall refer to it as $\overrightarrow{AB}$.

It is an obvious conclusion that if we move from point A to point B along the directed line segment $\overrightarrow{AB}$ and then from point B to point C along the directed line segment $\overrightarrow{BC}$, then the same displacement is achieved as if we move directly from point A to point C along the directed line segment $\overrightarrow{AC}$

†Vectors are a special case of more general objects called **quaternions**. The algebra of quaternions was invented by the Irish mathematician W. R. Hamilton in 1843. However, the algebra was found to be unnecessarily complicated for most applications and was simplified in both content and notation by the American mathematician J. W. Gibbs and independently by the English electrical engineer O. Heaviside from about 1880 onwards. See Appendix 3.

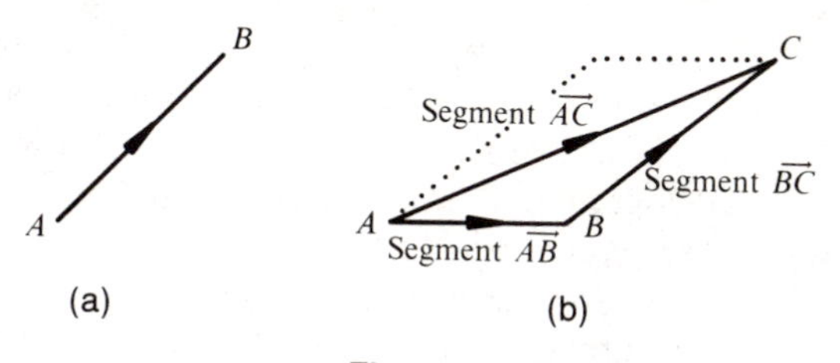

Figure 1.1

(see Figure 1.1b). The 'resultant' displacement $\overrightarrow{AC}$ is the diagonal of the parallelogram with sides parallel to $\overrightarrow{AB}$ and $\overrightarrow{BC}$, and this way of producing a new directed line segment $\overrightarrow{AC}$ from given segments $\overrightarrow{AB}$ and $\overrightarrow{BC}$ is known as the parallelogram law.

The directed line segment is thus a vector since it obeys the parallelogram law of combination and is characterized by both a magnitude (the length of the line) and (obviously) a direction in space.

Forces acting at a point

Consider a simple spring. We shall assume the spring is 'linear', so that if the spring extends through a distance l when a weight W is attached then it will extend by kl when a weight kW is attached. The extension of the spring is then a direct measure of the magnitude of the applied force.

Now imagine two forces are applied to spring c in Figure 1.2a by using a combination of weights which are free to slide in the vertical direction.

Each of the three springs a, b and c is assumed to have equal 'stiffness'; that is, each extends by the same amount if subjected to the same force.

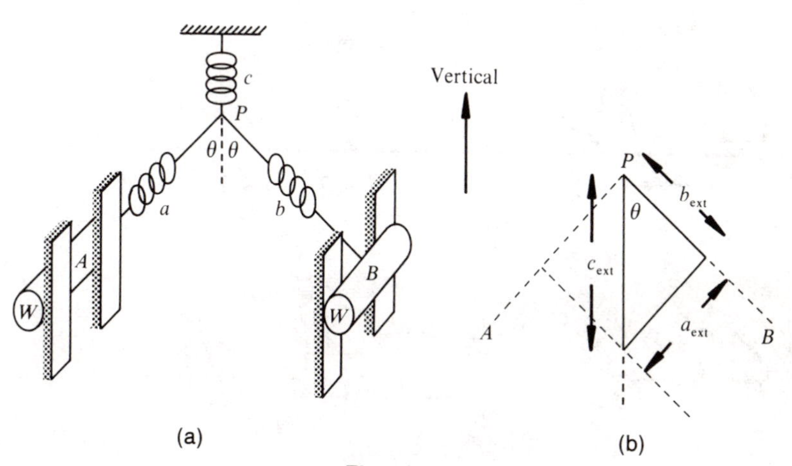

Figure 1.2

The force developed in each spring has a magnitude (directly proportional to the extension) and a direction (the axis of the spring) and so enjoys the first two characteristics of a vector. But do forces combine according to the parallelogram law? It is an *experimental fact* that no matter what weight W one uses (provided that W is not so large as to take the spring into non-linear behaviour), the extension of spring c is just sufficient to be the diagonal of a parallelogram formed by sides parallel to PB and PA with lengths as the extensions of springs b and a respectively (see Figure 1.2b). Forces acting at a point P thus combine together to produce a resultant force according to the parallelogram law, and so can be properly classified as vectors.

Velocity

The velocity of a moving object clearly has the first two characteristics of a vector: a magnitude (which we call the speed) and a direction in space which is simply the direction of motion. To discover how velocities combine consider the following simple experiment.

An observer on the bank of a river notes the position of a passenger who is walking across a barge which is floating downstream. If the speed of the stream is $v\ \mathrm{m\,s^{-1}}$, then the velocity of the barge can be described by an arrowed line of length v (see Figure 1.3a). If the passenger moves across the barge at a uniform speed of $u\ \mathrm{m\,s^{-1}}$ from point A to point B, then the velocity of the passenger, with respect to the barge, may be described by an arrowed line in the direction of $\overrightarrow{AB}$ of length equal to u.

In a time t seconds it is an *observed fact* that the final position of the passenger is at B', and it is clear that for each second of the journey the

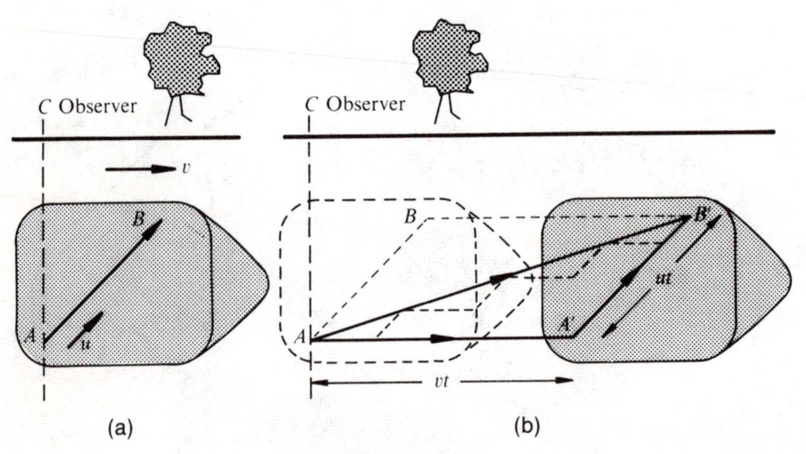

Figure 1.3

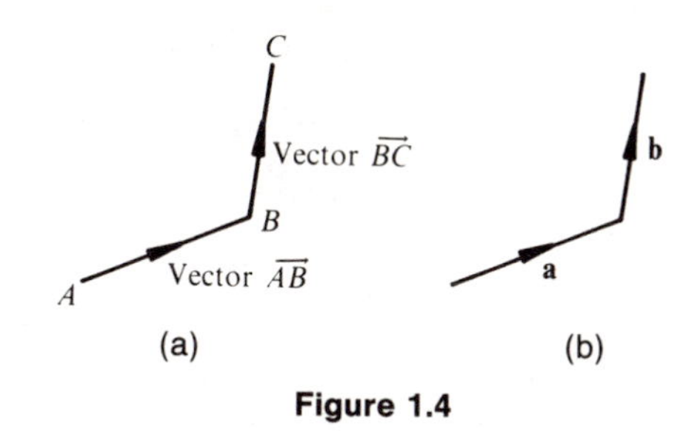

Figure 1.4

actual velocity (with respect to the fixed observer on the bank) of the passenger is along the diagonal of a parallelogram with sides formed from the separate velocities of stream and passenger (see Figure 1.3b). Clearly the velocities are combining together according to the parallelogram law, and so velocity enjoys all three characteristics required of a vector.

We could continue in this way and check that other quantities of interest have vector characteristics. However, we shall find it more convenient to consider vectors first from an (almost) abstract point of view and then, after gaining sufficient experience, return to the 'real world' to check that our abstract formulation has useful applications to physical systems. The abstraction will not be complete, as we shall rely heavily on using the directed line segment as the archetypal vector, which will ensure that the 'abstract' definitions introduced have an immediate physical interpretation.

Consider the directed line segment $\overrightarrow{AB}$ shown in Figure 1.4a. We shall often use the notation of a single bold letter, **a** say, as an alternative to an arrow above two letters to denote a vector. With this understanding, Figures 1.4a and 1.4b are completely equivalent.

1.2 Elementary operations with vectors

The magnitude of a vector

We remember that if s is a scalar (either positive or negative) then the magnitude (or modulus) of s is written $|s|$ and is defined by

$$|s| = \begin{cases} s & \text{if } s > 0 \\ -s & \text{if } s < 0 \end{cases} \tag{1.1}$$

We use a similar notation to denote the magnitude of a vector. If $\overrightarrow{AB}$ (or **a**) is a vector, then its magnitude is denoted by $|\overrightarrow{AB}|$ (or $|\mathbf{a}|$) and is often called the **modulus** of the vector. As far as the directed line segment is concerned, the modulus of this vector is simply its length.

The modulus of a vector is a scalar (a number).

A **null** or **zero vector** is any vector with zero modulus. (The 'direction' in this case is a meaningless concept.)

Equality of vectors

We say that two vectors **a** and **b** are **equal** if and only if both vectors have the same magnitude and both are pointing in the same direction; when this is the case, we write

$$\mathbf{a} = \mathbf{b}$$

Note that we do not demand that both vectors are in exactly the same position in space; hence, all of the vectors shown in Figure 1.5 are equal.

This definition is strictly only valid for what are called **free vectors**. We shall soon find out that not all vectors – quantities that satisfy the three basic characteristics – are of the same kind. As well as free vectors, we shall also encounter **bound vectors** (such as forces and vector fields), **sliding vectors** (such as torques) and **position vectors**.

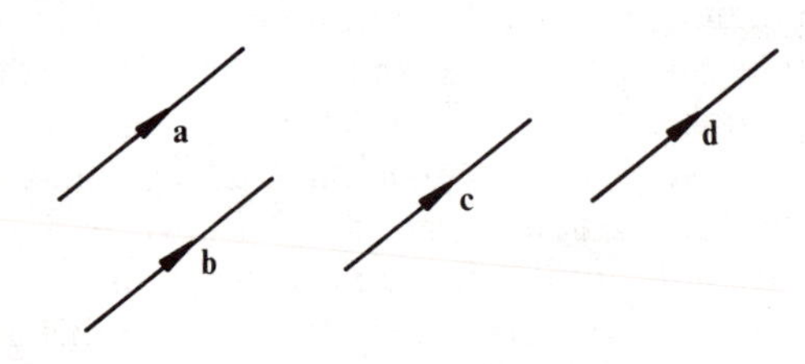

Figure 1.5

Addition of vectors

The sum of two vectors **a** and **b** is written **a** + **b** and is equal to a vector **c**:

$$\mathbf{a} + \mathbf{b} = \mathbf{c}$$

such that **c** is the diagonal of a parallelogram with sides represented by **a** and **b** (see Figure 1.6a). However, for free vectors we can slide **b** (keeping the direction fixed) so that it originates at O and similarly slide **a** so that it extends out from the endpoint of **b**. The result is clearly the same (see Figure 1.6b); that is,

$$\mathbf{b} + \mathbf{a} = \mathbf{c} = \mathbf{a} + \mathbf{b} \tag{1.2}$$

Vector addition is thus **commutative** (order is not important).

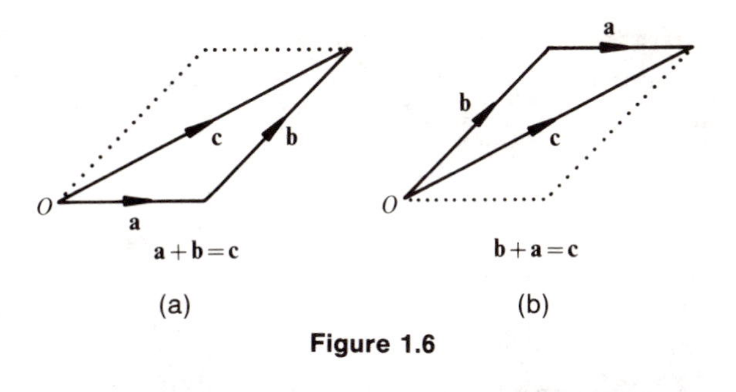

Figure 1.6

A simple extension of this construction will show further that if we have three vectors **a**, **b** and **c**, then

$$\mathbf{a} + (\mathbf{b} + \mathbf{c}) = (\mathbf{a} + \mathbf{b}) + \mathbf{c}$$

(the reader is invited to obtain this result). Therefore vector addition is **associative** (grouping is not important) and we can write, without ambiguity, $\mathbf{a} + \mathbf{b} + \mathbf{c}$.

Multiplication by a scalar

If **a** is a vector and $s > 0$ is a scalar (a positive number), then we define the product of s with **a**, written $s\mathbf{a}$, to be the vector in the same direction as **a** but with magnitude $s|\mathbf{a}|$. If $s < 0$ we define the product $s\mathbf{a}$ to be a vector in the opposite direction to **a** and of magnitude $|s||\mathbf{a}|$ (see Figure 1.7).

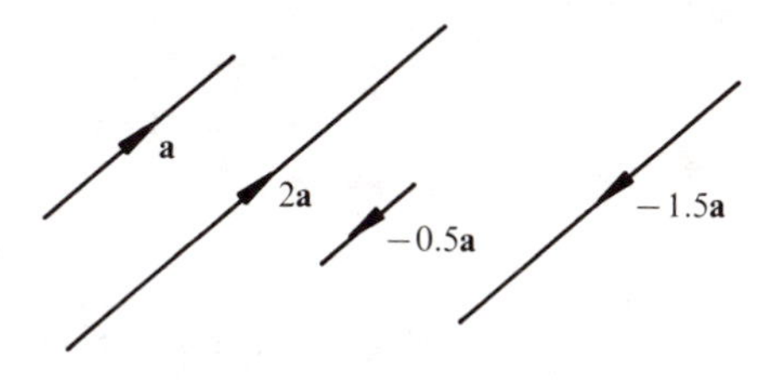

Figure 1.7

Unit vectors

In this book, a unit vector will *always* be denoted by a letter with a circumflex, for example $\hat{\mathbf{a}}$; and is such that it has a magnitude of 1, that is,

$$|\hat{\mathbf{a}}| = 1 \qquad \textbf{(1.3)}$$

If a vector **a** and a unit vector $\hat{\mathbf{a}}$ are in the same direction, then clearly we may write

$$\mathbf{a} = |\mathbf{a}|\hat{\mathbf{a}} \qquad \textbf{(1.4)}$$

since the vectors **a** and $|\mathbf{a}|\hat{\mathbf{a}}$ are in the same direction and have the same magnitude. Alternatively, if we wish to find a unit vector in the direction of **a**, we simply write

$$\hat{\mathbf{a}} = \frac{\mathbf{a}}{|\mathbf{a}|} \tag{1.5}$$

where we have multiplied (1.4) through by $1/|\mathbf{a}|$, a scalar.

Cartesian coordinates and the unit vectors $\hat{\mathbf{i}}$, $\hat{\mathbf{j}}$ and $\hat{\mathbf{k}}$

Before delving further into vectors, we shall remind the reader of the Cartesian description of three-dimensional space. We first choose the position of an origin O. From O we draw two perpendicular lines (or axes), the x- and y-axes. The origin splits the axes into positive and negative parts, the positive side being indicated by an arrow (see Figure 1.8a).

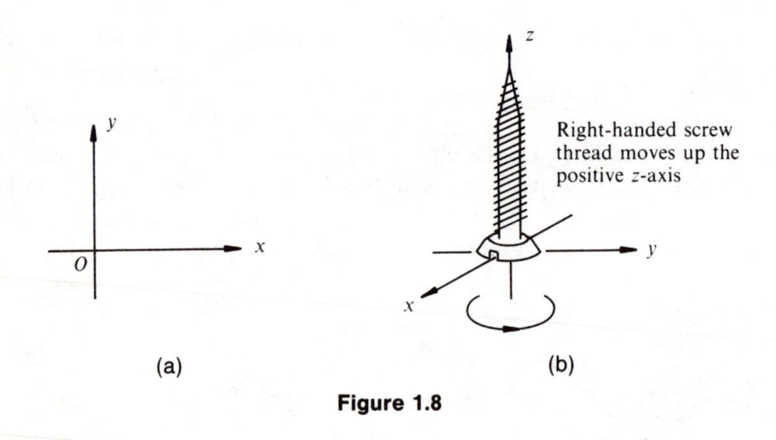

Figure 1.8

We choose a third axis – the z-axis – through O and perpendicular to Ox and Oy. The positive direction of Oz is chosen in such a way that if a screw with a right-handed thread (most screws are of this type!) were placed at O with its threaded end pointing along Oz then the screw would move in the positive direction when turned from Ox to Oy through 90° (see Figure 1.8b).

This choice of coordinate axes allows any point P in space to be uniquely defined by three numbers, x_P, y_P and z_P (often called coordinates); see Figure 1.9.

We now introduce three mutually perpendicular unit vectors $\hat{\mathbf{i}}$, $\hat{\mathbf{j}}$ and $\hat{\mathbf{k}}$, pointing along the positive Ox, Oy and Oz axes respectively (see Figure 1.10). These vectors are often called **basis vectors** since, as we shall now show, any arbitrary vector **a** may be expressed uniquely in terms of them.

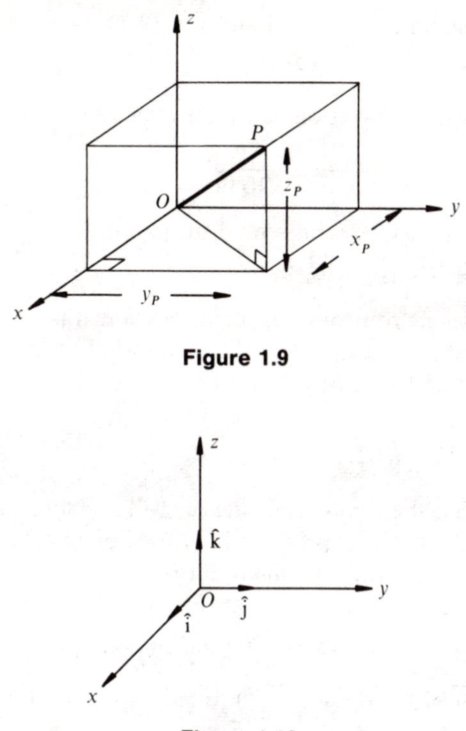

Figure 1.9

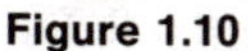

Figure 1.10

We can always slide the vector **a** so that it points directly out of the origin (see Figure 1.11). Using the properties of vector addition,

$$\overrightarrow{OB} + \overrightarrow{BE} = \mathbf{a}$$

But $\overrightarrow{OB} = \overrightarrow{OA} + \overrightarrow{AB}$ and $\overrightarrow{BE} = \overrightarrow{OG}$ while $\overrightarrow{AB} = \overrightarrow{OC}$, so for any vector

$$\mathbf{a} = \overrightarrow{OA} + \overrightarrow{OC} + \overrightarrow{OG}$$

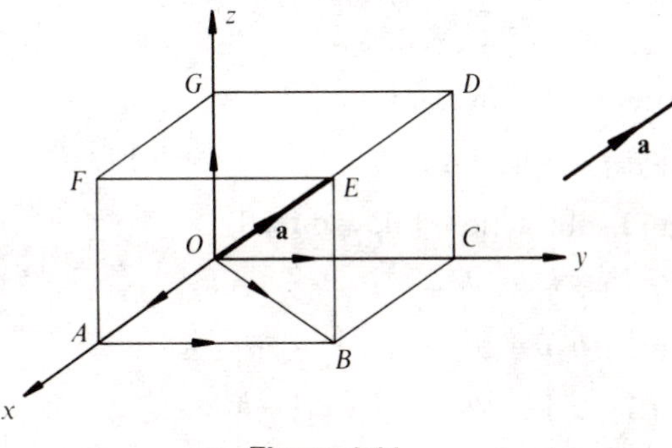

Figure 1.11

However, using the unit vectors $\hat{\mathbf{i}}$, $\hat{\mathbf{j}}$ and $\hat{\mathbf{k}}$ we have

$$\overrightarrow{OA} = |\overrightarrow{OA}|\hat{\mathbf{i}}$$

since $\overrightarrow{OA}$ is in the same direction as $\hat{\mathbf{i}}$ (along the x-axis) and has magnitude $|\overrightarrow{OA}|$. Similarly,

$$\overrightarrow{OC} = |\overrightarrow{OC}|\hat{\mathbf{j}} \quad \text{and} \quad \overrightarrow{OG} = |\overrightarrow{OG}|\hat{\mathbf{k}}$$

Hence *any* vector **a** may be represented in the form

$$\mathbf{a} = |\overrightarrow{OA}|\hat{\mathbf{i}} + |\overrightarrow{OC}|\hat{\mathbf{j}} + |\overrightarrow{OG}|\hat{\mathbf{k}} \tag{1.6}$$

Remembering, from our earlier discussion, that the numbers $|\overrightarrow{OA}|$, $|\overrightarrow{OC}|$ and $|\overrightarrow{OG}|$ are the coordinates of the endpoint of the vector **a** (with its other endpoint at the origin) allows us an alternative expression for the vector, namely

$$\mathbf{a} = (|\overrightarrow{OA}|, |\overrightarrow{OC}|, |\overrightarrow{OG}|) \tag{1.7}$$

We must realize that the right-hand side of (1.7) is a shorthand notation for the expansion in (1.6). The expression in (1.6) is often called the **component form** of **a**, whilst (1.7) is the **coordinate form**.

We should also note, again from Figure 1.11, that

$$\begin{aligned} |\mathbf{a}| &= \sqrt{(|\overrightarrow{OB}|^2 + |\overrightarrow{BE}|^2)} && \text{by Pythagoras' theorem} \\ &= \sqrt{(|\overrightarrow{OA}|^2 + |\overrightarrow{AB}|^2 + |\overrightarrow{BE}|^2)} && \text{again applying Pythagoras} \\ &= \sqrt{(|\overrightarrow{OA}|^2 + |\overrightarrow{OC}|^2 + |\overrightarrow{OG}|^2)} && \end{aligned} \tag{1.8}$$

In summary, any vector **a** may be written in the Cartesian form

$$\mathbf{a} = a_x\hat{\mathbf{i}} + a_y\hat{\mathbf{j}} + a_z\hat{\mathbf{k}} \qquad (a_x = |\overrightarrow{OA}|, a_y = |\overrightarrow{OC}|, a_z = |\overrightarrow{OG}|) \tag{1.9}$$

where a_x, a_y and a_z are called the **components** of **a**, and, with this representation,

$$|\mathbf{a}| = \sqrt{(a_x^2 + a_y^2 + a_z^2)} \tag{1.10}$$

A unit vector in the direction of **a** is

$$\hat{\mathbf{a}} = \frac{\mathbf{a}}{|\mathbf{a}|} = \frac{a_x\hat{\mathbf{i}} + a_y\hat{\mathbf{j}} + a_z\hat{\mathbf{k}}}{\sqrt{(a_x^2 + a_y^2 + a_z^2)}} \tag{1.11}$$

If two vectors are written in $\hat{\mathbf{i}}$, $\hat{\mathbf{j}}$, $\hat{\mathbf{k}}$ form as

$$\mathbf{a} = a_x\hat{\mathbf{i}} + a_y\hat{\mathbf{j}} + a_z\hat{\mathbf{k}} \quad \text{and} \quad \mathbf{b} = b_x\hat{\mathbf{i}} + b_y\hat{\mathbf{j}} + b_z\hat{\mathbf{k}}$$

then addition and subtraction follow directly:

$$\begin{aligned} \mathbf{a} + \mathbf{b} &= a_x\hat{\mathbf{i}} + a_y\hat{\mathbf{j}} + a_z\hat{\mathbf{k}} + b_x\hat{\mathbf{i}} + b_y\hat{\mathbf{j}} + b_z\hat{\mathbf{k}} \\ &= (a_x + b_x)\hat{\mathbf{i}} + (a_y + b_y)\hat{\mathbf{j}} + (a_z + b_z)\hat{\mathbf{k}} \end{aligned} \tag{1.12a}$$

$$\begin{aligned} \mathbf{a} - \mathbf{b} &= a_x\hat{\mathbf{i}} + a_y\hat{\mathbf{j}} + a_z\hat{\mathbf{k}} - b_x\hat{\mathbf{i}} - b_y\hat{\mathbf{j}} - b_z\hat{\mathbf{k}} \\ &= (a_x - b_x)\hat{\mathbf{i}} + (a_y - b_y)\hat{\mathbf{j}} + (a_z - b_z)\hat{\mathbf{k}} \end{aligned} \tag{1.12b}$$

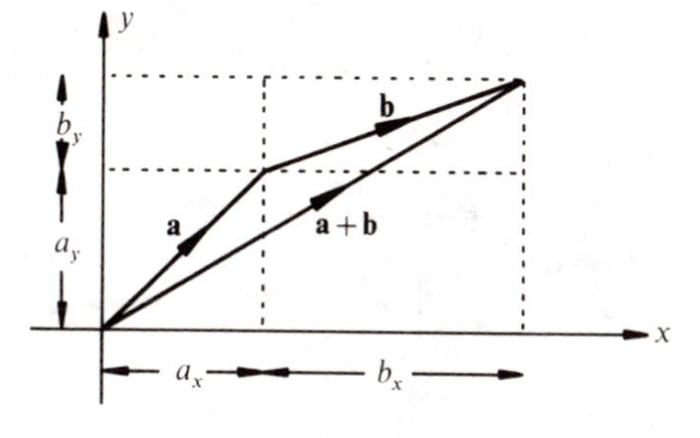

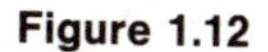
Figure 1.12

That is, we add (or subtract) corresponding components. We refer the reader to Figure 1.12, which illustrates (for two-dimensional vectors only) the connection between (1.12a) and the parallelogram law of addition.

Further, if we wish to multiply a vector by a scalar, we simply multiply each component by the scalar: if

$$\mathbf{a} = a_x\hat{\mathbf{i}} + a_y\hat{\mathbf{j}} + a_z\hat{\mathbf{k}} \quad \text{then} \quad 3\mathbf{a} = 3a_x\hat{\mathbf{i}} + 3a_y\hat{\mathbf{j}} + 3a_z\hat{\mathbf{k}}, \quad \text{etc.}$$

The test for the equality of two vectors is now clear. Two vectors **a** and **b** will be equal if and only if all of their corresponding components are equal: if

$$\mathbf{a} = a_x\hat{\mathbf{i}} + a_y\hat{\mathbf{j}} + a_z\hat{\mathbf{k}} \quad \text{and} \quad \mathbf{b} = b_x\hat{\mathbf{i}} + b_y\hat{\mathbf{j}} + b_z\hat{\mathbf{k}}$$

then

$$\mathbf{a} = \mathbf{b} \quad \text{if and only if} \quad a_x = b_x, \quad a_y = b_y \quad \text{and} \quad a_z = b_z \tag{1.13}$$

The use of the $\hat{\mathbf{i}}, \hat{\mathbf{j}}, \hat{\mathbf{k}}$ or coordinate notation is essentially the computational part of vector algebra. It allows for the quick evaluation of numerical results concerning vectors, whereas the use of single bold letters (or coordinate-free notation) allows a more general discussion of vector properties to be carried out.

Example 1.1

A vector **a** has magnitude 15.3 and is in the same direction as the line $\overrightarrow{BC}$, where B has coordinates $(1, 2, -4)$ and C has coordinates $(-1, 0, 3)$. Determine the component form of **a**.

Solution

It is always useful to draw some kind of diagram when solving vector problems, especially in the early stages. We can draw two vectors $\overrightarrow{OB}$ and $\overrightarrow{OC}$ from the origin, as shown in Figure 1.13. Then, by the

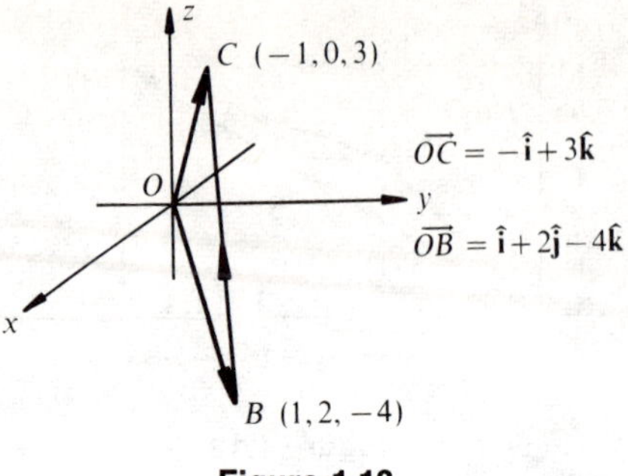

Figure 1.13

parallelogram rule of combination,

$$\overrightarrow{OB} + \overrightarrow{BC} = \overrightarrow{OC}$$

$$\therefore \quad \overrightarrow{BC} = \overrightarrow{OC} - \overrightarrow{OB}$$

$$= (-\hat{\mathbf{i}} + 3\hat{\mathbf{k}}) - (\hat{\mathbf{i}} + 2\hat{\mathbf{j}} - 4\hat{\mathbf{k}}) = -2\hat{\mathbf{i}} - 2\hat{\mathbf{j}} + 7\hat{\mathbf{k}}$$

(This means that if vector $\overrightarrow{BC}$ was slid so that point B was at the origin, then the coordinates of C would be $(-2, -2, 7)$.)

We have now found a vector in the direction $\overrightarrow{BC}$. However, its magnitude is $\sqrt{57}$, not 15.3. We must therefore construct a unit vector $\hat{\mathbf{a}}$ in the direction of $\overrightarrow{BC}$ using equation (1.11):

$$\hat{\mathbf{a}} = \frac{-2\hat{\mathbf{i}} - 2\hat{\mathbf{j}} + 7\hat{\mathbf{k}}}{\sqrt{(4+4+49)}} = \frac{1}{\sqrt{57}}(-2\hat{\mathbf{i}} - 2\hat{\mathbf{j}} + 7\hat{\mathbf{k}})$$

Hence the required vector $\mathbf{a}$ is given by

$$\mathbf{a} = |\mathbf{a}|\hat{\mathbf{a}} = 15.3\hat{\mathbf{a}} = 15.3\left(\frac{1}{\sqrt{57}}\right)(-2\hat{\mathbf{i}} - 2\hat{\mathbf{j}} + 7\hat{\mathbf{k}})$$

$$\therefore \quad \mathbf{a} = -4.053\hat{\mathbf{i}} - 4.053\hat{\mathbf{j}} + 14.186\hat{\mathbf{k}}$$

EXERCISES

1.1 A vector $\mathbf{a}$ is parallel to the line through A (1, 2, 3) and B (−1, 0, 2), whilst a vector $\mathbf{b}$ is parallel to the line through C (0, 0, 1) and D (−1, −2, −3). Both vectors have magnitude 4. Find $\mathbf{a} + \mathbf{b}$ and $\mathbf{a} - \mathbf{b}$ in component form.

1.2 Using an appropriate scale draw the vectors $\mathbf{a} = (1, 2, 0)$ and $\mathbf{b} = (2, 0.5, 0)$. On your diagram show the vector $\mathbf{a} + \mathbf{b}$ and verify that $\mathbf{a} + \mathbf{b} = (3, 2.5, 0)$.

1.3 Find the sum of the vectors $\mathbf{a} = 3\hat{\mathbf{i}} + 7\hat{\mathbf{j}} - 4\hat{\mathbf{k}}$, $\mathbf{b} = -\hat{\mathbf{j}} + 12\hat{\mathbf{k}}$ and $\mathbf{c} = 3\hat{\mathbf{i}} - 5\hat{\mathbf{j}} + 8\hat{\mathbf{k}}$. Calculate the moduli of $\mathbf{a}$, $\mathbf{b}$ and $\mathbf{c}$ and also the modulus of the vector sum. (Note that the sum of the moduli is not equal to the modulus of the sum.)

1.4 If the vertices of a triangle are the points $(1, 0, -1)$, $(-1.3, 0, 2)$ and $(2.4, 1, 2)$, what are the vectors that represent the sides? Determine the length of each side.

1.5 Sketch the vectors $\mathbf{a} = (1, 1, 1)$ and $\mathbf{b} = (2, 0, 1)$. On your sketch draw $\mathbf{a} + \mathbf{b}$, $2\mathbf{a}$ and $\mathbf{a} - \mathbf{b}$.

1.6 The vectors $\mathbf{a}$ and $\mathbf{b}$ are drawn from the origin to two points, A and B respectively. Show that the vector drawn from the origin to the midpoint of AB is $(\mathbf{a} + \mathbf{b})/2$.

1.7 The position vectors of two points A and B are $\hat{\mathbf{i}} + \hat{\mathbf{j}} - \hat{\mathbf{k}}$ and $3\hat{\mathbf{i}} - 2\hat{\mathbf{j}} - \hat{\mathbf{k}}$ respectively. Show that $\overrightarrow{AB}$ is parallel to the xy-plane and has length 3.605.

1.8 Using vectors, show that the lines joining midpoints of opposite sides of a tetrahedron all meet and bisect one another.

1.9 $ABCD$ is a quadrilateral with $\overrightarrow{AB} = \overrightarrow{DC}$. Prove that $ABCD$ is a parallelogram.

1.10 If $\mathbf{a} = 3\hat{\mathbf{i}} + 4\hat{\mathbf{j}} + 5\hat{\mathbf{k}}$, $\mathbf{b} = -\hat{\mathbf{i}} + \hat{\mathbf{j}}$ and $\mathbf{c} = \hat{\mathbf{i}} + \hat{\mathbf{j}} - 3\hat{\mathbf{k}}$, find the values of α and β such that $\mathbf{a} + \alpha\mathbf{b} + \beta\mathbf{c}$ is parallel to the y-axis.

1.11 Show that, for any two vectors $\mathbf{a}$ and $\mathbf{b}$:

(a) $|\mathbf{a} + \mathbf{b}| \leqslant |\mathbf{a}| + |\mathbf{b}|$ (b) $|\mathbf{a} - \mathbf{b}| \geqslant |\mathbf{a}| - |\mathbf{b}|$

1.12 $ABCD$ is a parallelogram and P is the midpoint of BC. Show that the intersection of AP with the diagonal DB is a point of trisection of DB.

1.13 ABC is a triangle. P and P' are the midpoints of AB and BC. Show that $\overrightarrow{PP'}$ is parallel to $\overrightarrow{AC}$ and that $|\overrightarrow{PP'}| = \frac{1}{2}|\overrightarrow{AC}|$.

1.3 Elementary applications of vectors

Centroids

If a particle of mass m is positioned at the point (x_0, y_0, z_0) then we say it has a **moment** about the zy-plane of value mx_0, a moment of value my_0 about the xz-plane and a moment of value mz_0 about the xy-plane (see Figure 1.14a). Now imagine a situation in which we have particles of mass

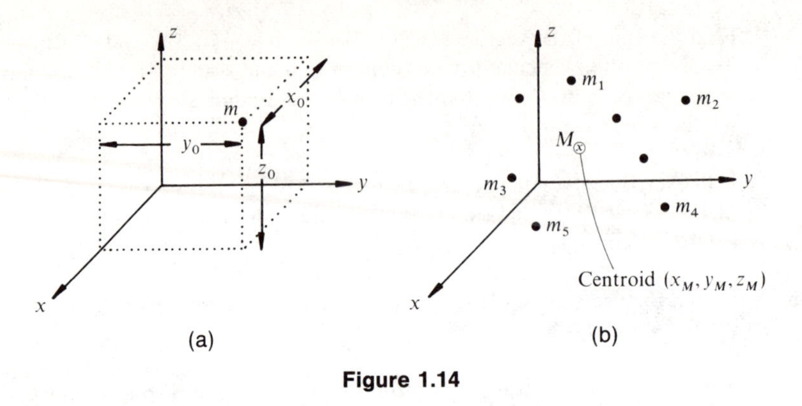

Figure 1.14

$m_1, m_2, \ldots, m_n$ situated at positions $(x_1, y_1, z_1), (x_2, y_2, z_2), \ldots, (x_n, y_n, z_n)$ respectively. If these particles are fixed relative to each other (as in a rigid body, for example) then for many applications they can be replaced by a single mass M equal to the sum of the masses of all the separate particles, placed in a position called the **centroid** (see Figure 1.14b).

The position of the centroid is such that the moment of M about any particular coordinate plane is equal to the sum of the moments of all the particles taken together about that coordinate plane. Hence, if the coordinates of the centroid are (x_M, y_M, z_M) we have, from this definition,

$$Mx_M = m_1x_1 + m_2x_2 + \cdots + m_nx_n$$

$$My_M = m_1y_1 + m_2y_2 + \cdots + m_ny_n$$

$$Mz_M = m_1z_1 + m_2z_2 + \cdots + m_nz_n$$

These three scalar equations may be combined in one vector equation as follows. Multiplying the three equations by the unit vectors $\hat{\mathbf{i}}$, $\hat{\mathbf{j}}$ and $\hat{\mathbf{k}}$ respectively and adding, we obtain

$$M\mathbf{r}_M = m_1\mathbf{r}_1 + m_2\mathbf{r}_2 + \cdots + m_n\mathbf{r}_n$$

where $\mathbf{r}_i = x_i\hat{\mathbf{i}} + y_i\hat{\mathbf{j}} + z_i\hat{\mathbf{k}}$ is the **position vector** of the ith particle $(i = 1, 2, \ldots, n)$ and $\mathbf{r}_M = x_M\hat{\mathbf{i}} + y_M\hat{\mathbf{j}} + z_M\hat{\mathbf{k}}$ is the position vector of the centroid. Hence

$$\mathbf{r}_M = \frac{m_1\mathbf{r}_1 + m_2\mathbf{r}_2 + \cdots + m_n\mathbf{r}_n}{M} = \frac{1}{M}\sum_{i=1}^{n} m_i\mathbf{r}_i \qquad \textbf{(1.14)}$$

Forces

The effect of a force acting on a body is dependent not only on the magnitude and direction of the force but also, generally, on the point of application.

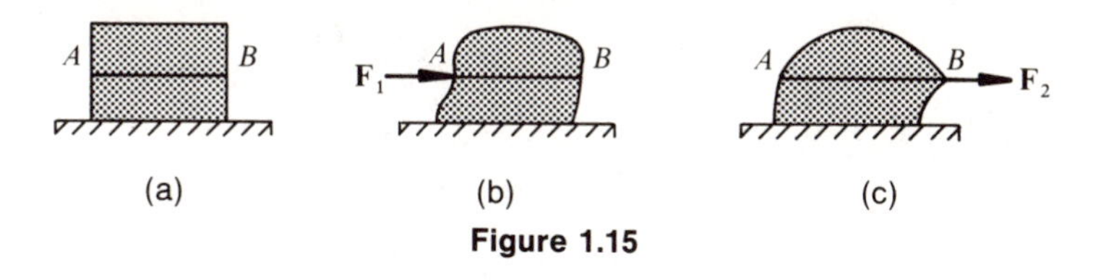

Figure 1.15

In some physical systems (such as rigid bodies), however, only the line of action of the force, as opposed to its actual point of application, is of importance. Two simple examples will illustrate the point.

Consider a block of rubber held fixed in some way (see Figure 1.15a). If we apply a force $\mathbf{F}_1$ at point A in the direction AB, then the deformation of the solid would be as indicated in Figure 1.15b. However, if a force $\mathbf{F}_2$ of the same magnitude and direction as $\mathbf{F}_1$ were applied at point B (that is along the same line of action), then the deformed shape would be similar to that shown in Figure 1.15c: a completely distinct deformation has resulted simply by changing the point of application of the force.

As a second example, consider the mechanical configuration shown in Figure 1.16. If we are interested in the moment developed at point A due to the force $\mathbf{F}$, then it is irrelevant whether the point of application of $\mathbf{F}$ is at B or at C or indeed at any point on the line BC (the line of action of $\mathbf{F}$), because exactly the same moment (which is $|\mathbf{F}|l$) is obtained in all cases.

It is unfortunate, but nonetheless a fact, that forces cannot generally be considered as 'free' vectors. However, in the special case where forces are concurrent (that is, where they have a common point of application), then they may be regarded as 'free' vectors of the type being considered in this text. This special application is not quite as artificial as it seems, since we shall discover later in this chapter that we can often arrange matters so that a complicated system of forces is replaced by an equivalent system of concurrent forces (plus, to be precise, a torque).

Two concurrent forces may always be replaced by a single force, acting at the common point of application. For example, if $\mathbf{F}_1$ and $\mathbf{F}_2$ are concurrent at point A (see Figure 1.17), then they can be replaced by a single force $\mathbf{F} = \mathbf{F}_1 + \mathbf{F}_2$ as shown, $\mathbf{F}$ being known as the **resultant**. The concept of the resultant can be extended to any number of concurrent forces. We define a system of concurrent forces to be in **equilibrium** if their resultant is the zero vector.

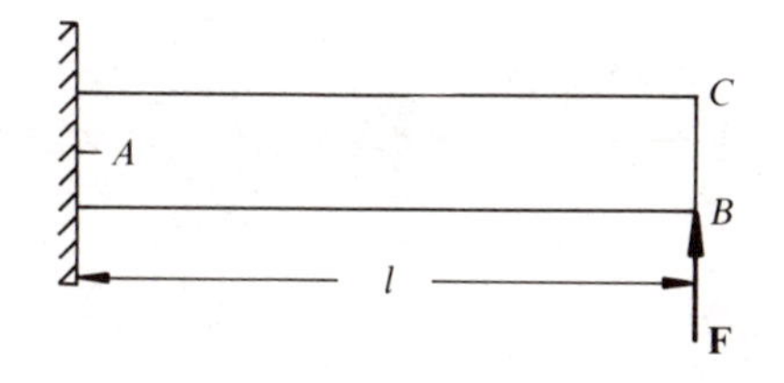

Figure 1.16

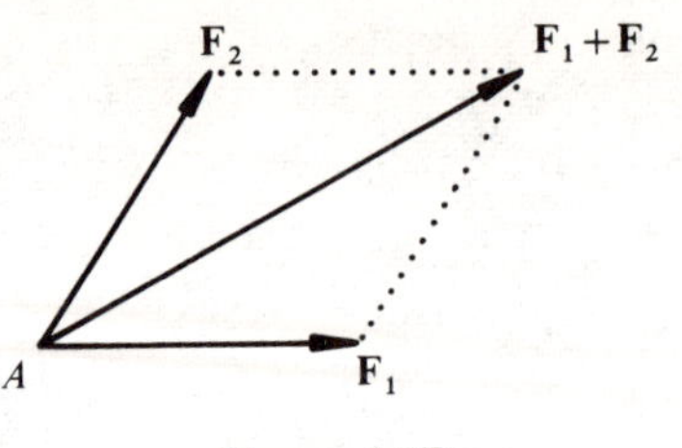

Figure 1.17

Geometry – the vector equation of a straight line

Consider two points A and B with coordinates (x_1, y_1, z_1) and (x_2, y_2, z_2). By an argument using the properties of similar triangles, we can deduce that if (x, y, z) is a general point on the straight line AB (see Figure 1.18a), then

$$\frac{x-x_1}{x_2-x_1}=\frac{y-y_1}{y_2-y_1}=\frac{z-z_1}{z_2-z_1}=t, \quad \text{say} \tag{1.15}$$

where t is a parameter denoting the common ratio. Here we require three (scalar) equations to define the straight line and a point (x, y, z) will lie on this line only if these three equations are satisfied.

We shall now show how this same line may be described (far more elegantly) in terms of vectors. Again we consider the situation shown in Figure 1.18a, but this time we introduce vectors $\mathbf{a}$ $(\equiv \overrightarrow{OA})$ and $\mathbf{b}$ $(\equiv \overrightarrow{OB})$ as shown in Figure 1.18b. If point P is a general point on the line with coordinates (x, y, z), then P is said to have **position vector**

$$\mathbf{r} = x\hat{\mathbf{i}} + y\hat{\mathbf{j}} + z\hat{\mathbf{k}} \tag{1.16}$$

(throughout this book we shall always use $\mathbf{r}$ to denote this position vector).

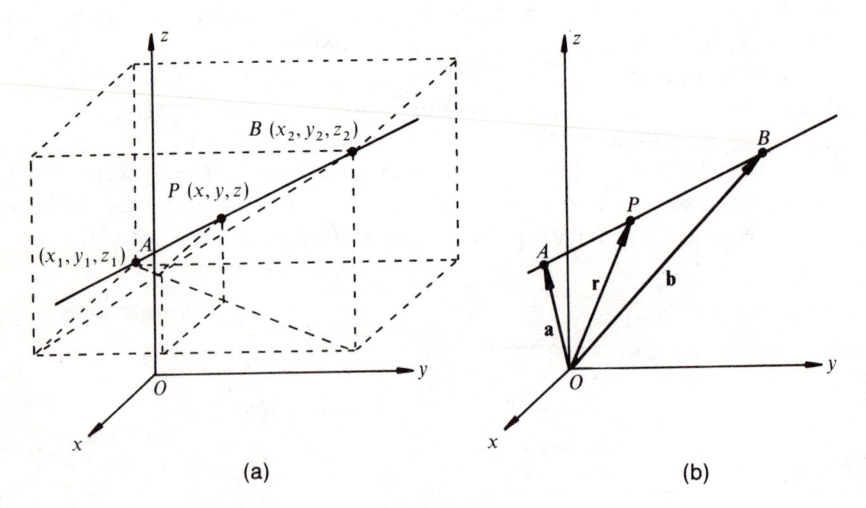

Figure 1.18

Now, using the rules of vector algebra,

$$\overrightarrow{OA} + \overrightarrow{AP} = \overrightarrow{OP}$$

that is,

$$\mathbf{a} + \overrightarrow{AP} = \mathbf{r} \qquad \textbf{(1.17)}$$

Clearly, vector $\overrightarrow{AP}$ is in the same direction as vector $\overrightarrow{AB}$, since both are parts of the same straight line. Hence, we may write

$$\overrightarrow{AP} = t\,\overrightarrow{AB}$$

where t is some scalar. Now

$$\overrightarrow{OA} + \overrightarrow{AB} = \overrightarrow{OB}$$

that is,

$$\overrightarrow{AB} = \overrightarrow{OB} - \overrightarrow{OA} = \mathbf{b} - \mathbf{a}$$

Therefore $\overrightarrow{AP} = t(\mathbf{b} - \mathbf{a})$ and, substituting into (1.17),

$$\mathbf{r} = \mathbf{a} + t(\mathbf{b} - \mathbf{a}) \qquad \textbf{(1.18a)}$$

This is the vector equation of a straight line passing through two points (the endpoints of the vectors **a** and **b**). To reach different points on the line we give different values to t (see Figure 1.19).

More generally, the vector equation of a line through point **a** and parallel to a vector **c** is

$$\mathbf{r} = \mathbf{a} + t\mathbf{c} \qquad \textbf{(1.18b)}$$

The scalar (or, as it is sometimes called, the Cartesian) form of the equation of a straight line may be easily recovered from (1.18a) by expressing all the vectors in component form:

$$\begin{aligned} x\hat{\mathbf{i}} + y\hat{\mathbf{j}} + z\hat{\mathbf{k}} &= x_1\hat{\mathbf{i}} + y_1\hat{\mathbf{j}} + z_1\hat{\mathbf{k}} + t[(x_2 - x_1)\hat{\mathbf{i}} + (y_2 - y_1)\hat{\mathbf{j}} + (z_2 - z_1)\hat{\mathbf{k}}] \\ &= [x_1 + t(x_2 - x_1)]\hat{\mathbf{i}} + [y_1 + t(y_2 - y_1)]\hat{\mathbf{j}} + [z_1 + t(z_2 - z_1)]\hat{\mathbf{k}} \end{aligned}$$

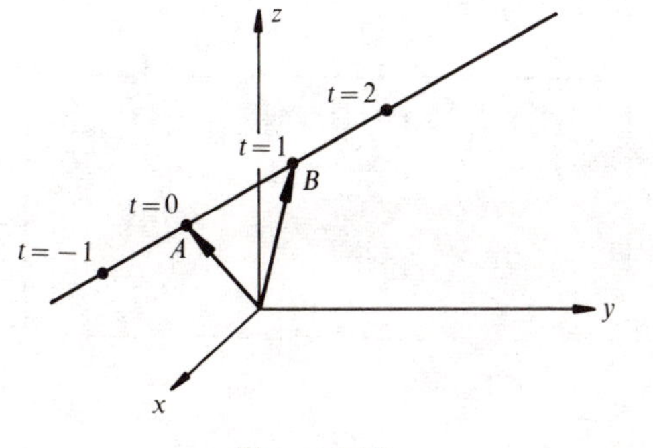

Figure 1.19

Then, using the definition of equality of vectors (equation (1.13)), we can conclude that

$$x = x_1 + t(x_2 - x_1), \qquad y = y_1 + t(y_2 - y_1), \qquad z = z_1 + t(z_2 - z_1)$$

or

$$\frac{x - x_1}{x_2 - x_1} = \frac{y - y_1}{y_2 - y_1} = \frac{z - z_1}{z_2 - z_1} = t \quad \text{as in (1.15)}$$

The information contained in the single vector equation (1.18a) is equivalent to that contained in the three scalar equations (1.15).

Example 1.2

Determine the vector equation of the straight line parallel to $\overrightarrow{CD}$, where C is the point (1, 2, 0) and D is (−1, 2, 3), and which passes through the point A (0, 1, 2). Find the point at which this line intersects the plane $x = 2$ (see Figure 1.20).

Solution

From (1.18b) with

$$\mathbf{a} = \hat{\mathbf{j}} + 2\hat{\mathbf{k}}$$

and

$$\mathbf{c} = \overrightarrow{OD} - \overrightarrow{OC} = (-\hat{\mathbf{i}} + 2\hat{\mathbf{j}} + 3\hat{\mathbf{k}}) - (\hat{\mathbf{i}} + 2\hat{\mathbf{j}}) = -2\hat{\mathbf{i}} + 3\hat{\mathbf{k}}$$

we obtain

$$\mathbf{r} = \hat{\mathbf{j}} + 2\hat{\mathbf{k}} + t(-2\hat{\mathbf{i}} + 3\hat{\mathbf{k}})$$

as the vector equation of the given straight line.

In Cartesian form, with $\mathbf{r} = x\hat{\mathbf{i}} + y\hat{\mathbf{j}} + z\hat{\mathbf{k}}$, we obtain (after equating coefficients of $\hat{\mathbf{i}}$, $\hat{\mathbf{j}}$ and $\hat{\mathbf{k}}$)

$$x = -2t, \qquad y = 1, \qquad z = 2 + 3t$$

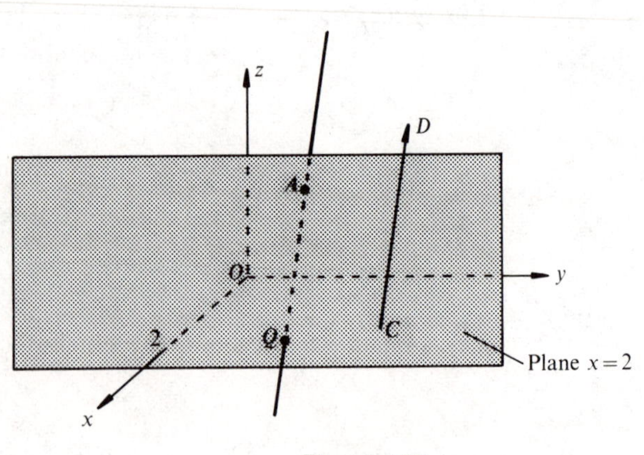

Figure 1.20

This line thus intersects the plane $x = 2$ when $t = -1$. Having determined the value of t, the values of y and z may be determined. We find $y = 1$ (always) and $z = -1$. Thus the point of intersection of the line with the plane $x = 2$ is the point Q with coordinates $(2, 1, -1)$.

Example 1.3

A particle moves uniformly in the xy-plane around a circle of radius 3 m with a rotational speed of 0.1 rad s^{-1}. If the initial position of the particle has coordinates (3, 0) and the motion is anticlockwise, find the position of the particle after 12 s and determine the velocity vector of the particle at this time.

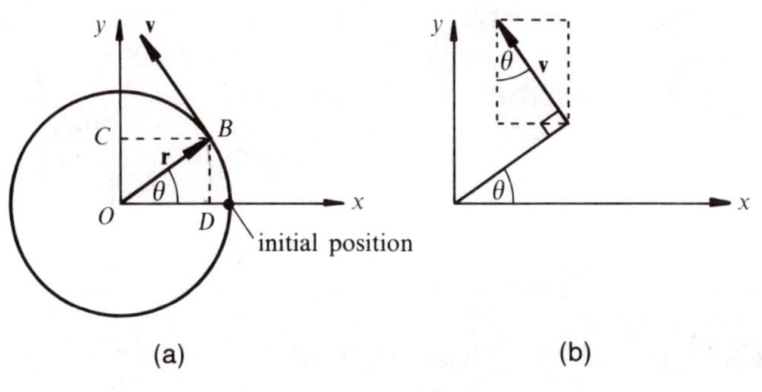

Figure 1.21

Solution

After 12 s the particle is at B (see Figure 1.21a) with $\theta = 1.2$ rad. Therefore

$$OD = 3\cos\theta = 1.087, \qquad OC = 3\sin\theta = 2.796$$

Therefore the position vector $\mathbf{r}$ of the particle after 12 s is

$$\mathbf{r} = OD\hat{\mathbf{i}} + OC\hat{\mathbf{j}} = 1.087\hat{\mathbf{i}} + 2.796\hat{\mathbf{j}}$$

The velocity is clearly tangential to the circle. In a time δt the particle moves through a small angle $\delta\theta$. The distance moved through in this time is $r\,\delta\theta$ where $r = |\mathbf{r}|$ is the radius of the circle, and so the speed of the particle is $r\,\delta\theta/\delta t$, that is the radius r multiplied by the rotational speed. Therefore, if $\mathbf{v}$ is the velocity vector,

$$|\mathbf{v}| = 3(0.1) \quad \text{m s}^{-1}$$

From Figure 1.21b, the components of this vector are $-|\mathbf{v}|\sin\theta$ in the x-direction and $|\mathbf{v}|\cos\theta$ in the y-direction; that is

$$\begin{aligned}\mathbf{v} = -|\mathbf{v}|\sin\theta\,\hat{\mathbf{i}} + |\mathbf{v}|\cos\theta\,\hat{\mathbf{j}} &= -3(0.1)\sin\theta\,\hat{\mathbf{i}} + 3(0.1)\cos\theta\,\hat{\mathbf{j}} \\ &= -0.2796\hat{\mathbf{i}} + 0.1087\hat{\mathbf{j}}\end{aligned}$$

We might note that, just as in scalar analysis speed is the rate of change of distance with time, so in vector analysis velocity is the rate of change of displacement with time:

$$\mathbf{v}=\frac{\mathrm{d}\mathbf{r}}{\mathrm{d}t}=\frac{\mathrm{d}}{\mathrm{d}t}(3\cos\theta\,\hat{\mathbf{i}}+3\sin\theta\,\hat{\mathbf{j}})=-3\sin\theta\frac{\mathrm{d}\theta}{\mathrm{d}t}\hat{\mathbf{i}}+3\cos\theta\frac{\mathrm{d}\theta}{\mathrm{d}t}\hat{\mathbf{j}}$$

$$=-3\sin\theta\,(0.1)\hat{\mathbf{i}}+3\cos\theta\,(0.1)\hat{\mathbf{j}}$$

as above. This derivative relationship between the velocity vector and the displacement vector will be more fully developed in Chapter 2.

Relative velocity

Suppose that two observers α and β are moving with velocities $\mathbf{v}_\alpha$ and $\mathbf{v}_\beta$ with respect to a fixed coordinate system. If we know their initial positions then we can plot the path that each observer takes in space.

If both observers are moving along the x-axis, then we can write

$$\mathbf{v}_\alpha=v_{\alpha x}\hat{\mathbf{i}} \quad \text{and} \quad \mathbf{v}_\beta=v_{\beta x}\hat{\mathbf{i}}$$

where $v_{\alpha x}$ and $v_{\beta x}$ are the respective speeds of each observer. In one second, therefore, observer β moves a distance $v_{\beta x}$ and observer α moves a distance $v_{\alpha x}$. Both movements are along the x-axis, so the observers' distance apart changes by $(v_{\beta x}-v_{\alpha x})$ in this time. Thus, from the viewpoint of observer α, observer β appears to be moving with a velocity

$$(v_{\beta x}-v_{\alpha x})\hat{\mathbf{i}}=\mathbf{v}_\beta-\mathbf{v}_\alpha$$

This result actually holds for any direction of motion, because if we write

$$\mathbf{v}_\alpha=v_{\alpha x}\hat{\mathbf{i}}+v_{\alpha y}\hat{\mathbf{j}}+v_{\alpha z}\hat{\mathbf{k}} \quad \text{and} \quad \mathbf{v}_\beta=v_{\beta x}\hat{\mathbf{i}}+v_{\beta y}\hat{\mathbf{j}}+v_{\beta z}\hat{\mathbf{k}}$$

then for each second that passes, β appears to be moving towards or away from α by amounts $(v_{\beta x}-v_{\alpha x})$ in the x-direction, $(v_{\beta y}-v_{\alpha y})$ in the y-direction and $(v_{\beta z}-v_{\alpha z})$ in the z-direction. Thus from the point of view of observer α, observer β appears to have a velocity

$$(v_{\beta x}-v_{\alpha x})\hat{\mathbf{i}}+(v_{\beta y}-v_{\alpha y})\hat{\mathbf{j}}+(v_{\beta z}-v_{\alpha z})\hat{\mathbf{k}}=\mathbf{v}_\beta-\mathbf{v}_\alpha \tag{1.19}$$

and we say that the **relative velocity** of observer β with respect to observer α is

$$\mathbf{v}_\beta-\mathbf{v}_\alpha$$

Example 1.4

Two particles P and Q are instantaneously at points A and B respectively at time $t=0$, A and B being 15 m apart. The particles are moving with uniform velocities, the former towards B at a speed of

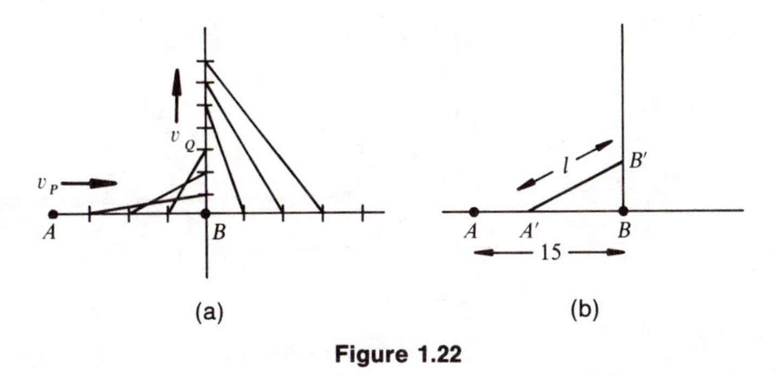

Figure 1.22

v_P m s^{-1} and the latter perpendicular to AB at v_Q m s^{-1}. Find the shortest distance apart of the two particles.

Solution

Figure 1.22a shows the positions of the particles at equal intervals of time (for the purposes of illustration we have assumed, in the diagram, that $v_Q < v_P$). If we consider the positions after a time t seconds, the particle P has moved a distance $v_P t$ from A to point A' and particle Q has moved a distance $v_Q t$ from B to B'. The distance between the particles at this instant is l, where

$$l^2 = (15 - v_P t)^2 + v_Q^2 t^2 \quad \text{(see Figure 1.22b)}$$

We can readily minimize l^2 using elementary calculus:

$$\frac{\mathrm{d}(l^2)}{\mathrm{d}t} = -2v_P(15 - v_P t) + 2v_Q^2 t$$

implying that l^2 (and hence l) is a minimum ($\mathrm{d}(l^2)/\mathrm{d}t = 0$) when

$$t = \frac{15v_P}{v_P^2 + v_Q^2}$$

so that the shortest distance apart is

$$l = \sqrt{\left[\left(15 - \frac{15v_P^2}{v_P^2 + v_Q^2}\right)^2 + \frac{225v_P^2 v_Q^2}{(v_P^2 + v_Q^2)^2}\right]} = \frac{15v_Q}{\sqrt{(v_P^2 + v_Q^2)}}$$

(If, as a specific example, we take $v_P = 4$ m s^{-1} and $v_Q = 3$ m s^{-1}, then $l = 9$ m and at the time of minimum separation P is 9.6 m from A and Q is 7.2 m from B.)

Alternatively, we can solve this problem using a vector approach. We choose a coordinate system so that point A is the origin and AB is in the direction of the positive x-axis. Then the velocity of particle P is $\mathbf{v}_P = v_P \hat{\mathbf{i}}$ and the velocity of particle Q is $\mathbf{v}_Q = v_Q \hat{\mathbf{j}}$. The relative velocity of particle Q with respect to particle P is the velocity that Q appears to have to an observer moving with P. Such an observer

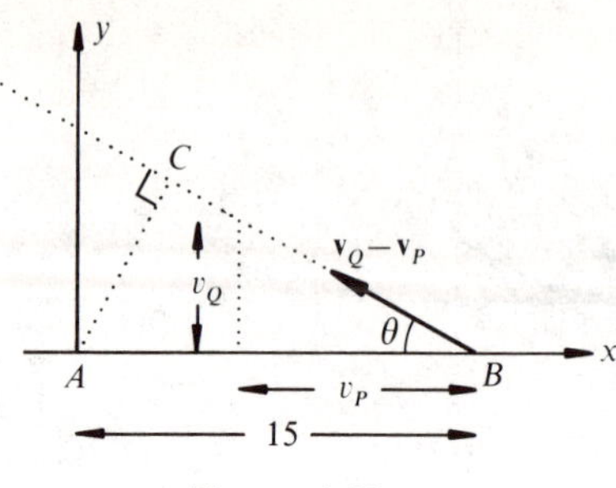

Figure 1.23

would see Q moving towards P with speed v_P as well as observing the motion of Q in the positive y-direction. In other words, the relative velocity of Q with respect to P is $-v_P\hat{\mathbf{i}}+v_Q\hat{\mathbf{j}}=\mathbf{v}_Q-\mathbf{v}_P$, as in the general theory. Hence we can solve the problem by imagining that P is stationary at the point A and that Q is moving with the relative velocity along the path shown in Figure 1.23. Clearly, the shortest distance between the particles is AC, where

$$AC = 15\sin\theta = \frac{15}{\operatorname{cosec}\theta} = \frac{15}{\sqrt{(1+\cot^2\theta)}} = \frac{15}{\sqrt{(1+v_P^2/v_Q^2)}}$$

$$= \frac{15v_Q}{\sqrt{(v_P^2+v_Q^2)}}, \quad \text{as before}$$

The vector approach is more direct and does not require the use of calculus to determine the solution.

Example 1.5

A weight W is supported by two rigid rods AB and BC as shown in Figure 1.24. Determine the forces developed in the two rods.

Solution

Although, in reality, there will be a slight vertical deflection of point B, this is generally so small that it can be ignored. The geometry of the

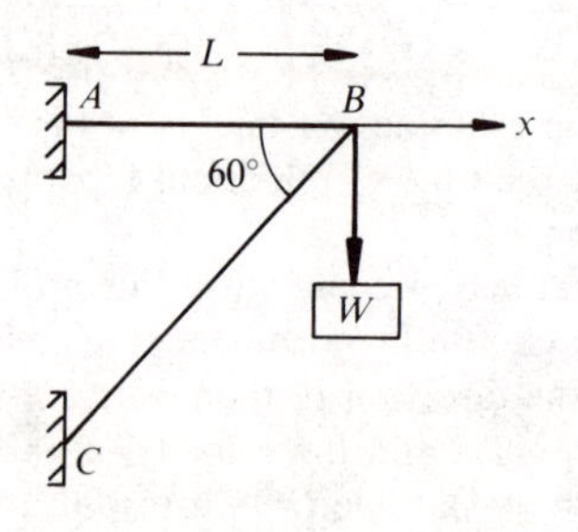

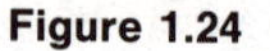

Figure 1.24

deformed structure is therefore assumed to be exactly the same as the undeformed structure, that is, before the weight is attached.

Let the forces developed in AB and BC be denoted by $\mathbf{F}_1$ and $\mathbf{F}_2$ respectively. Choosing a Cartesian coordinate system with origin at B and the positive y-axis as vertically upward, the force applied to joint B due to the weight W is $-W\hat{\mathbf{j}}$. Now, after the weight W is attached, the forces acting through B are in equilibrium. Therefore the forces through point B sum to zero:

$$\mathbf{F}_1 + \mathbf{F}_2 - W\hat{\mathbf{j}} = 0 \tag{1.20}$$

But clearly $\mathbf{F}_1$ is in the direction of AB, so

$$\mathbf{F}_1 = \alpha\hat{\mathbf{i}}, \quad \text{where } \alpha \text{ is to be determined}$$

and $\mathbf{F}_2$ is in the direction of BC so

$$\mathbf{F}_2 = \beta(-\cos 60^\circ\,\hat{\mathbf{i}} - \sin 60^\circ\,\hat{\mathbf{j}}), \quad \text{where } \beta \text{ is to be determined.}$$

(Note that $(-\cos 60^\circ\,\hat{\mathbf{i}} - \sin 60^\circ\,\hat{\mathbf{j}})$ is a unit vector in the direction of $\overrightarrow{BC}$.) Now, from (1.20),

$$\alpha\hat{\mathbf{i}} + \beta(-\cos 60^\circ\,\hat{\mathbf{i}} - \sin 60^\circ\,\hat{\mathbf{j}}) - W\hat{\mathbf{j}} = 0$$

Therefore equating coefficients of $\hat{\mathbf{i}}$ and $\hat{\mathbf{j}}$ to zero gives

$$\alpha - \beta\cos 60^\circ = 0 \quad \text{and} \quad -\beta\sin 60^\circ - W = 0$$

Therefore

$$\beta = -\frac{2}{\sqrt{3}}W = -\tfrac{2}{3}\sqrt{3}W \quad \text{and} \quad \alpha = -\frac{2}{\sqrt{3}}W\cos 60^\circ = -\tfrac{1}{3}\sqrt{3}W$$

Hence the required forces are (see Figure 1.25)

$$\mathbf{F}_1 = -\tfrac{1}{3}\sqrt{3}W\hat{\mathbf{i}} \quad \text{and} \quad \mathbf{F}_2 = \tfrac{2}{3}\sqrt{3}W(\tfrac{1}{2}\hat{\mathbf{i}} + \tfrac{1}{2}\sqrt{3}\hat{\mathbf{j}})$$

that is, $\mathbf{F}_1$ is directed away from B and $\mathbf{F}_2$ is directed towards B.

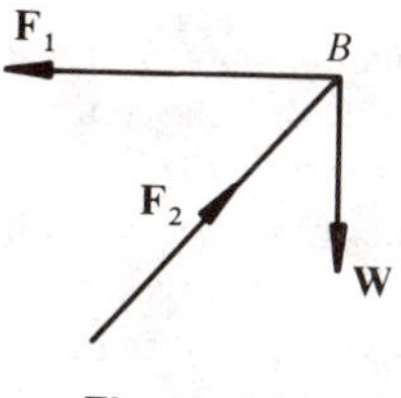

Figure 1.25

EXERCISES

1.14 Forces $\mathbf{F}_1$, $\mathbf{F}_2$ and $\mathbf{F}_3$ have magnitudes 1 N, 2.6 N and 3 N respectively and act concurrently through a point O. The forces are parallel to the vectors $\hat{\mathbf{i}} + \hat{\mathbf{j}}$, $3\hat{\mathbf{k}}$ and $\hat{\mathbf{i}} + \hat{\mathbf{j}} + \hat{\mathbf{k}}$ respectively. Find the resultant force and determine its magnitude.

1.15 Four particles of mass m_1, $2m_1$, $3m_1$ and $4m_1$ are placed at the points $(1, 0, 0)$, $(0, 0, 0)$, $(1, 2, -3.5)$ and $(4, 1, 0)$ respectively. Find the position of the centre of mass of the system.

1.16 Find the vector equation of the straight line through the points $(1, 2, -1)$ and $(2, 0, 1)$. Determine the Cartesian form of the line.

1.17 Deduce whether or not the three points A $(1, 2, 3)$, B $(5, 4, -2)$ and C $(-2, -1, 5)$ are collinear.

1.18 Two boats sailing at the Equator are initially three nautical miles apart. Boat A has a damaged rudder and damaged engine controls and so is unable to change its course or speed. It is steaming in a direction due north at a speed of 15 knots. What is the shortest time a rescue craft (boat B) can reach the crippled boat if its maximum speed is 20 knots? (*Note*: a 'knot' is a speed of one nautical mile per hour.)

1.19 A person travelling due east at 4 km h^{-1} observes that the wind appears to blow directly from the north. On doubling his speed, he finds the wind appearing to come from the north-east. Show that the wind is blowing from the north-west with a speed of 5.657 km h^{-1}.

1.20 Find the vector equation of the straight line which passes through the point $(3, -2, 5)$ and is parallel to the vector $4\hat{\mathbf{i}} + 6\hat{\mathbf{j}} - 7\hat{\mathbf{k}}$.

1.21 Two projectiles A and B have velocities $\mathbf{v}_A = a\hat{\mathbf{i}} + (b - ct)\hat{\mathbf{j}}$ and $\mathbf{v}_B = d\hat{\mathbf{i}} + (e - ct)\hat{\mathbf{j}}$, where a, b, c, d and e are constants and t is time. Determine the relation between a, b, d and e if one projectile is always at an angle of 45° to the other.

1.22 Show that the vectors drawn from the centre of a regular pentagon to its vertices sum to zero.

1.23 Masses $m, 2m, \ldots, 6m$ are placed at the vertices of a regular hexagon $ABCDEF$ of side h. Show that the centre of mass is a distance $1.113h$ from AB.

1.24 A, B, C and D are the midpoints of consecutive sides of a quadrilateral. Show that $ABCD$ is a parallelogram. Confirm that this result is not dependent on the quadrilateral being planar.

1.25 (a) Show that the vectors $3\hat{\mathbf{i}} + \hat{\mathbf{j}} - 2\hat{\mathbf{k}}$, $-\hat{\mathbf{i}} + 3\hat{\mathbf{j}} + 4\hat{\mathbf{k}}$ and $4\hat{\mathbf{i}} - 2\hat{\mathbf{j}} - 6\hat{\mathbf{k}}$ can form the sides of a triangle.
(b) Show that the medians of any triangle can form a triangle.

1.26 Show that the medians of a triangle intersect at a point which is a point of trisection of each median.

1.4 Orthogonal and perpendicular projections of a vector

Consider a vector **Q**. If we imagine **Q** as a directed line segment then this line segment could be the hypotenuse of many (in fact, infinitely many) right-angled triangles (see Figure 1.26). If we concentrate on just one of these

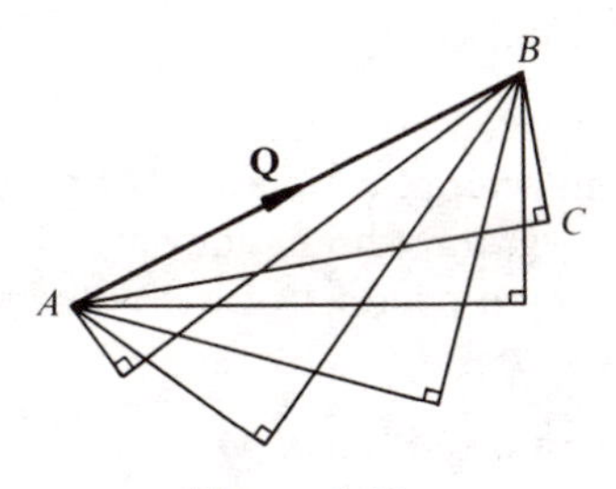

Figure 1.26

triangles (say ABC), then by the rules of vector addition we have

$$\overrightarrow{AC} + \overrightarrow{CB} = \mathbf{Q}$$

Clearly **Q** has been **resolved** into two mutually perpendicular vectors, $\overrightarrow{AC}$ and $\overrightarrow{CB}$.

In a similar fashion, if we have two vectors **Q** and **P**, then we can imagine resolving **Q** onto **P** by the construction shown in Figure 1.27. We say that AC is the **orthogonal projection** of **Q** onto **P**, and clearly

$$AC = |\mathbf{Q}| \cos \theta$$

Similarly, if BC is the **perpendicular projection** of **Q** onto **P**, then

$$BC = |\mathbf{Q}| \sin \theta$$

The vector **P** makes an implicit contribution to both projections through the angle θ, which is the angle between **Q** and **P**.

We might naturally consider two further expressions,

$$|\mathbf{Q}| \cos \theta \, |\mathbf{P}| \quad \text{and} \quad |\mathbf{Q}| \sin \theta \, |\mathbf{P}|$$

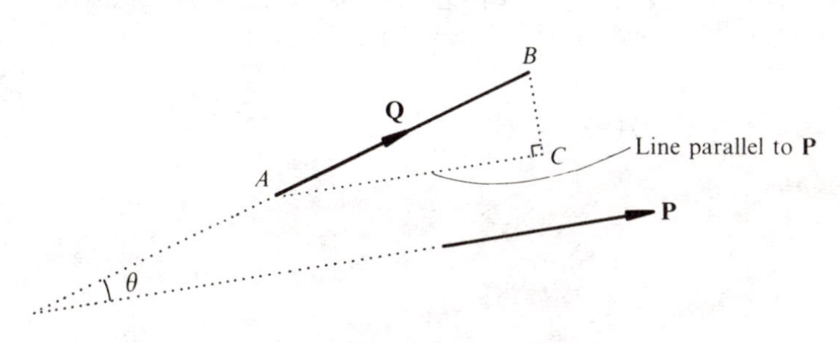

Figure 1.27

in which both **Q** and **P** are treated symmetrically. These two types of 'mutual projection' play a significant role in the further development of vectors and, as we shall see, have numerous applications in mathematics, physics and engineering. In physics, for example, the work done by a force **F** as it moves its point of application along a directed line segment **d** is *defined* as the product of the orthogonal projection of **F** onto **d** with $|\mathbf{d}|$; that is,

$$\text{work done} = |\mathbf{F}| \cos\theta \, |\mathbf{d}|$$

where θ is the angle between **F** and **d**.

A second elementary application of projections is the determination of the area of a parallelogram whose sides are the vectors **P** and **Q** (see Figure 1.28). The area is the product of the perpendicular projection of **P** onto **Q** with $|\mathbf{Q}|$; that is

$$\text{area of parallelogram} = |\mathbf{P}| \sin\theta \, |\mathbf{Q}|$$

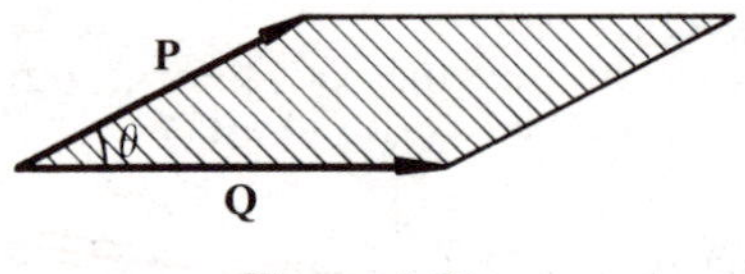

Figure 1.28

Still considering these two examples, however, whilst work is essentially a *scalar* quantity, an area is more conveniently regarded as a *vector*. As with a straight line segment, a plane area has both a magnitude and a direction associated with it. If **A** denotes the vector area, then $|\mathbf{A}|$ is the purely numerical value of the area, whilst the direction of **A** is in the direction of the normal to the plane area (see Figure 1.29). Thus we can write $\mathbf{A} = |\mathbf{A}| \, \hat{\mathbf{n}}$. There still remains some ambiguity about the direction of the normal – does it point outwards or inwards? The sense to be chosen will be defined in Section 1.6, after our discussion of the vector product.

Rather than checking that **A** enjoys the final characteristic required of a vector (the parallelogram law of addition), we shall regard vector area as a **derived quantity** – derived from 'simpler vectors', such as the directed line segment. This will be more fully discussed in Section 1.6.

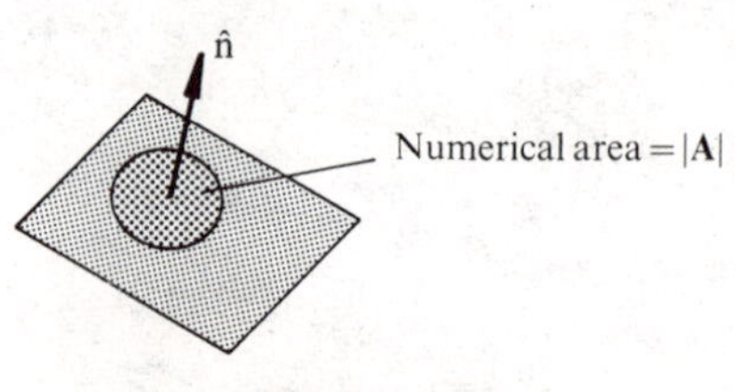

Figure 1.29

Thus far, two vectors have been combined through the process of addition to produce a new vector. In the next two sections we shall define two further distinct ways of combining vectors. The definitions we shall use will formalize the above ideas concerning the mutual projections of two vectors.

1.5 The scalar product

The **scalar product** of two vectors $\mathbf{a}$ and $\mathbf{b}$ is written $\mathbf{a}\cdot\mathbf{b}$ (often called the 'dot product') and is defined as the scalar quantity

$$\mathbf{a}\cdot\mathbf{b} = |\mathbf{a}||\mathbf{b}| \cos\theta \tag{1.21}$$

Here θ is the smaller angle between the vectors $\mathbf{a}$ and $\mathbf{b}$ when they are directed away from a common point, and so $0 \leqslant \theta \leqslant \pi$ (see Figure 1.30).

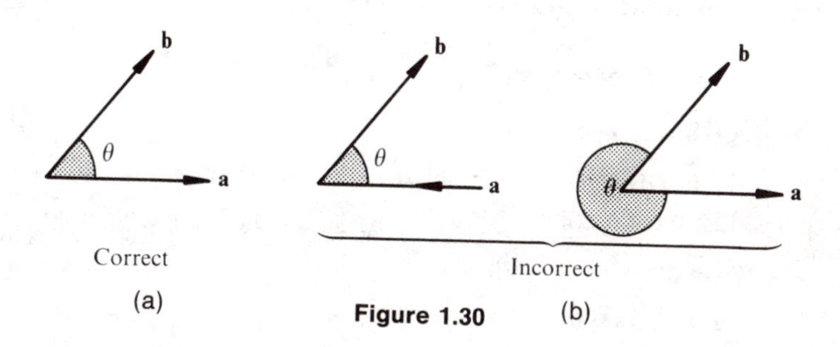

Figure 1.30

It is clear from the definition that

$$\mathbf{b}\cdot\mathbf{a} = |\mathbf{b}||\mathbf{a}| \cos\theta = \mathbf{a}\cdot\mathbf{b} \tag{1.22}$$

so that the scalar product is commutative.

It is easily shown that the distributive law for scalar multiplication,

$$s(t+u) = st + su \tag{1.23}$$

also holds for the scalar product. This is done by realizing, from the definition (1.21), that $\mathbf{a}\cdot\mathbf{b}$ can be regarded as the product of the orthogonal projection of $\mathbf{b}$ onto $\mathbf{a}$ (namely $|\mathbf{b}| \cos\theta$) multiplied by $|\mathbf{a}|$. Now consider any three vectors $\mathbf{a}$, $\mathbf{b}$ and $\mathbf{c}$. They can always be arranged (by sliding if necessary) as in Figure 1.31. From this figure it is clear that the sum of the orthogonal projections of $\mathbf{b}$ and $\mathbf{c}$ onto $\mathbf{a}$, that is $OP'' + P'Q$ is equal to $OP'' + P''Q'$, which is the orthogonal projection of $(\mathbf{b}+\mathbf{c})$ onto $\mathbf{a}$. Therefore

$$\mathbf{a}\cdot(\mathbf{b}+\mathbf{c}) = \mathbf{a}\cdot\mathbf{b} + \mathbf{a}\cdot\mathbf{c} \tag{1.24}$$

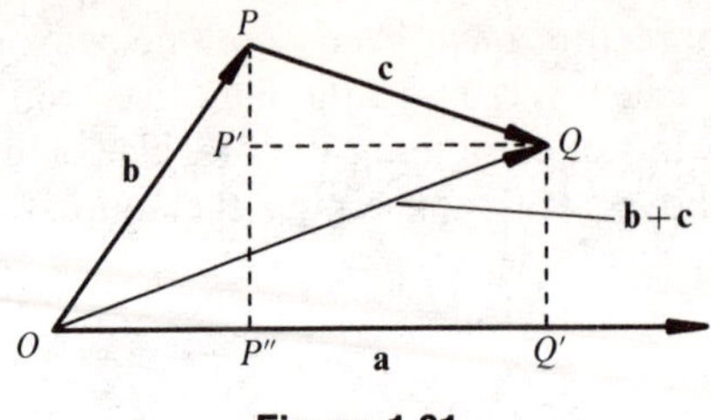

Figure 1.31

By repeatedly applying this basic result it is clear that the scalar product of two vector sums may be expanded as in scalar algebra. Thus

$$(\mathbf{a}+\mathbf{b}+\cdots)\cdot(\mathbf{p}+\mathbf{q}+\cdots)=\mathbf{a}\cdot\mathbf{p}+\mathbf{a}\cdot\mathbf{q}+\cdots+\mathbf{b}\cdot\mathbf{p}+\mathbf{b}\cdot\mathbf{q}+\cdots \tag{1.25}$$

Elementary properties of the scalar product

(1) If **a** and **b** are perpendicular vectors, then

$$\mathbf{a}\cdot\mathbf{b}=|\mathbf{a}||\mathbf{b}|\cos\tfrac{1}{2}\pi=0 \tag{1.26}$$

Conversely, if $\mathbf{a}\cdot\mathbf{b}=0$ and $|\mathbf{a}|\neq 0$, $|\mathbf{b}|\neq 0$, then **a** and **b** must be perpendicular vectors.

(2) If **a** and **b** are parallel, then

$$\mathbf{a}\cdot\mathbf{b}=\pm|\mathbf{a}||\mathbf{b}|$$

the positive sign being taken if **a** and **b** are in the same direction ($\theta=0$) and the negative sign if **a** and **b** are in opposite directions ($\theta=\pi$). In particular, if $\mathbf{b}\equiv\mathbf{a}$, then

$$\mathbf{a}\cdot\mathbf{a}=|\mathbf{a}|^2 \tag{1.27}$$

(3) It follows from the definition (1.21) that if s is a scalar, then

$$\mathbf{a}\cdot(s\mathbf{b})=s\mathbf{a}\cdot\mathbf{b}=(\mathbf{a}\cdot\mathbf{b})s \tag{1.28}$$

Component form of the scalar product

Again using the definition of the scalar product, we have immediately

$$\begin{aligned}
&\hat{\mathbf{i}}\cdot\hat{\mathbf{i}}=1, \quad &&\hat{\mathbf{i}}\cdot\hat{\mathbf{j}}=0, \quad &&\hat{\mathbf{i}}\cdot\hat{\mathbf{k}}=0\\
&\hat{\mathbf{j}}\cdot\hat{\mathbf{i}}=0, &&\hat{\mathbf{j}}\cdot\hat{\mathbf{j}}=1, &&\hat{\mathbf{j}}\cdot\hat{\mathbf{k}}=0\\
&\hat{\mathbf{k}}\cdot\hat{\mathbf{i}}=0, &&\hat{\mathbf{k}}\cdot\hat{\mathbf{j}}=0, &&\hat{\mathbf{k}}\cdot\hat{\mathbf{k}}=1
\end{aligned}$$

Hence, using the distributive property (1.25), if

$$\mathbf{a} = a_x\hat{\mathbf{i}} + a_y\hat{\mathbf{j}} + a_z\hat{\mathbf{k}} \quad \text{and} \quad \mathbf{b} = b_x\hat{\mathbf{i}} + b_y\hat{\mathbf{j}} + b_z\hat{\mathbf{k}}$$

then

$$\begin{aligned}\mathbf{a}\cdot\mathbf{b} &= (a_x\hat{\mathbf{i}} + a_y\hat{\mathbf{j}} + a_z\hat{\mathbf{k}})\cdot(b_x\hat{\mathbf{i}} + b_y\hat{\mathbf{j}} + b_z\hat{\mathbf{k}}) \\ &= a_xb_x + a_yb_y + a_zb_z \end{aligned} \tag{1.29}$$

which is the component form of the scalar product.

The result (1.29), in conjunction with (1.21), provides a convenient method for determining the angle θ between two vectors:

$$\mathbf{a}\cdot\mathbf{b} = |\mathbf{a}||\mathbf{b}|\cos\theta = a_xb_x + a_yb_y + a_zb_z$$

$$\therefore \qquad \theta = \cos^{-1}\left[\frac{a_xb_x + a_yb_y + a_zb_z}{|\mathbf{a}||\mathbf{b}|}\right] \tag{1.30}$$

Example 1.6

A particle, acted on by constant forces $\mathbf{F}_1 = 4\hat{\mathbf{i}} + \hat{\mathbf{j}} - 3\hat{\mathbf{k}}$ and $\mathbf{F}_2 = 3\hat{\mathbf{i}} + \hat{\mathbf{j}} - \hat{\mathbf{k}}$ (both measured in newtons), is displaced from the point (1, 2, 3) to the point (5, 4, 1) (measured in metres). Find the total work done by the forces.

Solution

Figure 1.32 shows the displacement of the particle and the forces acting on it. Although the forces are shown acting at the initial point A, they

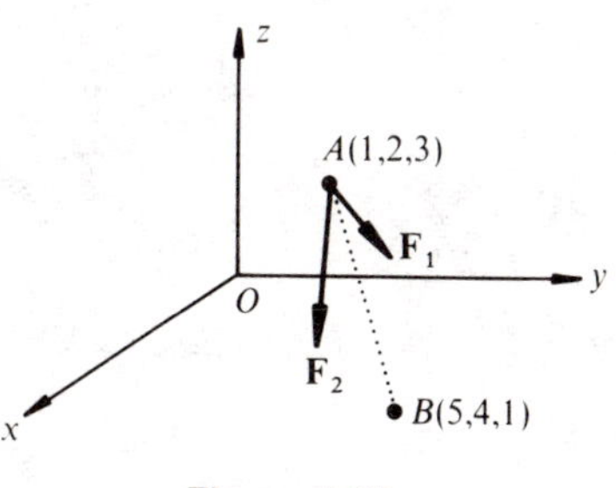

Figure 1.32

are assumed to act on the particle throughout the displacement from A to B. The resultant force is

$$\mathbf{F} = \mathbf{F}_1 + \mathbf{F}_2 = 7\hat{\mathbf{i}} + 2\hat{\mathbf{j}} - 4\hat{\mathbf{k}}$$

The displacement is the vector $\mathbf{d} = \overrightarrow{AB}$. But

$$\overrightarrow{OA} + \overrightarrow{AB} = \overrightarrow{OB}$$

$$\therefore \qquad \overrightarrow{AB} = \overrightarrow{OB} - \overrightarrow{OA} = (5\hat{\mathbf{i}} + 4\hat{\mathbf{j}} + \hat{\mathbf{k}}) - (\hat{\mathbf{i}} + 2\hat{\mathbf{j}} + 3\hat{\mathbf{k}})$$

and so

$$\mathbf{d} = 4\hat{\mathbf{i}} + 2\hat{\mathbf{j}} - 2\hat{\mathbf{k}}$$

The work done, W, is given by $\mathbf{F} \cdot \mathbf{d}$ (see Section 1.4). Therefore

$$W = (7\hat{\mathbf{i}} + 2\hat{\mathbf{j}} - 4\hat{\mathbf{k}}) \cdot (4\hat{\mathbf{i}} + 2\hat{\mathbf{j}} - 2\hat{\mathbf{k}}) = 28 + 4 + 8 = 40 \text{ joules}$$

Example 1.7

Derive the cosine rule for a triangle ABC; that is, show that

$$(BC)^2 = (AB)^2 + (AC)^2 - 2(AB)(AC)\cos\theta$$

Solution

Let $\mathbf{a}$ and $\mathbf{b}$ represent two sides of a triangle and θ the angle between them. Then, as shown in Figure 1.33, the third side is the vector $\mathbf{a} + \mathbf{b}$. If we form the scalar product of $\mathbf{a} + \mathbf{b}$ with itself (in order to determine the length of this side), then we find

$$(\mathbf{a} + \mathbf{b}) \cdot (\mathbf{a} + \mathbf{b}) = \mathbf{a} \cdot \mathbf{a} + \mathbf{a} \cdot \mathbf{b} + \mathbf{b} \cdot \mathbf{a} + \mathbf{b} \cdot \mathbf{b}$$

that is,

$$\begin{aligned} |\mathbf{a} + \mathbf{b}|^2 &= |\mathbf{a}|^2 + |\mathbf{b}|^2 + 2\mathbf{a} \cdot \mathbf{b} \\ &= |\mathbf{a}|^2 + |\mathbf{b}|^2 + 2|\mathbf{a}||\mathbf{b}|\cos(180^\circ - \theta) \end{aligned}$$

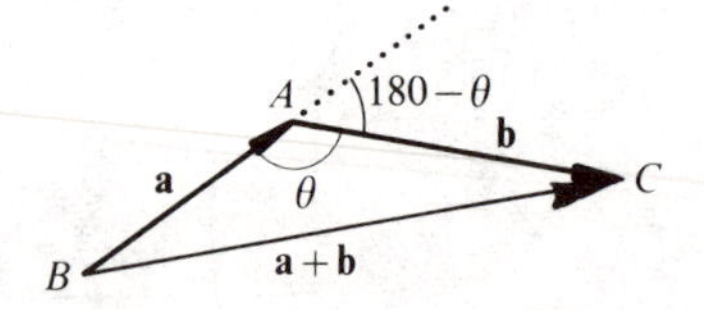

Figure 1.33

We have intentionally drawn Figure 1.33 to illustrate the proper use of the definition (1.21) concerning the measurement of the angle between two vectors. Hence, noting that $\cos(180^\circ - \theta) = -\cos\theta$,

$$|\mathbf{a} + \mathbf{b}|^2 = |\mathbf{a}|^2 + |\mathbf{b}|^2 - 2|\mathbf{a}||\mathbf{b}|\cos\theta \qquad \textbf{(1.31)}$$

which is the cosine rule of trigonometry. Note that Pythagoras' theorem is obtained for right-angled triangles ($\theta = \pi/2$).

Example 1.8

(a) Determine the equation of a plane which is perpendicular to the vector $3\hat{\mathbf{i}} + 2\hat{\mathbf{j}} + \hat{\mathbf{k}}$ and which passes through the point (2, 0, 1).
(b) Determine the shortest distance from the origin to this plane.

Solution

(a) Before solving this problem we shall derive the general vector form for the equation of a plane which is perpendicular to a given vector and which passes through a given point.

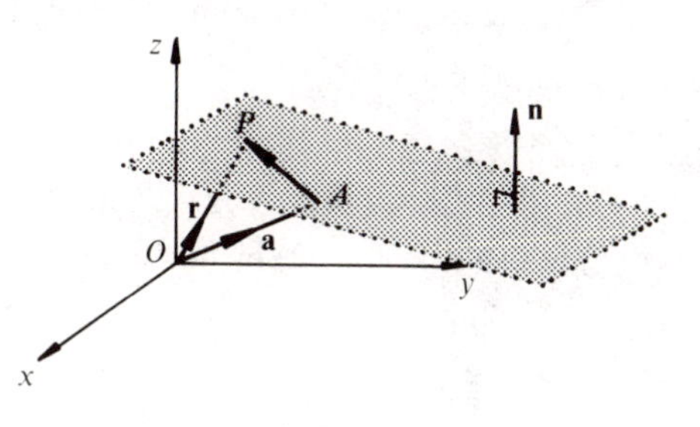

Figure 1.34

Let A (with position vector $\mathbf{a}$) be a given point on the plane and $\mathbf{n}$ a vector (not necessarily of unit length) perpendicular to the plane. If P is a general point on the plane with position vector $\mathbf{r}$ (see Figure 1.34), then the vector $\overrightarrow{AP} = \mathbf{r} - \mathbf{a}$ lies in the plane. All vectors lying in the plane are perpendicular to $\mathbf{n}$, and so

$$(\mathbf{r} - \mathbf{a}) \cdot \mathbf{n} = 0$$

$$\therefore \quad \mathbf{r} \cdot \mathbf{n} = \mathbf{a} \cdot \mathbf{n} \tag{1.32}$$

Equation (1.32) is the vector equation of the required plane. The equation may be expressed in the more usual Cartesian form by writing each vector in it in component form. As usual, we let $\mathbf{r} = x\hat{\mathbf{i}} + y\hat{\mathbf{j}} + z\hat{\mathbf{k}}$. If we let $\mathbf{n} = n_x\hat{\mathbf{i}} + n_y\hat{\mathbf{j}} + n_z\hat{\mathbf{k}}$ and $\mathbf{a} = a_x\hat{\mathbf{i}} + a_y\hat{\mathbf{j}} + a_z\hat{\mathbf{k}}$, then (1.32) becomes

$$xn_x + yn_y + zn_z = a_xn_x + a_yn_y + a_zn_z$$

In the particular example given,

$$\mathbf{n} = 3\hat{\mathbf{i}} + 2\hat{\mathbf{j}} + \hat{\mathbf{k}} \quad \text{and} \quad \mathbf{a} = 2\hat{\mathbf{i}} + \hat{\mathbf{k}}$$

therefore the equation of the plane is

$$3x + 2y + z = 6 + 1 = 7$$

(b) The shortest distance from the origin to the plane must be in a direction which is normal to the plane, that is, in the direction of $\mathbf{n}$.

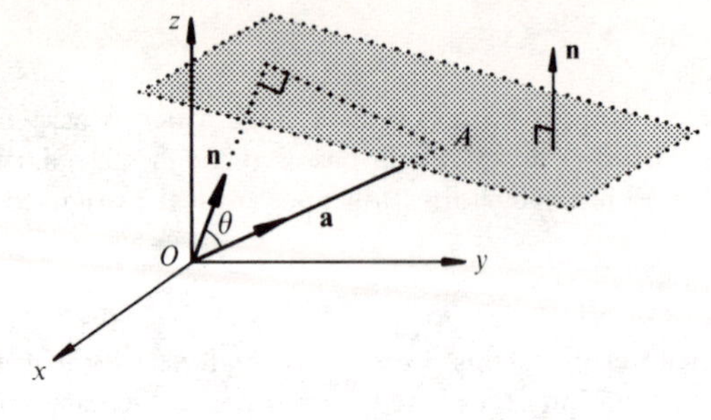

Figure 1.35

From Figure 1.35, the shortest distance is $||\mathbf{a}|\cos\theta|$. Now

$$\mathbf{a}\cdot\mathbf{n} = |\mathbf{a}|\,|\mathbf{n}|\cos\theta$$

$$\therefore \qquad |\mathbf{a}|\cos\theta = \frac{\mathbf{a}\cdot\mathbf{n}}{|\mathbf{n}|} = \mathbf{a}\cdot\hat{\mathbf{n}}$$

But on dividing both sides of (1.32) by $|\mathbf{n}|$ we find

$$\mathbf{r}\cdot\frac{\mathbf{n}}{|\mathbf{n}|} = \mathbf{a}\cdot\frac{\mathbf{n}}{|\mathbf{n}|}$$

$$\therefore \qquad \mathbf{r}\cdot\hat{\mathbf{n}} = \mathbf{a}\cdot\hat{\mathbf{n}} \tag{1.33}$$

Hence, if the equation of a plane is written such that the normal vector that is used is a *unit* vector, then the shortest distance from the origin to the plane is the modulus of the constant term $\mathbf{a}\cdot\hat{\mathbf{n}}$ on the right-hand side of (1.33).

In the particular example, the equation of the plane

$$3x + 2y + z = 7$$

is rearranged by dividing by $|\mathbf{n}|$ to obtain the required form (1.33):

$$\frac{3x + 2y + z}{\sqrt{(9 + 4 + 1)}} = \frac{7}{\sqrt{(9 + 4 + 1)}}$$

that is

$$\frac{3}{\sqrt{14}}x + \frac{2}{\sqrt{14}}y + \frac{1}{\sqrt{14}}z = \frac{7}{\sqrt{14}}$$

The shortest distance from the origin to the plane is then the right-hand side, $7/\sqrt{14}$ or 1.8708.

EXERCISES

1.27 If $\mathbf{a} = \hat{\mathbf{i}} + 3\hat{\mathbf{j}} - 4\hat{\mathbf{k}}$ and $\mathbf{b} = -2\hat{\mathbf{i}} - 3\hat{\mathbf{k}}$, find $|\mathbf{a}|$, $|\mathbf{b}|$, $\mathbf{a}\cdot\mathbf{b}$ and $|\mathbf{a}\cdot\mathbf{b}|$.

1.28 Find the equation of the plane P_1 through the point $2\hat{\mathbf{i}} + 3\hat{\mathbf{j}} - \hat{\mathbf{k}}$ and perpendicular to the vector $3\hat{\mathbf{i}} - 4\hat{\mathbf{j}} + \hat{\mathbf{k}}$. A plane P_2 parallel to P_1 passes through the point $(1, -1, -1)$. Find the perpendicular distance between the planes.

1.29 If $\mathbf{a} = 3\hat{\mathbf{i}} + 4\hat{\mathbf{j}} - 2\hat{\mathbf{k}}$, $\mathbf{b} = \hat{\mathbf{i}} - 2\hat{\mathbf{j}} + 5\hat{\mathbf{k}}$ and $\mathbf{c} = 2\hat{\mathbf{i}} + 6\hat{\mathbf{j}} - 7\hat{\mathbf{k}}$, find the angles between $\mathbf{a}$ and $\mathbf{b}$, $\mathbf{b}$ and $\mathbf{c}$, and $\mathbf{c}$ and $\mathbf{a}$, and show that the sum of two of these angles is equal to the third.

1.30 Find the work done by a force $\mathbf{F} = (3\hat{\mathbf{i}} + 4\hat{\mathbf{j}})$N as it moves its point of application from A $(1, 0, -1)$m to B $(6, 6, 1)$m.

1.31 Suppose that $\mathbf{a}\cdot\mathbf{c} = \mathbf{b}\cdot\mathbf{c}$ for *all* $\mathbf{c}$. By choosing $\mathbf{c}$ appropriately, show that $\mathbf{a} = \mathbf{b}$.

1.32 Show that the vectors $\mathbf{a} = 2\hat{\mathbf{i}} - \hat{\mathbf{j}} + \hat{\mathbf{k}}$, $\mathbf{b} = \hat{\mathbf{i}} - 3\hat{\mathbf{j}} - 5\hat{\mathbf{k}}$ and $\mathbf{c} = 3\hat{\mathbf{i}} - 4\hat{\mathbf{j}} - 4\hat{\mathbf{k}}$ form the sides of a right-angled triangle.

1.33 Three forces, one of magnitude 8 N in the direction of $\hat{\mathbf{i}}$, a second of magnitude 10 N in the direction of $3\hat{\mathbf{i}} + 4\hat{\mathbf{j}}$ and a third of magnitude 13 N in the direction of $5\hat{\mathbf{i}} + 12\hat{\mathbf{k}}$ displace a particle from A $(3, 2, -6)$m to B $(5, -1, 0)$m. Find the work done on the particle.

1.34 If $\mathbf{a} = \cos\theta\,\hat{\mathbf{i}} + \sin\theta\,\hat{\mathbf{j}}$ and $\mathbf{b} = \cos\phi\,\hat{\mathbf{i}} + \sin\phi\,\hat{\mathbf{j}}$ show, using the dot product, that

$$\cos(\theta - \phi) = \cos\theta\cos\phi + \sin\theta\sin\phi$$

1.35 Find the angle between the planes P_1, $4x - 2y + 2z - 11 = 0$, and P_2, $x + 0.666y - 0.333z = 4$.

1.36 If $\mathbf{a} = 3\hat{\mathbf{i}} + 4\hat{\mathbf{j}} - 7\hat{\mathbf{k}}$, $\mathbf{b} = 7\hat{\mathbf{i}} + 7\hat{\mathbf{j}} - 3\hat{\mathbf{k}}$ and $\mathbf{c} = \hat{\mathbf{i}} - \hat{\mathbf{j}}$, evaluate $\mathbf{a}\cdot\mathbf{b}$ and $\mathbf{a}\cdot\mathbf{c}$ and verify that $\mathbf{a}\cdot(\mathbf{b} + \mathbf{c}) = \mathbf{a}\cdot\mathbf{b} + \mathbf{a}\cdot\mathbf{c}$.

1.37 Prove that the four points $(3, 0, 0)$, $(4, -1, 0)$, $(0, -2, 10)$ and $(3.5, 0, -1)$ are coplanar.

1.38 A particle moves under the action of a constant force $\mathbf{F}$ around a *closed* polygon. Show that the work done on the particle is zero.

1.39 (a) Show that the perpendicular bisectors of the sides of a triangle meet in a point.
(b) ABC is a triangle. Show that the lines AD, BE and CF, which are perpendicular to BC, AC and AB respectively, intersect in a point.

1.6 The vector product

The **vector product** of two vectors $\mathbf{a}$ and $\mathbf{b}$ is written $\mathbf{a} \wedge \mathbf{b}$ (often called the 'cross product' and alternatively written as $\mathbf{a} \times \mathbf{b}$). It is defined as the vector quantity

$$\mathbf{a} \wedge \mathbf{b} = |\mathbf{a}|\,|\mathbf{b}| \sin\theta\,\hat{\mathbf{n}} \qquad \textbf{(1.34)}$$

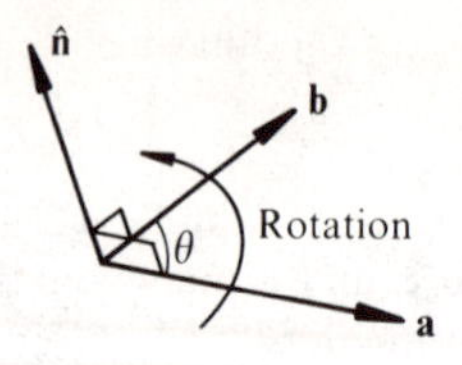

Figure 1.36

where θ is the smaller angle between $\mathbf{a}$ and $\mathbf{b}$ (as used in the scalar product) and $\hat{\mathbf{n}}$ is a unit vector perpendicular to both $\mathbf{a}$ and $\mathbf{b}$. The sense of $\hat{\mathbf{n}}$ is such that $\mathbf{a}, \mathbf{b}, \hat{\mathbf{n}}$ taken *in this order* form a right-handed system of vectors; that is, as we rotate from $\mathbf{a}$ to $\mathbf{b}$ through the angle θ, a screw with a right-handed thread would move in the direction of $\hat{\mathbf{n}}$ (see Figure 1.36).

If we reverse the order of $\mathbf{a}$ and $\mathbf{b}$ then, according to the definition,

$$\mathbf{a} \wedge \mathbf{b} = |\mathbf{a}|\,|\mathbf{b}| \sin\theta\,\hat{\mathbf{m}}$$

where $\hat{\mathbf{m}}$ is perpendicular to both $\mathbf{b}$ and $\mathbf{a}$ and such that $\mathbf{b}, \mathbf{a}, \hat{\mathbf{m}}$ taken in this order form a right-handed system. But clearly, as we rotate from $\mathbf{b}$ to $\mathbf{a}$, a screw with a right-handed thread would move in the direction of $-\hat{\mathbf{n}}$, that is $\hat{\mathbf{m}} = -\hat{\mathbf{n}}$. Hence,

$$\mathbf{b} \wedge \mathbf{a} = -|\mathbf{b}|\,|\mathbf{a}| \sin\theta\,\hat{\mathbf{n}} = -\mathbf{a} \wedge \mathbf{b} \qquad \textbf{(1.35)}$$

Therefore the vector product is *not* commutative and order *is* important.

We now state that the distributive law also holds for the vector product:

$$\mathbf{a} \wedge (\mathbf{b} + \mathbf{c}) = \mathbf{a} \wedge \mathbf{b} + \mathbf{a} \wedge \mathbf{c} \qquad \textbf{(1.36)}$$

The proof of this statement is not quite so immediate as was the case for the scalar product. Nevertheless we shall prove it after a brief discussion on area and volume.

We have already introduced the idea that area may be regarded as a vector quantity in Section 1.4 and with the definition of the vector product this idea can now be formalized. A parallelogram with sides represented by the vectors $\mathbf{a}$ and $\mathbf{b}$ (see Figure 1.37a) is *defined* to have the vector area $\mathbf{A}$, given by the expression

$$\mathbf{A} = \mathbf{a} \wedge \mathbf{b} = |\mathbf{a}|\,|\mathbf{b}| \sin\theta\,\hat{\mathbf{n}} \qquad \textbf{(1.37)}$$

Clearly $|\mathbf{A}|$ is the correct numerical value of the area $(= |\mathbf{a}|\,|\mathbf{b}| \sin\theta)$, but now we are giving a direction to the area, namely the direction of $\mathbf{a} \wedge \mathbf{b}$.

A prism (parallelepiped) will be formed if the parallelogram shape is 'projected' into the third dimension along a vector $\mathbf{c}$ (see Figure 1.37b). As with any prism, the volume of the solid is simply the cross-sectional area of the base multiplied by the vertical height. But if $\hat{\mathbf{n}}$ is a unit vector normal to the plane containing $\mathbf{a}$ and $\mathbf{b}$, then the vertical height is $\mathbf{c} \cdot \hat{\mathbf{n}}$. Therefore,

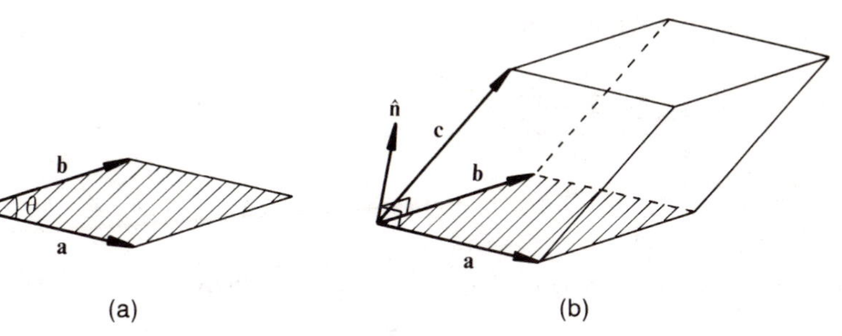

(a) (b)

Figure 1.37

utilizing (1.28),

$$\text{volume of prism} = (\mathbf{c}\cdot\hat{\mathbf{n}})(|\mathbf{a}|\ |\mathbf{b}| \sin\theta) = \mathbf{c}\cdot(|\mathbf{a}|\ |\mathbf{b}| \sin\theta\ \hat{\mathbf{n}}) = \mathbf{c}\cdot(\mathbf{a}\wedge\mathbf{b}) \tag{1.38}$$

But we could have taken the base of the prism to be formed by **b** and **c** or by **c** and **a**, giving, by symmetry arguments,

$$\text{volume of prism} = \mathbf{c}\cdot(\mathbf{a}\wedge\mathbf{b}) = \mathbf{a}\cdot(\mathbf{b}\wedge\mathbf{c}) = \mathbf{b}\cdot(\mathbf{c}\wedge\mathbf{a}) \tag{1.39}$$

Each of the products in (1.39) is called a **scalar triple product** and the equalities shown are actually true for any three vectors **a**, **b** and **c**.

We are now in a position to prove the distributive law (1.36) for vector products. Consider forming the scalar product of the expression $[\mathbf{a}\wedge(\mathbf{b}+\mathbf{c})-\mathbf{a}\wedge\mathbf{b}-\mathbf{a}\wedge\mathbf{c}]$ with an arbitrary non-zero vector **q**:

$$\begin{aligned}\mathbf{q}\cdot[\mathbf{a}\wedge(\mathbf{b}+\mathbf{c})-\mathbf{a}\wedge\mathbf{b}-\mathbf{a}\wedge\mathbf{c}]&\\ =\mathbf{q}\cdot[\mathbf{a}\wedge(\mathbf{b}+\mathbf{c})]-\mathbf{q}\cdot(\mathbf{a}\wedge\mathbf{b})-\mathbf{q}\cdot(\mathbf{a}\wedge\mathbf{c})&\end{aligned}$$

using the distributive law for scalar products. Now each of the three terms on the right-hand side of this equation may be rearranged using (1.39), and so we can write

$$\begin{aligned}\mathbf{q}\cdot[\mathbf{a}\wedge(\mathbf{b}+\mathbf{c})-\mathbf{a}\wedge\mathbf{b}-\mathbf{a}\wedge\mathbf{c}]&=(\mathbf{b}+\mathbf{c})\cdot(\mathbf{q}\wedge\mathbf{a})-\mathbf{b}\cdot(\mathbf{q}\wedge\mathbf{a})-\mathbf{c}\cdot(\mathbf{q}\wedge\mathbf{a})\\ &=[\mathbf{b}+\mathbf{c}-\mathbf{b}-\mathbf{c}]\cdot(\mathbf{q}\wedge\mathbf{a})=0\end{aligned} \tag{1.40}$$

again using the distributive law for scalar products.

But since **q** is an arbitrary non-zero vector, we must conclude from (1.40) that

$$[\mathbf{a}\wedge(\mathbf{b}+\mathbf{c})-\mathbf{a}\wedge\mathbf{b}-\mathbf{a}\wedge\mathbf{c}]=0$$

that is,

$$\mathbf{a}\wedge(\mathbf{b}+\mathbf{c})=\mathbf{a}\wedge\mathbf{b}+\mathbf{a}\wedge\mathbf{c} \tag{1.41}$$

which is the distributive law for vector products.

We may now deduce that for any vectors **a**, **b**, . . . , **p**, **q**, . . .

$$(\mathbf{a}+\mathbf{b}+\cdots)\wedge(\mathbf{p}+\mathbf{q}+\cdots)=\mathbf{a}\wedge\mathbf{p}+\mathbf{a}\wedge\mathbf{q}+\cdots+\mathbf{b}\wedge\mathbf{p}+\mathbf{b}\wedge\mathbf{q}+\cdots \tag{1.42}$$

Elementary properties of the vector product

(1) If $\mathbf{a}$ and $\mathbf{b}$ are parallel vectors ($\theta = 0$ or $\theta = \pi$), then

$$\mathbf{a} \wedge \mathbf{b} = |\mathbf{a}|\,|\mathbf{b}| \sin\theta\, \hat{\mathbf{n}} = 0 \tag{1.43}$$

Conversely, if $\mathbf{a} \wedge \mathbf{b} = 0$ and $|\mathbf{a}| \neq 0$, $|\mathbf{b}| \neq 0$, then $\mathbf{a}$ and $\mathbf{b}$ must be parallel vectors.

(2) If $\hat{\mathbf{a}}$ and $\hat{\mathbf{b}}$ are perpendicular unit vectors, then

$$\hat{\mathbf{a}} \wedge \hat{\mathbf{b}} = |\hat{\mathbf{a}}|\,|\hat{\mathbf{b}}| \sin\tfrac{1}{2}\pi\, \hat{\mathbf{n}} = \hat{\mathbf{n}}$$

where $\hat{\mathbf{a}}$, $\hat{\mathbf{b}}$ and $\hat{\mathbf{n}}$ form a right-handed system of unit vectors (similar to $\hat{\mathbf{i}}$, $\hat{\mathbf{j}}$ and $\hat{\mathbf{k}}$).

Component form of the vector product

From the above properties we can quickly deduce that

$$\begin{array}{lll} \hat{\mathbf{i}} \wedge \hat{\mathbf{i}} = 0, & \hat{\mathbf{i}} \wedge \hat{\mathbf{j}} = \hat{\mathbf{k}}, & \hat{\mathbf{i}} \wedge \hat{\mathbf{k}} = -\hat{\mathbf{j}} \\ \hat{\mathbf{j}} \wedge \hat{\mathbf{i}} = -\hat{\mathbf{k}}, & \hat{\mathbf{j}} \wedge \hat{\mathbf{j}} = 0, & \hat{\mathbf{j}} \wedge \hat{\mathbf{k}} = \hat{\mathbf{i}} \\ \hat{\mathbf{k}} \wedge \hat{\mathbf{i}} = \hat{\mathbf{j}}, & \hat{\mathbf{k}} \wedge \hat{\mathbf{j}} = -\hat{\mathbf{i}}, & \hat{\mathbf{k}} \wedge \hat{\mathbf{k}} = 0 \end{array} \tag{1.44}$$

An easy way to remember these results is to write the vectors $\hat{\mathbf{i}}$, $\hat{\mathbf{j}}$ and $\hat{\mathbf{k}}$ in a list

$$\hat{\mathbf{i}} \quad \hat{\mathbf{j}} \quad \hat{\mathbf{k}} \quad \hat{\mathbf{i}} \quad \hat{\mathbf{j}} \quad \hat{\mathbf{k}}$$

Now the vector product of any two consecutive vectors (reading from left to right) in the list is equal to the vector immediately following:

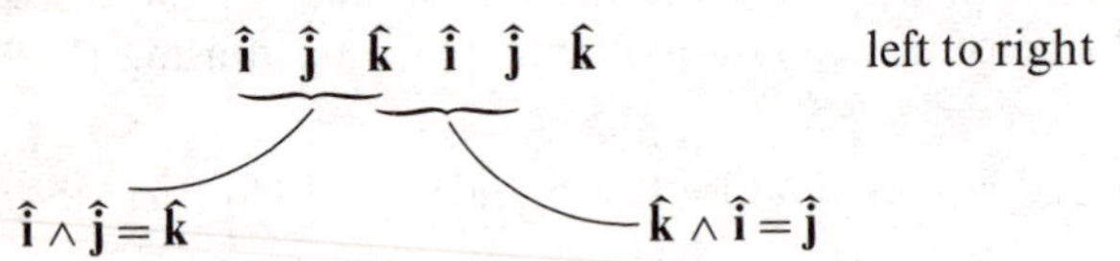

If, instead, we read from right to left, then the vector product of any two consecutive vectors in the list is equal to *minus* the immediately preceding vector:

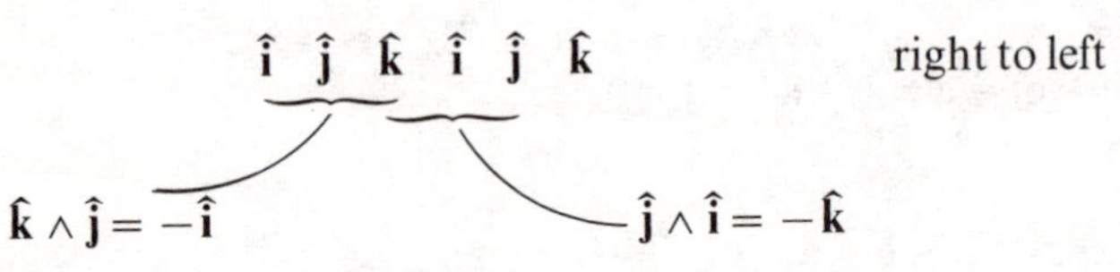

We are now in a position to derive the component form of the vector product. If

$$\mathbf{a} = a_x\hat{\mathbf{i}} + a_y\hat{\mathbf{j}} + a_z\hat{\mathbf{k}} \quad \text{and} \quad \mathbf{b} = b_x\hat{\mathbf{i}} + b_y\hat{\mathbf{j}} + b_z\hat{\mathbf{k}}$$

then $\quad \mathbf{a} \wedge \mathbf{b} = (a_x\hat{\mathbf{i}} + a_y\hat{\mathbf{j}} + a_z\hat{\mathbf{k}}) \wedge (b_x\hat{\mathbf{i}} + b_y\hat{\mathbf{j}} + b_z\hat{\mathbf{k}})$

and by the distributive law for the vector product we find

$$\begin{aligned}\mathbf{a}\wedge\mathbf{b} &= a_x b_x \hat{\mathbf{i}}\wedge\hat{\mathbf{i}} + a_x b_y \hat{\mathbf{i}}\wedge\hat{\mathbf{j}} + a_x b_z \hat{\mathbf{i}}\wedge\hat{\mathbf{k}} \\ &\quad + a_y b_x \hat{\mathbf{j}}\wedge\hat{\mathbf{i}} + a_y b_y \hat{\mathbf{j}}\wedge\hat{\mathbf{j}} + a_y b_z \hat{\mathbf{j}}\wedge\hat{\mathbf{k}} \\ &\quad + a_z b_x \hat{\mathbf{k}}\wedge\hat{\mathbf{i}} + a_z b_y \hat{\mathbf{k}}\wedge\hat{\mathbf{j}} + a_z b_z \hat{\mathbf{k}}\wedge\hat{\mathbf{k}}\end{aligned}$$

Using (1.44) and grouping parallel components together, we find

$$\mathbf{a}\wedge\mathbf{b} = (a_y b_z - a_z b_y)\hat{\mathbf{i}} + (a_z b_x - a_x b_z)\hat{\mathbf{j}} + (a_x b_y - a_y b_x)\hat{\mathbf{k}} \tag{1.45}$$

For those who have experience of determinants, this may be more formally expressed as

$$\mathbf{a}\wedge\mathbf{b} = \begin{vmatrix} \hat{\mathbf{i}} & \hat{\mathbf{j}} & \hat{\mathbf{k}} \\ a_x & a_y & a_z \\ b_x & b_y & b_z \end{vmatrix} \tag{1.46}$$

where expansion about the top row gives us back (1.45). (An introduction to determinants, suitable for use in vector analysis, is given in Appendix 1.)

Example 1.9

Find a unit vector perpendicular to the vectors $\mathbf{a} = 3\hat{\mathbf{i}} + \hat{\mathbf{j}}$ and $\mathbf{b} = -\hat{\mathbf{i}} + 2\hat{\mathbf{j}} + 2\hat{\mathbf{k}}$.

Solution

A vector perpendicular to **a** and **b** is $\mathbf{a}\wedge\mathbf{b}$. Using equation (1.46),

$$\begin{aligned}\mathbf{a}\wedge\mathbf{b} &= \begin{vmatrix} \hat{\mathbf{i}} & \hat{\mathbf{j}} & \hat{\mathbf{k}} \\ 3 & 1 & 0 \\ -1 & 2 & 2 \end{vmatrix} \\ &= \hat{\mathbf{i}}(2-0) - \hat{\mathbf{j}}(6-0) + \hat{\mathbf{k}}(6+1) \\ &= 2\hat{\mathbf{i}} - 6\hat{\mathbf{j}} + 7\hat{\mathbf{k}}\end{aligned}$$

A unit vector, perpendicular to **a** and **b**, in this direction is obtained by simply dividing $\mathbf{a}\wedge\mathbf{b}$ by its magnitude. Thus

$$\hat{\mathbf{c}} = \frac{2\hat{\mathbf{i}} - 6\hat{\mathbf{j}} + 7\hat{\mathbf{k}}}{\sqrt{[2^2 + (-6)^2 + 7^2]}} = 0.212\hat{\mathbf{i}} - 0.636\hat{\mathbf{j}} + 0.742\hat{\mathbf{k}}$$

is the required vector. The vector $-\hat{\mathbf{c}}$ is also acceptable, of course.

Interpretation of the components of an area vector

If **A** represents a directed line segment

$$\mathbf{A} = A_x\hat{\mathbf{i}} + A_y\hat{\mathbf{j}} + A_z\hat{\mathbf{k}}$$

then the components A_x, A_y, A_z have an obvious interpretation (noted in equation (1.9)) in terms of the resolution of **A** along the three coordinate axes.

If **A** represents an area, however, then the interpretation of the three components is different. The x-component $A_x = \mathbf{A}\cdot\hat{\mathbf{i}}$ is interpreted as the numerical value of the area **A** obtained by projection of this area onto the zy-plane (with normal $\hat{\mathbf{i}}$). Similarly, A_y $(=\mathbf{A}\cdot\hat{\mathbf{j}})$ and A_z $(=\mathbf{A}\cdot\hat{\mathbf{k}})$ are interpreted as the areas of projection of **A** onto the zx- and xy-planes respectively.

To illustrate this, we consider a parallelogram with sides **a** and **b**. If $\mathbf{a} = a_x\hat{\mathbf{j}} + a_y\hat{\mathbf{j}} + a_z\hat{\mathbf{k}}$ and $\mathbf{b} = b_x\hat{\mathbf{i}} + b_y\hat{\mathbf{j}} + b_z\hat{\mathbf{k}}$, then, using (1.37) and (1.46), the area of this parallelogram is the vector **A**, where

$$\mathbf{A} = \mathbf{a} \wedge \mathbf{b} = \begin{vmatrix} \hat{\mathbf{i}} & \hat{\mathbf{j}} & \hat{\mathbf{k}} \\ a_x & a_y & a_z \\ b_x & b_y & b_z \end{vmatrix}$$

That is, the components of **A** are

$$A_x = a_y b_z - a_z b_y, \qquad A_y = a_z b_x - a_x b_z, \qquad A_z = a_x b_y - a_y b_x$$

If this parallelogram is projected onto the xy-plane (see Figure 1.38), then it will form a parallelogram with sides $\mathbf{a}'$ and $\mathbf{b}'$ such that

$$\mathbf{a}' = a_x\hat{\mathbf{i}} + a_y\hat{\mathbf{j}} \quad \text{and} \quad \mathbf{b}' = b_x\hat{\mathbf{i}} + b_y\hat{\mathbf{j}}$$

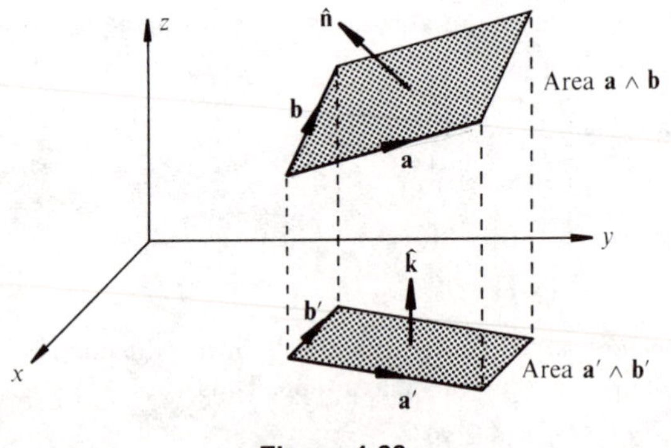

Figure 1.38

The area of this second parallelogram is

$$\mathbf{a}' \wedge \mathbf{b}' = \hat{\mathbf{k}}(a_x b_y - a_y b_x) = \hat{\mathbf{k}} A_z$$

which agrees with our interpretation of A_z outlined above. We can obtain similar verifications of our interpretations of A_y and A_z if **A** is projected onto the other two coordinate planes.

Rotation

Consider a rigid body rotating with a constant angular speed ω radians per second about an axis. Just as a body undergoing linear motion may be described by a velocity vector, so may rotational motion be described by a vector called the **angular velocity**. The magnitude of this vector is the rotational speed ω, whilst its direction is parallel to the axis of rotation (with the positive direction being taken as the direction in which a right-handed screw would advance if it was undergoing the rotation). We denote this angular velocity by $\boldsymbol{\omega}$. Although the body is rotating about a *fixed* axis in space, we shall see in Chapter 8 that the angular velocity vector $\boldsymbol{\omega}$ is a *free vector* and is not necessarily located on this axis. If we look directly along the rotational axis of the body, we simply see a body rotating in a circle with an angular speed ω, rather similar to Figure 1.21 in Example 1.3.

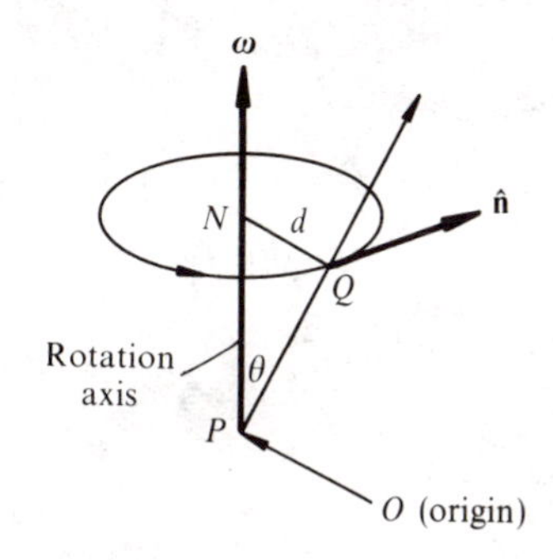

Figure 1.39

From Figure 1.39, the speed at Q is clearly $|\boldsymbol{\omega}|d$ in the direction of the unit vector $\hat{\mathbf{n}}$ (into the page). If P is an arbitrary point on the axis, then the velocity is perpendicular to the plane containing PN and PQ. Clearly,

$$d = |\overrightarrow{PQ}| \sin \theta$$

so $$\overrightarrow{PN} \wedge \overrightarrow{PQ} = |\overrightarrow{PN}|\,|\overrightarrow{PQ}| \sin \theta\, \hat{\mathbf{n}} = d|\overrightarrow{PN}|\hat{\mathbf{n}} \quad \text{or} \quad \frac{\overrightarrow{PN}}{|\overrightarrow{PN}|} \wedge \overrightarrow{PQ} = d\hat{\mathbf{n}}$$

But, if O is an arbitrarily chosen origin, $\overrightarrow{PQ} + \overrightarrow{OP} = \mathbf{r}$, the position vector of Q. Therefore

$$\frac{\overrightarrow{PN}}{|\overrightarrow{PN}|} \wedge (\mathbf{r} - \overrightarrow{OP}) = d\hat{\mathbf{n}}$$

Therefore the linear velocity at point Q is

$$(d\hat{\mathbf{n}})|\boldsymbol{\omega}| = \boldsymbol{\omega} \wedge (\mathbf{r} - \overrightarrow{OP}) \qquad \textbf{(1.47)}$$

(since $\overrightarrow{PN}/|\overrightarrow{PN}|$ is a unit vector in the direction of $\boldsymbol{\omega}$).

Example 1.10

A rigid body is rotating with an angular velocity of 3 rad s^{-1} about an axis passing through the point P (1, 1, 0) and parallel to the vector $2\hat{\mathbf{i}}+\hat{\mathbf{j}}-\hat{\mathbf{k}}$ (see Figure 1.40). Determine the velocity of the rigid body at Q (2, 1, 1).

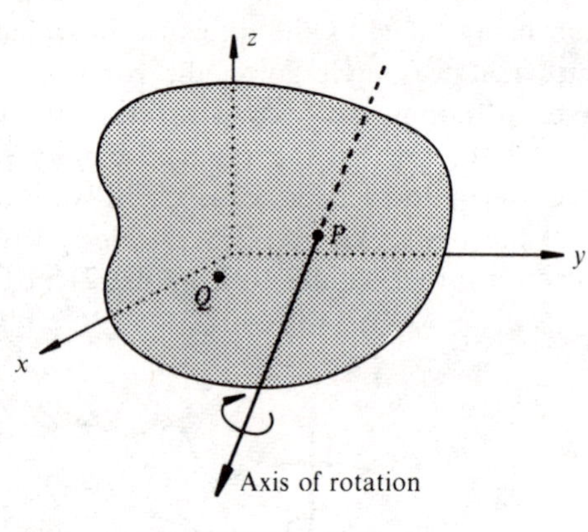

Figure 1.40

Solution

In this example,

$$\boldsymbol{\omega}=3\left(\frac{2\hat{\mathbf{i}}+\hat{\mathbf{j}}-\hat{\mathbf{k}}}{\sqrt{(4+1+1)}}\right)=\frac{3}{\sqrt{6}}(2\hat{\mathbf{i}}+\hat{\mathbf{j}}-\hat{\mathbf{k}})$$

An arbitrary point on the axis is P (1, 1, 0), and the point of interest is Q (2, 1, 1). Therefore the velocity at Q is, from (1.47),

$$\boldsymbol{\omega}\wedge(\mathbf{r}-\overrightarrow{OP})=\frac{(6\hat{\mathbf{i}}+3\hat{\mathbf{j}}-3\hat{\mathbf{k}})}{\sqrt{6}}\wedge(2\hat{\mathbf{i}}+\hat{\mathbf{j}}+\hat{\mathbf{k}}-\hat{\mathbf{i}}-\hat{\mathbf{j}})$$

$$=\frac{(6\hat{\mathbf{i}}+3\hat{\mathbf{j}}-3\hat{\mathbf{k}})\wedge(\hat{\mathbf{i}}+\hat{\mathbf{k}})}{\sqrt{6}}$$

$$=\frac{1}{\sqrt{6}}(3\hat{\mathbf{j}}\wedge\hat{\mathbf{i}}-3\hat{\mathbf{k}}\wedge\hat{\mathbf{i}}+6\hat{\mathbf{i}}\wedge\hat{\mathbf{k}}+3\hat{\mathbf{j}}\wedge\hat{\mathbf{k}})$$

$$=\frac{1}{\sqrt{6}}(-3\hat{\mathbf{k}}-3\hat{\mathbf{j}}-6\hat{\mathbf{j}}+3\hat{\mathbf{i}})$$

$$=\frac{1}{\sqrt{6}}(3\hat{\mathbf{i}}-9\hat{\mathbf{j}}-3\hat{\mathbf{k}})$$

Alternatively, using the determinant form of the cross product,

$$\text{velocity} = \frac{1}{\sqrt{6}}\begin{vmatrix} \hat{\mathbf{i}} & \hat{\mathbf{j}} & \hat{\mathbf{k}} \\ 6 & 3 & -3 \\ 1 & 0 & 1 \end{vmatrix} = \frac{1}{\sqrt{6}}[\hat{\mathbf{i}}(3) - \hat{\mathbf{j}}(3+6) + \hat{\mathbf{k}}(-3)]$$

$$= \frac{1}{\sqrt{6}}(3\hat{\mathbf{i}} - 9\hat{\mathbf{j}} - 3\hat{\mathbf{k}})$$

Mechanisms

A **mechanism** is an assembly of linkages (rigid rods) connected together. Its purpose is to transmit power and motion from one point in a machine to another. Perhaps the simplest example is the **slider–crank** mechanism shown in Figure 1.41. This mechanism can operate in two ways. Firstly, the slider can be made to execute linear motion (possibly by exploding gases, as in a piston engine). This induces the crank to rotate about O. Secondly, the crank can be made to rotate, thereby inducing the slider to execute oscillatory linear motion (as in a press).

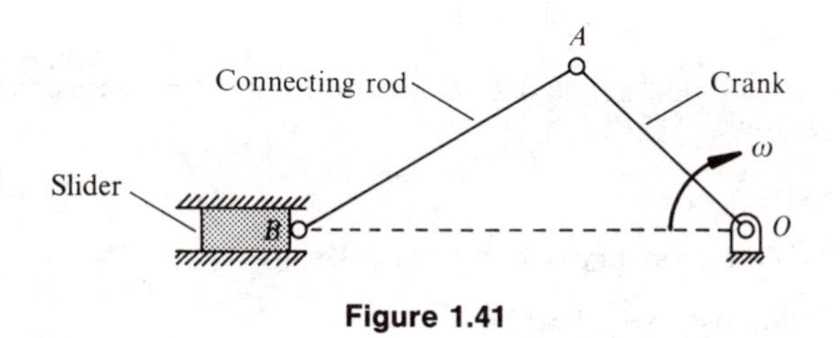

Figure 1.41

If ω is the angular speed of the crank (not necessarily a constant), then the speed of the joint A will be $(OA)\omega$. The velocity of A will therefore be in a direction perpendicular to OA and of magnitude $(OA)\omega$. The velocity of B is, obviously, always directed along the slide. Because AB is a rigid link, an observer at A will see B rotating about A with some angular speed Ω. Clearly, the velocity of B relative to A has magnitude $(AB)\Omega$.

Example 1.11

The crank OA of an engine rotates with an angular speed 100 rad s^{-1}. Find the velocity of the piston B, the angular speed of the rod AB and the velocity of the point C on the rod, a distance of 0.1 m from B, at the instant when the crank is 45° from the line BO. The length of OA is 0.1 m and that of AB is 0.25 m (see Figure 1.42).

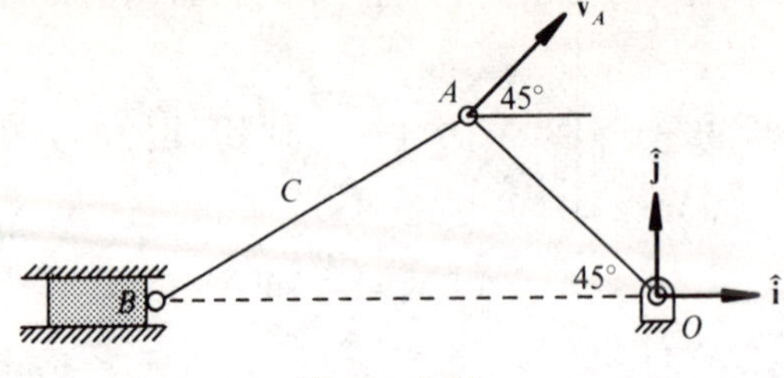

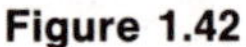

Figure 1.42

Solution

Choosing an origin at O and axes as shown, it is clear that the velocity of B is $\mathbf{v}_B = v_B\hat{\mathbf{i}}$ where v_B is to be determined and $|\mathbf{v}_A| = 0.1 \times 100 = 10 \text{ m s}^{-1}$. Therefore

$$\mathbf{v}_A = 10\left(\frac{1}{\sqrt{2}}\hat{\mathbf{i}} + \frac{1}{\sqrt{2}}\hat{\mathbf{j}}\right) = 7.07(\hat{\mathbf{i}} + \hat{\mathbf{j}})$$

The velocity of B relative to A is $\mathbf{v}_B - \mathbf{v}_A$:

$$\mathbf{v}_B - \mathbf{v}_A = v_B\hat{\mathbf{i}} - 7.07(\hat{\mathbf{i}} + \hat{\mathbf{j}})$$

Since AB is rigid, the relative velocity vector must be perpendicular to the vector $\overrightarrow{BA}$; that is,

$$(\mathbf{v}_B - \mathbf{v}_A) \cdot \overrightarrow{BA} = 0$$

But from the geometry of the triangle ABO,

$$\overrightarrow{BA} = 0.239\hat{\mathbf{i}} + 0.0707\hat{\mathbf{j}}$$

Hence, taking the scalar product indicated,

$$v_B(0.239) - 7.07(0.239 + 0.0707) = 0$$

leading to $v_B = 9.16 \text{ m s}^{-1}$. Therefore $\mathbf{v}_B = 9.16\hat{\mathbf{i}}$ and so

$$\mathbf{v}_B - \mathbf{v}_A = 2.09\hat{\mathbf{i}} - 7.07\hat{\mathbf{j}}$$

Also, $|\mathbf{v}_B - \mathbf{v}_A| = 0.25\Omega$. Therefore

$$\Omega = \frac{\sqrt{[(2.09)^2 + (7.07)^2]}}{0.25} = 29.49 \text{ rad s}^{-1}$$

is the angular speed of the rod AB.

The velocity of C relative to A is parallel to $(\mathbf{v}_B - \mathbf{v}_A)$ but has magnitude 0.15Ω; that is,

$$\mathbf{v}_C - \mathbf{v}_A = \frac{2.09\hat{\mathbf{i}} - 7.07\hat{\mathbf{j}}}{\sqrt{[(2.09)^2 + (7.07)^2]}} \times (0.15\Omega)$$

$$= 1.25\hat{\mathbf{i}} - 4.24\hat{\mathbf{j}}$$

Hence the velocity of C is

$$\mathbf{v}_C = 1.25\hat{\mathbf{i}} - 4.24\hat{\mathbf{j}} + 7.07(\hat{\mathbf{i}} + \hat{\mathbf{j}})$$
$$= (8.32\hat{\mathbf{i}} + 2.83\hat{\mathbf{j}}) \text{ m s}^{-1}$$

EXERCISES

1.40 If $\mathbf{a} = \hat{\mathbf{i}} + \hat{\mathbf{j}} + \hat{\mathbf{k}}$ and $\mathbf{b} = -3\hat{\mathbf{i}} + 4\hat{\mathbf{k}}$, find $|\mathbf{a}|$, $|\mathbf{b}|$, $\mathbf{a} \wedge \mathbf{b}$ and $|\mathbf{a} \wedge \mathbf{b}|$, and verify that $(\mathbf{a} - \mathbf{b}) \wedge (\mathbf{a} + \mathbf{b}) = 2\mathbf{a} \wedge \mathbf{b}$.

1.41 Find a unit vector perpendicular to the two vectors $\hat{\mathbf{i}} + 3\hat{\mathbf{j}} + 4\hat{\mathbf{k}}$ and $2\hat{\mathbf{i}} + \hat{\mathbf{k}}$.

1.42 A rigid body rotates with constant angular velocity of 60 rad s^{-1} about an axis parallel to $2\hat{\mathbf{i}} + 3\hat{\mathbf{j}} + 6\hat{\mathbf{k}}$ passing through the point (0, 0, 0). Find the speed at the point (1, 0, 0) in the body. (Displacement units are metres.)

1.43 Find the area of a triangle whose vertices are at (1, 1, −1), (2, 1, 0) and (0, 0, 0).

1.44 Determine the shortest distance between the two skew lines

$$\frac{x-1}{2} = \frac{y}{3} = \frac{z+1}{1} \quad \text{and} \quad \frac{x}{3} = \frac{y-3}{4} = \frac{z-1}{2}$$

1.45 If two points A and B have coordinates (3, 7, −2) and (1, 5, −1) respectively, and $\overrightarrow{AC} = \hat{\mathbf{i}} + 2\hat{\mathbf{j}}$ and $\overrightarrow{BD} = -\hat{\mathbf{i}} + \hat{\mathbf{k}}$, show that the lines AC and BD intersect. (*Hint:* determine the vector equations of the two lines with parameters s and t. Show that there is a possible solution for s and t giving a common value of (x, y, z) on each of the two lines.)

1.46 If the position vectors of three points A, B and C are $\mathbf{a}$, $\mathbf{b}$ and $\mathbf{c}$ respectively, show that the vector

$$\mathbf{a} \wedge \mathbf{b} + \mathbf{b} \wedge \mathbf{c} + \mathbf{c} \wedge \mathbf{a}$$

is perpendicular to the plane ABC.

1.47 A four-bar mechanism is shown in Figure 1.43. AB rotates about A at

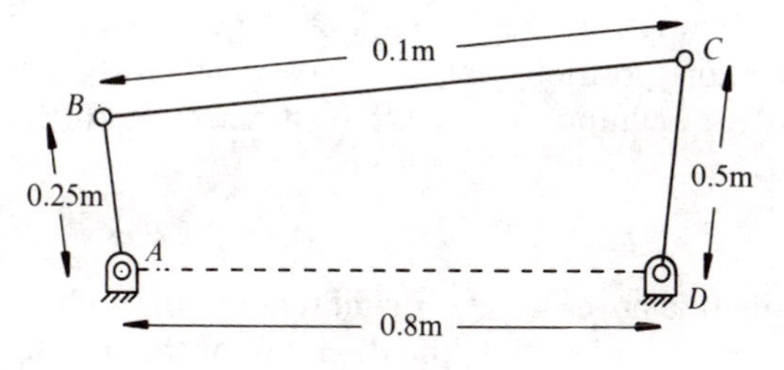

Figure 1.43

5 rev s^{-1}. DC oscillates about D. Show that the speed of the joint C is 0.915 m s^{-1} when the rod is perpendicular to AD.

1.48 A triangle has vertices at (0, 1, 3), (3, 4, 5) and (2, 0, −1). Determine its vector area. Find the magnitudes of the areas of projection onto the xy-plane and onto a plane with normal $2\hat{\mathbf{i}} - 3\hat{\mathbf{j}} + 4\hat{\mathbf{k}}$.

1.49 Show that the perpendicular distance between a corner of a unit cube and a diagonal which does not pass through it is 0.816.

1.7 Applications of vectors to forces

The moment of a force

When a force **F** (with a given line of action) acts on a body, then we say that the force has a **moment** about a point O equal to the magnitude of **F** times the perpendicular distance from the line of action of **F** to O; referring to Figure 1.44 for example,

$$\text{moment of } \mathbf{F} \text{ about } O = |\mathbf{F}|d$$

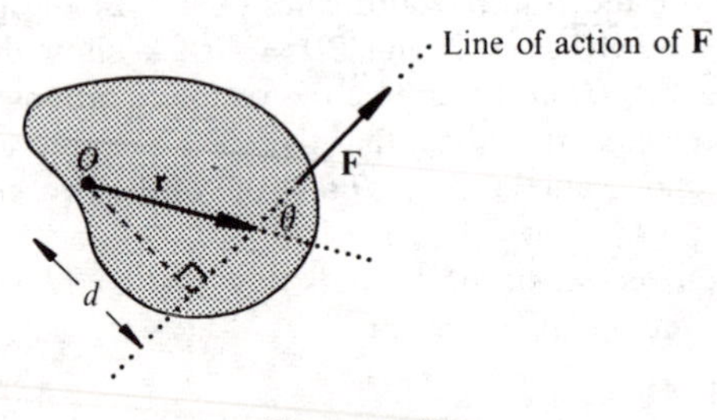

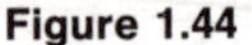

Figure 1.44

If the body is pivoted at O, then the physical effect of the moment is to induce a 'turning' about an axis through O perpendicular to the plane of **r** and **F**, where **r** is the position vector of *any* point on the line of action of **F**. This suggests the definition of the vector moment **M** (or the **torque**) of **F** about O as

$$\mathbf{M} = \mathbf{r} \wedge \mathbf{F} \tag{1.48}$$

This concise equation gives us the magnitude of the moment (this being $|\mathbf{M}| = |\mathbf{r}|\,|\mathbf{F}| \sin\theta = |\mathbf{F}|d$) as well as the direction of the turning axis referred to above, this being the direction of $\mathbf{r} \wedge \mathbf{F}$.

Couples

If we have a pair of forces, equal in magnitude but opposite in direction and with distinct lines of action, then this system of forces constitutes a **couple** (see Figure 1.45).

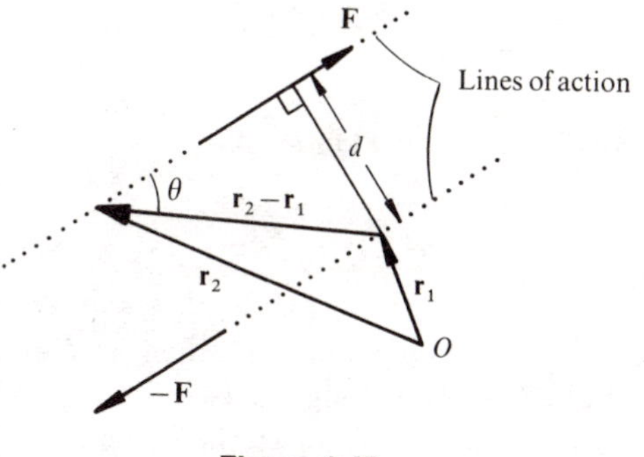

Figure 1.45

Let $\mathbf{r}_1$ and $\mathbf{r}_2$ be the position vectors of *any* two points on the lines of action of the forces with respect to an origin at O. The resultant torque of the forces about O is defined simply as the vector sum of the two vectors $\mathbf{r}_1 \wedge (-\mathbf{F})$ and $\mathbf{r}_2 \wedge (\mathbf{F})$, these being the separate torques of the forces about O. The resultant torque is

$$\mathbf{r}_1 \wedge (-\mathbf{F}) + \mathbf{r}_2 \wedge (\mathbf{F}) = (\mathbf{r}_2 - \mathbf{r}_1) \wedge \mathbf{F}$$

(using the distributive law for vector products). But $\mathbf{r}_2 - \mathbf{r}_1$ is *any* vector joining the lines of action of the two forces, so this resultant is independent of the choice of origin. The direction of the resultant torque is perpendicular to the plane containing the two lines of action. From the definition of the vector product we see that the magnitude of the torque is $|\mathbf{r}_2 - \mathbf{r}_1|\,|\mathbf{F}| \sin\theta$. But from Figure 1.45, $d = |\mathbf{r}_2 - \mathbf{r}_1| \sin\theta$, where d is the perpendicular distance between the parallel lines of action. Hence the torque has the simple magnitude $|\mathbf{F}|d$.

The resolution of forces

We are now in a position to prove that, suitably treated, forces may be regarded as free vectors. We show that *any* system of forces may be reduced to a single force acting through a convenient point O and a single torque about O.

Let $\mathbf{F}_1, \mathbf{F}_2, \ldots, \mathbf{F}_n$ be the forces and $\mathbf{r}_1, \mathbf{r}_2, \ldots, \mathbf{r}_n$ be the position vectors from an origin O to any point on the lines of action of the forces (see Figure 1.46).

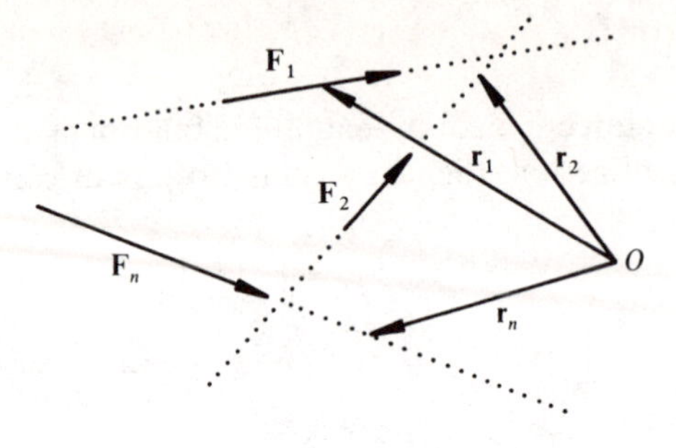

Figure 1.46

If we consider a particular force $\mathbf{F}_i$, then we can add and subtract a force parallel to $\mathbf{F}_i$ but acting through O (see Figure 1.47). In other words, we have simply added the zero vector to the system. Now this collection of three forces is equivalent to a single force $\mathbf{F}_i$ acting through O, together with a single couple about O with torque $\mathbf{r}_i \wedge \mathbf{F}_i$. Repeating this construction for each force in the system, we see that we can reduce the original system to

- a single resultant force $\sum_{i=1}^{n} \mathbf{F}_i$ (vector sum) through O,
- a single resultant couple with torque $\sum_{i=1}^{n} \mathbf{r}_i \wedge \mathbf{F}_i$ (another vector sum) about O.

A body acted on by a system of forces is said to be in equilibrium if *both* the resultant force *and* the resultant couple are zero.

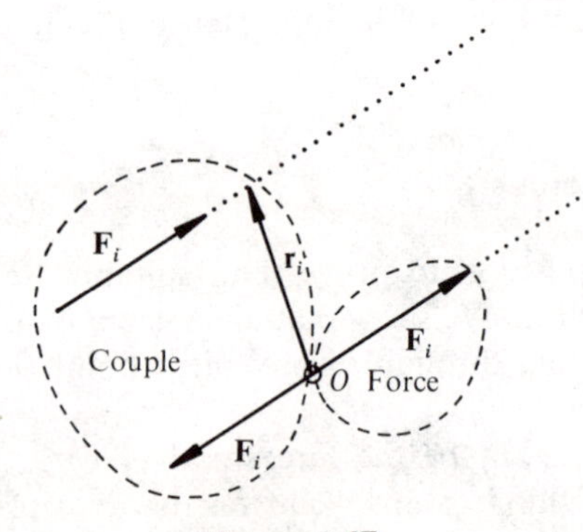

Figure 1.47

1.8 Introduction to the vector analysis of stress *(optional)*

If we apply equal and opposite forces to either end of a straight bar of uniform (though arbitrary) cross-section, then it is a matter of common experience that after a very short time (during which it suffers a slight elongation) the bar will be in equilibrium (see Figure 1.48). The result of applying these forces is to 'stress' the bar. If we touch the exterior curved part of the bar, we feel no force. Even if we cut the bar in the direction of F, no change in the configuration of the bar will be observed. However, if we cut the bar transversely at any point, then it will immediately begin to tear apart. The forces have been transmitted through the bar to the point of the cut.

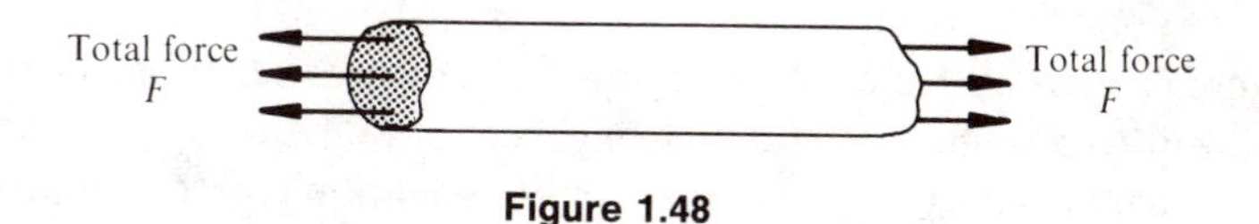

Figure 1.48

We are also familiar with the fact that if this same force F is applied to a bar of larger area of cross-section, then its effect in deforming the bar is reduced in proportion. Indeed, the deformation produced (elongation and thinning in this case) is directly related to the 'normal' stress σ which is defined, for this simple system, as the magnitude of the applied force divided by the cross-sectional area:

$$\sigma = \frac{F}{A}$$

Now consider a non-uniform bar of the type shown in Figure 1.49. It is again a matter of common experience that if the force F is large enough and the bar ruptures, then it is very likely to rupture at a point near P. It is at this point that the normal stress is largest, since the same force F is transmitted across each cross-section but the area through which it is transmitted is much smaller in the vicinity of P.

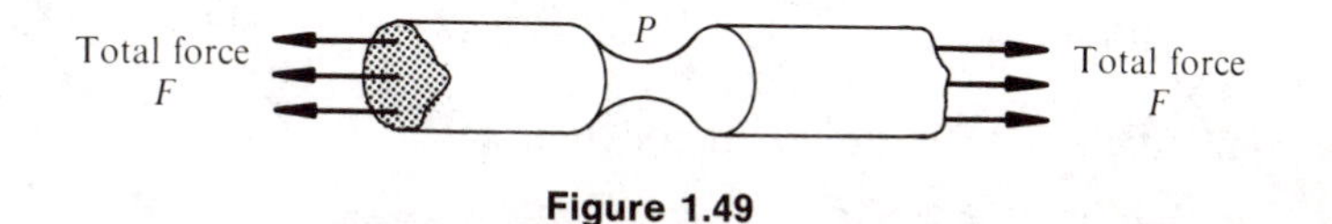

Figure 1.49

We now extend these ideas to more generally shaped bodies. Consider a continuous body separated into two parts, 1 and 2, by a dividing plane surface S and let Q be any point on S (see Figure 1.50). By Newton's third law, the forces exerted by region 1 on region 2 across S are equal and opposite

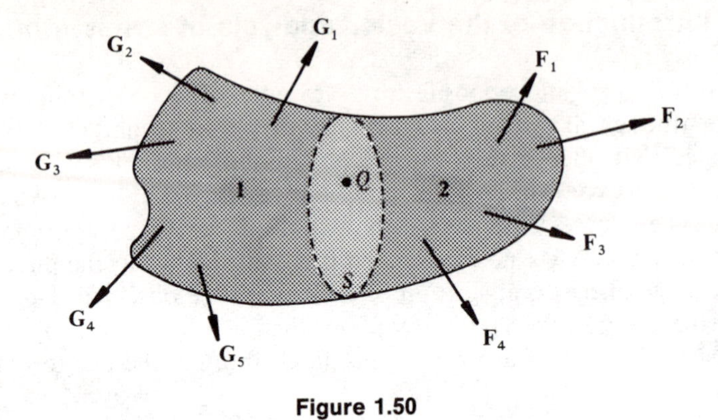

Figure 1.50

to those exerted by region 2 on region 1. The forces acting in region 2 are equivalent to a single force **f** and a single couple **h** acting at Q. Now shrink S but always keeping Q within S. It is a basic assumption of stress analysis that

$$\lim_{S\to 0}\frac{\mathbf{f}}{S}=\mathbf{P} \quad \text{and} \quad \lim_{S\to 0}\frac{\mathbf{h}}{S}=0 \tag{1.49}$$

P is called the stress vector at Q and its components have the dimensions of force/area. **P** will depend not only on the coordinates of Q but also on the orientation of S. If $\hat{\mathbf{n}}$ is the unit normal to S, then we emphasize this dependence by writing $\mathbf{P}^n$ instead of just **P**.

The fundamental result of stress analysis is that each component of $\mathbf{P}^n$ is linearly related to the normal $\hat{\mathbf{n}}$ via three vectors $\boldsymbol{\sigma}_x$, $\boldsymbol{\sigma}_y$ and $\boldsymbol{\sigma}_z$ which are independent of $\hat{\mathbf{n}}$. More precisely, it can be shown that

$$P^n_x=\boldsymbol{\sigma}_x\cdot\hat{\mathbf{n}}, \qquad P^n_y=\boldsymbol{\sigma}_y\cdot\hat{\mathbf{n}}, \qquad P^n_z=\boldsymbol{\sigma}_z\cdot\hat{\mathbf{n}} \tag{1.50}$$

The three vectors are

$$\boldsymbol{\sigma}_x=\sigma_{xx}\hat{\mathbf{i}}+\sigma_{xy}\hat{\mathbf{j}}+\sigma_{xz}\hat{\mathbf{k}}, \qquad \boldsymbol{\sigma}_y=\sigma_{yx}\hat{\mathbf{i}}+\sigma_{yy}\hat{\mathbf{j}}+\sigma_{yz}\hat{\mathbf{k}},$$

$$\boldsymbol{\sigma}_z=\sigma_{zx}\hat{\mathbf{i}}+\sigma_{zy}\hat{\mathbf{j}}+\sigma_{zz}\hat{\mathbf{k}}$$

or, in coordinate form,

$$\boldsymbol{\sigma}_x=(\sigma_{xx},\sigma_{xy},\sigma_{xz}), \qquad \boldsymbol{\sigma}_y=(\sigma_{yx},\sigma_{yy},\sigma_{yz}), \qquad \boldsymbol{\sigma}_z=(\sigma_{zx},\sigma_{zy},\sigma_{zz}) \tag{1.51}$$

The components satisfy the symmetry conditions

$$\sigma_{xy}=\sigma_{yx}, \qquad \sigma_{xz}=\sigma_{zx}, \qquad \sigma_{yz}=\sigma_{zy} \tag{1.52}$$

so that, at each point on a small plane within the material, six numbers, σ_{xx}, σ_{xy}, σ_{xz}, σ_{yy}, σ_{yz} and σ_{zz}, may be found that relate the orientation of the plane to the stress vector at the point.

We should note that the direction of $\mathbf{P}^n$ is not necessarily the direction of $\hat{\mathbf{n}}$. However, $\mathbf{P}^n$ has a component along $\hat{\mathbf{n}}$ called the **normal** (or **tensile**)

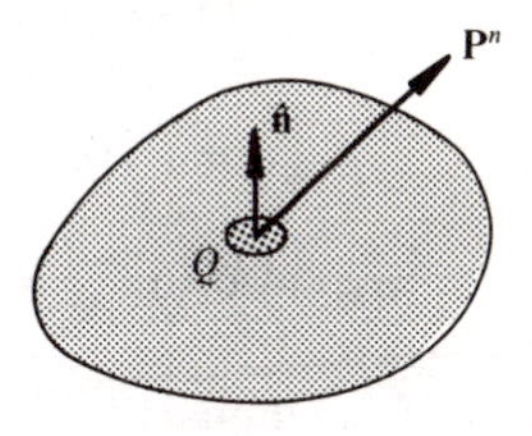

Figure 1.51

stress at Q and a component tangential to the plane area called the **shear** stress at Q (see Figure 1.51). Clearly, the component of $\mathbf{P}^n$ in the direction of $\hat{\mathbf{n}}$ is $\mathbf{P}^n \cdot \hat{\mathbf{n}}$ and so, using vector addition,

$$\mathbf{P}^n = (\mathbf{P}^n \cdot \hat{\mathbf{n}})\hat{\mathbf{n}} + [\mathbf{P}^n - (\mathbf{P}^n \cdot \hat{\mathbf{n}})\hat{\mathbf{n}}] \tag{1.53}$$

The first component on the right-hand side of (1.53) is parallel to $\hat{\mathbf{n}}$ and the second is perpendicular to $\hat{\mathbf{n}}$.

The magnitude of the normal stress is $\mathbf{P}^n \cdot \hat{\mathbf{n}}$ and that of the shear stress is (by Pythagoras)

$$[|\mathbf{P}^n|^2 - (\mathbf{P}^n \cdot \hat{\mathbf{n}})^2]^{1/2}$$

Part of the purpose of stress analysis is to determine the three vectors $\boldsymbol{\sigma}_x$, $\boldsymbol{\sigma}_y$ and $\boldsymbol{\sigma}_z$ at each point within a body that is deformed by forces. The analysis of stress applies to all continua and depends only on geometry and on Newton's laws. The problem in general is quite complicated, requiring the solution of systems of partial differential equations. In this text we shall only consider some of the simpler aspects.

Example 1.12

A uniform bar of cross-sectional area A has its axis aligned along the x-axis and is subjected to equal and opposite forces F as shown in Figure 1.52. If

$$\sigma_{xx} = \frac{F}{A}, \qquad \sigma_{xy} = \sigma_{xz} = \sigma_{yy} = \sigma_{yz} = \sigma_{zz} = 0$$

determine the stress vector at any point on a plane at an angle α to the axis of the bar. Determine the normal and shear stress on this plane and, finally, verify that the total force on this plane is F.

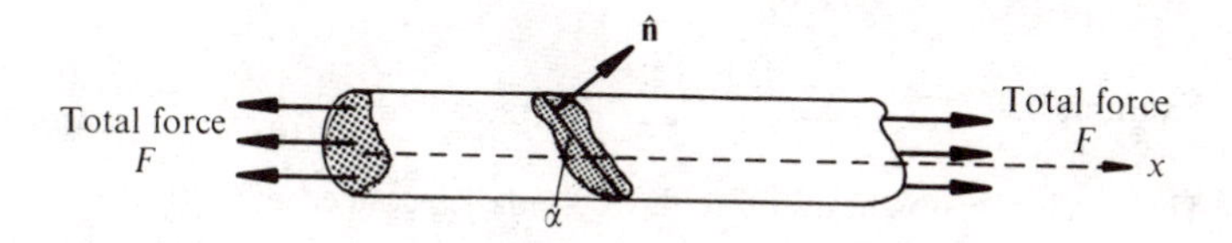

Figure 1.52

Solution

In this problem, clearly (using equation (1.51))

$$\boldsymbol{\sigma}_x = \left(\frac{F}{A}, 0, 0\right), \qquad \boldsymbol{\sigma}_y = (0, 0, 0), \qquad \boldsymbol{\sigma}_z = (0, 0, 0)$$

Now $\hat{\mathbf{n}} = \sin\alpha\,\hat{\mathbf{i}} + \cos\alpha\,\hat{\mathbf{j}}$ and so (from (1.50))

$$P_x^n = \frac{F}{A}\sin\alpha, \qquad P_y^n = 0, \qquad P_z^n = 0$$

Hence the stress vector (at any point on the plane) is

$$\mathbf{P}^n = \frac{F}{A}\sin\alpha\,\hat{\mathbf{i}}$$

The normal component of $\mathbf{P}^n$ is

$$\mathbf{P}^n \cdot \hat{\mathbf{n}} = \frac{F}{A}\sin^2\alpha$$

which is the normal stress, and the tangential component of $\mathbf{P}^n$ is

$$|\mathbf{P}^n \wedge \hat{\mathbf{n}}| = \frac{F}{A}\sin\alpha\cos\alpha$$

which is the shear stress on this plane.

The total force on the inclined plane is

$$\frac{F}{A}\sin\alpha\, A'$$

in the $\hat{\mathbf{i}}$ direction, where A' is the area of the inclined element. Now $\mathbf{A}' = A'\hat{\mathbf{n}}$ is the vector area of this inclined element. But the projection of $\mathbf{A}'$ onto the zy-plane is $\mathbf{A}' \cdot \hat{\mathbf{i}}$, and this is equal to A. Therefore

$$\begin{aligned} A = \mathbf{A}' \cdot \hat{\mathbf{i}} &= A'(\sin\alpha\,\hat{\mathbf{i}} + \cos\alpha\,\hat{\mathbf{j}}) \cdot \hat{\mathbf{i}} \\ &= A'\sin\alpha \end{aligned}$$

Therefore the total force on the plane element is

$$\frac{F}{A}\sin\alpha \times \frac{A}{\sin\alpha} = F$$

as expected.

Example 1.13

The torsion problem of a circular cylinder is concerned with the deformation produced when the two ends of the cylinder are twisted by equal and opposite couples.

A circular cylinder of radius a has its axis aligned along the z-axis. Verify that if the three vectors $\boldsymbol{\sigma}_x$, $\boldsymbol{\sigma}_y$ and $\boldsymbol{\sigma}_z$ at each point (x, y, z) are given by

$$\boldsymbol{\sigma}_x = (0, 0, -\mu y), \qquad \boldsymbol{\sigma}_y = (0, 0, \mu x), \qquad \boldsymbol{\sigma}_z = (-\mu y, \mu x, 0)$$

where μ is a constant, then this fully specifies a situation in which the cylinder is subjected to torsion by equal and opposite couples applied to either end.

Solution

We shall simply verify that the forces which result from the given values for $\boldsymbol{\sigma}_x$, $\boldsymbol{\sigma}_y$ and $\boldsymbol{\sigma}_z$ are compatible with the torsion problem. First we shall show that the lateral or curved surface of the cylinder is stress-free, that is, that no forces are applied to this part of the boundary. Secondly we shall show that the forces applied to the ends of the cylinder constitute a couple.

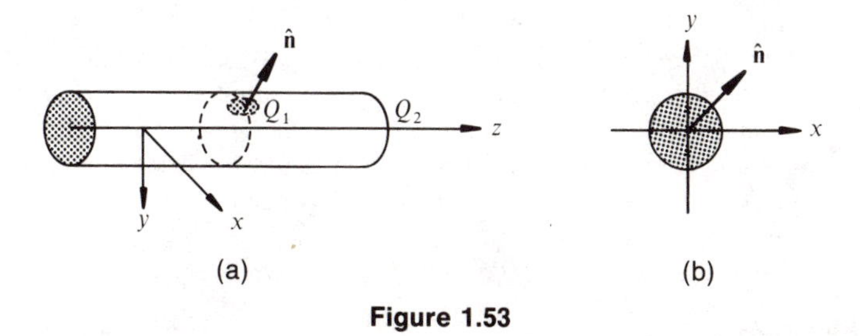

Figure 1.53

From Figure 1.53a we can see that a normal to the lateral surface has the form $\hat{\mathbf{n}} = \alpha\hat{\mathbf{i}} + \beta\hat{\mathbf{j}}$, so clearly it has a zero z-component. Hence, if we consider a point Q_1 on the lateral surface, then the stress vector at that point has components

$$P_x^n = \boldsymbol{\sigma}_x \cdot \hat{\mathbf{n}} = 0, \qquad P_y^n = \boldsymbol{\sigma}_y \cdot \hat{\mathbf{n}} = 0, \qquad P_z^n = \boldsymbol{\sigma}_z \cdot \hat{\mathbf{n}} = -\mu y\alpha + \mu x\beta$$

But on a circle in the xy-plane (centred on the z-axis), the unit normal at any point on the boundary $x^2 + y^2 = a^2$ has components $\hat{\mathbf{n}} = (x/a)\hat{\mathbf{i}} + (y/a)\hat{\mathbf{j}}$ since the normal is always in the same direction as the position vector of the point under consideration (see Figure 1.53b). Therefore

$$P_z^n = -\mu y(x/a) + \mu x(y/a) = 0$$

so that the lateral surface is indeed stress-free.

If we now choose a point Q_2 on the right-hand end of the cylinder, then the normal at Q_2 is $\hat{\mathbf{n}} = \hat{\mathbf{k}}$. Thus at this point

$$P_x^n = \boldsymbol{\sigma}_x \cdot \hat{\mathbf{n}} = -\mu y, \qquad P_y^n = \boldsymbol{\sigma}_y \cdot \hat{\mathbf{n}} = \mu x, \qquad P_z^n = \boldsymbol{\sigma}_z \cdot \hat{\mathbf{n}} = 0$$

and so the stress vector at this point is $\mathbf{P}^n = (-\mu y, \mu x, 0)$. But this

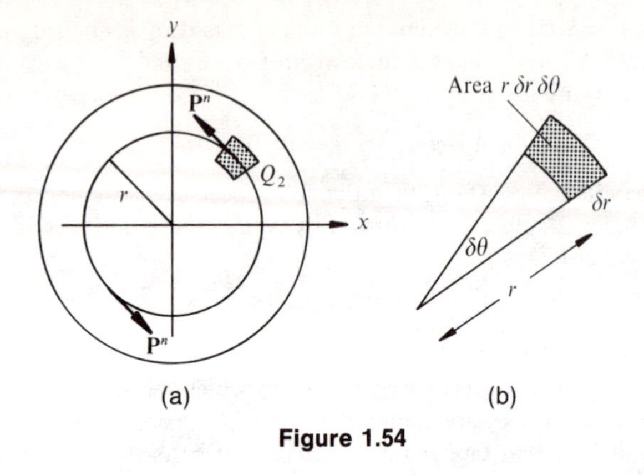

Figure 1.54

vector is always tangential to the circle $x^2 + y^2 = r^2$ since $\mathbf{P}^n \cdot \mathbf{m} = 0$, where $\mathbf{m} = x\hat{\mathbf{i}} + y\hat{\mathbf{j}}$ is a normal to the circle. It is in the direction shown in Figure 1.54a since $\mathbf{m} \wedge \mathbf{P}^n = \mu(x^2 + y^2)\hat{\mathbf{k}}$, a vector in the positive z-direction. Of course, had $\mathbf{P}^n$ been in the opposite direction to that shown, then $\mathbf{m} \wedge \mathbf{P}^n$ would have been a vector in the negative z-direction.

It is clear that the resultant of all these forces is a couple in the direction of $\hat{\mathbf{k}}$. We can readily determine the magnitude of the couple as follows. An elementary area at Q_2 (see Figure 1.54b) has an area $r\,\delta r\,\delta\theta$ and so the force acting over this area has magnitude

$$|\mathbf{P}^n|\, r\,\delta r\,\delta\theta = \mu\sqrt{(y^2 + x^2)}\, r\,\delta r\,\delta\theta = \mu r^2\,\delta r\,\delta\theta$$

This force has a couple about the origin of magnitude $\mu r^3\,\delta r\,\delta\theta$, and adding all forces at a distance r from the axis we obtain a couple of magnitude $\mu r^3\,\delta r(2\pi)$ due to the 'ring' of forces. Now, summing all couples as r changes from 0 to a, we find the total couple acting over the end of the cylinder to be

$$\hat{\mathbf{k}}\left(\int_0^a \mu r^3(2\pi)\,\mathrm{d}r\right) = \tfrac{1}{2}\pi\mu a^4\hat{\mathbf{k}}$$

The calculation at the other end of the cylinder is identical except that we must use $\hat{\mathbf{n}} = (0, 0, -1)$. For this end, the resultant couple is of equal magnitude but in the opposite direction, that is, it is given by $-\tfrac{1}{2}\pi\mu a^4\hat{\mathbf{k}}$, which is required for the overall equilibrium of the body.

Example 1.14

A square plate is acted on by a shearing stress τ on each of its edges as shown in Figure 1.55. Determine the tensile and shearing stresses on planar elements at Q_1 and Q_2. Assume that the vectors $\boldsymbol{\sigma}_x$, $\boldsymbol{\sigma}_y$ and $\boldsymbol{\sigma}_z$ are constant throughout the plate.

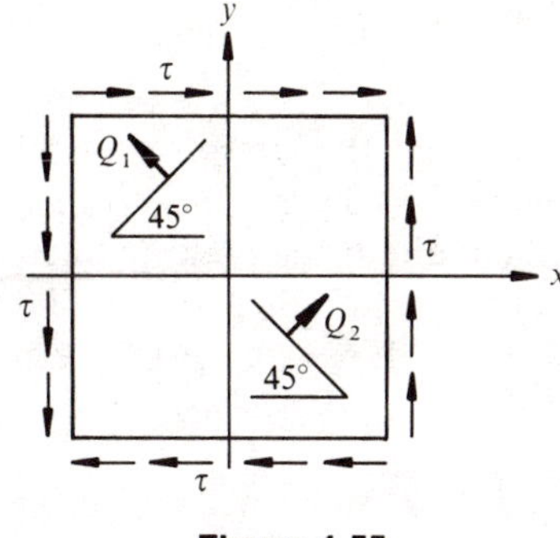

Figure 1.55

Solution

On the right-hand boundary we have $\hat{\mathbf{n}} = (1, 0, 0)$ and $\mathbf{P}^n = (0, \tau, 0)$ (shearing stress). Hence, from the relations in (1.50),

$$0 = \sigma_{xx}, \qquad \tau = \sigma_{yx}, \qquad 0 = \sigma_{zx}$$

On the top boundary we have $\hat{\mathbf{n}} = (0, 1, 0)$ and $\mathbf{P}^n = (\tau, 0, 0)$ and so, again using (1.50),

$$\tau = \sigma_{xy}, \qquad 0 = \sigma_{yy}, \qquad 0 = \sigma_{zy}$$

No forces are applied to the face of the plate for which $\hat{\mathbf{n}} = (0, 0, 1)$, so $\mathbf{P}^n = (0, 0, 0)$ and

$$0 = \sigma_{xz}, \qquad 0 = \sigma_{yz}, \qquad 0 = \sigma_{zz}$$

We can conclude that, throughout the body,

$$\boldsymbol{\sigma}_x = (0, \tau, 0), \qquad \boldsymbol{\sigma}_y = (\tau, 0, 0), \qquad \boldsymbol{\sigma}_z = (0, 0, 0)$$

On section Q_1, $\hat{\mathbf{n}} = (-\frac{1}{2}\sqrt{2}, \frac{1}{2}\sqrt{2}, 0)$ and so the stress vector on this plane has components

$$P_x^n = \boldsymbol{\sigma}_x \cdot \hat{\mathbf{n}} = \tfrac{1}{2}\sqrt{2}\,\tau, \qquad P_y^n = \boldsymbol{\sigma}_y \cdot \hat{\mathbf{n}} = -\tfrac{1}{2}\sqrt{2}\,\tau, \qquad P_z^n = \boldsymbol{\sigma}_z \cdot \hat{\mathbf{n}} = 0$$

hence the tensile stress $= \mathbf{P}^n \cdot \hat{\mathbf{n}} = -\frac{1}{2}\tau - \frac{1}{2}\tau = -\tau$

(The negative sign indicates a compression: if the material were cut along this plane, it would not tear.)

The shearing stress is

$$[|\mathbf{P}^n|^2 - (\mathbf{P}^n \cdot \hat{\mathbf{n}})^2]^{1/2} = 0$$

On section Q_2, $\hat{\mathbf{n}} = (\frac{1}{2}\sqrt{2}, \frac{1}{2}\sqrt{2}, 0)$, leading to a stress vector on this plane with components

$$P_x^n = \tfrac{1}{2}\sqrt{2}\,\tau, \qquad P_y^n = \tfrac{1}{2}\sqrt{2}\,\tau, \qquad P_z^n = 0$$

$\therefore$ tensile stress $= \mathbf{P}^n \cdot \hat{\mathbf{n}} = \tau$

(The positive sign indicates tension: if the material were cut along this plane, it would tend to tear apart.)

Again the shearing stress on this section is zero.

Principal stresses

We have already noted that the stress vector $\mathbf{P}^n$ at a point Q on a plane with normal $\hat{\mathbf{n}}$ is not necessarily parallel to $\hat{\mathbf{n}}$. However, we may imagine rotating the area element about the point Q (thereby changing the direction of $\hat{\mathbf{n}}$), and it is natural to ask whether we can find a particular area element for which $\mathbf{P}^n$ is parallel to $\hat{\mathbf{n}}$. If $\hat{\mathbf{n}} = (\alpha, \beta, \gamma)$, for example, we look for a scalar quantity λ such that

$$\mathbf{P}^n = \lambda \hat{\mathbf{n}} \tag{1.54}$$

Using equation (1.50), the vector equation (1.54) is equivalent to the three scalar equations

$$\begin{aligned}
\sigma_{xx}\alpha + \sigma_{xy}\beta + \sigma_{xz}\gamma &= \lambda\alpha \\
\sigma_{yx}\alpha + \sigma_{yy}\beta + \sigma_{yz}\gamma &= \lambda\beta \\
\sigma_{zx}\alpha + \sigma_{zy}\beta + \sigma_{zz}\gamma &= \lambda\gamma
\end{aligned} \tag{1.55}$$

This is a homogeneous system of three equations in the unknowns α, β and γ. A non-zero set of values of (α, β, γ) will exist only if

$$\begin{vmatrix} \sigma_{xx} - \lambda & \sigma_{xy} & \sigma_{xz} \\ \sigma_{yx} & \sigma_{yy} - \lambda & \sigma_{yz} \\ \sigma_{zx} & \sigma_{zy} & \sigma_{zz} - \lambda \end{vmatrix} = 0 \tag{1.56}$$

(Readers who are unfamiliar with determinants will need to refer to Appendix 1.)

This equation is a cubic polynomial in λ. For each root of this cubic, λ_1, λ_2 and λ_3, we may substitute back into (1.55) and solve for the corresponding values of (α, β, γ). It may be shown that each of the roots λ_1, λ_2 and λ_3 are real, so that (1.55) may always be used to determine three directions at the point Q for which the stress vector $\mathbf{P}^n$ is parallel to $\hat{\mathbf{n}}$.

Taking the scalar product of both sides of (1.54) with $\hat{\mathbf{n}}$ implies that

$$\mathbf{P}^n \cdot \hat{\mathbf{n}} = \lambda$$

or that the values of λ are just the normal stresses at Q. Because of the special relation between the stress vector and $\hat{\mathbf{n}}$ (they are parallel vectors), the values λ_1, λ_2 and λ_3 are called the **principal stresses** at Q and the directions associated with them are known as the **principal directions**. Principal stresses are of great importance because it may easily be shown that the maximum normal stress at a point within a body is one of the principal stresses.

We note, finally, that an area element at Q aligned with its normal parallel to a principal direction has zero shear stress. This follows since the magnitude of the shear stress at any point is $[|\mathbf{P}^n|^2 - (\mathbf{P} \cdot \hat{\mathbf{n}})^2]^{1/2}$. But $\mathbf{P}^n \cdot \hat{\mathbf{n}} = \lambda$ and so the shear stress magnitude is $[\lambda^2 \hat{\mathbf{n}} \cdot \hat{\mathbf{n}} - \lambda^2 (\hat{\mathbf{n}} \cdot \hat{\mathbf{n}})^2]^{1/2} = 0$.

EXERCISES

1.50 A particle positioned at the corner of a cube is acted on by three forces of magnitudes 1 N, 2 N and 3 N. The directions of the forces are along the diagonals of the faces of the cube which meet at the corner. Determine the resultant force.

1.51 A force of 12 N acts through the point A (3, 4, 5)m in the direction of the vector $3\hat{\mathbf{i}} + 4\hat{\mathbf{j}} - 5\hat{\mathbf{k}}$. What is the moment of the force about B (2, 1, 0)m?

1.52 Verify that the forces $\mathbf{F}_1 = 3\hat{\mathbf{i}} - \hat{\mathbf{k}}$ acting at (1, 2, 0) and $\mathbf{F}_2 = 2\hat{\mathbf{i}} + 3\hat{\mathbf{j}}$ acting at (0, −1, 3) reduce to a resultant force $5\hat{\mathbf{i}} + 3\hat{\mathbf{j}} - \hat{\mathbf{k}}$ at the origin together with a couple $-11\hat{\mathbf{i}} + 7\hat{\mathbf{j}} - 4\hat{\mathbf{k}}$. Show that this system of forces cannot be reduced to a single resultant force.

1.53 (a) OAB is a triangle with origin at O, and $\mathbf{a}$ and $\mathbf{b}$ are the position vectors of A and B respectively. Show that if $\mathbf{r}$ is the position vector of a point on AB which divides the line in the ratio m:n, then

$$(m+n)\mathbf{r} = n\mathbf{a} + m\mathbf{b}$$

(b) Two concurrent forces are represented by $p\,\overrightarrow{OA}$ and $q\,\overrightarrow{OB}$ respectively. Show that their resultant is $(p+q)\,\overrightarrow{OR}$, where R divides AB in the ratio p:q.

1.54 Three forces are represented by $\overrightarrow{BA}$, $\overrightarrow{CA}$ and $2\,\overrightarrow{BC}$, where A, B and C are the vertices of a triangle. Show that the resultant force is represented by $6\,\overrightarrow{DE}$, where D bisects BC and E is the point of trisection of CA, with E being nearer to C than to A.

1.55 A circular cylinder is subjected to torsional forces (see Example 1.13). Consider a point on the lateral surface. Verify that one of the principal stresses is zero and that the corresponding principal direction is perpendicular to the lateral surface.

1.56 A state of stress at a point on a plane with normal (0, 0, 1) is represented by the vectors

$$\boldsymbol{\sigma}_x = (0, 0, \alpha), \qquad \boldsymbol{\sigma}_y = (0, 0, \beta), \qquad \boldsymbol{\sigma}_z = (\alpha, \beta, \gamma)$$

If the tensile stress is T and the shear stress is S, show that the principal stresses are 0 and $\frac{1}{2}T \pm (\frac{1}{4}T^2 + S^2)^{1/2}$.

1.9 Triple products

There are two types of triple product (products involving three vectors): the **scalar triple product** $\mathbf{a}\cdot(\mathbf{b} \wedge \mathbf{c})$ and the **vector triple product** $\mathbf{a} \wedge (\mathbf{b} \wedge \mathbf{c})$. These products occur sufficiently often in practice to warrant a detailed examination.

Scalar triple product

We have already introduced the scalar triple product in our discussion of the vector product in Section 1.6. We saw that $|\mathbf{a}\cdot(\mathbf{b}\wedge\mathbf{c})|$ represented the volume of a parallelepiped of sides **a**, **b** and **c**. On the basis of symmetry we then deduced that

$$\mathbf{a}\cdot(\mathbf{b}\wedge\mathbf{c})=\mathbf{b}\cdot(\mathbf{c}\wedge\mathbf{a})=\mathbf{c}\cdot(\mathbf{a}\wedge\mathbf{b}) \tag{1.57}$$

A useful way of expressing the scalar triple product is in determinant form. Since

$$\mathbf{b}\wedge\mathbf{c}=\begin{vmatrix}\hat{\mathbf{i}} & \hat{\mathbf{j}} & \hat{\mathbf{k}}\\ b_x & b_y & b_z\\ c_x & c_y & c_z\end{vmatrix}$$

it readily follows from (1.29) that

$$\mathbf{a}\cdot(\mathbf{b}\wedge\mathbf{c})=\begin{vmatrix}a_x & a_y & a_z\\ b_x & b_y & b_z\\ c_x & c_y & c_z\end{vmatrix} \tag{1.58}$$

The equalities of (1.57) then follow from the properties of determinants.

Example 1.15

Determine the scalar triple product of $\mathbf{a}=2\hat{\mathbf{i}}-\hat{\mathbf{j}}+\hat{\mathbf{k}}$, $\mathbf{b}=\hat{\mathbf{i}}+2\hat{\mathbf{j}}-3\hat{\mathbf{k}}$ and $\mathbf{c}=6\hat{\mathbf{i}}-8\hat{\mathbf{j}}+10\hat{\mathbf{k}}$. Hence deduce that the vectors are coplanar.

Solution

Using (1.58),

$$\begin{aligned}\mathbf{a}\cdot(\mathbf{b}\wedge\mathbf{c}) &=\begin{vmatrix}2 & -1 & 1\\ 1 & 2 & -3\\ 6 & -8 & 10\end{vmatrix}\\ &=2\begin{vmatrix}2 & -3\\ -8 & 10\end{vmatrix}-(-1)\begin{vmatrix}1 & -3\\ 6 & 10\end{vmatrix}+1\begin{vmatrix}1 & 2\\ 6 & -8\end{vmatrix}\\ &=2(20-24)+(10+18)+(-8-12)\\ &=0\end{aligned}$$

Since the volume of the parallelepiped formed by **a**, **b** and **c** is zero, these three vectors must be coplanar.

Vector triple product

From the properties of the cross product it follows that the vector $\mathbf{a} \wedge (\mathbf{b} \wedge \mathbf{c})$ is perpendicular both to $\mathbf{a}$ and to $(\mathbf{b} \wedge \mathbf{c})$. Since $(\mathbf{b} \wedge \mathbf{c})$ is perpendicular to both $\mathbf{b}$ and $\mathbf{c}$, it follows that $\mathbf{a} \wedge (\mathbf{b} \wedge \mathbf{c})$ must lie in the plane containing both $\mathbf{b}$ and $\mathbf{c}$ (see Figure 1.56). Hence we may write

$$\mathbf{a} \wedge (\mathbf{b} \wedge \mathbf{c}) = \alpha\mathbf{b} + \beta\mathbf{c} \qquad \textbf{(1.59)}$$

where α and β are yet to be determined. We shall evaluate the vector product on the left-hand side and use the right-hand side in order to determine α and β.

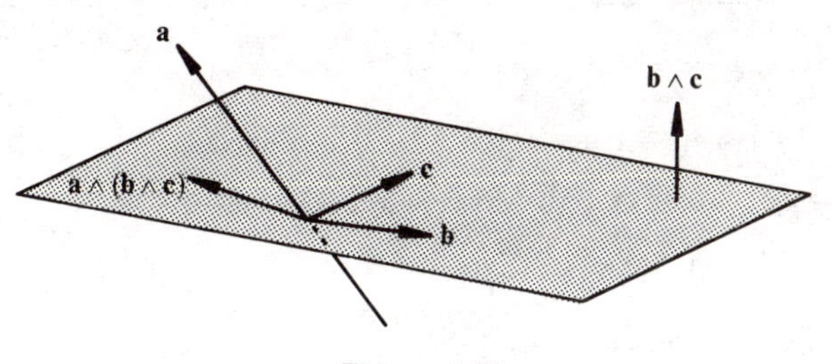

Figure 1.56

Now if $\mathbf{a} = a_x\hat{\mathbf{i}} + a_y\hat{\mathbf{j}} + a_z\hat{\mathbf{k}}$, and similarly for $\mathbf{b}$ and $\mathbf{c}$,

$$\mathbf{b} \wedge \mathbf{c} = \hat{\mathbf{i}}(b_yc_z - b_zc_y) - \hat{\mathbf{j}}(b_xc_z - c_xb_z) + \hat{\mathbf{k}}(b_xc_y - c_xb_y)$$

and so, using (1.46),

$$\begin{aligned}
\mathbf{a} \wedge (\mathbf{b} \wedge \mathbf{c}) &= \begin{vmatrix} \hat{\mathbf{i}} & \hat{\mathbf{j}} & \hat{\mathbf{k}} \\ a_x & a_y & a_z \\ b_yc_z - b_zc_y & c_xb_z - b_xc_z & b_xc_y - c_xb_y \end{vmatrix} \\
&= \hat{\mathbf{i}}[a_y(b_xc_y - c_xb_y) - a_z(c_xb_z - b_xc_z)] \\
&\quad - \hat{\mathbf{j}}[a_x(b_xc_y - c_xb_y) - a_z(b_yc_z - b_zc_y)] \\
&\quad + \hat{\mathbf{k}}[a_x(c_xb_z - b_xc_z) - a_y(b_yc_z - b_zc_y)]
\end{aligned}$$

Now, regrouping as far as possible according to the form of the right-hand side of (1.59), we find

$$\begin{aligned}
\mathbf{a} \wedge (\mathbf{b} \wedge \mathbf{c}) = {}& \hat{\mathbf{i}}b_x(a_yc_y + a_zc_z) + \hat{\mathbf{j}}b_y(a_xc_x + a_zc_z) + \hat{\mathbf{k}}b_z(a_xc_x + a_yc_y) \\
& - \hat{\mathbf{i}}c_x(a_yb_y + a_zb_z) - \hat{\mathbf{j}}c_y(a_xb_x + a_zb_z) - \hat{\mathbf{k}}c_z(a_xb_x + a_yb_y)
\end{aligned} \qquad \textbf{(1.60)}$$

Adding and subtracting

$$\hat{\mathbf{i}}(b_xa_xc_x) + \hat{\mathbf{j}}(b_ya_yc_y) + \hat{\mathbf{k}}(b_za_zc_z)$$

to the right-hand side of (1.60) finally gives

$$\mathbf{a} \wedge (\mathbf{b} \wedge \mathbf{c}) = (\mathbf{a} \cdot \mathbf{c})\mathbf{b} - (\mathbf{a} \cdot \mathbf{b})\mathbf{c} \tag{1.61}$$

By similar reasoning, the vector triple product $(\mathbf{a} \wedge \mathbf{b}) \wedge \mathbf{c}$ is a vector which lies in the plane containing $\mathbf{a}$ and $\mathbf{b}$. A detailed calculation shows that

$$(\mathbf{a} \wedge \mathbf{b}) \wedge \mathbf{c} = (\mathbf{a} \cdot \mathbf{c})\mathbf{b} - (\mathbf{b} \cdot \mathbf{c})\mathbf{a}$$

Clearly, $\mathbf{a} \wedge (\mathbf{b} \wedge \mathbf{c})$ is a vector with a different magnitude and direction from the vector $(\mathbf{a} \wedge \mathbf{b}) \wedge \mathbf{c}$. In other words, the associative law fails for the vector triple product.

EXERCISES

1.57 If $\mathbf{a} = \hat{\mathbf{i}} + \hat{\mathbf{j}}$, $\mathbf{b} = \hat{\mathbf{i}} + 2\hat{\mathbf{j}} + \hat{\mathbf{k}}$ and $\mathbf{c} = \hat{\mathbf{i}} - 3\hat{\mathbf{j}}$, evaluate:

(a) $\mathbf{a} \cdot [(\mathbf{a} - \mathbf{b}) \wedge \mathbf{c}]$ (b) $(\mathbf{a} \wedge \mathbf{c}) \cdot (\mathbf{b} \wedge \mathbf{c})$ (c) $\mathbf{a} \wedge (\mathbf{b} \wedge \mathbf{c})$

(d) $(\mathbf{a} \cdot \mathbf{c})\mathbf{b} - (\mathbf{a} \cdot \mathbf{b})\mathbf{c}$

1.58 The vector equations of two lines are

$$\mathbf{r} = \mathbf{a} + t\hat{\mathbf{g}} \quad \text{and} \quad \mathbf{r} = \mathbf{b} + s\hat{\mathbf{h}}$$

where $\hat{\mathbf{g}}$, $\hat{\mathbf{h}}$ are unit vectors and t, s are scalar parameters. Show that the shortest distance between the two lines is

$$\frac{|(\mathbf{a} - \mathbf{b}) \cdot (\hat{\mathbf{g}} \wedge \hat{\mathbf{h}})|}{|\hat{\mathbf{g}} \wedge \hat{\mathbf{h}}|}$$

Hence, show that 4.064 is the shortest distance between the lines AB and CD, where A is the point $(-1, -1, 1)$, B is $(0, 0, 0)$, C is $(0, 4, 2)$ and D is $(2, 0, 5)$.

1.59 By multiplying the equation

$$\mathbf{a} = \beta\mathbf{b} + \gamma\mathbf{c} + \delta\mathbf{d}$$

by terms of the form $\mathbf{b} \wedge \mathbf{c}$, show that

$$\beta = \frac{\mathbf{a} \cdot (\mathbf{c} \wedge \mathbf{d})}{\mathbf{b} \cdot (\mathbf{c} \wedge \mathbf{d})}, \qquad \gamma = \frac{\mathbf{a} \cdot (\mathbf{d} \wedge \mathbf{b})}{\mathbf{b} \cdot (\mathbf{c} \wedge \mathbf{d})}, \qquad \delta = \frac{\mathbf{a} \cdot (\mathbf{b} \wedge \mathbf{c})}{\mathbf{b} \cdot (\mathbf{c} \wedge \mathbf{d})}$$

provided that $\mathbf{b} \cdot (\mathbf{c} \wedge \mathbf{d}) \neq 0$.

1.60 Using properties of the vector triple product and the scalar triple product, prove that

$$(\mathbf{a} \wedge \mathbf{b}) \cdot (\mathbf{c} \wedge \mathbf{d}) = (\mathbf{a} \cdot \mathbf{c})(\mathbf{b} \cdot \mathbf{d}) - (\mathbf{b} \cdot \mathbf{c})(\mathbf{a} \cdot \mathbf{d})$$

1.61 Show that the three planes

$$\mathbf{r} \cdot \hat{\mathbf{n}}_1 = d_1, \quad \mathbf{r} \cdot \hat{\mathbf{n}}_2 = d_2 \quad \text{and} \quad \mathbf{r} \cdot \hat{\mathbf{n}}_3 = d_3$$

intersect at a point with position vector

$$\mathbf{r}_0 = \frac{d_1(\hat{\mathbf{n}}_2 \wedge \hat{\mathbf{n}}_3) + d_2(\hat{\mathbf{n}}_3 \wedge \hat{\mathbf{n}}_1) + d_3(\hat{\mathbf{n}}_1 \wedge \hat{\mathbf{n}}_2)}{\hat{\mathbf{n}}_1 \cdot (\hat{\mathbf{n}}_2 \wedge \hat{\mathbf{n}}_3)}$$

ADDITIONAL EXERCISES

1 (a) If $\hat{\mathbf{a}} = a_x\hat{\mathbf{i}} + a_y\hat{\mathbf{j}} + a_z\hat{\mathbf{k}}$ is a unit vector, show that a_x, a_y and a_z are the cosines of the angles that the vector makes with the positive x-, y- and z-axes respectively. (It is for this reason that a_x, a_y and a_z are called **direction cosines**.)
(b) Express the following vectors as products of a scalar and a unit vector:

(i) $\hat{\mathbf{i}} + \hat{\mathbf{j}} + \hat{\mathbf{k}}$ (ii) $3\hat{\mathbf{i}} + 4\hat{\mathbf{j}}$

Determine the direction cosines of each of the vectors (i) and (ii).

2 Three ships are observed from a lighthouse at 30-minute intervals. They have the following distance and velocity vectors:

$\mathbf{s}_1 = 2\hat{\mathbf{i}} + 6\hat{\mathbf{j}}$ and $\mathbf{v}_1 = 5\hat{\mathbf{i}} + 4\hat{\mathbf{j}}$ at 12 noon

$\mathbf{s}_2 = 6\hat{\mathbf{i}} + 9\hat{\mathbf{j}}$ and $\mathbf{v}_2 = 4\hat{\mathbf{i}} + 3\hat{\mathbf{j}}$ at 12.30 p.m.

$\mathbf{s}_3 = 11\hat{\mathbf{i}} + 6\hat{\mathbf{j}}$ and $\mathbf{v}_3 = 2\hat{\mathbf{i}} + 7\hat{\mathbf{j}}$ at 1 p.m.

Prove that, if the ships continue with the same velocities, two of them will collide, and find the time of collision. If at the instant of collision the third ship changes course and then proceeds directly to the scene of collision at its original speed, find the time at which it will arrive.

3 The **moment** of a bar magnet may be characterized by a vector **m**. (The direction of **m** is along the axis of the magnet from the negative (south) to the positive (north) pole; the magnitude depends on the size of the magnet and the material from which it is made.) It can be shown that the magnetic field **H** (a vector) experienced at a point whose position vector relative to the centre of the magnet is **r**, is given by

$$\mathbf{H} = -\frac{\mathbf{m}}{|\mathbf{r}|^3} + \frac{3(\mathbf{m}\cdot\mathbf{r})\mathbf{r}}{|\mathbf{r}|^5}$$

Find **H** at $(1, 0, 0)$ if the magnetic moment has a strength 5 in the direction of the vector $\hat{\mathbf{i}} + \hat{\mathbf{j}}$ and if the magnet is placed with its centre at the point $(-1, 2, 3)$. (A small magnet would tend to orientate in the direction of **H**.)

4 The equation of the surface of a sphere is $x^2 + y^2 + z^2 = 50$. Verify that the point P $(3, 4, 5)$ lies on the sphere and find the equation of the tangent plane through P.

5 An aircraft, A, has a speed of 1000 km h^{-1}. Taking the x-axis in an easterly direction and the z-axis vertically upwards, find the velocity vector $\mathbf{v}_A$ of the aircraft as it climbs steadily in a north-easterly direction and increases its height by 4 km h^{-1}.

A second aircraft, B, leaves at the same time and from the same point as A with a velocity vector $\mathbf{v}_B = 500\hat{\mathbf{i}} + 8\hat{\mathbf{k}}$ km h^{-1}. What is the relative velocity of the second aircraft with respect to the first? Determine the distance apart of the two aircraft two hours after take-off.

6 Three forces of magnitudes F, $2F$ and $3F$ act along the sides AB, BC and CA respectively of an equilateral triangle ABC. Determine the resultant of these forces and show that the line of action of the resultant cuts BC in a point P such that $2BP = 3BC$.

7 Prove that the four vector areas (with outward-pointing normals) of the faces of a tetrahedron sum to zero.

8 (a) The volume of a tetrahedron is given by $\frac{1}{3}$(perpendicular height × area of base). If $\mathbf{d}$, $\mathbf{a}$ and $\mathbf{f}$ are vectors representing the three sides of a tetrahedron which intersect at a corner, show that the volume of the tetrahedron may be expressed in the form

$$V = \tfrac{1}{6}|\mathbf{d} \cdot (\mathbf{a} \wedge \mathbf{f})|$$

(b) If $\mathbf{a}$, $\mathbf{b}$, $\mathbf{c}$ and $\mathbf{d}$ are the position vectors of four points A, B, C and D respectively, show that the condition for the four points to be coplanar is

$$\mathbf{a} \cdot (\mathbf{b} \wedge \mathbf{c}) + \mathbf{a} \cdot (\mathbf{c} \wedge \mathbf{d}) + \mathbf{a} \cdot (\mathbf{d} \wedge \mathbf{b}) = \mathbf{b} \cdot (\mathbf{c} \wedge \mathbf{d})$$

SUMMARY

- A **scalar** is a quantity which is characterized by a magnitude only.

- A **vector** is a quantity which is characterized by both a magnitude and a direction and is such that two vectors combine ('add up') according to a rule known as the **parallelogram law**.

- A vector may be written in component form or in coordinate form:

$$\mathbf{a} = a_x\hat{\mathbf{i}} + a_y\hat{\mathbf{j}} + a_z\hat{\mathbf{k}} \quad \text{or} \quad (a_x, a_y, a_z)$$

- The modulus of a vector is its scalar magnitude $|\mathbf{a}|$, where $|\mathbf{a}| = \sqrt{(a_x^2 + a_y^2 + a_z^2)}$.

- The **resultant** of two vectors $\mathbf{a}$ and $\mathbf{b}$ is their **vector sum**, $\mathbf{a} + \mathbf{b}$.

- Vector addition satisfies the commutative and associative laws:

$$\mathbf{a}+\mathbf{b}=\mathbf{b}+\mathbf{a} \quad \text{and} \quad \mathbf{a}+(\mathbf{b}+\mathbf{c})=(\mathbf{a}+\mathbf{b})+\mathbf{c}$$

- The product of a scalar s with a vector $\mathbf{a}$ is written $s\mathbf{a}$ and is such that

$$s\mathbf{a}=sa_x\hat{\mathbf{i}}+sa_y\hat{\mathbf{j}}+sa_z\hat{\mathbf{k}}$$

- A unit vector in the direction of $\mathbf{a}$ is $\hat{\mathbf{a}}=\dfrac{\mathbf{a}}{|\mathbf{a}|}$

- The **scalar product** of two vectors $\mathbf{a}$ and $\mathbf{b}$ is written $\mathbf{a}\cdot\mathbf{b}$ and is defined by

$$\mathbf{a}\cdot\mathbf{b}=|\mathbf{a}|\,|\mathbf{b}|\cos\theta$$

or, in component form,

$$\mathbf{a}\cdot\mathbf{b}=a_xb_x+a_yb_y+a_zb_z$$

The scalar product satisfies the commutative and distributive laws:

$$\mathbf{a}\cdot\mathbf{b}=\mathbf{b}\cdot\mathbf{a} \quad \text{and} \quad \mathbf{a}\cdot(\mathbf{b}+\mathbf{c})=\mathbf{a}\cdot\mathbf{b}+\mathbf{a}\cdot\mathbf{c}$$

- The **vector product** of two vectors $\mathbf{a}$ and $\mathbf{b}$ is written $\mathbf{a}\wedge\mathbf{b}$ and is defined by

$$\mathbf{a}\wedge\mathbf{b}=|\mathbf{a}|\,|\mathbf{b}|\sin\theta\,\hat{\mathbf{n}}$$

where $\mathbf{a}$, $\mathbf{b}$, $\hat{\mathbf{n}}$, taken in this order, form a right-handed system of vectors. In component form,

$$\mathbf{a}\wedge\mathbf{b}=(a_yb_z-a_zb_y)\hat{\mathbf{i}}+(a_zb_x-a_xb_z)\hat{\mathbf{j}}+(a_xb_y-a_yb_x)\hat{\mathbf{k}}$$

or, using the determinant formalism,

$$\mathbf{a}\wedge\mathbf{b}=\begin{vmatrix}\hat{\mathbf{i}} & \hat{\mathbf{j}} & \hat{\mathbf{k}}\\ a_x & a_y & a_z\\ b_x & b_y & b_z\end{vmatrix}$$

The vector product is not commutative but does satisfy the distributive law:

$$\mathbf{a}\wedge\mathbf{b}=-\mathbf{b}\wedge\mathbf{a}, \qquad \mathbf{a}\wedge(\mathbf{b}+\mathbf{c})=\mathbf{a}\wedge\mathbf{b}+\mathbf{a}\wedge\mathbf{c}$$

- The **scalar triple product** $\mathbf{a}\cdot(\mathbf{b}\wedge\mathbf{c})$ satisfies

$$\mathbf{a}\cdot(\mathbf{b}\wedge\mathbf{c})=\mathbf{b}\cdot(\mathbf{c}\wedge\mathbf{a})=\mathbf{c}\cdot(\mathbf{a}\wedge\mathbf{b})$$

- The **vector triple product** $\mathbf{a}\wedge(\mathbf{b}\wedge\mathbf{c})$ satisfies

$$\mathbf{a}\wedge(\mathbf{b}\wedge\mathbf{c})=(\mathbf{a}\cdot\mathbf{c})\mathbf{b}-(\mathbf{a}\cdot\mathbf{b})\mathbf{c}$$

SUPPLEMENT

1S.1 Index notation

The notation we have adopted in this text in order to describe vectors is perhaps the most widely accepted. Certainly, in our view, it is in many ways the 'best' notation with which to *introduce* vectors. However, there are a number of alternative notations and in this supplement we shall record the main results in terms of one of the most useful of these alternatives – the index notation.

The index notation for the description of vectors enjoys a number of advantages. The first is *conciseness*: it allows standard results to be written in an immediately transparent though compact form. A second, less obvious, advantage is that the index approach often allows a surprisingly quick proof of the results that we need. There are many other advantages (and one or two disadvantages) which readers will discover for themselves. We would also point out that the index approach to the description of vectors is the most 'natural' notation if one wishes to consider a generalization of the vector which has found many applications in mathematics, physics and engineering – the **tensor**.

We shall find it convenient to relabel the coordinates (x, y, z) as (x_1, x_2, x_3) or simply as x_i, with the agreement that the index i can take any of the values 1, 2 and 3. Using this index notation, a vector $\mathbf{a}$ may be denoted by a_i. Thus the components of the vector a_i are (a_1, a_2, a_3).

Equality

Two vectors a_i and b_i are equal if and only if

$$a_i = b_i \tag{1S.1}$$

We shall take it for granted that if the index is unspecified then the equation in which it occurs is assumed to be valid for each of the three possible values that the index can take. Thus (1S.1) is shorthand for

$$a_1 = b_1, \qquad a_2 = b_2, \qquad a_3 = b_3 \tag{1S.2}$$

The summation convention

If an index is repeated in a term, then it is assumed that the term is summed as the repeated index takes the values 1, 2 and 3. Thus

$$a_i b_i \quad \text{means} \quad \sum_{i=1}^{3} a_i b_i \equiv a_1 b_1 + a_2 b_2 + a_3 b_3 \tag{1S.3}$$

Using this convention, if a_i and b_i are two vectors (representing **a** and **b**), then the scalar product $\mathbf{a}\cdot\mathbf{b}$ is represented by $a_i b_i$ (see equation (1.29)).

A repeated index is called a *dummy index*, since the precise letter used is irrelevant. For example, $a_i b_i$ has exactly the same meaning as $a_m b_m$, $a_j b_j$, $a_s b_s$, etc., that is, $a_1 b_1 + a_2 b_2 + a_3 b_3$.

The Kronecker delta

We define a two-indexed quantity δ_{ij} called the Kronecker delta by

$$\delta_{ij} = \begin{cases} 1 & \text{if } i = j \\ 0 & \text{otherwise} \end{cases} \qquad \textbf{(1S.4)}$$

Thus $\delta_{11} = \delta_{22} = \delta_{33} = 1$ and $\delta_{12} = \delta_{21} = \delta_{13} = \delta_{31} = \delta_{23} = \delta_{32} = 0$. Another name for this quantity is the **substitution operator**, since it can be used to substitute one index for another. This follows from the easily verified result

$$a_j = \delta_{ji} a_i \qquad \textbf{(1S.5)}$$

(The right-hand side is $\delta_{j1}a_1 + \delta_{j2}a_2 + \delta_{j3}a_3$, so if $j = 1$ in (1S.5) $a_1 = \delta_{11}a_1 + \delta_{12}a_2 + \delta_{13}a_3 = a_1$, with similar results for $j = 2$ and $j = 3$.)

An index that is not repeated (in a term) is called a *free index*.

Just as in ordinary algebra, an equation will be formed from a collection of terms added together and equated to zero. Each term will comprise one or more indexed objects multiplied together. It is an inherent rule when using the index notation that the number and type of free indices in any term of a given equation must be the same. Thus

$$a_i c_j + g_{ij} d_k e_k - h_i c_j = 0$$

is meaningful (i and j are free indices in each term, and k is a dummy index), but

$$a_i c_k + d_{ij} a_t b_t = 0$$

is not meaningful (i and k are free in the first term, i and j are free in the second term, and t is a dummy index).

EXERCISES

1S.1 If a_i and b_i correspond to vectors **a** and **b**, determine the index form of the vector equation

$$\mathbf{a} + (\mathbf{a}\cdot\mathbf{b})\mathbf{b} = 0$$

1S.2 Which of the following equations are meaningful indexed expressions?

(a) $a_i b_j + c_i d_j a_k a_k = 0$ (b) $a_i b_j + a_i = 0$ (c) $a_k b_k c_i + a_k d_i = 0$

(d) $\delta_{ii} - 3 = 0$ (e) $c_i + \varepsilon_{ijk} b_j d_k = 0$

1S.3 Verify that:

(a) $\delta_{ij}\delta_{jk} = \delta_{ik}$ (b) $\delta_{ij}\delta_{ij} = 3$

The permutation symbol

This is defined by

$$\varepsilon_{ijk} = \begin{cases} 1 & \text{if } i, j, k \text{ is a cyclic permutation of } 1, 2, 3 \\ -1 & \text{if } i, j, k \text{ is a non-cyclic permutation of } 1, 2, 3 \\ 0 & \text{otherwise} \end{cases} \tag{1S.6}$$

Each subscript can only take on values 1, 2 or 3, so ε_{ijk} has 27 'components'. From the definition, it follows that

$$\varepsilon_{123} = \varepsilon_{231} = \varepsilon_{312} = +1, \qquad \varepsilon_{132} = \varepsilon_{321} = \varepsilon_{213} = -1 \tag{1S.7a}$$

and all other $\varepsilon_{ijk} = 0$, for example $\varepsilon_{223} = \varepsilon_{233} = \varepsilon_{311} = 0$ **(1S.7b)**

These results imply the following identities:

$$\varepsilon_{ijk} = \varepsilon_{jki} = \varepsilon_{kij} \tag{1S.7c}$$

That is, a cyclic permutation of the indices on the permutation symbol does not alter its value.

Also,

$$\varepsilon_{ijk} = -\varepsilon_{kji}, \quad \varepsilon_{ijk} = -\varepsilon_{jik}, \quad \varepsilon_{ijk} = -\varepsilon_{ikj} \tag{1S.7d}$$

that is, interchange of any two indices on the permutation symbol introduces a minus sign. This corresponds to an anticyclic permutation of indices.

With the introduction of this symbol we can now show that the ith component of the vector product of two vectors **a** and **b** (or a_i and b_i) is

$$(\mathbf{a} \wedge \mathbf{b})_i = \varepsilon_{ijk} a_j b_k \tag{1S.8}$$

The right-hand side of (1S.8) involves two repeated indices and so involves a double summation: firstly

$$\varepsilon_{ijk} a_j b_k = \varepsilon_{i1k} a_1 b_k + \varepsilon_{i2k} a_2 b_k + \varepsilon_{i3k} a_3 b_k \quad \text{(summing over the index } j\text{)}$$

Now summing over the index k (and using the result implicit in (1S.7b) that

ε_{ijk} is zero if it contains two identical indices),

$$\varepsilon_{ijk}a_jb_k = \varepsilon_{i12}a_1b_2 + \varepsilon_{i13}a_1b_3 + \varepsilon_{i21}a_2b_1 + \varepsilon_{i23}a_2b_3 + \varepsilon_{i31}a_3b_1 + \varepsilon_{i32}a_3b_2$$

Therefore if index i takes the value 1,

$$\varepsilon_{1jk}a_jb_k = \varepsilon_{123}a_2b_3 + \varepsilon_{132}a_3b_2 = a_2b_3 - a_3b_2$$

Similarly, if $i = 2$,

$$\varepsilon_{2jk}a_jb_k = \varepsilon_{213}a_1b_3 + \varepsilon_{231}a_3b_1 = a_3b_1 - a_1b_3$$

and if $i = 3$,

$$\varepsilon_{3jk}a_jb_k = \varepsilon_{312}a_1b_2 + \varepsilon_{321}a_2b_1 = a_1b_2 - a_2b_1$$

But these three terms are the components of $\mathbf{a} \wedge \mathbf{b}$ (see equation (1.45)) and so (1S.8) is verified.

The ε–δ identity

There exists a basic identity connecting the ε symbol and the δ symbol. It is

$$\varepsilon_{ijk}\varepsilon_{ipr} = \delta_{jp}\delta_{kr} - \delta_{jr}\delta_{kp}$$

This identity is proved using the properties of determinants and is fully discussed in Appendix 1. It has numerous applications when developing vector identities; for example, let us re-prove, using the index approach, the identity (1.61), namely,

$$\mathbf{a} \wedge (\mathbf{b} \wedge \mathbf{c}) = (\mathbf{a}\cdot\mathbf{c})\mathbf{b} - (\mathbf{a}\cdot\mathbf{b})\mathbf{c}$$

Now the ith component of the left-hand side is, using (1S.8) twice,

$$\varepsilon_{ijk}a_j(\mathbf{b} \wedge \mathbf{c})_k = \varepsilon_{ijk}\varepsilon_{klm}a_jb_lc_m$$

After a cyclic permutation on the first ε symbol on the right-hand side, we find

$$\begin{aligned}\varepsilon_{kij}\varepsilon_{klm}a_jb_lc_m &= (\delta_{il}\delta_{jm} - \delta_{im}\delta_{jl})a_jb_lc_m \\ &= a_jb_ic_j - a_jb_jc_i \\ &= (\mathbf{a}\cdot\mathbf{c})b_i - (\mathbf{a}\cdot\mathbf{b})c_i \\ &= \{(\mathbf{a}\cdot\mathbf{c})\mathbf{b} - (\mathbf{a}\cdot\mathbf{b})\mathbf{c}\}_i\end{aligned}$$

which proves the result.

EXERCISES

1S.4 Show that:

(a) $\varepsilon_{iik}=0$ (b) $\varepsilon_{ijk}a_ja_k=0$ (c) $\varepsilon_{ijk}\varepsilon_{ijk}=6$

1S.5 Use the index notation to show that $\mathbf{a}\cdot(\mathbf{b}\wedge\mathbf{c})=(\mathbf{a}\wedge\mathbf{b})\cdot\mathbf{c}$.

Symmetry and skew-symmetry

As we have already explained, a 'term' comprises one or more indexed objects multiplied together. Such a term (or group of terms) may be denoted by $T_{ij\cdots k}$, where $ij\cdots k$ refer to the free indices in the term. Thus we might, for convenience, denote the collection $\varepsilon_{ijk}\varepsilon_{ipr}-\delta_{jp}\delta_{kr}+\delta_{jr}\delta_{kp}$ by T_{jkpr}, whilst ε_{iik} might be denoted by T_k.

Consider now a general term $T_{ij\cdots k}$. If when two particular indices are interchanged we obtain the result

$$T_{ij\cdots p\cdots q\cdots k}=T_{ij\cdots q\cdots p\cdots k}$$

then we say the term is **symmetric** in the indices p and q. If, instead, we obtain the result

$$T_{ij\cdots p\cdots q\cdots k}=-T_{ij\cdots q\cdots p\cdots k}$$

then we say the term is **skew-symmetric** in the indices p and q.

Typical examples of symmetric and skew-symmetric objects are the Kronecker delta and the permutation symbol respectively. Since $\delta_{ij}=\delta_{ji}$ (from (1S.4)) for each value of i and j, it follows that the Kronecker delta is symmetric in its indices (or simply symmetric). Also, from (1S.7d) it follows that the permutation symbol is skew-symmetric in *any* pair of indices. For this reason, it is often referred to as the completely skew-symmetric object.

Not every indexed object (with two or more indices) is either symmetric or skew-symmetric, although every indexed object (with two or more indices) can be written (for each pair of indices) as the sum of a symmetric and a skew-symmetric object. For example, if A_{ij} is a given two-indexed object, then by simple algebra

$$A_{ij}=\tfrac{1}{2}(A_{ij}+A_{ji})+\tfrac{1}{2}(A_{ij}-A_{ji})$$

and it is easily verified that the object $C_{ij}\equiv\tfrac{1}{2}(A_{ij}+A_{ji})$ is symmetric whilst the object $D_{ij}\equiv\tfrac{1}{2}(A_{ij}-A_{ji})$ is skew-symmetric.

Example 1S.1

Show that if $C_{ij\cdots p\cdots q\cdots k}$ is symmetric in p and q and $D_{lm\cdots p\cdots q\cdots n}$ is skew-symmetric in p and q, then

$$C_{ij\cdots p\cdots q\cdots k}D_{lm\cdots p\cdots q\cdots n}=0$$

This is a result that often arises.

Solution

Since both p and q are dummy indices we may use two different indices:

$$C_{ij\cdots p\cdots q\cdots k}D_{lm\cdots p\cdots q\cdots n}=C_{ij\cdots a\cdots b\cdots k}D_{lm\cdots a\cdots b\cdots n}$$

Now changing back from a, b to p, q but with $a \rightarrow q$ and $b \rightarrow p$ gives

$$C_{ij\cdots a\cdots b\cdots k}D_{lm\cdots a\cdots b\cdots n}=C_{ij\cdots q\cdots p\cdots k}D_{lm\cdots q\cdots p\cdots n}$$

But, since C is symmetric in p and q,

$$C_{ij\cdots q\cdots p\cdots k}=C_{ij\cdots p\cdots q\cdots k}$$

and since D is skew-symmetric in p and q,

$$D_{ij\cdots q\cdots p\cdots k}=-D_{ij\cdots p\cdots q\cdots k}$$

We can write

$$C_{ij\cdots p\cdots q\cdots k}D_{lm\cdots p\cdots q\cdots n}=-C_{ij\cdots p\cdots q\cdots k}D_{lm\cdots p\cdots q\cdots n}$$

which is only possible if both sides are zero (if $A=-A$, then $A=0$). This proves the result.

The stress tensor

In Section 1.8 we introduced three vectors $\boldsymbol{\sigma}_x$, $\boldsymbol{\sigma}_y$ and $\boldsymbol{\sigma}_z$ which governed the 'state of stress' at a point inside a body. In fact, these three vectors may be grouped together by use of the two-indexed object σ_{ij}:

$$\sigma_{1j}=(\boldsymbol{\sigma}_x)_j, \qquad \sigma_{2j}=(\boldsymbol{\sigma}_y)_j, \qquad \sigma_{3j}=(\boldsymbol{\sigma}_z)_j$$

Of course, σ_{ij} is symmetric: $\sigma_{ij}=\sigma_{ji}$. Using this notation, equation (1.50) may be written more elegantly as

$$P_i^n=\sigma_{ij}n_j$$

Contraction

If in a given equation, for example

$$C_{ij\cdots p\cdots q\cdots k}=0$$

we make two of the free indices the same (implying summation), for example

$$C_{ij\cdots p\cdots p\cdots k}=0$$

then we say we have **contracted** the equation over the indices p and q. The process of contraction reduces (or contracts) the number of free indices in an equation by two. This may be done for any pair of free indices and the equation will remain valid. As the reader might well have perceived, contraction is a shorthand procedure for recording the effect of multiplying both sides of the original equation by the Kronecker delta δ_{pq}:

$$0=\delta_{pq}C_{ij\cdots p\cdots q\cdots k}=C_{ij\cdots p\cdots p\cdots k}$$

Although a contracted equation is as valid as the original equation, the former contains less information, since we have two fewer indices. For example, it may be shown that

$$\varepsilon_{ijk}\delta_{mn}-\varepsilon_{mjk}\delta_{in}-\varepsilon_{imk}\delta_{jn}-\varepsilon_{ijm}\delta_{kn}=0$$

(There are 243 equations here.) However, if we contract k with m, we find

$$\varepsilon_{ijn}-\varepsilon_{mjm}\delta_{in}-\varepsilon_{imm}\delta_{jn}-\varepsilon_{ijn}=0$$

Then, using the properties of the permutation symbol $\varepsilon_{mjm}=0$ and $\varepsilon_{imm}=0$, we get

$$0=0$$

which, although valid, contains no information whatsoever!

A simple example of the use of contraction occurs in linear elasticity theory, where two two-indexed objects, the stress tensor σ_{ij} and the strain tensor η_{ij}, arise. (The actual meaning of the term 'tensor' need not concern us here.) Both objects are symmetric, that is $\sigma_{ij}=\sigma_{ji}$ and $\eta_{ij}=\eta_{ji}$. It may be shown that the relationship between stress and strain is elegantly described in the 'generalized Hooke's law',

$$\sigma_{ij}=2\mu\eta_{ij}+\lambda\delta_{ij}\eta_{kk} \qquad \textbf{(1S.9)}$$

in which μ and λ are constants (known as the Lamé constants) directly related to Young's modulus and the Poisson ratio). In this form, the stress tensor is explicitly expressed in terms of the strain tensor. For some applications we need to invert this relation and express the strain tensor explicitly in terms of the stress tensor. To do this we contract (1S.9) (obviously over the indices i and j) to give

$$\sigma_{ii}=2\mu\eta_{ii}+\lambda\delta_{ii}\eta_{kk}$$

But $\delta_{ii}=\delta_{11}+\delta_{22}+\delta_{33}=3$ and η_{ii} is identical to η_{kk}, both being equal to $\eta_{11}+\eta_{22}+\eta_{33}$. Therefore

$$\sigma_{ii}=2\mu\eta_{ii}+3\lambda\eta_{ii}=(2\mu+3\lambda)\eta_{ii} \quad \text{and} \quad \eta_{ii}=\frac{\sigma_{ij}}{(2\mu+3\lambda)}$$

Hence (1S.9) may be re-expressed as

$$\sigma_{ij} = 2\mu\eta_{ij} + \lambda\delta_{ij}\frac{\sigma_{kk}}{(2\mu + 3\lambda)} \tag{1S.10}$$

where (to avoid ambiguity) we have used k instead of i as the dummy index in the expression for η_{ii}. Rearranging (1S.10) so that all the stress 'components' appear on one side of the equation, we find

$$\eta_{ij} = \frac{1}{2\mu}\left(\sigma_{ij} - \frac{\lambda}{(2\mu + 3\lambda)}\delta_{ij}\sigma_{kk}\right) \tag{1S.11}$$

and this completes our 'inversion'.

As we extend our knowledge of vectors we shall attempt (in the chapters' supplements) to reformulate the major results using the index notation.

Parametric Equations of Curves and Surfaces

PREVIEW

In this chapter we show how vectors can be used to obtain useful descriptions of curves in space. We introduce the concept of the derivative of a vector, which enables us readily to find tangents and normals to curves. We then extend the discussion to surfaces and show how to calculate a unit vector normal to a given surface. The material, particularly that dealing with surfaces, is fundamental to later chapters.

2.1 Curves

You are probably familiar with the Cartesian form of a curve,

$$y = f(x) \qquad \textbf{(2.1)}$$

where f is a given function of the variable x. This relation has a geometrical interpretation as a curve in the xy-plane (see Figure 2.1). In this form, the variables x and y are of different types. For each given value of x, the relation $y = f(x)$ enables us to determine the corresponding y-value. We normally say that x is the **independent** variable and y is the **dependent** variable.

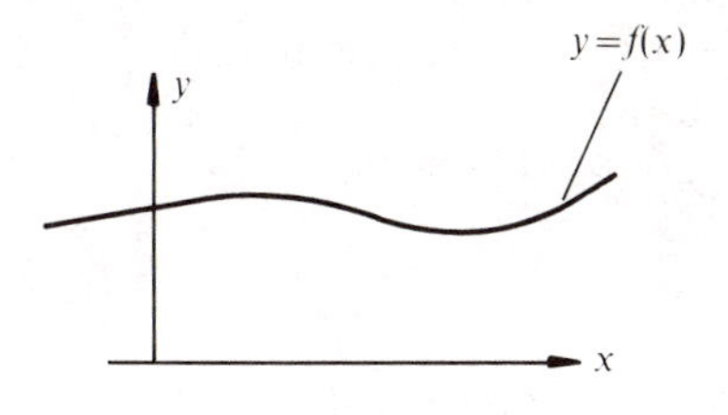

Figure 2.1

There is an alternative way of describing a curve in space in which both the variables x and y are on the same 'level' in that both are treated as dependent variables. This is the **parametric** form, in which we introduce a parameter t such that the equation of the curve has the form

$$x = g(t), \qquad y = h(t) \qquad \textbf{(2.2)}$$

This appears to be more complicated than (2.1), but in fact the parametric form can prove advantageous in many problems.

Of course, as well as giving the equation of a curve, we should also state the range of values that the independent variable can assume. We should also emphasize that the functions f, g and h must be single valued: each value of the independent variable must give rise to just one value of the dependent variable.

Example 2.1

Determine parametric equations for the straight line which connects the points (1, 1) and (3, 5).

Solution

This straight line has the Cartesian form

$$\frac{y-1}{5-1} = \frac{x-1}{3-1}, \qquad 1 \leqslant x \leqslant 3$$

or $$y=2x-1, \qquad 1 \leqslant x \leqslant 3 \tag{2.3}$$

We could write this in parametric form as

$$y=t, \quad x=\tfrac{1}{2}(t+1), \qquad 1 \leqslant t \leqslant 5 \tag{2.4}$$

(2.4) is of course equivalent to (2.3) because when we eliminate t from the equations of (2.4) we obtain (2.3).

One important point is that there is no unique parametric form corresponding to the equation of a given curve – the straight line of (2.3) could also be parametrized by, for example,

$$y=t^2+t, \quad x=\tfrac{1}{2}(t^2+t+1), \qquad \tfrac{1}{2}(-1+\sqrt{5}) \leqslant t \leqslant \tfrac{1}{2}(-1+\sqrt{21}) \tag{2.5}$$

Example 2.2

Determine a parametric representation for the segment of the circle shown in Figure 2.2a.

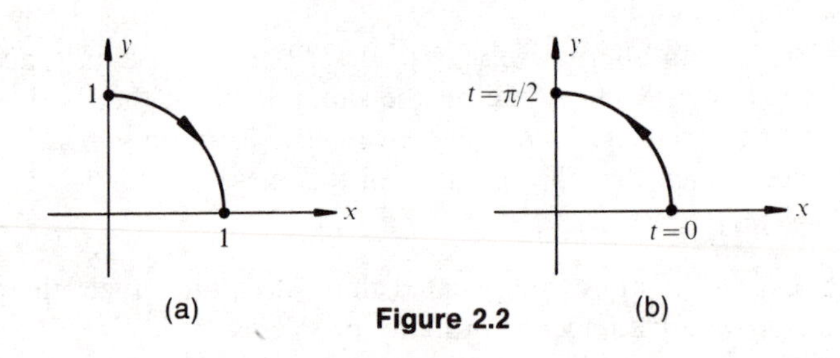

Figure 2.2

Solution

The Cartesian form is

$$y=+\sqrt{(1-x^2)}, \qquad 0 \leqslant x \leqslant 1 \tag{2.6}$$

and the arrow in Figure 2.2a indicates the direction in which the curve is followed as the independent variable, in this case x, increases.

There are numerous possible parametric representations, of which the most popular is

$$x=\cos t, \quad y=\sin t, \qquad 0 \leqslant t \leqslant \pi/2 \tag{2.7}$$

because, clearly, $x^2+y^2=\cos^2 t+\sin^2 t=1$, from which we can recover (2.6).

With the parametrization of (2.7), although the curve described is indeed the circle of Figure 2.2a, we would have to reverse the direction of the arrow. This is because the curve described by (2.7) is followed in an anticlockwise sense as the independent variable, which is now t, increases from 0 to $\pi/2$ (see Figure 2.2b).

If we wanted a parametrization for which the arrow direction of Figure 2.2a was preserved, we could use, for example,

$$x = \sin t, \quad y = \cos t, \qquad 0 \leqslant t \leqslant \pi/2 \tag{2.8}$$

instead of (2.7).

These subtle points can be quite important when parametric representations are used in, for example, the evaluation of line integrals.

Curves in three dimensions are more naturally expressed in parametric form than in Cartesian form. We write

$$x = x(t), \quad y = y(t), \quad z = z(t), \qquad t_0 \leqslant t \leqslant t_1 \tag{2.9}$$

that is, the dependent variables x, y and z are each given functions of an independent variable t as t varies over a range $t_0 \leqslant t \leqslant t_1$. For example, a single turn of a helix (see Figure 2.3) might have a parametric representation

$$x = \cos t, \quad y = \sin t, \quad z = t, \qquad 0 \leqslant t \leqslant 2\pi \tag{2.10}$$

whilst two turns of the helix would be described by exactly the same parametric equations, but the range of the parameter t would be $0 \leqslant t \leqslant 4\pi$.

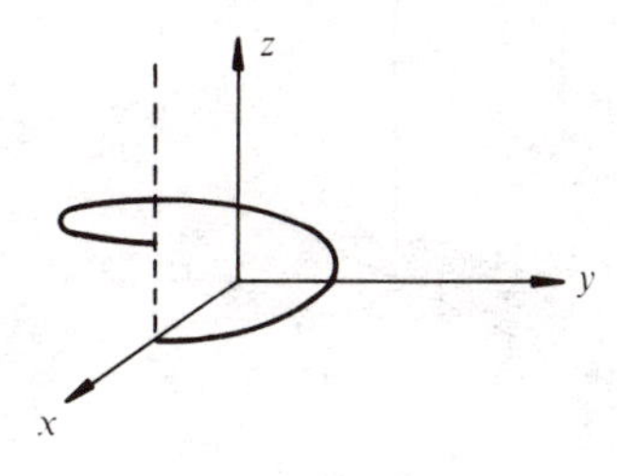

Figure 2.3

2.2 Surfaces

As was the case with curves, the equation of a surface may be written in many different ways. Again, perhaps the most common method is the explicit Cartesian form,

$$z = f(x, y), \qquad (x, y) \text{ within a domain } D \tag{2.11}$$

where f is a given function of x and y (see Figure 2.4). As the point $P_1(x, y)$ ranges through the domain D, the point P_2 with coordinates $(x, y, z = f(x, y))$ ranges over part of the surface S defined by equation (2.11). For example, the equation

$$z = +\sqrt{(4 - x^2 - y^2)}, \quad (x, y) \text{ within the domain } x^2 + y^2 \leqslant 4 \tag{2.12}$$

defines the upper half of a sphere centred on the origin and of radius 2. This

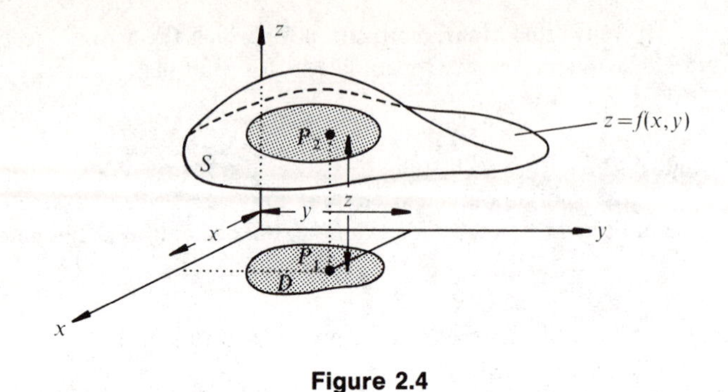

Figure 2.4

follows since, obviously, z is positive for all permissible x, y and

$$z^2 = 4 - x^2 - y^2 \quad \text{or} \quad x^2 + y^2 + z^2 = 4 \tag{2.13}$$

which is the familiar Cartesian equation in implicit form for the surface of a sphere (see Figure 2.5).

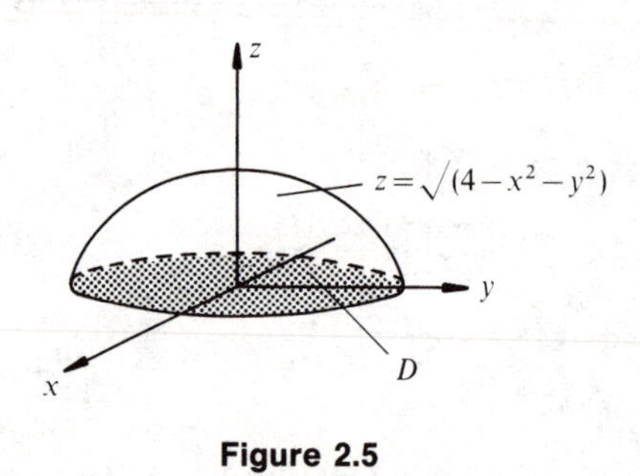

Figure 2.5

In a similar manner, the equation of a plane may be written in explicit form as

$$z = (-ax - by + d)/c, \qquad \text{all } (x, y) \tag{2.14}$$

The implicit form of this equation is

$$ax + by + cz = d \tag{2.15}$$

and, as we have seen in Example 1.8, this is the equation of a plane with normal $a\hat{\mathbf{i}} + b\hat{\mathbf{j}} + c\hat{\mathbf{k}}$ and at a distance $d/\sqrt{(a^2 + b^2 + c^2)}$ from the origin.

In the examples above (and in normal practice), x and y have been chosen as the independent variables and z as the dependent variable. We could, of course, have chosen any two of the variables x, y and z as independent and the remaining one would then have been dependent. The choice is usually dictated by the example we have in mind. Thus, if we wished

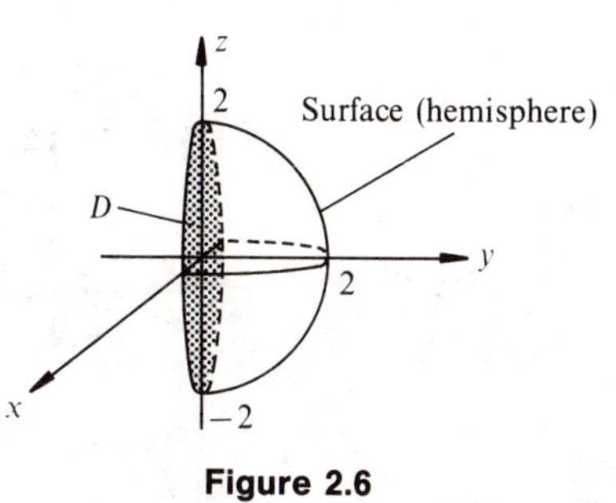

Figure 2.6

to describe the hemispherical surface in Figure 2.6 using z as the dependent variable we would need two statements if we restricted ourselves to single-valued functions. (On the surface shown, each pair of values of (x, y) gives rise to *two* values of z.) We would have

$$z = \begin{cases} +\sqrt{(4 - x^2 - y^2)} & \text{above the } xy\text{-plane} \\ -\sqrt{(4 - x^2 - y^2)} & \text{below the } xy\text{-plane} \end{cases} \quad \textbf{(2.16)}$$

whereas if we make y the dependent variable, then the equation of the surface is more simply expressed as

$$y = +\sqrt{(4 - x^2 - z^2)}, \qquad (x, z) \text{ within } D \quad \textbf{(2.17)}$$

The slight drawback with an explicit Cartesian representation of a surface, namely that all variables are not on the same level, can be removed by writing the equation in implicit form,

$$G(x, y, z) = 0 \quad \textbf{(2.18)}$$

where G is a given function of the variables x, y and z. Here all the variables x, y and z are treated in a symmetrical way and (2.18) may be solved to obtain any one variable explicitly in terms of the other two. Unfortunately, this process may not lead to single-valued functions and an alternative representation for surfaces is desirable. This is the parametric form,

$$x = x(u, v), \qquad y = y(u, v), \qquad z = z(u, v) \quad \textbf{(2.19)}$$

where $x(u, v)$, $y(u, v)$ and $z(u, v)$ are given functions of the two parameters u and v. As u and v range over a domain in a 'uv-plane', so x, y and z range over some surface S, the precise surface depending on the choice of the three functions in (2.19). Note that if we move along a line on which u is a constant in the uv-plane, or one on which v is constant, then motion along a curve on the surface S is implied. Such a curve on S is called a **parametric** curve and, unless the choice of the parametric equations is somewhat perverse, the drawing of these parametric curves gives a pictorial realization of the 'shape' of the surface (see Figure 2.7).

The parametric description of a surface clearly encompasses the previous methods, as we can imagine solving two of the equations for u and

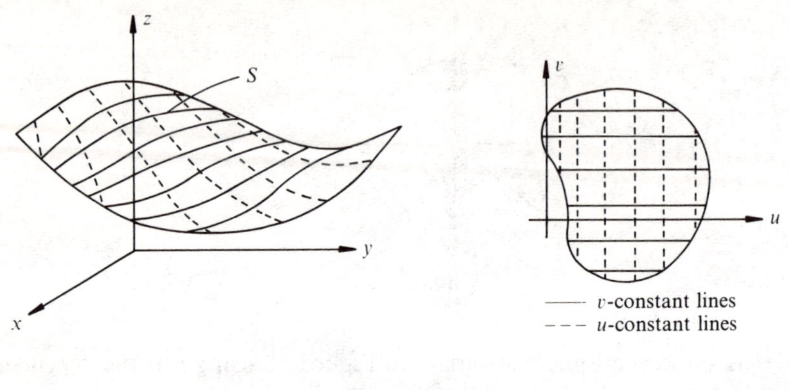

Figure 2.7

v in terms of the Cartesian coordinates. For example, solving the first two equations of (2.19) for u and v in terms of x and y gives

$$u = u(x, y), \qquad v = v(x, y) \tag{2.20}$$

in which case the third implies

$$z = z(u(x, y), v(x, y)) = f(x, y) \tag{2.21}$$

which is an explicit equation for a surface.

As an example, the parametric equations for the surface of a sphere centre the origin and of radius a are

$$x = a \sin u \cos v, \quad y = a \sin u \sin v, \quad z = a \cos u,$$

$$0 \leqslant v \leqslant 2\pi, \quad 0 \leqslant u \leqslant \pi \tag{2.22}$$

where u and v are the angles shown in Figure 2.8. (These angles are usually represented by the symbols θ and ϕ respectively in spherical coordinates, as we shall discuss more fully in Chapter 4.) To find the Cartesian equation for the sphere, the parameters u and v need to be eliminated from these three

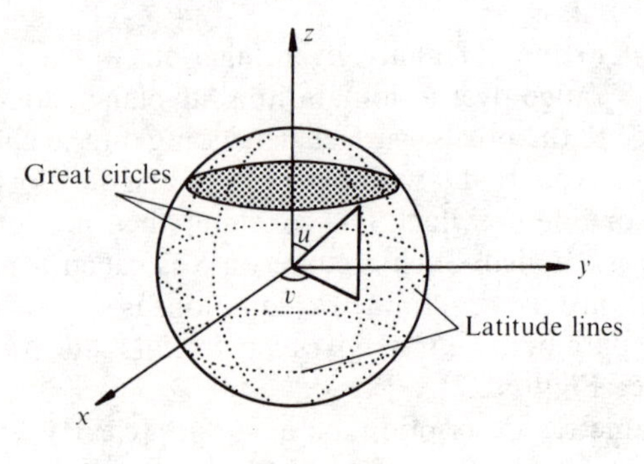

Figure 2.8

equations. Clearly, from (2.22),

$$x^2 + y^2 = a^2 \sin^2 u \cos^2 v + a^2 \sin^2 u \sin^2 v$$
$$= a^2 \sin^2 u$$

$$\therefore \quad x^2 + y^2 + z^2 = a^2 \sin^2 u + a^2 \cos^2 u = a^2$$

In this particular example, u-constant lines are latitude lines on the sphere and v-constant lines are great circles.

In solving problems mathematically, flexibility of mind is a considerable asset as it would be extremely difficult to solve all problems using the same approach. If it is convenient to change representation, then this should be done. The description of curves and surfaces is a case in point: for some problems the explicit Cartesian form is the most appropriate approach, for others the implicit Cartesian form might be more suitable, and for still others a parametric representation might lead to a simple solution.

2.3 The vector equation of a curve

In this section we shall normally deal with three-dimensional curves, but our discussion is also relevant for planar curves.

We imagine moving along a given curve whose parametric representation is

$$x = x(t), \quad y = y(t), \quad z = z(t), \qquad t_0 \leqslant t \leqslant t_1 \tag{2.23}$$

Each value of t specifies a different position on the curve. The position vector $\mathbf{r}$ at a general value of t is the vector drawn from the origin to meet the curve (see Figure 2.9). The magnitude and direction of $\mathbf{r}$ depend on the value of t and to emphasize this, we will often write $\mathbf{r}(t)$ instead of simply $\mathbf{r}$. Hence we can write

$$\mathbf{r}(t) = x(t)\hat{\mathbf{i}} + y(t)\hat{\mathbf{j}} + z(t)\hat{\mathbf{k}} \tag{2.24}$$

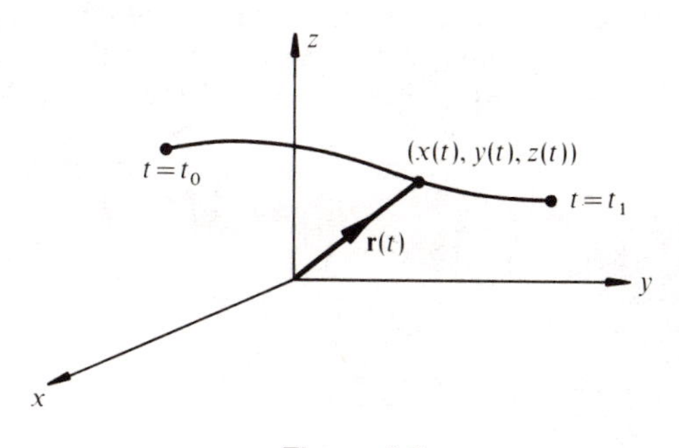

Figure 2.9

so that instead of using the three equations (2.23), we can simply write the equation of the curve in the vector form

$$\mathbf{r} = \mathbf{r}(t), \qquad t_0 \leqslant t \leqslant t_1 \tag{2.25}$$

For example, the helix described earlier by equation (2.10),

$$x = \cos t, \quad y = \sin t, \quad z = t, \qquad 0 \leqslant t \leqslant 2\pi$$

would have vector representation

$$\mathbf{r} = \cos t\,\hat{\mathbf{i}} + \sin t\,\hat{\mathbf{j}} + t\hat{\mathbf{k}}, \qquad 0 \leqslant t \leqslant 2\pi \tag{2.26}$$

whilst the straight line of (2.4) would have the vector equation

$$\mathbf{r} = \frac{(t+1)}{2}\hat{\mathbf{i}} + t\hat{\mathbf{j}}, \qquad 1 \leqslant t \leqslant 5 \tag{2.27}$$

It can be seen that a vector function of a single independent parameter t may be interpreted geometrically as a curve in three dimensions.

EXERCISES

2.1 Determine the Cartesian form of each of the following curves, and describe them geometrically:

(a) $\mathbf{r} = 3\cos\theta\,\hat{\mathbf{i}} + 4\sin\theta\,\hat{\mathbf{j}} + 3\hat{\mathbf{k}}$ (b) $\mathbf{r} = 2p^2\hat{\mathbf{i}} + 4p\hat{\mathbf{j}}$

(c) $\mathbf{r} = 3\sec\theta\,\hat{\mathbf{i}} + 3\hat{\mathbf{j}} + 4\tan\theta\,\hat{\mathbf{k}}$

2.2 Part of a circle has radius a and centre at (a, a) (see Figure 2.10). Determine the vector equation of the circle using θ as the parameter. What is the range of θ?

2.3 (a) Show that the vector equation of a sphere of radius r_0 with its centre at the endpoint of the position vector $\mathbf{a}$ is

$$|\mathbf{r} - \mathbf{a}| = r_0$$

where $\mathbf{r}$ is the position vector of an arbitrary point on the sphere.
(b) Determine the equation of the tangent plane at the point (0.5, 1.5, 0.707) on the surface of the sphere with centre at (1, 2, 0) and radius 1.

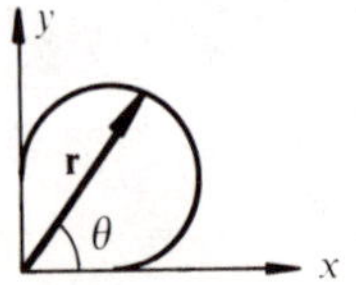

Figure 2.10

2.4 Show that every point on the curve

$$\mathbf{r} = t \cos t\, \hat{\mathbf{i}} + t \sin t\, \hat{\mathbf{j}} + t\hat{\mathbf{k}}, \qquad t > 0$$

lies on a cone with vertex at the origin. Show that straight lines on the cone passing through the vertex are aligned at 45° to the z-axis.

2.5 Sketch the parametric curves on the surface defined parametrically by

$$x = v \cos u, \quad y = v \sin u, \quad z = v^2, \qquad 0 \leqslant u \leqslant 2\pi, \quad v \geqslant 0$$

2.6 Show that the two curves

$$x = r \cos \theta + r, \quad y = r \sin \theta, \qquad 0 \leqslant \theta \leqslant 2\pi$$

and

$$x = r \cos \phi, \quad y = r \sin \phi + r, \qquad 0 \leqslant \phi \leqslant 2\pi$$

intersect at two points, $(0, 0)$ and (r, r).

2.4 The derivative of a vector

If we consider an ordinary function of a single variable,

$$y = f(x) \tag{2.28}$$

then we know how to find the derivative of y with respect to x. For example, if $f(x) = \cos x$, then

$$\frac{\mathrm{d}y}{\mathrm{d}x} = -\sin x$$

Now consider the situation in which we are given a vector function of a single independent variable, specifically a position vector:

$$\mathbf{r}(t) = x(t)\hat{\mathbf{i}} + y(t)\hat{\mathbf{j}} + z(t)\hat{\mathbf{k}} \tag{2.29}$$

We define the derivative of $\mathbf{r}$ with respect to t to be

$$\frac{\mathrm{d}\mathbf{r}}{\mathrm{d}t} = \frac{\mathrm{d}x}{\mathrm{d}t}\hat{\mathbf{i}} + \frac{\mathrm{d}y}{\mathrm{d}t}\hat{\mathbf{j}} + \frac{\mathrm{d}z}{\mathrm{d}t}\hat{\mathbf{k}} \tag{2.30}$$

(the notation $\dot{\mathbf{r}}$ is sometimes used instead of $\mathrm{d}\mathbf{r}/\mathrm{d}t$). For example, for the straight line in (2.27),

$$\mathbf{r} = \frac{(t+1)}{2}\hat{\mathbf{i}} + t\hat{\mathbf{j}}$$

we have

$$\frac{\mathrm{d}\mathbf{r}}{\mathrm{d}t} = \tfrac{1}{2}\hat{\mathbf{i}} + \hat{\mathbf{j}}$$

and for the helix of (2.26),

$$\mathbf{r} = \cos t\,\hat{\mathbf{i}} + \sin t\,\hat{\mathbf{j}} + t\hat{\mathbf{k}}$$

we have

$$\frac{\mathrm{d}\mathbf{r}}{\mathrm{d}t} = -\sin t\,\hat{\mathbf{i}} + \cos t\,\hat{\mathbf{j}} + \hat{\mathbf{k}}$$

The definition (2.30) implies that the operation of finding the derivative of a vector function of a single variable simply involves the straightforward process of calculating three ordinary derivatives. The more difficult problem is the interpretation of the vector $\mathrm{d}\mathbf{r}/\mathrm{d}t$. To help us obtain an interpretation, we recall the significance of the derivative of a scalar function of one variable, $y = f(x)$.

The definition of $\mathrm{d}y/\mathrm{d}x$ at a point is in terms of a limit:

$$\frac{\mathrm{d}y}{\mathrm{d}x} = \lim_{\delta x \to 0} \frac{f(x + \delta x) - f(x)}{\delta x} \tag{2.31}$$

The geometrical interpretation of $\mathrm{d}y/\mathrm{d}x$ is obtained from Figure 2.11, where we consider two points P $(x, f(x))$ and Q $(x + \delta x, f(x + \delta x))$ on the plane curve $y = f(x)$. The ratio $(f(x + \delta x) - f(x))/\delta x$ gives us the slope of the line PQ. As $\delta x \to 0$, the point Q approaches the point P and the line PQ approaches the tangent line at P. Hence from (2.31), in the limiting case,

$$\frac{\mathrm{d}y}{\mathrm{d}x} = \tan\theta \tag{2.32}$$

that is, the value of $\mathrm{d}y/\mathrm{d}x$ at P gives us the slope of the tangent line at P.

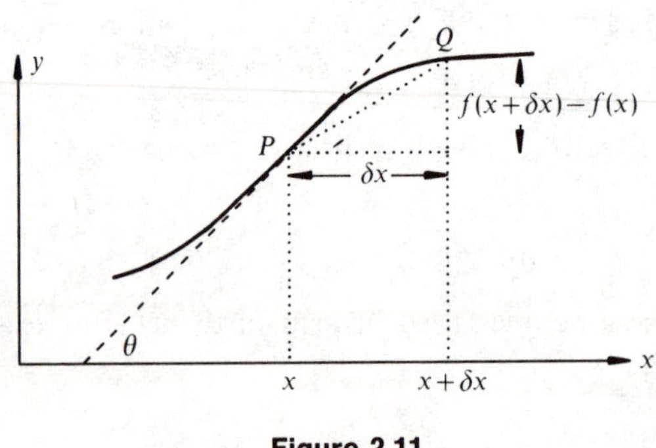

Figure 2.11

To obtain a similar interpretation of the vector derivative $\mathrm{d}\mathbf{r}/\mathrm{d}t$, we consider two points on the curve $\mathbf{r} = \mathbf{r}(t)$, namely P (position vector $\mathbf{r}(t)$) and Q (position vector $\mathbf{r}(t + \delta t)$), as shown in Figure 2.12. Now, from the

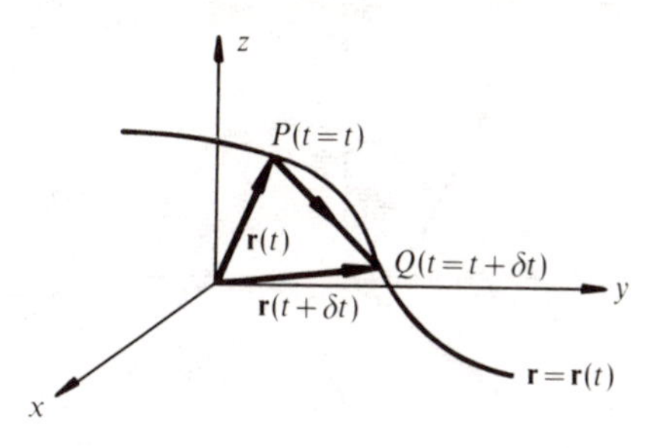

Figure 2.12

parallelogram rule for vector addition,

$$\mathbf{r}(t) + \overrightarrow{PQ} = \mathbf{r}(t + \delta t)$$

$$\therefore \qquad \overrightarrow{PQ} = \mathbf{r}(t + \delta t) - \mathbf{r}(t) \tag{2.33}$$

If δt is small then so is the magnitude of the vector $\overrightarrow{PQ}$ and, using an obvious notation, we can replace $\overrightarrow{PQ}$ by $\delta\mathbf{r}$, that is,

$$\delta\mathbf{r} = \mathbf{r}(t + \delta t) - \mathbf{r}(t)$$

Multiplying this vector by the scalar $1/\delta t$,

$$\frac{\delta\mathbf{r}}{\delta t} = \frac{\mathbf{r}(t + \delta t) - \mathbf{r}(t)}{\delta t}$$

or, in component form,

$$\frac{\delta\mathbf{r}}{\delta t} = \left(\frac{x(t + \delta t) - x(t)}{\delta t}\right)\hat{\mathbf{i}} + \left(\frac{y(t + \delta t) - y(t)}{\delta t}\right)\hat{\mathbf{j}} + \left(\frac{z(t + \delta t) - z(t)}{\delta t}\right)\hat{\mathbf{k}}$$

If we let $\delta t \to 0$, we obtain

$$\lim_{\delta t \to 0} \frac{\delta\mathbf{r}}{\delta t} = \frac{\mathrm{d}x}{\mathrm{d}t}\hat{\mathbf{i}} + \frac{\mathrm{d}y}{\mathrm{d}t}\hat{\mathbf{j}} + \frac{\mathrm{d}z}{\mathrm{d}t}\hat{\mathbf{k}}$$

which is our definition (2.30) of the derivative $\mathrm{d}\mathbf{r}/\mathrm{d}t$.

We see that

$$\frac{\mathrm{d}\mathbf{r}}{\mathrm{d}t} = \lim_{\delta t \to 0} \frac{\delta\mathbf{r}}{\delta t} = \lim_{\delta t \to 0} \frac{\mathbf{r}(t + \delta t) - \mathbf{r}(t)}{\delta t} \tag{2.34}$$

But, from Figure 2.12, as $\delta t \to 0$ the vector $\delta\mathbf{r}$ or $\overrightarrow{PQ}$ becomes closer and closer to the tangent vector to the curve $\mathbf{r} = \mathbf{r}(t)$ at P. We can interpret $\mathrm{d}\mathbf{r}/\mathrm{d}t$, therefore, as a **tangent vector** to the curve $\mathbf{r} = \mathbf{r}(t)$ at the point given by parameter value t.

For the straight line (2.27),

$$\mathbf{r} = \frac{(t + 1)}{2}\hat{\mathbf{i}} + t\hat{\mathbf{j}} \quad \text{and} \quad \frac{\mathrm{d}\mathbf{r}}{\mathrm{d}t} = \tfrac{1}{2}\hat{\mathbf{i}} + \hat{\mathbf{j}} \quad \text{(a constant vector)}$$

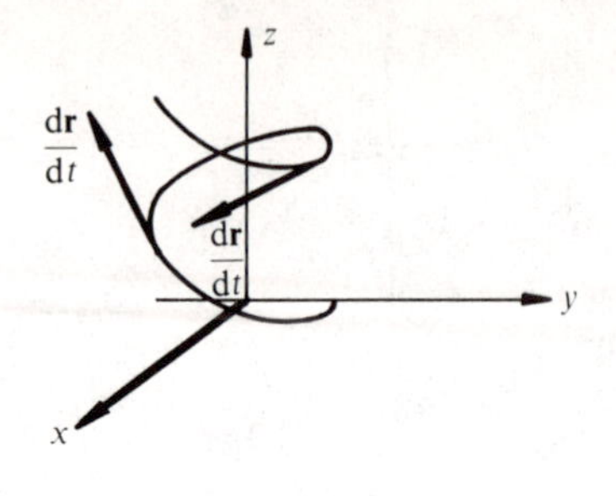

Figure 2.13

we obtain the obvious result that the tangent vector to a straight line has the same direction at all points. For a helix (different from the helix of (2.10)),

$$\mathbf{r} = \sin t\,\hat{\mathbf{i}} + \cos t\,\hat{\mathbf{j}} + t\hat{\mathbf{k}} \quad \text{and} \quad \frac{\mathrm{d}\mathbf{r}}{\mathrm{d}t} = \cos t\,\hat{\mathbf{i}} - \sin t\,\hat{\mathbf{j}} + \hat{\mathbf{k}}$$

we see that $\mathbf{dr}/\mathrm{d}t$ changes in direction with t, which agrees with obvious geometrical ideas (see Figure 2.13).

2.5 Properties of the derivatives of a vector

Although we have considered only the derivative of a position vector with respect to a parameter it is clear that *any* vector function of a single parameter t, say $\mathbf{a}(t)$, may be regarded as a position vector, since any vector may be translated so as to emanate from the origin of coordinates. It follows that the derivative as defined in Section 2.4 holds for any such vector function of a single parameter.

The familiar properties of the derivatives of scalar functions $f(t)$ and $g(t)$,

(1) $$\frac{\mathrm{d}}{\mathrm{d}t}(c) = 0 \quad \text{if } c \text{ is a constant}$$

(2) $$\frac{\mathrm{d}}{\mathrm{d}t}(f(t) + g(t)) = \frac{\mathrm{d}f}{\mathrm{d}t} + \frac{\mathrm{d}g}{\mathrm{d}t}$$

(3) $$\frac{\mathrm{d}}{\mathrm{d}t}(f(t)g(t)) = \frac{\mathrm{d}f}{\mathrm{d}t}g + \frac{\mathrm{d}g}{\mathrm{d}t}f$$

generalize in an obvious way to vector functions. For example, if $\mathbf{c}$ is a constant vector $\mathbf{c} = c_x\hat{\mathbf{i}} + c_y\hat{\mathbf{j}} + c_z\hat{\mathbf{k}}$, where c_x, c_y and c_z are constants (implying that the direction and magnitude of $\mathbf{c}$ remain fixed), then

$$\frac{\mathrm{d}\mathbf{c}}{\mathrm{d}t} = \frac{\mathrm{d}c_x}{\mathrm{d}t}\hat{\mathbf{i}} + \frac{\mathrm{d}c_y}{\mathrm{d}t}\hat{\mathbf{j}} + \frac{\mathrm{d}c_z}{\mathrm{d}t}\hat{\mathbf{k}} = 0$$

(or, more properly, equals the zero vector). Also, if

$$\mathbf{a}(t) = a_x(t)\hat{\mathbf{i}} + a_y(t)\hat{\mathbf{j}} + a_z(t)\hat{\mathbf{k}} \quad \text{and} \quad \mathbf{b}(t) = b_x(t)\hat{\mathbf{i}} + b_y(t)\hat{\mathbf{j}} + b_z(t)\hat{\mathbf{k}}$$

then
$$\frac{\mathrm{d}}{\mathrm{d}t}(\mathbf{a}+\mathbf{b}) = \frac{\mathrm{d}}{\mathrm{d}t}[(a_x+b_x)\hat{\mathbf{i}} + (a_y+b_y)\hat{\mathbf{j}} + (a_z+b_z)\hat{\mathbf{k}}]$$
$$= \left(\frac{\mathrm{d}a_x}{\mathrm{d}t} + \frac{\mathrm{d}b_x}{\mathrm{d}t}\right)\hat{\mathbf{i}} + \left(\frac{\mathrm{d}a_y}{\mathrm{d}t} + \frac{\mathrm{d}b_y}{\mathrm{d}t}\right)\hat{\mathbf{j}} + \left(\frac{\mathrm{d}a_z}{\mathrm{d}t} + \frac{\mathrm{d}b_z}{\mathrm{d}t}\right)\hat{\mathbf{k}}$$
$$= \frac{\mathrm{d}\mathbf{a}}{\mathrm{d}t} + \frac{\mathrm{d}\mathbf{b}}{\mathrm{d}t}$$

Rule (3) above may be shown to be applicable to all the products that we have defined in Chapter 1 between scalars and vectors and between vectors and vectors. Thus, for a scalar $f(t)$ and vectors $\mathbf{a}(t)$ and $\mathbf{b}(t)$,

$$\frac{\mathrm{d}}{\mathrm{d}t}(f(t)\mathbf{a}(t)) = \frac{\mathrm{d}f}{\mathrm{d}t}\mathbf{a} + f\frac{\mathrm{d}\mathbf{a}}{\mathrm{d}t}, \qquad \text{a vector equation}$$

$$\frac{\mathrm{d}}{\mathrm{d}t}(\mathbf{a}\cdot\mathbf{b}) = \frac{\mathrm{d}\mathbf{a}}{\mathrm{d}t}\cdot\mathbf{b} + \mathbf{a}\cdot\frac{\mathrm{d}\mathbf{b}}{\mathrm{d}t}, \qquad \text{a scalar equation}$$

$$\frac{\mathrm{d}}{\mathrm{d}t}(\mathbf{a}\wedge\mathbf{b}) = \frac{\mathrm{d}\mathbf{a}}{\mathrm{d}t}\wedge\mathbf{b} + \mathbf{a}\wedge\frac{\mathrm{d}\mathbf{b}}{\mathrm{d}t}, \qquad \text{a vector equation}$$

We should emphasize that in this last result (as in all formulae involving the vector product), the order of the vectors is important.

Other, perhaps less obvious, formulae follow from these results. For example,

$$\frac{\mathrm{d}}{\mathrm{d}t}(\mathbf{a}\cdot\mathbf{a}) = \frac{\mathrm{d}\mathbf{a}}{\mathrm{d}t}\cdot\mathbf{a} + \mathbf{a}\cdot\frac{\mathrm{d}\mathbf{a}}{\mathrm{d}t} = 2\mathbf{a}\cdot\frac{\mathrm{d}\mathbf{a}}{\mathrm{d}t}$$

since in scalar products the order is irrelevant. But we know that if $|\mathbf{a}|$ is the modulus of $\mathbf{a}$, then

$$\mathbf{a}\cdot\mathbf{a} = |\mathbf{a}|^2$$

so
$$\frac{\mathrm{d}}{\mathrm{d}t}(\mathbf{a}\cdot\mathbf{a}) = \frac{\mathrm{d}}{\mathrm{d}t}(|\mathbf{a}|^2) = 2|\mathbf{a}|\frac{\mathrm{d}}{\mathrm{d}t}(|\mathbf{a}|)$$

Therefore we must have

$$\mathbf{a}\cdot\frac{\mathrm{d}\mathbf{a}}{\mathrm{d}t} = |\mathbf{a}|\frac{\mathrm{d}|\mathbf{a}|}{\mathrm{d}t}$$

This result is often useful in the simplification of formulae, but a particularly useful form with a direct geometrical application may be deduced for constant magnitude vectors. If $\mathbf{a}$ is a vector with constant magnitude, then $|\mathbf{a}|$ is a

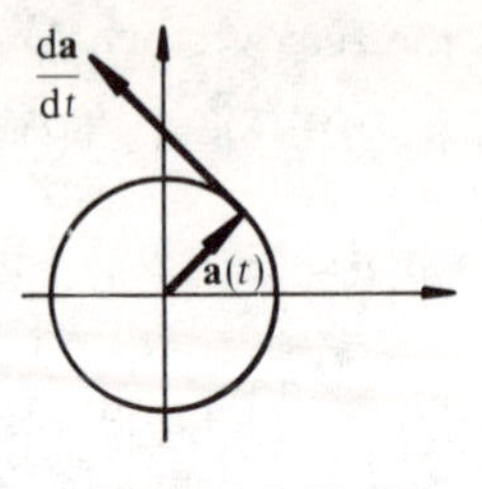

Figure 2.14

constant, so that $\mathrm{d}|\mathbf{a}|/\mathrm{d}t = 0$. It follows that for vectors with this property,

$$\mathbf{a} \cdot \frac{\mathrm{d}\mathbf{a}}{\mathrm{d}t} = 0 \tag{2.35}$$

that is, $\mathbf{a}$ and its derivative $\mathrm{d}\mathbf{a}/\mathrm{d}t$ are mutually perpendicular vectors in this case.

A constant magnitude vector $\mathbf{a}(t)$ can be represented as a radius vector from some origin O with the endpoint of the vector $\mathbf{a}$ tracing out a circle of radius $|\mathbf{a}|$ as the parameter t changes (see Figure 2.14). It is clear from this diagram that $\mathrm{d}\mathbf{a}/\mathrm{d}t$ (which is a tangent vector to the circle) is perpendicular to the radius vector $\mathbf{a}$, thus confirming (2.35).

The operation of taking a vector drivative may be extended in two ways. Firstly, we can obtain the second derivative $\mathbf{a}(t)$ with respect to t; this is denoted by $\mathrm{d}^2\mathbf{a}/\mathrm{d}t^2$ and is defined in an obvious way:

$$\frac{\mathrm{d}^2\mathbf{a}}{\mathrm{d}t^2} = \frac{\mathrm{d}}{\mathrm{d}t}\left(\frac{\mathrm{d}\mathbf{a}}{\mathrm{d}t}\right)$$

Higher derivatives are similarly defined, for example

$$\frac{\mathrm{d}^3\mathbf{a}}{\mathrm{d}t^3} = \frac{\mathrm{d}}{\mathrm{d}t}\left(\frac{\mathrm{d}^2\mathbf{a}}{\mathrm{d}t^2}\right)$$

Secondly, we can reverse the differentiation operation by the process of **vector integration**. If we have two vector functions $\mathbf{a}(t)$ and $\mathbf{A}(t)$, related to each other by

$$\frac{\mathrm{d}\mathbf{A}}{\mathrm{d}t} = \mathbf{a} \tag{2.36}$$

then $\mathbf{A}$ is called the **indefinite integral** of $\mathbf{a}$ and we write

$$\mathbf{A} = \int \mathbf{a}\,\mathrm{d}t \tag{2.37}$$

Now, since $\mathrm{d}\mathbf{c}/\mathrm{d}t = 0$ for any constant vector $\mathbf{c}$, the integral is only unique up to the addition of a constant vector. For example, if

$$\mathbf{a} = t\hat{\mathbf{i}} + t^2\hat{\mathbf{j}} + \cos t\,\hat{\mathbf{k}}, \quad \text{then} \quad \mathbf{A} = \frac{t^2}{2}\hat{\mathbf{i}} + \frac{t^3}{3}\hat{\mathbf{j}} + \sin t\,\hat{\mathbf{k}} + \mathbf{c}$$

since $\mathrm{d}\mathbf{A}/\mathrm{d}t = \mathbf{a}$ for any choice of constant vector $\mathbf{c}$.

The most obvious application of the derivative of a vector occurs, as we have seen, when the parameter t is the time and when $\mathbf{r}(t)$ describes the path of a moving particle in space. In this case, $\mathrm{d}\mathbf{r}/\mathrm{d}t$ represents the **velocity** of the particle in both magnitude and direction and $\mathrm{d}^2\mathbf{r}/\mathrm{d}t^2$ represents the **acceleration** of the particle, again in both magnitude and direction.

Example 2.3

Determine the path of a particle falling freely (not necessarily vertically) under gravity.

Solution

We choose axes so that the z-axis points upwards (see Figure 2.15). If the particle has mass m, then applying Newton's second law (mass × acceleration equals applied force) to the particle we get

$$m\frac{\mathrm{d}^2\mathbf{r}}{\mathrm{d}t^2} = -mg\hat{\mathbf{k}}, \quad \text{where } g \text{ is the acceleration due to gravity}$$

Integrating,

$$m\frac{\mathrm{d}\mathbf{r}}{\mathrm{d}t} = -mgt\hat{\mathbf{k}} + \mathbf{d}$$

Now if $\mathrm{d}\mathbf{r}/\mathrm{d}t = \mathbf{v}_0$ when $t = 0$ (that is, the particle is set in motion with an initial velocity $\mathbf{v}_0$), then $\mathbf{d} = m\mathbf{v}_0$. Therefore

$$m\frac{\mathrm{d}\mathbf{r}}{\mathrm{d}t} = -mgt\hat{\mathbf{k}} + m\mathbf{v}_0$$

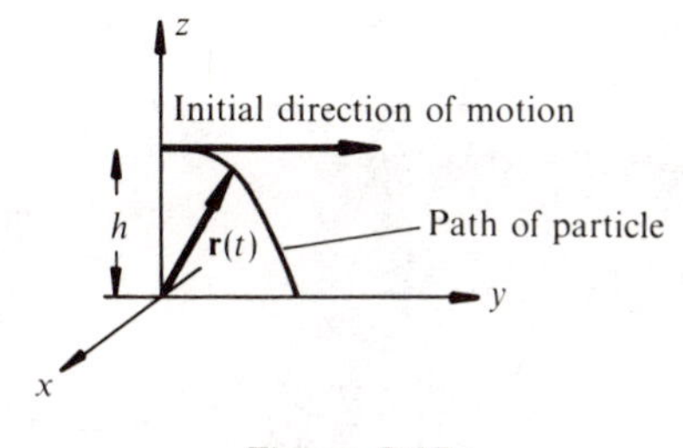

Figure 2.15

Integrating again,

$$m\mathbf{r} = -\tfrac{1}{2}mgt^2\hat{\mathbf{k}} + m\mathbf{v}_0 t + \mathbf{e}$$

If $\mathbf{r} = \mathbf{r}_0$ when $t = t_0$ (that is, the particle starts its motion from a point with position vector $\mathbf{r}_0$), then $\mathbf{e} = m\mathbf{r}_0$, giving

$$\mathbf{r} = -\tfrac{1}{2}gt^2\hat{\mathbf{k}} + \mathbf{v}_0 t + \mathbf{r}_0$$

We now show that the path of the particle is a parabola. For simplicity, we consider the case where $\mathbf{r}_0 = h\hat{\mathbf{k}}$ and $\mathbf{v}_0 = v_0\hat{\mathbf{j}}$ (that is, the particle is initially projected horizontally with speed v_0 at a height h above ground level). For this situation,

$$\mathbf{r} = (-\tfrac{1}{2}gt^2 + h)\hat{\mathbf{k}} + v_0 t\hat{\mathbf{j}}$$

or, equating components,

$$x = 0, \qquad y = v_0 t, \qquad z = -\tfrac{1}{2}gt^2 + h$$

which give the coordinates of the particle at any time t. Eliminating t gives

$$x = 0, \qquad z = -\tfrac{1}{2}gv_0^2 y^2 + h$$

which is the equation of a parabola in the zy-plane.

2.6 Unit tangent vectors

The vector derivative $\mathrm{d}\mathbf{r}/\mathrm{d}t$, as we have just seen, is a tangent vector to a curve. However, it is not necessarily a unit vector. The magnitude of $\mathrm{d}\mathbf{r}/\mathrm{d}t$ is dependent both on the shape of the curve $\mathbf{r} = \mathbf{r}(t)$ and, more importantly, on the choice of parametrization. For example, for the straight line whose parametrization is (2.27),

$$\mathbf{r} = \frac{(t+1)}{2}\hat{\mathbf{i}} + t\hat{\mathbf{j}}$$

we have

$$\frac{\mathrm{d}\mathbf{r}}{\mathrm{d}t} = \tfrac{1}{2}\hat{\mathbf{i}} + \hat{\mathbf{j}} \quad \text{and} \quad \left|\frac{\mathrm{d}\mathbf{r}}{\mathrm{d}t}\right| = \tfrac{1}{2}\sqrt{5}$$

which is independent of t. For the alternative parametrization of this same straight line (2.5),

$$\mathbf{r} = \tfrac{1}{2}(t^2 + t + 1)\hat{\mathbf{i}} + (t^2 + t)\hat{\mathbf{j}}$$

$$\therefore \qquad \frac{\mathrm{d}\mathbf{r}}{\mathrm{d}t} = (t + \tfrac{1}{2})\hat{\mathbf{i}} + (2t + 1)\hat{\mathbf{j}}$$

and $$\left|\frac{\mathrm{d}\mathbf{r}}{\mathrm{d}t}\right| = \sqrt{[(t + \tfrac{1}{2})^2 + (2t + 1)^2]} = \sqrt{(5t^2 + 5t + \tfrac{5}{4})}$$

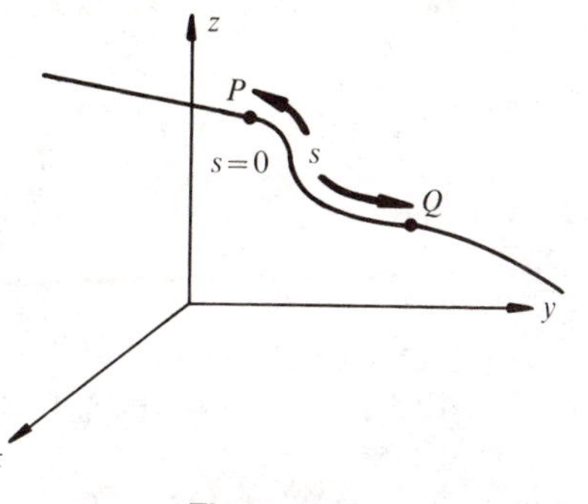

Figure 2.16

so that the magnitude *is* dependent on t. On the other hand, the direction of $\mathbf{dr}/dt$ for this parametrization is parallel to the vector $\hat{\mathbf{i}} + 2\hat{\mathbf{j}}$ (since $(t + \frac{1}{2})$ is a factor) and is thus clearly independent of t, as we would expect for a straight line.

It is sometimes important to choose a parametrization which is, in a sense, characteristic of the curve itself. Such a choice is the arc length s. Considering a curve $\mathbf{r} = \mathbf{r}(t)$ (see Figure 2.16), we can choose any point P on the curve as the origin for s. The value of s at a point Q on the curve is simply the arc length as measured from P. Since each point on the curve is characterized by a unique value of s, then we can indeed use s as a parameter for the curve. With this choice of parameter, we can see from Figure 2.16 that, if Q is close to P, the magnitude of $\overrightarrow{PQ}$, or $\delta\mathbf{r}$, is $|\delta\mathbf{r}| \approx \delta s$, and so the vector derivative $\mathbf{dr}/ds$ will be a *unit* vector. For example, consider the circle $x^2 + y^2 = 4$. The arc length on this curve is $s = 2\theta$, where θ is the angle measured in radians, as shown in Figure 2.17. The parametrization of the

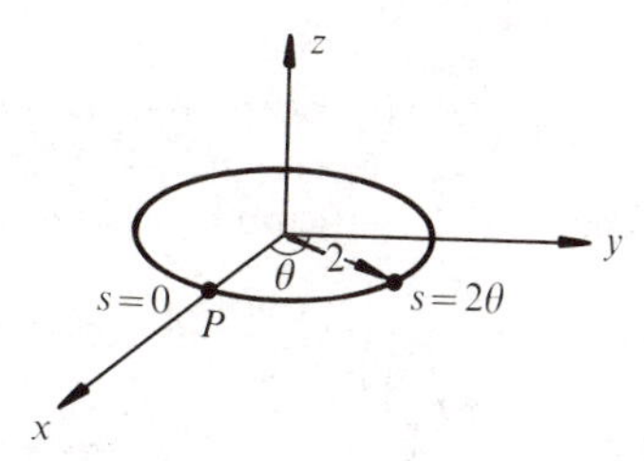

Figure 2.17

circle in terms of s would be

$$x = 2\cos\tfrac{1}{2}s, \quad y = 2\sin\tfrac{1}{2}s, \quad z = 0, \qquad 0 \leqslant s \leqslant 4\pi$$

Thus the position vector at any point of the circle is

$$\mathbf{r} = 2\cos\tfrac{1}{2}s\,\hat{\mathbf{i}} + 2\sin\tfrac{1}{2}s\,\hat{\mathbf{j}}$$

$$\therefore \qquad \frac{d\mathbf{r}}{ds} = -\sin\tfrac{1}{2}s\,\hat{\mathbf{i}} + \cos\tfrac{1}{2}s\,\hat{\mathbf{j}}$$

which clearly has magnitude 1.

The artifice of using arc length as the parameter is generally only convenient in the theoretical development of the subject. For general curves, unlike the example given above, it is often very difficult to obtain the arc length s in terms of some more convenient parameter.

EXERCISES

2.7 Show, for any vector $\mathbf{r}(t)$, that

$$\frac{d}{dt}\left(\mathbf{r}\wedge\frac{d\mathbf{r}}{dt}\right) = \mathbf{r}\wedge\frac{d^2\mathbf{r}}{dt^2}$$

2.8 Verify the following results for the vectors $\mathbf{a} = t^2\hat{\mathbf{i}} + (t^2 - t^3)\hat{\mathbf{j}} + \sin t\,\hat{\mathbf{k}}$ and $\mathbf{b} = 2t^2\hat{\mathbf{i}} + 4t\hat{\mathbf{j}}$:

(a) $\dfrac{d}{dt}(\mathbf{a}\cdot\mathbf{b}) = \mathbf{a}\cdot\dfrac{d\mathbf{b}}{dt} + \dfrac{d\mathbf{a}}{dt}\cdot\mathbf{b}$ (b) $\dfrac{d}{dt}(\mathbf{a}\wedge\mathbf{b}) = \mathbf{a}\wedge\dfrac{d\mathbf{b}}{dt} + \dfrac{d\mathbf{a}}{dt}\wedge\mathbf{b}$

2.9 Determine the indefinite integrals of the following vectors:

(a) $\mathbf{a}(\theta) = 3\sin\theta\,\hat{\mathbf{i}} + 4\sin\theta\cos\theta\,\hat{\mathbf{j}} + \theta^3\hat{\mathbf{k}}$ (b) $\mathbf{a}(t) = \tfrac{2}{3}t^{3/2}\hat{\mathbf{i}} + t\hat{\mathbf{j}} + t\hat{\mathbf{k}}$

2.10 A particle moves along the helical path

$$\mathbf{r}(t) = \cos t\,\hat{\mathbf{i}} + \sin t\,\hat{\mathbf{j}} + t\hat{\mathbf{k}}$$

where the parameter t is the time in seconds. At $t = 3$ s the particle flies off the path along a tangent. Determine the velocity and acceleration of the particle at the time of separation and find its position one second later, assuming that it continues along the tangent line at the same speed.

2.11 An object is moving around a circle of radius r with constant angular speed ω. Show that the acceleration has magnitude $\omega^2 r$ and is directed towards the centre of the circle.

2.12 Show that the curve $\mathbf{r} = a\left(\dfrac{1-t^2}{1+t^2}\right)\hat{\mathbf{i}} + \left(\dfrac{2at}{1+t^2}\right)\hat{\mathbf{j}}$ represents a circle.

2.13 The vector equation of a curve is given by

$$\mathbf{r} = (2 + 5\sin\theta)\hat{\mathbf{i}} + (3 + 4\cos\theta)\hat{\mathbf{j}} + (4 + 3\cos\theta)\hat{\mathbf{k}}$$

Show that the curve is a circle with centre at (2, 3, 4) and of radius 5. Show that a normal to the plane containing the circle is $15\hat{\mathbf{j}} - 20\hat{\mathbf{k}}$.

2.14 A particle moves with constant velocity $\mathbf{v}$, starting from a point with position vector $\mathbf{r}_0$ at time $t = 0$. Show that the particle will be nearest to the origin at a time $t = -\mathbf{r}_0 \cdot \mathbf{v}/\mathbf{v} \cdot \mathbf{v}$.

2.15 Let the path of a projectile be

$$\mathbf{r} = (u \cos \theta)t\hat{\mathbf{i}} + [(u \sin \theta)t - \tfrac{1}{2}gt^2]\hat{\mathbf{j}}$$

where u is the initial speed, θ is the angle of projection and t is the time measured from the instant of projection. Verify the following statements:

(a) The time for the projectile to reach its highest point is $(u \sin \theta)/g$.
(b) The maximum height of the projectile is $(u \sin \theta)^2/2g$.
(c) The velocity of the projectile at its highest point is $u \cos \theta\, \hat{\mathbf{i}}$.
(d) The velocity of the projectile on returning to ground level is $u \cos \theta\, \hat{\mathbf{i}} - u \sin \theta\, \hat{\mathbf{j}}$.

2.7 Simple geometry of curves

We have seen in Section 2.6 that the vector equation of a curve may be parametrized by $\mathbf{r} = \mathbf{r}(s)$, where s is arc length, and that, for each value of s, we can calculate a unit tangent vector as $\mathrm{d}\mathbf{r}/\mathrm{d}s$. We give this unit tangent vector a special symbol $\hat{\mathbf{t}}$ (not to be confused with the parameter t):

$$\hat{\mathbf{t}} = \frac{\mathrm{d}\mathbf{r}}{\mathrm{d}s} \qquad \textbf{(2.38)}$$

We now consider the vector $\mathrm{d}\hat{\mathbf{t}}/\mathrm{d}s$. From (2.35) it follows, since $\hat{\mathbf{t}}$ has constant magnitude, that

$$\hat{\mathbf{t}} \cdot \frac{\mathrm{d}\hat{\mathbf{t}}}{\mathrm{d}s} = 0 \qquad \textbf{(2.39)}$$

that is, $\hat{\mathbf{t}}$ and $\mathrm{d}\hat{\mathbf{t}}/\mathrm{d}s$ are everywhere perpendicular. We can therefore express $\mathrm{d}\hat{\mathbf{t}}/\mathrm{d}s$ in the form $\kappa\hat{\mathbf{n}}$, where κ is some scalar and $\hat{\mathbf{n}}$ is a unit normal to the curve at a point P (see Figure 2.18). The quantity κ is called the **curvature**

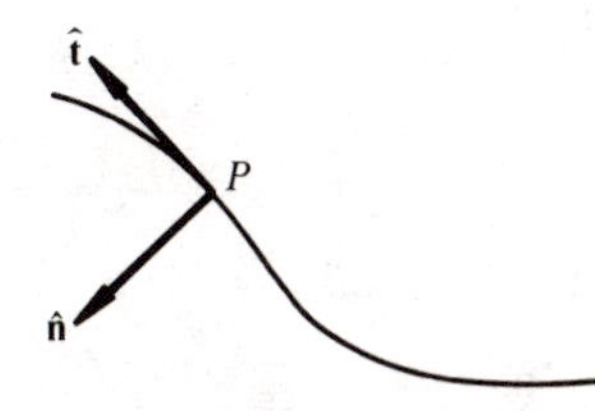

Figure 2.18

at P. By taking scalar products we have immediately that

$$\hat{\mathbf{n}}\cdot\frac{\mathrm{d}\hat{\mathbf{t}}}{\mathrm{d}s}=\kappa\hat{\mathbf{n}}\cdot\hat{\mathbf{n}}=\kappa \tag{2.40}$$

Note that since $\kappa=\left|\frac{\mathrm{d}\hat{\mathbf{t}}}{\mathrm{d}s}\right|$, then $\hat{\mathbf{n}}=\frac{\mathrm{d}\hat{\mathbf{t}}}{\mathrm{d}s}\Big/\left|\frac{\mathrm{d}\hat{\mathbf{t}}}{\mathrm{d}s}\right|$.

We now apply these ideas to planar curves lying in the xy-plane. We can easily find an explicit expression for the unit normal $\hat{\mathbf{n}}$ in this case. By definition,

$$\hat{\mathbf{t}}=\frac{\mathrm{d}\mathbf{r}}{\mathrm{d}s}=\frac{\mathrm{d}x}{\mathrm{d}s}\hat{\mathbf{i}}+\frac{\mathrm{d}y}{\mathrm{d}s}\hat{\mathbf{j}}$$

from which it follows that

$$\left(\frac{\mathrm{d}x}{\mathrm{d}s}\right)^2+\left(\frac{\mathrm{d}y}{\mathrm{d}s}\right)^2=1$$

Thus, if $\hat{\mathbf{n}}$ has the form $a\hat{\mathbf{i}}+b\hat{\mathbf{j}}$ then, since $\hat{\mathbf{n}}$ and $\hat{\mathbf{t}}$ are perpendicular vectors,

$$(a\hat{\mathbf{i}}+b\hat{\mathbf{j}})\cdot\left(\frac{\mathrm{d}x}{\mathrm{d}s}\hat{\mathbf{i}}+\frac{\mathrm{d}y}{\mathrm{d}s}\hat{\mathbf{j}}\right)=0$$

or
$$a\frac{\mathrm{d}x}{\mathrm{d}s}+b\frac{\mathrm{d}y}{\mathrm{d}s}=0 \tag{2.41}$$

Also $\hat{\mathbf{n}}$ is a unit normal, so that

$$a^2+b^2=1 \tag{2.42}$$

Finally, by convention, the direction of $\hat{\mathbf{n}}$ is chosen so that $\hat{\mathbf{t}}$, $\hat{\mathbf{n}}$ and $\hat{\mathbf{k}}$ form a right-handed system of vectors, or

$$\hat{\mathbf{t}}\wedge\hat{\mathbf{n}}=\hat{\mathbf{k}} \tag{2.43}$$

But
$$\hat{\mathbf{t}}\wedge\hat{\mathbf{n}}=\begin{vmatrix}\hat{\mathbf{i}} & \hat{\mathbf{j}} & \hat{\mathbf{k}}\\ \frac{\mathrm{d}x}{\mathrm{d}s} & \frac{\mathrm{d}y}{\mathrm{d}s} & 0\\ a & b & 0\end{vmatrix}=\hat{\mathbf{k}}\left(b\frac{\mathrm{d}x}{\mathrm{d}s}-a\frac{\mathrm{d}y}{\mathrm{d}s}\right)$$

Hence, from (2.43), we require

$$b\frac{\mathrm{d}x}{\mathrm{d}s}-a\frac{\mathrm{d}y}{\mathrm{d}s}=1 \tag{2.44}$$

The only solution to (2.41), (2.42) and (2.44) is

$$a=-\frac{\mathrm{d}y}{\mathrm{d}s},\qquad b=+\frac{\mathrm{d}x}{\mathrm{d}s} \tag{2.45}$$

so that

$$\hat{\mathbf{n}} = -\frac{dy}{ds}\hat{\mathbf{i}} + \frac{dx}{ds}\hat{\mathbf{j}} \tag{2.46}$$

is the required unit normal.

If we now introduce the parameter t as 'time', then the velocity of a particle moving on the curve is

$$\mathbf{v} = \frac{d\mathbf{r}}{dt} = \frac{d\mathbf{r}}{ds}\frac{ds}{dt} = \hat{\mathbf{t}}\frac{ds}{dt}$$

where ds/dt is the speed of the particle. The acceleration $\mathbf{a}$ of the particle is

$$\mathbf{a} = \frac{d^2\mathbf{r}}{dt^2} = \frac{d}{dt}\left(\hat{\mathbf{t}}\frac{ds}{dt}\right) = \frac{d\hat{\mathbf{t}}}{dt}\frac{ds}{dt} + \hat{\mathbf{t}}\frac{d^2s}{dt^2} = \frac{d\hat{\mathbf{t}}}{ds}\left(\frac{ds}{dt}\right)^2 + \hat{\mathbf{t}}\frac{d^2s}{dt^2}$$

or, in terms of the curvature κ,

$$\mathbf{a} = \kappa\hat{\mathbf{n}}\left(\frac{ds}{dt}\right)^2 + \hat{\mathbf{t}}\frac{d^2s}{dt^2} \tag{2.47}$$

This equation has the obvious interpretation that the acceleration of a particle moving on a curve $\mathbf{r} = \mathbf{r}(t)$ has two components – a tangential component of magnitude d^2s/dt^2 (the rate of change of speed) and a normal component of magnitude $\kappa(ds/dt)^2$ (the curvature times the square of the speed). Thus, to keep the particle of mass m moving on the curve with speed ds/dt, a force of magnitude $m\kappa(ds/dt)^2$ must be applied in the direction of $\hat{\mathbf{n}}$.

Equation (2.47) will only be of use if we can develop a method for determining the curvature κ in terms of a general parameter t (not necessarily time) rather than in terms of the special arc-length parameter s. Now for our planar curve we have, from (2.46),

$$\hat{\mathbf{n}} = -\frac{dy}{ds}\hat{\mathbf{i}} + \frac{dx}{ds}\hat{\mathbf{j}} = \left(-\frac{dy}{dt}\hat{\mathbf{i}} + \frac{dx}{dt}\hat{\mathbf{j}}\right)\frac{dt}{ds}$$

Using the definition of $\hat{\mathbf{t}}$,

$$\hat{\mathbf{t}} = \frac{d\mathbf{r}}{ds} = \frac{dx}{ds}\hat{\mathbf{i}} + \frac{dy}{ds}\hat{\mathbf{j}} = \left(\frac{dx}{dt}\hat{\mathbf{i}} + \frac{dy}{dt}\hat{\mathbf{j}}\right)\frac{dt}{ds}$$

$$\therefore \quad \frac{d\hat{\mathbf{t}}}{dt} = \left(\frac{d^2x}{dt^2}\hat{\mathbf{i}} + \frac{d^2y}{dt^2}\hat{\mathbf{j}}\right)\frac{dt}{ds} + \left(\frac{dx}{dt}\hat{\mathbf{i}} + \frac{dy}{dt}\hat{\mathbf{j}}\right)\left(\frac{d}{dt}\left(\frac{dt}{ds}\right)\right)$$

using the product rule for differentiating a product of a scalar and a vector. But using (2.40),

$$\kappa = \hat{\mathbf{n}}\cdot\frac{d\hat{\mathbf{t}}}{ds} = \hat{\mathbf{n}}\cdot\frac{d\hat{\mathbf{t}}}{dt}\left(\frac{dt}{ds}\right)$$

Substituting for $\hat{\mathbf{n}}$ and $\mathrm{d}\hat{\mathbf{t}}/\mathrm{d}t$ gives

$$\kappa = \left(-\frac{\mathrm{d}^2 x}{\mathrm{d}t^2}\frac{\mathrm{d}y}{\mathrm{d}t} + \frac{\mathrm{d}x}{\mathrm{d}t}\frac{\mathrm{d}^2 y}{\mathrm{d}t^2}\right)\left(\frac{\mathrm{d}t}{\mathrm{d}s}\right)^3$$

But $\quad \dfrac{\mathrm{d}\mathbf{r}}{\mathrm{d}t} = \dfrac{\mathrm{d}x}{\mathrm{d}t}\hat{\mathbf{i}} + \dfrac{\mathrm{d}y}{\mathrm{d}t}\hat{\mathbf{j}} \quad$ and $\quad \dfrac{\mathrm{d}s}{\mathrm{d}t} = \left|\dfrac{\mathrm{d}\mathbf{r}}{\mathrm{d}t}\right| = \sqrt{\left[\left(\dfrac{\mathrm{d}x}{\mathrm{d}t}\right)^2 + \left(\dfrac{\mathrm{d}y}{\mathrm{d}t}\right)^2\right]}$

Therefore, finally,

$$\kappa = \frac{-\dfrac{\mathrm{d}^2 x}{\mathrm{d}t^2}\dfrac{\mathrm{d}y}{\mathrm{d}t} + \dfrac{\mathrm{d}x}{\mathrm{d}t}\dfrac{\mathrm{d}^2 y}{\mathrm{d}t^2}}{\left[\left(\dfrac{\mathrm{d}x}{\mathrm{d}t}\right)^2 + \left(\dfrac{\mathrm{d}y}{\mathrm{d}t}\right)^2\right]^{3/2}} \tag{2.48}$$

which is an expression for κ in terms of the parameter t. An alternative form of (2.48) is

$$\kappa = \frac{\left|\dfrac{\mathrm{d}\mathbf{r}}{\mathrm{d}t} \wedge \dfrac{\mathrm{d}^2\mathbf{r}}{\mathrm{d}t^2}\right|}{\left|\dfrac{\mathrm{d}\mathbf{r}}{\mathrm{d}t}\right|^3} \tag{2.49}$$

Example 2.4

Deduce the curvature of

(a) the straight line $\mathbf{r} = \frac{1}{2}(t+1)\hat{\mathbf{i}} + t\hat{\mathbf{j}}$ (see equations (2.4) and (2.27)),

(b) the circle $\mathbf{r} = a\cos t\,\hat{\mathbf{i}} + a\sin t\,\hat{\mathbf{j}}$.

Solution

(a) We have

$$\frac{\mathrm{d}\mathbf{r}}{\mathrm{d}t} = \tfrac{1}{2}\hat{\mathbf{i}} + \hat{\mathbf{j}} \quad \text{so} \quad \frac{\mathrm{d}^2\mathbf{r}}{\mathrm{d}t^2} = 0$$

Hence, by (2.49),

$$\kappa = \frac{\left|\dfrac{\mathrm{d}\mathbf{r}}{\mathrm{d}t} \wedge \dfrac{\mathrm{d}^2\mathbf{r}}{\mathrm{d}t^2}\right|}{\left|\dfrac{\mathrm{d}\mathbf{r}}{\mathrm{d}t}\right|^3} = 0$$

as expected for a straight line.

The result (2.48) for the curvature is of course independent of the actual parametrization used. If we use, for this straight line, the parametrization (2.5),

$$\mathbf{r} = \tfrac{1}{2}(t^2 + t + 1)\hat{\mathbf{i}} + (t^2 + t)\hat{\mathbf{j}}$$

we obtain

$$\frac{d\mathbf{r}}{dt} = (t + \tfrac{1}{2})\hat{\mathbf{i}} + (2t + 1)\hat{\mathbf{j}} \quad \text{and} \quad \frac{d^2\mathbf{r}}{dt^2} = \hat{\mathbf{i}} + 2\hat{\mathbf{j}}$$

Therefore

$$\kappa = \frac{|[(t + \frac{1}{2})\hat{\mathbf{i}} + (2t + 1)\hat{\mathbf{j}}] \wedge [\hat{\mathbf{i}} + 2\hat{\mathbf{j}}]|}{[(t + \frac{1}{2})^2 + (2t + 1)^2]^{3/2}} = \frac{|2(t + \frac{1}{2})\hat{\mathbf{k}} + (2t + 1)(-\hat{\mathbf{k}})|}{[(t + \frac{1}{2})^2 + (2t + 1)^2]^{3/2}} = 0$$

as expected.

(b) Here

$$\frac{d\mathbf{r}}{dt} = -a \sin t\,\hat{\mathbf{i}} + a \cos t\,\hat{\mathbf{j}} \quad \text{and} \quad \frac{d^2\mathbf{r}}{dt^2} = -a \cos t\,\hat{\mathbf{i}} - a \sin t\,\hat{\mathbf{j}}$$

leading to

$$\left|\frac{d\mathbf{r}}{dt}\right| = \sqrt{(a^2 \sin^2 t + a^2 \cos^2 t)} = a$$

and

$$\frac{d\mathbf{r}}{dt} \wedge \frac{d^2\mathbf{r}}{dt^2} = \begin{vmatrix} \hat{\mathbf{i}} & \hat{\mathbf{j}} & \hat{\mathbf{k}} \\ -a \sin t & a \cos t & 0 \\ -a \cos t & -a \sin t & 0 \end{vmatrix} = \hat{\mathbf{k}}(a^2 \sin^2 t + a^2 \cos^2 t) = a^2\hat{\mathbf{k}}$$

Finally, by (2.49),

$$\kappa = \frac{a^2}{a^3} = \frac{1}{a}$$

that is, the larger is the radius a, the smaller is κ, and vice versa. It is for this reason that, for general curves, the term $1/\kappa$ is called the **radius of curvature**.

This topic will be developed more fully in Supplement 2S.1 at the end of this chapter.

Knowledge of curvature has practical application in the design of railway tracks. Consider, for example, the joining together of two tracks at P (see Figure 2.19), one track being part of a circle and the other a straight line segment.

The curves are continuous at P (as are the slopes), but their curvature is not. The curvature of the circle is $1/a$, whilst that of the straight line section is 0 (see Figure 2.20). The discontinuity in curvature at the join P will, by (2.47), cause a change in the normal component of the acceleration when a train passes through P. This may cause discomfort to passengers, who will

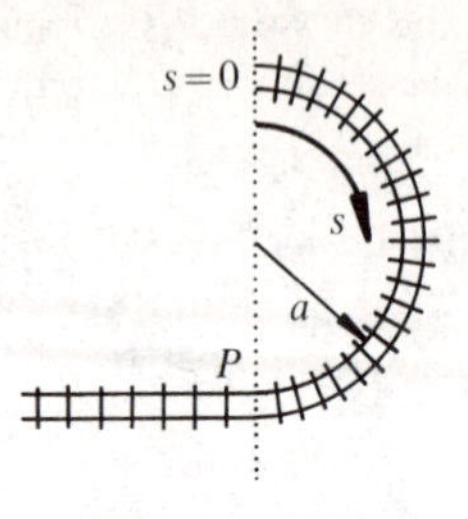

Figure 2.19

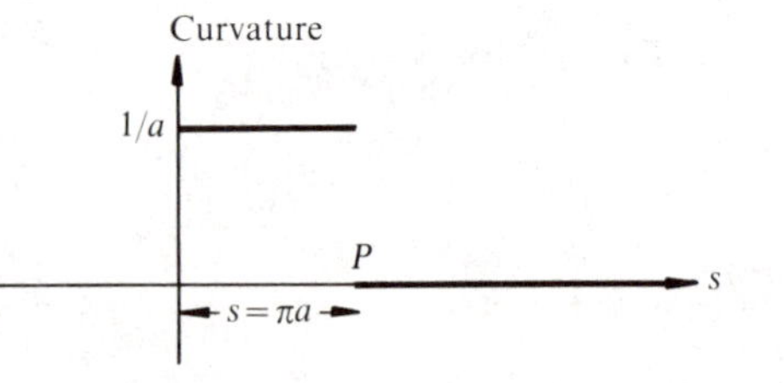

Figure 2.20

experience a 'jerk':

$$\text{'jerk'} \propto m\left(\frac{\mathrm{d}s}{\mathrm{d}t}\right)^2_{\text{at } P} (\kappa_{\text{before } P} - \kappa_{\text{after } P})$$

The 'jerk' is thus directly proportional to the 'tightness' of the circular section and to the square of the speed of the train. To eliminate this jerk one might use a transition track which gradually reduces the curvature from $1/a$ down to 0. Choosing such a curve is not a simple calculation. It may be shown that a circle can be smoothly matched to a straight line by using (amongst

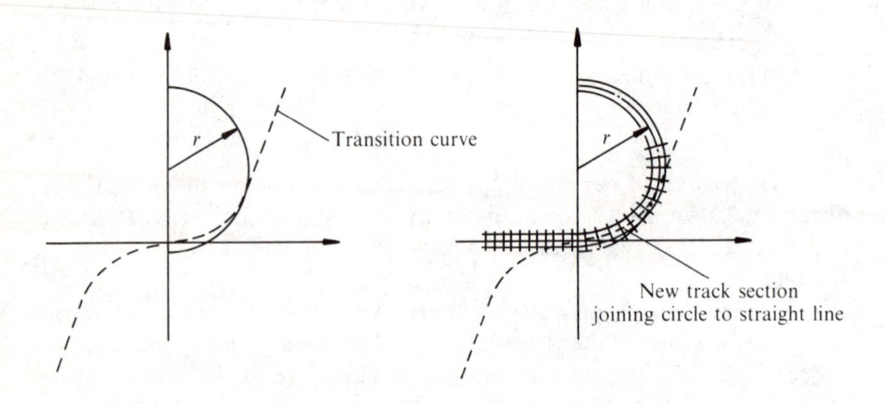

Figure 2.21

other alternatives) a section of a cubic curve (see Figure 2.21). The remaining details of this problem are omitted since no significant use of vectors is involved.

2.8 The normal to a planar curve

If we are given the equation of a curve in the xy-plane in Cartesian form, $y = f(x)$, then a tangent vector at any point is

$$\mathbf{t} = \hat{\mathbf{i}} + \hat{\mathbf{j}}\frac{\mathrm{d}y}{\mathrm{d}x} \tag{2.50}$$

This is easily shown since we know, by (2.38), that when the curve is represented in parametric form, $\mathbf{r} = \mathbf{r}(s)$, a unit tangent vector is given by

$$\hat{\mathbf{t}} = \frac{\mathrm{d}\mathbf{r}}{\mathrm{d}s} = \hat{\mathbf{i}}\frac{\mathrm{d}x}{\mathrm{d}s} + \hat{\mathbf{j}}\frac{\mathrm{d}y}{\mathrm{d}s} = \frac{\mathrm{d}x}{\mathrm{d}s}\left(\hat{\mathbf{i}} + \hat{\mathbf{j}}\frac{\mathrm{d}y}{\mathrm{d}x}\right)$$

Thus the vector $\hat{\mathbf{i}} + \hat{\mathbf{j}}\,\mathrm{d}y/\mathrm{d}x$ is parallel to $\hat{\mathbf{t}}$ and hence must be a tangent vector.

By a similar argument using (2.46), a normal vector to the curve is

$$\mathbf{n} = -\frac{\mathrm{d}y}{\mathrm{d}x}\hat{\mathbf{i}} + \hat{\mathbf{j}} \tag{2.51}$$

Note that $\mathbf{t}\cdot\mathbf{n} = 0$ as we would expect (see Figure 2.22) and that written in these forms $\mathbf{t}$ and $\mathbf{n}$ are not necessarily unit vectors.

We shall find it instructive to determine the normal in an alternative way. We begin again with the equation of the curve $y = f(x)$, but now written as

$$y - f(x) = 0 \tag{2.52}$$

or more generally,

$$G(x, y) = 0 \tag{2.53}$$

where G is a function of two variables. (2.53) is the implicit equation describing a planar curve, whereas (2.52) is the explicit equation of the curve.

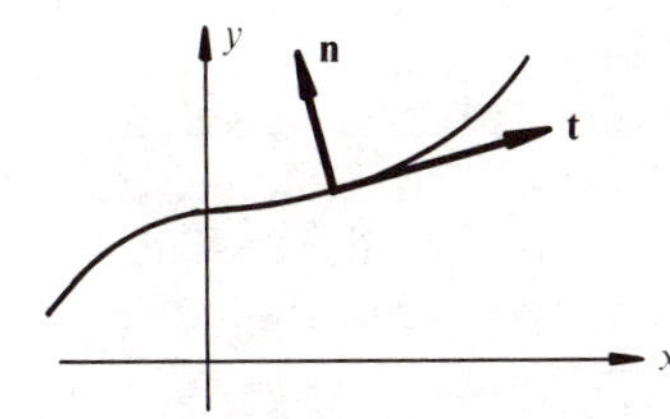

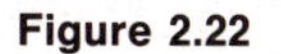

Figure 2.22

Now since G remains fixed at zero on the curve, if we make small changes of $\mathrm{d}x$ and $\mathrm{d}y$ in x and y but remain on the curve, then the change in G must be zero, that is,

$$\mathrm{d}G = 0 \tag{2.54}$$

But $$\mathrm{d}G = \frac{\partial G}{\partial x}\mathrm{d}x + \frac{\partial G}{\partial y}\mathrm{d}y \tag{2.55}$$

so (2.54) can be written in vector form as a scalar product:

$$\mathrm{d}G = \left(\frac{\partial G}{\partial x}\hat{\mathbf{i}} + \frac{\partial G}{\partial y}\hat{\mathbf{j}}\right)\cdot\left(\hat{\mathbf{i}}\,\mathrm{d}x + \hat{\mathbf{j}}\,\mathrm{d}y\right) = 0 \tag{2.56}$$

The second bracketed term is a tangent vector to the curve. The vector in the first bracket is perpendicular to this tangent vector so we have an alternative prescription for determining a normal vector to a curve when the equation is written in the implicit form (2.53)

$$\mathbf{n} = \hat{\mathbf{i}}\frac{\partial G}{\partial x} + \hat{\mathbf{j}}\frac{\partial G}{\partial y} \tag{2.57}$$

We can easily recover the previous form using

$$G(x, y) \equiv y - f(x)$$

$$\therefore \quad \frac{\partial G}{\partial x} = -\frac{\mathrm{d}f}{\mathrm{d}x} \quad \text{and} \quad \frac{\partial G}{\partial y} = 1$$

$$\therefore \quad \mathbf{n} = -\hat{\mathbf{i}}\frac{\mathrm{d}f}{\mathrm{d}x} + \hat{\mathbf{j}}$$

which is identical to the form (2.51) since on the curve, $y = f(x)$. The expression $\hat{\mathbf{i}}\,\partial G/\partial x + \hat{\mathbf{j}}\,\partial G/\partial y$ is important in vector field theory and will be discussed fully later.

2.9 The normal to a surface

The procedure described in Section 2.8 for finding a normal to a planar curve breaks down for three-dimensional curves. This is not surprising since, although a three-dimensional curve has a unique tangent at a point, there is no unique normal (see Figure 2.23). However, a *surface* does have a unique normal direction and a procedure similar to that used for planar curves may be applied to determine it. We write the equation of the surface in the implicit form

$$G(x, y, z) = 0 \tag{2.58}$$

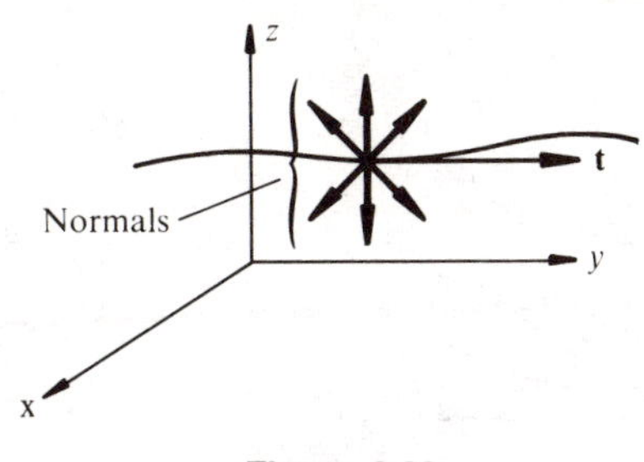

Figure 2.23

If we make small changes in x, y and z such that we remain on the surface, then

$$\mathrm{d}G = \frac{\partial G}{\partial x}\,\mathrm{d}x + \frac{\partial G}{\partial y}\,\mathrm{d}y + \frac{\partial G}{\partial z}\,\mathrm{d}z = 0 \tag{2.59}$$

As in (2.56), we can write this as a scalar product:

$$\mathrm{d}G = \left(\hat{\mathbf{i}}\frac{\partial G}{\partial x} + \hat{\mathbf{j}}\frac{\partial G}{\partial y} + \hat{\mathbf{k}}\frac{\partial G}{\partial z}\right)\cdot(\hat{\mathbf{i}}\,\mathrm{d}x + \hat{\mathbf{j}}\,\mathrm{d}y + \hat{\mathbf{k}}\,\mathrm{d}z) = 0 \tag{2.60}$$

But since the changes in x, y and z are such that we have remained on the surface, then $\hat{\mathbf{i}}\,\mathrm{d}x + \hat{\mathbf{j}}\,\mathrm{d}y + \hat{\mathbf{k}}\,\mathrm{d}z$ is a tangent vector to the surface. Hence, by (2.60), the vector

$$\hat{\mathbf{i}}\frac{\partial G}{\partial x} + \hat{\mathbf{j}}\frac{\partial G}{\partial y} + \hat{\mathbf{k}}\frac{\partial G}{\partial z} \tag{2.61}$$

must be a vector normal to the surface $G(x, y, z) = 0$. The vector (2.61) is, as we shall see in Chapter 3, fundamental in vector field theory; recall from the previous section that $\hat{\mathbf{i}}\,\partial G/\partial x + \hat{\mathbf{j}}\,\partial G/\partial y$ is a vector normal to the curve $G(x, y) = 0$.

If the surface is represented in the parametric form (2.19), then an alternative expression for the normal may be deduced. The position vector $\mathbf{r}$ of any point on the surface is given by

$$\mathbf{r} = \mathbf{r}(u, v) = x(u, v)\hat{\mathbf{i}} + y(u, v)\hat{\mathbf{j}} + z(u, v)\hat{\mathbf{k}} \tag{2.62}$$

Keeping u constant, say $u = c$, implies

$$\mathbf{r} = \mathbf{r}(c, v) = x(c, v)\hat{\mathbf{i}} + y(c, v)\hat{\mathbf{j}} + z(c, v)\hat{\mathbf{k}}$$

that is, $\mathbf{r}$ is a function of a single parameter v and describes a curve in space. The derivative of $\mathbf{r}$ with respect to v is denoted by $\partial\mathbf{r}/\partial v$ and means the rate of change of $\mathbf{r}$ with v for a constant value of u. Geometrically, $\partial\mathbf{r}/\partial v$ is a tangent vector to the curve $\mathbf{r} = \mathbf{r}(c, v)$ for exactly the same reasons that $\mathrm{d}\mathbf{r}/\mathrm{d}t$ is a tangent vector to the curve $\mathbf{r} = \mathbf{r}(t)$. Similarly, $\partial\mathbf{r}/\partial u$ is a tangent vector to a curve $\mathbf{r} = \mathbf{r}(u, c)$ on the surface corresponding to a v-constant line (see

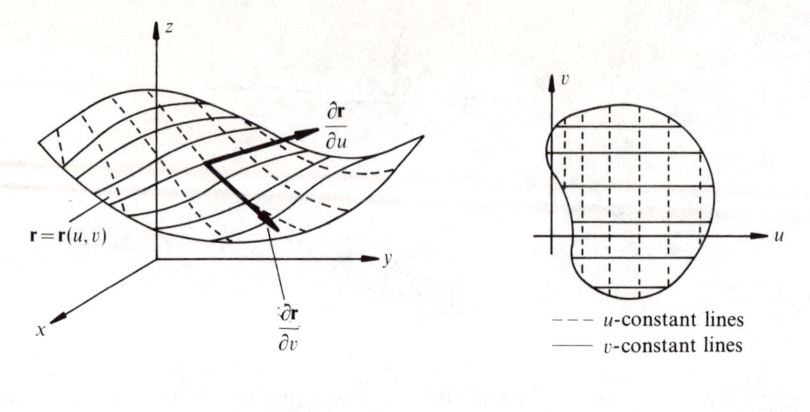

Figure 2.24

Figure 2.24). Thus a normal vector (perpendicular to the two tangent vectors) is given by the expression

$$\mathbf{n} = \frac{\partial \mathbf{r}}{\partial u} \wedge \frac{\partial \mathbf{r}}{\partial v} \tag{2.63}$$

Example 2.5

Find a vector normal to the surface of a sphere centre the origin and of unit radius.

Solution

The equation of the surface in implicit form is

$$G(x, y, z) = x^2 + y^2 + z^2 - 1$$

Hence, using (2.61),

$$\begin{aligned}\mathbf{n} &= \hat{\mathbf{i}}\frac{\partial G}{\partial x} + \hat{\mathbf{j}}\frac{\partial G}{\partial y} + \hat{\mathbf{k}}\frac{\partial G}{\partial z} \\ &= 2x\hat{\mathbf{i}} + 2y\hat{\mathbf{j}} + 2z\hat{\mathbf{k}}\end{aligned} \tag{2.64}$$

is a normal vector.

Alternatively, using the parametric representation (2.22) of the spherical surface,

$$x = \sin u \cos v, \qquad y = \sin u \sin v, \qquad z = \cos u$$

or $\quad \mathbf{r} = \sin u \cos v\,\hat{\mathbf{i}} + \sin u \sin v\,\hat{\mathbf{j}} + \cos u\,\hat{\mathbf{k}}$

gives $\quad \dfrac{\partial \mathbf{r}}{\partial v} = -\sin u \sin v\,\hat{\mathbf{i}} + \sin u \cos v\,\hat{\mathbf{j}}$

and $$\frac{\partial \mathbf{r}}{\partial u} = \cos u \cos v\,\hat{\mathbf{i}} + \cos u \sin v\,\hat{\mathbf{j}} - \sin u\,\hat{\mathbf{k}}$$

Therefore, by (2.63),

$$\begin{aligned}
\mathbf{n} &= \begin{vmatrix} \hat{\mathbf{i}} & \hat{\mathbf{j}} & \hat{\mathbf{k}} \\ \cos u \cos v & \cos u \sin v & -\sin u \\ -\sin u \sin v & \sin u \cos v & 0 \end{vmatrix} \\
&= \hat{\mathbf{i}}(\cos v \sin^2 u) - \hat{\mathbf{j}}(-\sin v \sin^2 u) \\
&\quad + \hat{\mathbf{k}}(\sin u \cos u \cos^2 v + \cos u \sin^2 v \sin u) \\
&= \sin u(\sin u \cos v\,\hat{\mathbf{i}} + \sin u \sin v\,\hat{\mathbf{j}} + \cos u\,\hat{\mathbf{k}}) \\
&= \sin u(x\hat{\mathbf{i}} + y\hat{\mathbf{j}} + z\hat{\mathbf{k}}) \qquad \textbf{(2.65)}
\end{aligned}$$

which is clearly parallel to the vector **n** in (2.64) and hence is also normal to the surface of the sphere.

EXERCISES

2.16 Determine unit normals to the following planar curves at the points indicated:

(a) $y = 3x^2 + x + 1, \quad x = 3$

(b) $x = 3\cos t, \quad y = 2\sin t, \quad t = \pi/4$

In each case, determine also the curvature at these points.

2.17 A surface has the vector equation

$$\mathbf{r} = \mathbf{r}(u, v) = v \cos u\,\hat{\mathbf{i}} + v \sin u\,\hat{\mathbf{j}} + v^2\hat{\mathbf{k}}$$

Determine a normal at the point $v = 3$, $u = \pi/8$. Hence determine the equation of the tangent plane to the surface at this point.

2.18 Find an expression for the unit normal to the ellipsoid

$$\frac{x^2}{2} + \frac{y^2}{4} + \frac{z^2}{12} = 1$$

at the point (0, 1, 3).

2.19 Find the angle between the surfaces $x^2 + y^2 + z^2 = 3$ and $x - z^2 - y^2 + 3 = 0$ at the point $(-1, 1, 1)$.

2.20 The vector equation of a surface is

$$\mathbf{r} = \mathbf{a} + u\mathbf{b} + v\mathbf{c}, \qquad u_0 \leqslant u \leqslant u_1, \quad v_0 \leqslant v \leqslant v_1$$

where **a**, **b** and **c** are constant vectors. Show that the surface is a plane and is at a distance

$$\frac{|\mathbf{a}\cdot(\mathbf{b}\wedge\mathbf{c})|}{|\mathbf{b}\wedge\mathbf{c}|}$$

from the origin.

2.21 Verify that, whenever the following two surfaces intersect, the tangent planes at the points of intersection are perpendicular:

$$\mathbf{r}=u\hat{\mathbf{i}}+v\hat{\mathbf{j}}+uv\hat{\mathbf{k}}, \quad \mathbf{r}=3\sqrt{2}\sin\phi\cos\theta\,\hat{\mathbf{i}}+6\sin\phi\sin\theta\,\hat{\mathbf{j}}+2\cos\phi\,\hat{\mathbf{k}}$$

2.22 Show that (a) the radius of curvature of the parabola $\mathbf{r}=at^2\hat{\mathbf{i}}+2at\hat{\mathbf{j}}$ is $2a(1+t^2)^{3/2}$ and (b) that of the rectangular hyperbola

$$\mathbf{r}=ct\hat{\mathbf{i}}+\frac{c}{t}\hat{\mathbf{j}} \quad \text{is} \quad \frac{c}{2}\left(t^2+\frac{1}{t^2}\right)^{3/2}$$

2.23 Show by a sketch that the surface

$$\mathbf{r}=u\hat{\mathbf{i}}+v\hat{\mathbf{j}}+\tfrac{1}{2}(u^2-v^2)\hat{\mathbf{k}}, \qquad -1\leqslant u\leqslant 1, \quad -1\leqslant v\leqslant 1$$

is saddle shaped and that it has a normal in the direction of the z-axis at the 'saddle point' $u=0$, $v=0$.

2.24 Show that, at the origin, the curve $\mathbf{r}=t\hat{\mathbf{i}}+t^2\hat{\mathbf{j}}+2t\hat{\mathbf{k}}$ has a unit tangent vector $\hat{\mathbf{t}}=\frac{1}{5}\sqrt{5}(\hat{\mathbf{i}}+2\hat{\mathbf{k}})$ and curvature $\kappa=\frac{2}{5}$.

2.25 P is a point on the curve $\mathbf{r}=t\hat{\mathbf{i}}+m\cosh(t/m)\hat{\mathbf{j}}$ (where m is a constant). The normal to the curve at P cuts the x-axis at a point Q. Show that PQ is numerically equal to the radius of curvature at P.

2.26 (a) Show that the surface

$$\mathbf{r}=\mathbf{r}_0=(R+a\cos\theta)\cos\phi\,\hat{\mathbf{i}}+(R+a\cos\theta)\sin\phi\,\hat{\mathbf{j}}+a\sin\theta\,\hat{\mathbf{k}},$$

$$0\leqslant\theta\leqslant 2\pi, \quad 0\leqslant\phi\leqslant 2\pi$$

is a torus.
(b) Show that the surface $\mathbf{r}=\mathbf{r}_0+\theta\hat{\mathbf{k}}$, where $\mathbf{r}_0$ is defined as in (a), represents one turn of a helical spring.

2.27 Realizing that, in index form, the ith component of $\mathrm{d}\mathbf{a}/\mathrm{d}t$ is $\mathrm{d}a_i/\mathrm{d}t$, prove, using the index notation, the results (a) and (b) of exercise 2.8.

ADDITIONAL EXERCISES

1 According to Newton's Law of Gravitation, the force between two particles of masses m and M is

$$\mathbf{F} = -GmM\frac{\mathbf{r}}{|\mathbf{r}|^3}$$

where G is the gravitational constant and $\mathbf{r}$ is the position vector separating the particles. Assuming the Earth to be a fixed body, determine the period of a satellite moving in a circular orbit 500 km above the Earth. (*Hint:* the centripetal force required by the satellite to move in a circular orbit is supplied by the gravitational force.)

($G = 6.67 \times 10^{-11}$ in SI units; mass of the Earth $= 5.98 \times 10^{24}$ kg; radius of the Earth $= 6.37 \times 10^6$ m.)

2 Imagine following a particle as it moves on a planar curve. Instead of describing its position at time t in terms of Cartesian coordinates $(x(t), y(t))$, suppose we describe its position in terms of polar coordinates $(\rho(t), \phi(t))$ as shown in Figure 2.25. Two unit vectors $\hat{\boldsymbol{\rho}}$ and $\hat{\boldsymbol{\phi}}$ are introduced at each point in the radial and transverse directions.

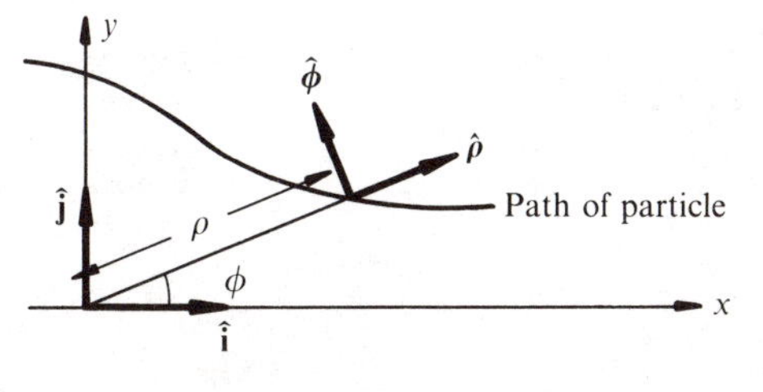

Figure 2.25

(a) Show that $\hat{\boldsymbol{\rho}} = \cos\phi\,\hat{\mathbf{i}} + \sin\phi\,\hat{\mathbf{j}}$ and $\hat{\boldsymbol{\phi}} = -\sin\phi\,\hat{\mathbf{i}} + \cos\phi\,\hat{\mathbf{j}}$

(b) Verify that $\dfrac{d\hat{\boldsymbol{\rho}}}{dt} = \hat{\boldsymbol{\phi}}\dfrac{d\phi}{dt}$ and $\dfrac{d\hat{\boldsymbol{\phi}}}{dt} = -\hat{\boldsymbol{\rho}}\dfrac{d\phi}{dt}$

(c) Show that the acceleration of the particle on the path is given by

$$\frac{d^2\boldsymbol{\rho}}{dt^2} = \left(\frac{d^2\rho}{dt^2} - \rho\left(\frac{d\phi}{dt}\right)^2\right)\hat{\boldsymbol{\rho}} + \left(2\frac{d\rho}{dt}\frac{d\phi}{dt} + \rho\frac{d^2\phi}{dt^2}\right)\hat{\boldsymbol{\phi}}, \quad \text{where } \boldsymbol{\rho} = \rho\hat{\boldsymbol{\rho}}$$

3 In the mechanism shown in Figure 2.26 the rod AB passes through a point O and moves round a circle of radius 1 m. If AB rotates about O at a constant angular velocity of 3 rad s^{-1}, show that A has a constant speed of 6 m s^{-1} and find the magnitude of its acceleration. (*Hint:* use the results of exercise 2.)

4 What altitude must a satellite have in order for it to appear stationary in the sky when viewed from the Earth?

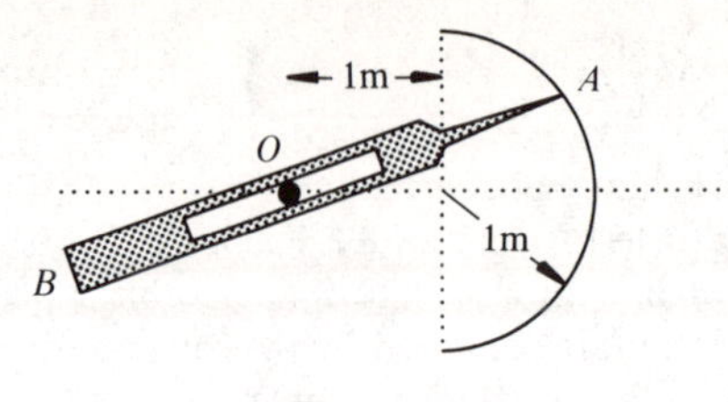

Figure 2.26

5 A small absorbent particle of mass m_0 is projected under gravity. Its mass increases (owing to condensation of moisture) at a constant rate $m_0 k$, where k is a constant. The particle is projected from the origin with an initial velocity $\mathbf{v}_0 = v \cos \alpha \,\hat{\mathbf{i}} + v \sin \alpha \,\hat{\mathbf{j}}$. The path of the particle lies entirely in the xy-plane with the y-axis taken vertically upwards. Show that the position vector of the particle at time t is

$$\mathbf{r} = \frac{-1}{4k^2}(kt+1)^2 g\hat{\mathbf{j}} + \left[\frac{v}{k}(\cos\alpha\,\hat{\mathbf{i}} + \sin\alpha\,\hat{\mathbf{j}}) + \frac{g}{2k^2}\hat{\mathbf{j}}\right]\log(kt+1) + \frac{g}{4k^2}\hat{\mathbf{j}}$$

(*Hint:* Newton's law for variable mass is that applied force equals rate of change of momentum.)

SUMMARY

- The **vector equation** of a curve in parametric form is

$$\mathbf{r} = \mathbf{r}(t) = x(t)\hat{\mathbf{i}} + y(t)\hat{\mathbf{j}} + z(t)\hat{\mathbf{k}}, \qquad t_0 \leqslant t \leqslant t_1$$

- The **derivative** of a vector function $\mathbf{a}(t)$ is

$$\frac{\mathrm{d}\mathbf{a}}{\mathrm{d}t} = \frac{\mathrm{d}a_x}{\mathrm{d}t}\hat{\mathbf{i}} + \frac{\mathrm{d}a_y}{\mathrm{d}t}\hat{\mathbf{j}} + \frac{\mathrm{d}a_z}{\mathrm{d}t}\hat{\mathbf{k}}$$

In particular,

$$\frac{\mathrm{d}}{\mathrm{d}t}(\mathbf{a}\cdot\mathbf{b}) = \mathbf{a}\cdot\frac{\mathrm{d}\mathbf{b}}{\mathrm{d}t} + \frac{\mathrm{d}\mathbf{a}}{\mathrm{d}t}\cdot\mathbf{b}$$

$$\frac{\mathrm{d}}{\mathrm{d}t}(\mathbf{a}\wedge\mathbf{b}) = \mathbf{a}\wedge\frac{\mathrm{d}\mathbf{b}}{\mathrm{d}t} + \frac{\mathrm{d}\mathbf{a}}{\mathrm{d}t}\wedge\mathbf{b}$$

$$\mathbf{a}\cdot\frac{\mathrm{d}\mathbf{a}}{\mathrm{d}t} = |\mathbf{a}|\,\frac{\mathrm{d}|\mathbf{a}|}{\mathrm{d}t}$$

- The **indefinite integral** of a vector function of a single variable, $\mathbf{a}(t)$, is

$$\int \mathbf{a}(t)\,\mathrm{d}t = \hat{\mathbf{i}}\int a_x(t)\,\mathrm{d}t + \hat{\mathbf{j}}\int a_y(t)\,\mathrm{d}t + \hat{\mathbf{k}}\int a_z(t)\,\mathrm{d}t$$

- A **tangent vector** to the curve $\mathbf{r} = \mathbf{r}(t)$ is given by $\dfrac{\mathrm{d}\mathbf{r}}{\mathrm{d}t}$

- A **unit tangent vector** is given by $\dfrac{\mathrm{d}\mathbf{r}}{\mathrm{d}t}\Big/\left|\dfrac{\mathrm{d}\mathbf{r}}{\mathrm{d}t}\right|$ or by $\dfrac{\mathrm{d}\mathbf{r}}{\mathrm{d}s}$, where s is the arc-length parameter.

- If the position vector of a particle at time t is given by $\mathbf{r} = \mathbf{r}(t)$ then its **velocity** at any instant is $\dfrac{\mathrm{d}\mathbf{r}}{\mathrm{d}t}$ and its **acceleration** is $\dfrac{\mathrm{d}^2\mathbf{r}}{\mathrm{d}t^2}$

- The **curvature** κ of a planar curve is defined by

$$\kappa^2 = \frac{\mathrm{d}\hat{\mathbf{t}}}{\mathrm{d}s}\cdot\frac{\mathrm{d}\hat{\mathbf{t}}}{\mathrm{d}s} \quad \text{or by} \quad \kappa^2 = \frac{\left[-\dfrac{\mathrm{d}^2x}{\mathrm{d}t^2}\dfrac{\mathrm{d}y}{\mathrm{d}t} + \dfrac{\mathrm{d}x}{\mathrm{d}t}\dfrac{\mathrm{d}^2y}{\mathrm{d}t^2}\right]^2}{\left[\left(\dfrac{\mathrm{d}x}{\mathrm{d}t}\right)^2 + \left(\dfrac{\mathrm{d}y}{\mathrm{d}t}\right)^2\right]^3}$$

- A **normal** to a planar curve $\mathbf{r} = \mathbf{r}(t) = x(t)\hat{\mathbf{i}} + y(t)\hat{\mathbf{j}}$ is given by

$$\mathbf{n} = -\frac{\mathrm{d}y}{\mathrm{d}t}\hat{\mathbf{i}} + \frac{\mathrm{d}x}{\mathrm{d}t}\hat{\mathbf{j}}$$

  If the curve is given in **implicit Cartesian form**, $G(x, y) = 0$, then a normal is

$$\mathbf{n} = \frac{\partial G}{\partial x}\hat{\mathbf{i}} + \frac{\partial G}{\partial y}\hat{\mathbf{j}}$$

- A **normal** to a **surface** given in **implicit Cartesian form** $G(x, y, z) = 0$ is

$$\mathbf{n} = \frac{\partial G}{\partial x}\hat{\mathbf{i}} + \frac{\partial G}{\partial y}\hat{\mathbf{j}} + \frac{\partial G}{\partial z}\hat{\mathbf{k}}$$

  whilst if the surface equation is given in parametric form, then a normal to the surface is

$$\mathbf{n} = \frac{\partial \mathbf{r}}{\partial u} \wedge \frac{\partial \mathbf{r}}{\partial v}$$

SUPPLEMENT

2S.1 Differential geometry

In Sections 2.7 and 2.8 we gave a brief discussion of the vector geometry of planar curves. We now briefly outline the necessary amendments and extensions that need to be made if the curve under consideration is not restricted to lie in a plane.

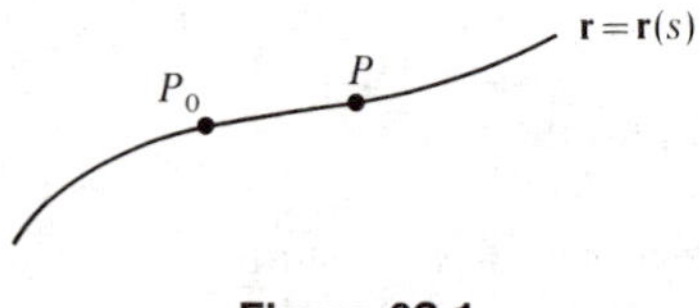

Figure 2S.1

Consider a curve $\mathbf{r} = \mathbf{r}(s)$ and a particular point P on this curve (see Figure 2S.1). If we consider a point P_0, close to P, then a plane through P_0 containing the unit tangent vector $\hat{\mathbf{t}}$ may be constructed. As $P_0 \to P$ this plane contains the point P and is called the **plane of curvature** or the **osculating plane** at P. A unit normal vector $\hat{\mathbf{n}}$ in the osculating plane is introduced at P and, as in Section 2.7, we can write

$$\frac{\mathrm{d}\hat{\mathbf{t}}}{\mathrm{d}s} = \kappa\hat{\mathbf{n}} \tag{2S.1}$$

where κ is the curvature at P. We can say that the curve in the close neighbourhood of the point P lies in the osculating plane. As far as its curvature properties are concerned, the curve close to the point P appears to be a circle of radius $1/\kappa$. This circle lies in the osculating plane and its centre (the centre of curvature) lies along $\hat{\mathbf{n}}$ (see Example 2.4). We know that the curve has an infinite collection of vectors perpendicular to $\hat{\mathbf{t}}$. In this context, $\hat{\mathbf{n}}$ is picked out and is known as the **principal normal** at P.

The plane perpendicular to the tangent vector at P is called the **normal plane** at P.

The straight line through P perpendicular to the osculating plane is called the **binormal** at P. The unit vector in this direction is denoted by $\hat{\mathbf{b}}$. The vectors $\hat{\mathbf{t}}$, $\hat{\mathbf{n}}$ and $\hat{\mathbf{b}}$ form a right-handed system when taken in this order (see Figure 2S.2).

Now

$$\frac{\mathrm{d}}{\mathrm{d}s}(\hat{\mathbf{t}}\cdot\hat{\mathbf{b}}) = \frac{\mathrm{d}\hat{\mathbf{t}}}{\mathrm{d}s}\cdot\hat{\mathbf{b}} + \hat{\mathbf{t}}\cdot\frac{\mathrm{d}\hat{\mathbf{b}}}{\mathrm{d}s} = 0$$

but $\quad \dfrac{\mathrm{d}\hat{\mathbf{t}}}{\mathrm{d}s} = \kappa\hat{\mathbf{n}} \quad$ and so $\quad \dfrac{\mathrm{d}\hat{\mathbf{t}}}{\mathrm{d}s}\cdot\hat{\mathbf{b}} = 0$

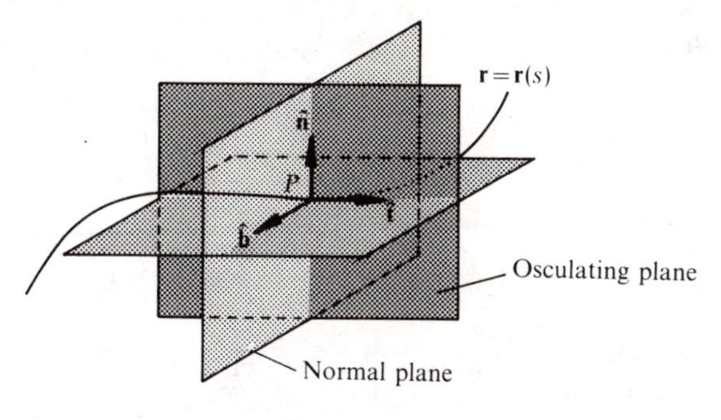

Figure 2S.2

Therefore

$$\hat{\mathbf{t}}\cdot\frac{\mathrm{d}\hat{\mathbf{b}}}{\mathrm{d}s}=0$$

which implies that $\mathrm{d}\hat{\mathbf{b}}/\mathrm{d}s$ is perpendicular to $\hat{\mathbf{t}}$. Also, since $\hat{\mathbf{b}}$ is a unit vector, $\mathrm{d}\hat{\mathbf{b}}/\mathrm{d}s$ is, by (2.35), perpendicular to $\hat{\mathbf{b}}$. Therefore $\mathrm{d}\hat{\mathbf{b}}/\mathrm{d}s$ must be parallel to $\hat{\mathbf{n}}$. We write

$$\frac{\mathrm{d}\hat{\mathbf{b}}}{\mathrm{d}s}=-\tau\hat{\mathbf{n}} \tag{2S.2}$$

where τ is called the **torsion** of the curve at P. It is a measure of the arc rate of change of $\hat{\mathbf{b}}$. The larger is τ, the 'quicker' the curve moves away from the osculating plane at P. In fact, it can be shown that the condition for a curve to be planar is that it has zero torsion.

Since $\hat{\mathbf{n}}=\hat{\mathbf{b}}\wedge\hat{\mathbf{t}}$ then

$$\begin{aligned}\frac{\mathrm{d}\hat{\mathbf{n}}}{\mathrm{d}s}&=\frac{\mathrm{d}\hat{\mathbf{b}}}{\mathrm{d}s}\wedge\hat{\mathbf{t}}+\hat{\mathbf{b}}\wedge\frac{\mathrm{d}\hat{\mathbf{t}}}{\mathrm{d}s}\\&=-\tau(\hat{\mathbf{n}}\wedge\hat{\mathbf{t}})+\hat{\mathbf{b}}\wedge(\kappa\hat{\mathbf{n}})=-\tau(-\hat{\mathbf{b}})+\kappa(-\hat{\mathbf{t}})\\&=\tau\hat{\mathbf{b}}-\kappa\hat{\mathbf{t}}\end{aligned} \tag{2S.3}$$

The three formulae

$$\frac{\mathrm{d}\hat{\mathbf{t}}}{\mathrm{d}s}=\kappa\hat{\mathbf{n}},\qquad\frac{\mathrm{d}\hat{\mathbf{b}}}{\mathrm{d}s}=-\tau\hat{\mathbf{n}},\qquad\frac{\mathrm{d}\hat{\mathbf{n}}}{\mathrm{d}s}=\tau\hat{\mathbf{b}}-\kappa\hat{\mathbf{t}} \tag{2S.4}$$

are called the **Serret–Frenet equations** and are fundamental to the study of the differential geometry of curves.

Example 2S.1

Determine the unit vectors $\hat{\mathbf{t}}$, $\hat{\mathbf{n}}$ and $\hat{\mathbf{b}}$ and the constants κ and τ for the helix

$$\mathbf{r} = \mathbf{r}(t) = \cos t\,\hat{\mathbf{i}} + \sin t\,\hat{\mathbf{j}} + t\hat{\mathbf{k}}$$

Solution

$$\hat{\mathbf{t}} = \frac{\mathrm{d}\mathbf{r}}{\mathrm{d}s} = \frac{\mathrm{d}\mathbf{r}}{\mathrm{d}t}\frac{\mathrm{d}t}{\mathrm{d}s} \quad \text{where} \quad \frac{\mathrm{d}\mathbf{r}}{\mathrm{d}t} = -\sin t\,\hat{\mathbf{i}} + \cos t\,\hat{\mathbf{j}} + \hat{\mathbf{k}}$$

It follows that $|\mathrm{d}\mathbf{r}/\mathrm{d}t| = \sqrt{2}$ and, for $\hat{\mathbf{t}}$ to be a unit vector, $|\mathrm{d}s/\mathrm{d}t| = \sqrt{2}$. Also, since the rate of change of arc length with t is positive, $\mathrm{d}s/\mathrm{d}t = +\sqrt{2}$. Therefore

$$\hat{\mathbf{t}} = \frac{1}{\sqrt{2}}[-\sin t\,\hat{\mathbf{i}} + \cos t\,\hat{\mathbf{j}} + \hat{\mathbf{k}}]$$

Also, $$\frac{\mathrm{d}\hat{\mathbf{t}}}{\mathrm{d}s} = \frac{\mathrm{d}\hat{\mathbf{t}}}{\mathrm{d}t}\frac{\mathrm{d}t}{\mathrm{d}s} = \frac{1}{\sqrt{2}}\frac{\mathrm{d}\hat{\mathbf{t}}}{\mathrm{d}t} = \tfrac{1}{2}(-\cos t\,\hat{\mathbf{i}} - \sin t\,\hat{\mathbf{j}})$$

so using $\mathrm{d}\hat{\mathbf{t}}/\mathrm{d}s = \kappa\hat{\mathbf{n}}$ we find

$$\kappa = \left|\frac{\mathrm{d}\hat{\mathbf{t}}}{\mathrm{d}s}\right| = \tfrac{1}{2} \quad \text{and} \quad \hat{\mathbf{n}} = -\cos t\,\hat{\mathbf{i}} - \sin t\,\hat{\mathbf{j}}$$

Hence

$$\hat{\mathbf{b}} = \hat{\mathbf{t}} \wedge \hat{\mathbf{n}} = \frac{1}{\sqrt{2}}\begin{vmatrix} \hat{\mathbf{i}} & \hat{\mathbf{j}} & \hat{\mathbf{k}} \\ -\sin t & \cos t & 1 \\ -\cos t & -\sin t & 0 \end{vmatrix} = \frac{1}{\sqrt{2}}(\sin t\,\hat{\mathbf{i}} - \cos t\,\hat{\mathbf{j}} + \hat{\mathbf{k}})$$

therefore

$$\frac{\mathrm{d}\hat{\mathbf{b}}}{\mathrm{d}s} = \frac{\mathrm{d}\hat{\mathbf{b}}}{\mathrm{d}t}\frac{\mathrm{d}t}{\mathrm{d}s} = \tfrac{1}{2}(\cos t\,\hat{\mathbf{i}} + \sin t\,\hat{\mathbf{j}})$$

Finally, using (2S.2),

$$\tau = \left|\frac{\mathrm{d}\hat{\mathbf{b}}}{\mathrm{d}s}\right| = \tfrac{1}{2}$$

We note here that both the curvature and the torsion are constant for every point on the curve. It can be shown that the circular helix is the *only* curve for which this is true.

We conclude this supplement by noting that the calculations leading to equation (2.47) are still valid even for non-planar curves and so its application to the dynamics of moving particles remains unaltered.

2S.2 Impulsive forces

The **impulse J** applied to a particle of mass m in an interval of time from t_1 to t_2 is defined as the change in momentum of the particle during the interval. If $\mathbf{v}_1$ is the velocity of the particle at t_1 and $\mathbf{v}_2$ is the velocity at t_2, then

$$\mathbf{J} = m(\mathbf{v}_2 - \mathbf{v}_1) \tag{2S.5}$$

which, of course, is a vector equation.

To change the momentum of a particle, a force must be applied. If $\mathbf{a}(t)$ is the acceleration of the particle at time t and $\mathbf{F}(t)$ is the force acting, then

$$\mathbf{F}(t) = m\mathbf{a}(t)$$

Integrating from t_1 to t_2,

$$\int_{t_1}^{t_2} \mathbf{F}(t)\,\mathrm{d}t = m\int_{t_1}^{t_2} \mathbf{a}(t)\,\mathrm{d}t = m\int_{t_1}^{t_2} \frac{\mathrm{d}\mathbf{v}}{\mathrm{d}t}\,\mathrm{d}t = m\int_{t_1}^{t_2} \mathrm{d}(\mathbf{v}) = m[\mathbf{v}]_{t_1}^{t_2}$$
$$= m(\mathbf{v}_2 - \mathbf{v}_1) = \mathbf{J}$$

In other words, the impulse **J** is the time integral of the force vector. If the time interval $t_2 - t_1$ tends to zero and the magnitude of **F** remains finite, then $\int_{t_1}^{t_2} \mathbf{F}(t)\,\mathrm{d}t \to 0$. However, if the magnitude of **F** tends to infinity as $t_2 - t_1$ tends to zero, then $\int_{t_1}^{t_2} \mathbf{F}(t)\,\mathrm{d}t$ may tend to a definite value. In this case, we say $\lim_{t_2 \to t_1} \int_{t_1}^{t_2} \mathbf{F}(t)\,\mathrm{d}t$ is an **impulse** or **impulsive force**, and we write

$$\mathbf{J} = \lim_{t_2 \to t_1} \int_{t_1}^{t_2} \mathbf{F}(t)\,\mathrm{d}t \tag{2S.6}$$

(Note that we are modelling the situation which arises when a very large force is applied over a very short time interval.)

The most obvious application of impulse is to the collision of particles. When two particles collide the impulsive force experienced by one particle will, by Newton's third law, be equal and opposite to the impulsive force experienced by the other particle. If the two particles have masses m_1 and m_2, velocities $\mathbf{v}_1$ and $\mathbf{v}_2$ before collision and velocities $\mathbf{u}_1$ and $\mathbf{u}_2$ after collision, then the impulse on m_1 is

$$\mathbf{J}_1 = m_1(\mathbf{u}_1 - \mathbf{v}_1)$$

and that on mass m_2 is

$$J_2 = m_2(\mathbf{u}_2 - \mathbf{v}_2)$$

But $\mathbf{J}_2 = -\mathbf{J}_1$ and so

$$m_2\mathbf{u}_2 - m_2\mathbf{v}_2 = -m_1\mathbf{u}_1 + m_1\mathbf{v}_1$$

$$\therefore \quad m_1\mathbf{v}_1 + m_2\mathbf{v}_2 = m_1\mathbf{u}_1 + m_2\mathbf{u}_2 \tag{2S.7}$$

that is, the combined momentum before collision is equal to the combined momentum after collision.

If the bodies coalesce together after impact then this 'conservation of linear momentum' is sufficient to determine the motion after impact. For if the final velocity of $(m_1 + m_2)$ after impact is $\mathbf{u}$, then

$$(m_1 + m_2)\mathbf{u} = m_1\mathbf{v}_1 + m_2\mathbf{v}_2$$

$$\therefore \qquad \mathbf{u} = \frac{m_1\mathbf{v}_1 + m_2\mathbf{v}_2}{m_1 + m_2} \tag{2S.8}$$

If the particles bounce off each other, then the principle of conservation of linear momentum is not sufficient to determine the ensuing motion as there are six unknown components – three from $\mathbf{v}_1$ and three from $\mathbf{v}_2$ – but only three equations. The theory is now supplemented by an experimental law (supposedly discovered by Newton).

We suppose that on contact the bodies have a common tangent plane with normal $\hat{\mathbf{n}}$ (see Figure 2S.3). The experimental law states that the relative

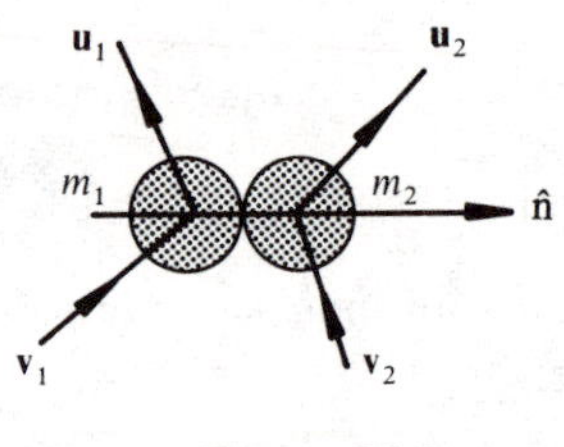

Figure 2S.3

velocity perpendicular to $\hat{\mathbf{n}}$, before and after impact, is conserved but that the relative velocity in the direction of $\hat{\mathbf{n}}$ after impact is $(-e)$ times the relative velocity in the direction of $\hat{\mathbf{n}}$ before impact. The constant e is called the **coefficient of restitution**. Before expressing this experimental law in mathematical form we note that we can express *any* vector $\mathbf{b}$ in the form

$$\mathbf{b} = (\mathbf{b}\cdot\hat{\mathbf{n}})\hat{\mathbf{n}} + [\mathbf{b} - (\mathbf{b}\cdot\hat{\mathbf{n}})\hat{\mathbf{n}}] \tag{2S.9}$$

simply by adding and subtracting $(\mathbf{b}\cdot\hat{\mathbf{n}})\hat{\mathbf{n}}$ to $\mathbf{b}$. This shows that $\mathbf{b}$ may be rewritten as the sum of a vector parallel to $\hat{\mathbf{n}}$ and a vector perpendicular to $\hat{\mathbf{n}}$. That the vector $\mathbf{b} - (\mathbf{b}\cdot\hat{\mathbf{n}})\hat{\mathbf{n}}$ is perpendicular to $\hat{\mathbf{n}}$ is easily checked by noting that its dot product with $\hat{\mathbf{n}}$ is zero.

Since the relative velocity before impact is $\mathbf{v}_2 - \mathbf{v}_1$ and the relative velocity after impact is $\mathbf{u}_2 - \mathbf{u}_1$, the experimental law takes the form

$$(\mathbf{u}_2 - \mathbf{u}_1) - [(\mathbf{u}_2 - \mathbf{u}_1)\cdot\hat{\mathbf{n}}]\hat{\mathbf{n}} = (\mathbf{v}_2 - \mathbf{v}_1) - [(\mathbf{v}_2 - \mathbf{v}_1)\cdot\hat{\mathbf{n}}]\hat{\mathbf{n}} \tag{2S.10}$$

and $$(\mathbf{u}_2 - \mathbf{u}_1)\cdot\hat{\mathbf{n}} = -e(\mathbf{v}_2 - \mathbf{v}_1)\cdot\hat{\mathbf{n}} \tag{2S.11}$$

These two equations, together with the conservation of linear momentum, provide just enough information to determine $\mathbf{u}_2$ and $\mathbf{u}_1$. (Since (2S.11) is a scalar equation, it contains a single piece of information, and (2S.10), although a vector equation, contains only two pieces of information, because the normal component of the vector has been subtracted.)

Example 2S.2

Two particles A and B, of mass $2m$ and $1m$ respectively and moving with velocities $\mathbf{v}_1 = \hat{\mathbf{i}} + 2\hat{\mathbf{j}} + 3\hat{\mathbf{k}}$ and $\mathbf{v}_2 = 2\hat{\mathbf{i}} + \hat{\mathbf{j}} - 3\hat{\mathbf{k}}$, collide. If the plane of contact has a unit normal $\hat{\mathbf{n}} = \dfrac{1}{\sqrt{3}}(\hat{\mathbf{i}} - \hat{\mathbf{j}} + \hat{\mathbf{k}})$ and the coefficient of restitution between the particles is 0.75, determine the velocities of the two particles after collision.

Solution

Let $\mathbf{u}_1$ and $\mathbf{u}_2$ be the velocities after the impact of A and B respectively. The relative velocity before impact is $\mathbf{v}_2 - \mathbf{v}_1 = \hat{\mathbf{i}} - \hat{\mathbf{j}} - 6\hat{\mathbf{k}}$ and so

$$(\mathbf{v}_2 - \mathbf{v}_1)\cdot\hat{\mathbf{n}} = \frac{1}{\sqrt{3}}(-4)$$

Conservation of linear momentum implies that

$$2m\mathbf{u}_1 + m\mathbf{u}_2 = 2m\mathbf{v}_1 + m\mathbf{v}_2$$

$$\therefore \quad 2\mathbf{u}_1 + \mathbf{u}_2 = 4\hat{\mathbf{i}} + 5\hat{\mathbf{j}} + 3\hat{\mathbf{k}}$$

$$\therefore \quad \mathbf{u}_2 = 4\hat{\mathbf{i}} + 5\hat{\mathbf{j}} + 3\hat{\mathbf{k}} - 2\mathbf{u}_1 \tag{2S.12}$$

Newton's experimental law (2S.10) implies that

$$(\mathbf{u}_2 - \mathbf{u}_1) - [(\mathbf{u}_2 - \mathbf{u}_1)\cdot\hat{\mathbf{n}}]\hat{\mathbf{n}} = \hat{\mathbf{i}} - \hat{\mathbf{j}} - 6\hat{\mathbf{k}} - \left(-\frac{4}{\sqrt{3}}\right)\frac{1}{\sqrt{3}}(\hat{\mathbf{i}} - \hat{\mathbf{j}} + \hat{\mathbf{k}})$$

$$= \hat{\mathbf{i}}(\tfrac{7}{3}) + \hat{\mathbf{j}}(-\tfrac{7}{3}) + \hat{\mathbf{k}}(-\tfrac{14}{3}) \tag{2S.13}$$

But, using (2S.12), the left-hand side of this equation may be expressed in terms of $\mathbf{u}_1$ alone:

$$(\mathbf{u}_2 - \mathbf{u}_1)\cdot\hat{\mathbf{n}} = \frac{1}{\sqrt{3}}(2) - \frac{3}{\sqrt{3}}\mathbf{u}_1\cdot(\hat{\mathbf{i}} - \hat{\mathbf{j}} + \hat{\mathbf{k}})$$

$$\therefore \quad [(\mathbf{u}_2 - \mathbf{u}_1)\cdot\hat{\mathbf{n}}]\hat{\mathbf{n}} = \tfrac{1}{3}[2 - 3(u_{1x} - u_{1y} + u_{1z})](\hat{\mathbf{i}} - \hat{\mathbf{j}} + \hat{\mathbf{k}}) \tag{2S.14}$$

where (u_{1x}, u_{1y}, u_{1z}) are the components of $\mathbf{u}_1$. Substituting into (2S.13) and expanding the single vector equation into its three component parts, we find

$$-6u_{1x} - 3u_{1y} + 3u_{1z} = -3$$

$$-3u_{1x} - 6u_{1y} - 3u_{1z} = -24$$

$$3u_{1x} - 3u_{1y} - 6u_{1z} = -21 \tag{2S.15}$$

The three equations (2S.15) are not independent, in that the third equation can be obtained by subtracting the first from the second. Hence we cannot solve for the three unknowns by these equations alone. This is related to the remark made earlier, that even though this set has arisen from a vector equation it contains only two pieces of information, namely

$$2u_{1x} + u_{1y} - u_{1z} = 1 \tag{2S.16a}$$

$$u_{1x} + 2u_{1y} + u_{1z} = 8 \tag{2S.16b}$$

An additional equation is obtained using the second part of Newton's experimental law (2S.11):

$$\frac{1}{\sqrt{3}}(2) - \frac{3}{\sqrt{3}}\mathbf{u}_1 \cdot (\hat{\mathbf{i}} - \hat{\mathbf{j}} + \hat{\mathbf{k}}) = -e\left(-\frac{4}{\sqrt{3}}\right) = \frac{3}{\sqrt{3}}$$

that is,

$$-3u_{1x} + 3u_{1y} - 3u_{1z} = 1 \tag{2S.16c}$$

The system of equations in (2S.16) may be solved immediately to obtain

$$\mathbf{u}_1 = \tfrac{2}{9}\hat{\mathbf{i}} + \tfrac{25}{9}\hat{\mathbf{j}} + \tfrac{20}{9}\hat{\mathbf{k}}$$

and hence, using (2S.12),

$$\mathbf{u}_2 = 4\hat{\mathbf{i}} + 5\hat{\mathbf{j}} + 3\hat{\mathbf{k}} - 2\mathbf{u}_1 = \tfrac{32}{9}\hat{\mathbf{i}} - \tfrac{5}{9}\hat{\mathbf{j}} - \tfrac{27}{9}\hat{\mathbf{k}}$$

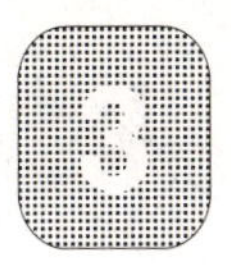

Scalar and Vector Fields

PREVIEW

In this chapter we define the concept of a field; in particular, scalar fields and vector fields are studied. The fundamental operator in vector field theory – the gradient – is introduced and, by analogy with the scalar and vector products of vector algebra, used to define the divergence operator and the curl operator. The major identities of vector differential calculus are obtained and utilized, and the Laplacian operator is considered. Applications to elasticity theory are discussed.

3.1 Scalar fields

Consider the temperature distribution within a tank of fluid. At each point in the fluid, one can measure the temperature (a scalar), $T(x, y, z)$. This is an example of a **scalar field**. More formally, a scalar field is a rule which

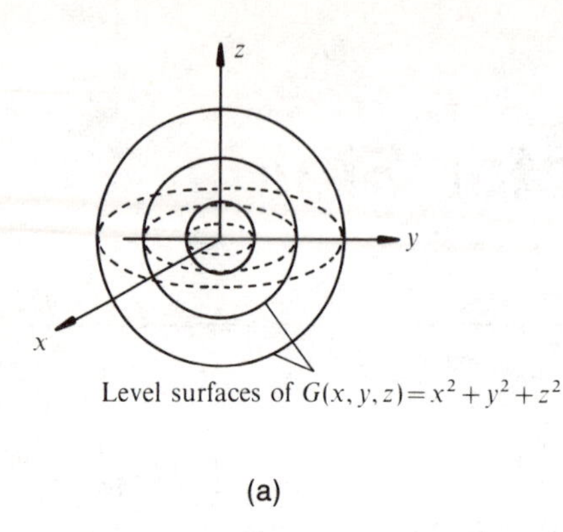

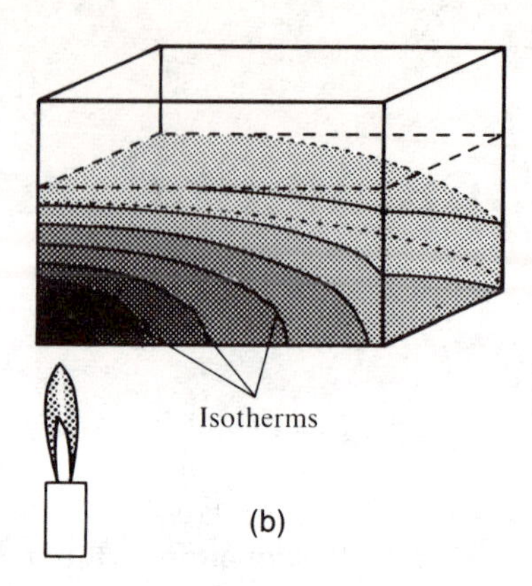

Figure 3.1

associates a number with each point in a region of space. In mathematical terms, a scalar field is simply a function of the variables x, y and z (or x and y in two dimensions). For example, $G(x, y, z) \equiv x^2 + y^2 + z^2$ is a scalar field: every value of (x, y, z) has associated with it the number $x^2 + y^2 + z^2$.

The **level surfaces** of a given scalar field $G(x, y, z)$ are those surfaces on which the scalar field has a constant value. Explicitly, these surfaces take the form

$$G(x, y, z) = k \tag{3.1}$$

where k is the constant value on the surface. The level surfaces of the scalar field $x^2 + y^2 + z^2$ are spheres centred on the origin (see Figure 3.1a), whilst those of the temperature distribution in the tank of fluid mentioned above might take the form shown in Figure 3.1b. Level surfaces are sometimes given special names for particular scalar fields G: **isotherms** if G represents temperature, **isobars** if G represents pressure, **equipotentials** if G represents a scalar potential.

If $G(x, y, z)$ is a scalar field, then we define the gradient of G, written ∇G (or **grad** G) by

$$\nabla G = \hat{\mathbf{i}}\frac{\partial G}{\partial x} + \hat{\mathbf{j}}\frac{\partial G}{\partial y} + \hat{\mathbf{k}}\frac{\partial G}{\partial z} \tag{3.2}$$

(∇ is pronounced as 'del' or 'nabla'). We have already met ∇G (without giving it a name) in Section 2.9, where we showed that ∇G evaluated at a point P is a vector that is normal to the level surface $G(x, y, z) = \text{constant}$ passing through P. In Section 2.8 we showed that the two-dimensional form

$$\nabla G = \hat{\mathbf{i}}\frac{\partial G}{\partial x} + \hat{\mathbf{j}}\frac{\partial G}{\partial y}$$

evaluated at a point P is normal to the level 'curve' $G(x, y) = 0$.

Example 3.1

Determine a unit normal to the surface S defined by $x^2 + y^2 + z^2 = a^2$.

Solution

Here $G(x, y, z) = x^2 + y^2 + z^2$, so a normal to S is

$$\nabla G = \hat{\mathbf{i}}\frac{\partial G}{\partial x} + \hat{\mathbf{j}}\frac{\partial G}{\partial y} + \hat{\mathbf{k}}\frac{\partial G}{\partial z} = \hat{\mathbf{i}}(2x) + \hat{\mathbf{j}}(2y) + \hat{\mathbf{k}}(2z)$$

Therefore a unit normal is

$$\hat{\mathbf{n}} = \frac{(2x\hat{\mathbf{i}} + 2y\hat{\mathbf{j}} + 2z\hat{\mathbf{k}})}{\sqrt{(4x^2 + 4y^2 + 4z^2)}} = \frac{x\hat{\mathbf{i}} + y\hat{\mathbf{j}} + z\hat{\mathbf{k}}}{\sqrt{(x^2 + y^2 + z^2)}}$$

But $x^2 + y^2 + z^2 = a^2$ on S, so

$$\hat{\mathbf{n}} = \left(\frac{x}{a}\right)\hat{\mathbf{i}} + \left(\frac{y}{a}\right)\hat{\mathbf{j}} + \left(\frac{z}{a}\right)\hat{\mathbf{k}} = \frac{\mathbf{r}}{a}$$

where $\mathbf{r}$ is the usual position vector of a point on the surface. The vector $-\hat{\mathbf{n}}$ is the other unit normal to S.

It will sometimes be convenient to consider the vector operator ∇ alone:

$$\nabla \equiv \hat{\mathbf{i}}\frac{\partial}{\partial x} + \hat{\mathbf{j}}\frac{\partial}{\partial y} + \hat{\mathbf{k}}\frac{\partial}{\partial z} \tag{3.3}$$

so that the gradient of G, ∇G, can be thought of as the operator ∇ applied to G to produce

$$\hat{\mathbf{i}}\frac{\partial G}{\partial x} + \hat{\mathbf{j}}\frac{\partial G}{\partial y} + \hat{\mathbf{k}}\frac{\partial G}{\partial z}$$

The operator ∇ has properties similar to the ordinary derivative operator d/dt in that, if G_1 and G_2 are any two scalar functions, then

(1) $$\nabla(G_1 + G_2) = \nabla G_1 + \nabla G_2 \tag{3.4}$$

(2) $$\nabla(G_1 G_2) = G_1 \nabla G_2 + G_2 \nabla G_1 \tag{3.5}$$

which are direct generalizations of properties (2) and (3) for ordinary derivatives in Section 2.5.

3.2 The directional derivative

As well as being a convenient method for determining normals to surfaces (which, as we shall see later, are needed in integrating vectors over surfaces), the gradient operator also enables us to calculate the derivative of a scalar field along a particular curve – the so-called **directional derivative**. The directional derivative is of considerable importance in that it measures the rate of change of the scalar quantity of interest along a particular direction in space. For example, if we imagine standing on a hill, then for the scalar field

$$\Phi(x, y, z) = \text{height above sea level}$$

the value of $d\Phi/ds$ (the directional derivative) is a measure of the steepness of the hill in the direction in which s is measured.

To see how we can calculate the directional derivative, consider a scalar field $\Phi(x, y, z)$, so that the level surfaces of Φ are $\Phi(x, y, z) = k$ for each value of the constant k (see Figure 3.2). Consider a point P on the curve Γ and let Q be a point on the curve close to P. Let δs be the arc-length distance between P and Q. Suppose that P has coordinates (x, y, z) and Q has coordinates $(x + \delta x, y + \delta y, z + \delta z)$, so that

$$(\delta s)^2 = (\delta x)^2 + (\delta y)^2 + (\delta z)^2$$

Now the change in Φ between P and Q is

$$\begin{aligned}\delta\Phi &= \Phi(x + \delta x, y + \delta y, z + \delta z) - \Phi(x, y, z) \\ &\approx \frac{\partial\Phi}{\partial x}\delta x + \frac{\partial\Phi}{\partial y}\delta y + \frac{\partial\Phi}{\partial z}\delta z\end{aligned}$$

Dividing through by δs and then taking the limit as $\delta s \to 0$, we obtain

$$\begin{aligned}\frac{d\Phi}{ds} &= \frac{\partial\Phi}{\partial x}\frac{dx}{ds} + \frac{\partial\Phi}{\partial y}\frac{dy}{ds} + \frac{\partial\Phi}{\partial z}\frac{dz}{ds} \\ &= \left(\hat{\mathbf{i}}\frac{\partial\Phi}{\partial x} + \hat{\mathbf{j}}\frac{\partial\Phi}{\partial y} + \hat{\mathbf{k}}\frac{\partial\Phi}{\partial z}\right)\cdot\left(\hat{\mathbf{i}}\frac{dx}{ds} + \hat{\mathbf{j}}\frac{dy}{ds} + \hat{\mathbf{k}}\frac{dz}{ds}\right) \\ &= \nabla\Phi\cdot\hat{\mathbf{t}} \qquad \textbf{(3.6)}\end{aligned}$$

where $\hat{\mathbf{t}}$ is a unit tangent vector to Γ at P (see Section 2.7). Equation (3.6) tells us that the rate of change of a scalar field in the direction of the unit tangent vector $\hat{\mathbf{t}}$ (that is, along Γ) is given by the scalar product of $\nabla\Phi$ with

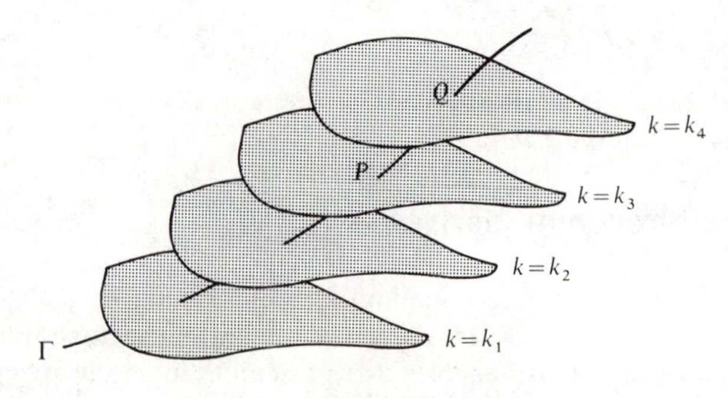

Figure 3.2

$\hat{\mathbf{t}}$. If $\hat{\mathbf{t}}$ lies in the level surface (if Γ is a curve within the surface at P), then we would expect $\mathrm{d}\Phi/\mathrm{d}s$ to be zero, since Φ is not changing within the level surface. This expectation is confirmed, since we know that $\nabla\Phi$ is normal to the surface $\Phi = \text{constant}$ and so, for this particular choice of $\hat{\mathbf{t}}$, $\nabla\Phi\cdot\hat{\mathbf{t}} = 0$.

It is of importance in applications to determine the maximum rate of change of Φ and also to find the direction in which this maximum occurs, the so-called steepest-descent direction. Now

$$\frac{\mathrm{d}\Phi}{\mathrm{d}s} = \nabla\Phi\cdot\hat{\mathbf{t}} = |\nabla\Phi|\,|\hat{\mathbf{t}}|\cos\theta \qquad \mathbf{(3.7)}$$

where θ is the angle between the vectors $\nabla\Phi$ and $\hat{\mathbf{t}}$ at the point of calculation of $\mathrm{d}\Phi/\mathrm{d}s$. Now at a particular point in space, $|\nabla\Phi|$ is fixed by Φ itself and $|\hat{\mathbf{t}}| = 1$ since $\hat{\mathbf{t}}$ is a unit vector. Hence the magnitude of $\mathrm{d}\Phi/\mathrm{d}s$ depends only on the value of $\cos\theta$. Clearly, $\mathrm{d}\Phi/\mathrm{d}s$ is a maximum when $\theta = 0$ or π, that is, when $\hat{\mathbf{t}}$ and $\nabla\Phi$ are aligned parallel or antiparallel in a direction normal to the surface $\Phi = \text{constant}$. Therefore, when $\mathrm{d}\Phi/\mathrm{d}s$ is a maximum,

$$\hat{\mathbf{t}} = \frac{\nabla\Phi}{|\nabla\Phi|} \qquad \mathbf{(3.8)}$$

$$\therefore \qquad \max\left|\frac{\mathrm{d}\Phi}{\mathrm{d}s}\right| = \frac{|\nabla\Phi\cdot\nabla\Phi|}{|\nabla\Phi|} = |\nabla\Phi| \qquad \mathbf{(3.9)}$$

(Alternatively, this can be obtained directly from (3.7) with $\cos\theta = +1$ or -1.)

As a specific example, suppose that $\Phi(x, y, z)$ represents the potential in an electrostatic field. A charge placed in that field will move in the direction of maximum $\mathrm{d}\Phi/\mathrm{d}s$, that is, normal to the equipotentials. Thus the **electric field lines** (lines on which charges move) will be everywhere perpendicular to the equipotentials. Figure 3.3 shows a simple configuration of charges; the arrowed lines are field lines and the broken lines are equipotentials.

Equation (3.7) tells us that the directional derivative along a direction from P specified by the unit vector $\hat{\mathbf{t}}$ is simply the component of $\nabla\Phi$ along that direction, while equation (3.9) tells us that the scalar field at P has a maximum rate of change of $|\nabla\Phi|$ and that this occurs if we travel from P along the direction of $\nabla\Phi$.

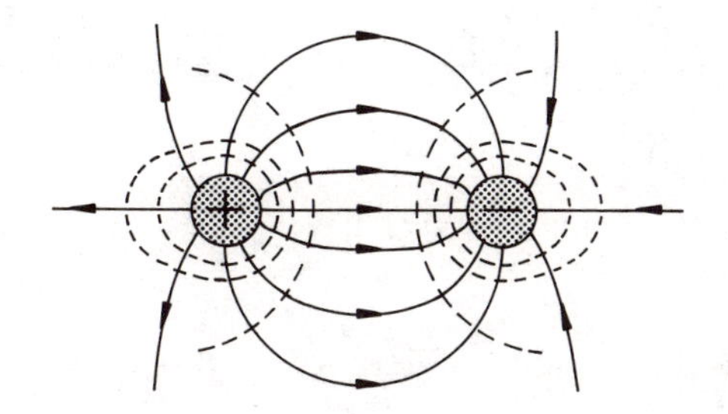

Figure 3.3

We see that a knowledge of $\nabla\Phi$ gives us all the information we need about the directional derivative at a point P in a scalar field. We note that, if $\mathrm{d}\Phi/\mathrm{d}n$ denotes the rate of change of Φ in the direction normal to a level surface $\Phi = \text{constant}$,

$$\frac{\mathrm{d}\Phi}{\mathrm{d}n} = \max\left|\frac{\mathrm{d}\Phi}{\mathrm{d}s}\right| = |\nabla\Phi|$$

so
$$\nabla\Phi = |\nabla\Phi|\hat{\mathbf{n}} = \frac{\mathrm{d}\Phi}{\mathrm{d}n}\hat{\mathbf{n}}$$

Example 3.2

Find the directional derivative of $\Phi = x^2y^2z^2 + x + y + z$ in the direction of the vector $2\hat{\mathbf{i}} - \hat{\mathbf{j}} + 2\hat{\mathbf{k}}$ at the point (1, 1, 1). In what direction does Φ change most quickly at this point?

Solution

Here
$$\begin{aligned}\nabla\Phi &= \nabla(x^2y^2z^2 + x + y + z)\\ &= (2xy^2z^2 + 1)\hat{\mathbf{i}} + (2yx^2z^2 + 1)\hat{\mathbf{j}} + (2zx^2y^2 + 1)\hat{\mathbf{k}}\\ &= 3\hat{\mathbf{i}} + 3\hat{\mathbf{j}} + 3\hat{\mathbf{k}} \quad \text{at } (1, 1, 1)\end{aligned}$$

A unit vector in the direction of $2\hat{\mathbf{i}} - \hat{\mathbf{j}} + 2\hat{\mathbf{k}}$ is

$$\hat{\mathbf{t}} = \tfrac{2}{3}\hat{\mathbf{i}} - \tfrac{1}{3}\hat{\mathbf{j}} + \tfrac{2}{3}\hat{\mathbf{k}}$$

Thus the required directional derivative is

$$\nabla\Phi\cdot\hat{\mathbf{t}} = (3\hat{\mathbf{i}} + 3\hat{\mathbf{j}} + 3\hat{\mathbf{k}})\cdot(\tfrac{2}{3}\hat{\mathbf{i}} - \tfrac{1}{3}\hat{\mathbf{j}} + \tfrac{2}{3}\hat{\mathbf{k}}) = 3$$

The positive value indicates that the scalar field Φ is *increasing* in the direction $2\hat{\mathbf{i}} - \hat{\mathbf{j}} + 2\hat{\mathbf{k}}$ at the point (1, 1, 1). The maximum value of $\mathrm{d}\Phi/\mathrm{d}s$ at this point is given by the value of $|\nabla\Phi|$, namely $\sqrt{27}$, and the direction at which this maximum rate of change would be observed is parallel to $\nabla\Phi$.

EXERCISES

3.1 Sketch the level surfaces of the following scalar fields and determine unit normals to these surfaces:

(a) $\dfrac{x^2}{1} + \dfrac{y^2}{2} + \dfrac{z^2}{9}$ (b) $z - \sqrt{(x^2 + y^2)}$

(c) $z - (x^2 + y^2)$ (d) $z^2 - (x^2 + y^2)$

3.2 Compute the gradient $\nabla\Phi$ at the point (1, 1, 1) for the scalar fields:

(a) $\Phi = xy + yz + xz$ (b) $\Phi = xyz \cos \pi y$

3.3 Determine the rate of change of the scalar field $\Phi = x + 2xy - 3z^2$ at the point P (1, 1, 2) in the direction of the vector $\mathbf{a} = 3\hat{\mathbf{i}} + 4\hat{\mathbf{j}} + \sqrt{24}\,\hat{\mathbf{k}}$. In what direction from P is the rate of change a maximum?

3.4 Let $\Phi(x, y) = -x^2 - \frac{1}{2}y^2 + 9$ denote the height on a mountain at position (x, y). In what direction from (1, 0) is the steepest descent? Use your knowledge of the gradient to determine the position of the top of the mountain.

3.5 If $\mathbf{a}$ is a constant vector, show that:

(a) $\nabla(\mathbf{a}\cdot\mathbf{r}) = \mathbf{a}$ (b) $\nabla(\mathbf{r}\cdot\mathbf{r}) = 2\mathbf{r}$ (c) $\nabla|\mathbf{r} - \mathbf{a}|^2 = 2(\mathbf{r} - \mathbf{a})$

3.6 The directional derivative of a function $\Phi(x, y, z)$ at the point (2, 0, 3) in the direction towards (3, −2, 3) is 1.789; in the direction towards (2, 4, 4) it is 0.243, whilst in the direction towards (4, −1, 2) it is zero. By showing that the three first-order partial derivatives of Φ are 2.0, −1.0 and 5.0 respectively, verify that the value of the directional derivative of Φ at (2, 0, 3) in the direction towards (0, 2, 14) ia 4.314.

3.7 For each of the following surfaces, determine the equations of the tangent plane and the normal line at the point indicated:

(a) $x^2 + y^2 + z^2 = 12$, (2, 2, −2) (b) $2x^2 + 3y^2 + z^2 = 9$, (1, 1, 2)

3.8 Prove the properties of the gradient operator stated in equations (3.4) and (3.5).

3.9 Show that the surfaces $x^2 + y^2 + z^2 = 9$ and $x^2 + y^2 - z = 3$ intersect in the circle $x^2 + y^2 = 5$ and also at the single point (0, 0, −3). Show that on the curve of intersection the surfaces intersect at an angle of 54° 25′.

3.10 (a) By defining the symbol $\mathbf{a}\cdot\nabla$ as the operator

$$a_x \frac{\partial}{\partial x} + a_y \frac{\partial}{\partial y} + a_z \frac{\partial}{\partial z}$$

show that if Φ is a scalar field then

$$(\mathbf{a}\cdot\nabla)\Phi \equiv \mathbf{a}\cdot\nabla\Phi$$

(b) If $\mathbf{b}$ is a vector field, give a meaning to $(\mathbf{a}\cdot\nabla)\mathbf{b}$.

3.3 Introduction to the vector analysis of strain (*optional*)

The gradient operator finds considerable application in the analysis of strain in an elastic body. Consider a body as shown in Figure 3.4. We focus attention on an elementary line AB within the body. Suppose that A has position

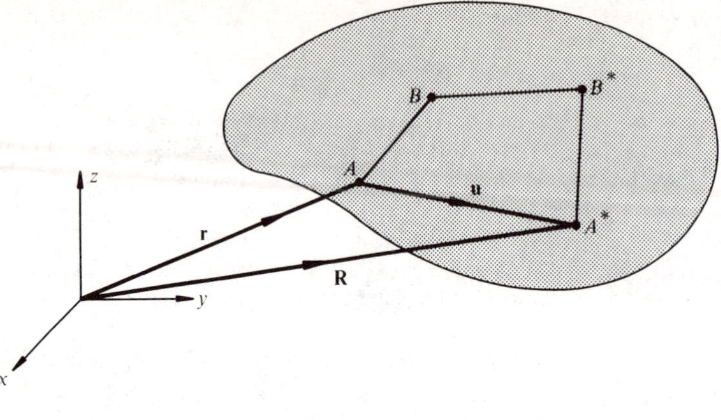

Figure 3.4

vector $\mathbf{r} = x\hat{\mathbf{i}} + y\hat{\mathbf{j}} + z\hat{\mathbf{k}}$ and that B has position vector $\mathbf{r} + \mathrm{d}\mathbf{r}$. The length of AB is then $\mathrm{d}s = |\mathrm{d}\mathbf{r}|$. When forces are applied to the body (surface forces on the boundary and possibly body forces, such as gravity, within the material) it deforms and the line AB moves, say to the line A^*B^*. For most materials these deformations are exceedingly small, even when the body is subjected to large forces, and the assumption that A^*B^* is still linear is a reasonable one. Suppose that A^* has position vector $\mathbf{R} = X\hat{\mathbf{i}} + Y\hat{\mathbf{j}} + Z\hat{\mathbf{k}}$ and that B^* has position vector $\mathbf{R} + \mathrm{d}\mathbf{R}$, so that the length of A^*B^* is $\mathrm{d}s^* = |\mathrm{d}\mathbf{R}|$.

There are many measures of strain that can be used to characterize the deformation within the body. The measure that is generally accepted is $(\mathrm{d}s^*)^2 - \mathrm{d}s^2$ because the more obvious measure $(\mathrm{d}s^* - \mathrm{d}s)$ involves the difference between two square roots and is thus, mathematically, more difficult to deal with. Clearly,

$$(\mathrm{d}s^*)^2 - \mathrm{d}s^2 = \mathrm{d}\mathbf{R}\cdot\mathrm{d}\mathbf{R} - \mathrm{d}\mathbf{r}\cdot\mathrm{d}\mathbf{r} \qquad \textbf{(3.10)}$$

It is not unreasonable to demand that for each particle of the body the final coordinates (X, Y, Z) are functionally related to the original coordinates (x, y, z), that is,

$$X = X(x, y, z), \qquad Y = Y(x, y, z), \qquad Z = Z(x, y, z)$$

We assume that the functions are sufficiently well-behaved for partial derivatives to be taken as required. Now

$$\begin{aligned}\mathrm{d}X &= \frac{\partial X}{\partial x}\,\mathrm{d}x + \frac{\partial X}{\partial y}\,\mathrm{d}y + \frac{\partial X}{\partial z}\,\mathrm{d}z \\ &= \mathrm{d}\mathbf{r}\cdot\nabla X\end{aligned}$$

where we have introduced the gradient operator. Similarly,

$$\mathrm{d}Y = \mathrm{d}\mathbf{r}\cdot\nabla Y \quad \text{and} \quad \mathrm{d}Z = \mathrm{d}\mathbf{r}\cdot\nabla Z$$

and so

$$\begin{aligned}
\mathrm{d}\mathbf{R} &= \mathrm{d}X\,\hat{\mathbf{i}} + \mathrm{d}Y\hat{\mathbf{j}} + \mathrm{d}Z\,\hat{\mathbf{k}} \\
&= \mathrm{d}x\frac{\partial}{\partial x}(\hat{\mathbf{i}}X) + \mathrm{d}y\frac{\partial}{\partial y}(\hat{\mathbf{i}}X) + \mathrm{d}z\frac{\partial}{\partial z}(\hat{\mathbf{i}}X) \\
&\qquad + \mathrm{d}x\frac{\partial}{\partial x}(\hat{\mathbf{j}}Y) + \mathrm{d}y\frac{\partial}{\partial y}(\hat{\mathbf{j}}Y) + \mathrm{d}z\frac{\partial}{\partial z}(\hat{\mathbf{j}}Y) \\
&\qquad + \mathrm{d}x\frac{\partial}{\partial x}(\hat{\mathbf{k}}Z) + \mathrm{d}y\frac{\partial}{\partial y}(\hat{\mathbf{k}}Z) + \mathrm{d}z\frac{\partial}{\partial z}(\hat{\mathbf{k}}Z) \\
&= \mathrm{d}x\frac{\partial}{\partial x}(\mathbf{R}) + \mathrm{d}y\frac{\partial}{\partial y}(\mathbf{R}) + \mathrm{d}z\frac{\partial}{\partial z}(\mathbf{R}) \\
&= (\mathrm{d}\mathbf{r}\cdot\nabla)\mathbf{R}
\end{aligned} \tag{3.11}$$

We have therefore, using (3.10),

$$(\mathrm{d}s^*)^2 - \mathrm{d}s^2 = [(\mathrm{d}\mathbf{r}\cdot\nabla)\mathbf{R}]\cdot[(\mathrm{d}\mathbf{r}\cdot\nabla)\mathbf{R}] - \mathrm{d}\mathbf{r}\cdot\mathrm{d}\mathbf{r} \tag{3.12}$$

This result applies to all continuous media, both solids and fluids. In solid mechanics, it is normal to express all quantities in terms of the original configuration. This is easily accomplished if we introduce the displacement vector $\mathbf{u}$,

$$\mathbf{u} = \mathbf{R} - \mathbf{r}$$

(see Figure 3.4). Hence

$$(\mathrm{d}s^*)^2 - \mathrm{d}s^2 = [(\mathrm{d}\mathbf{r}\cdot\nabla))(\mathbf{u}+\mathbf{r})]\cdot[(\mathrm{d}\mathbf{r}\cdot\nabla)(\mathbf{u}+\mathbf{r})] - \mathrm{d}\mathbf{r}\cdot\mathrm{d}\mathbf{r}$$

Now $$(\mathrm{d}\mathbf{r}\cdot\nabla)\mathbf{r} = \mathrm{d}x\frac{\partial}{\partial x}(\mathbf{r}) + \mathrm{d}y\frac{\partial}{\partial y}(\mathbf{r}) + \mathrm{d}z\frac{\partial}{\partial z}(\mathbf{r}) = \mathrm{d}\mathbf{r} \tag{3.13}$$

therefore

$$(\mathrm{d}s^*)^2 - \mathrm{d}s^2 = [(\mathrm{d}\mathbf{r}\cdot\nabla)\mathbf{u} + \mathrm{d}\mathbf{r}]\cdot[(\mathrm{d}\mathbf{r}\cdot\nabla)\mathbf{u} + \mathrm{d}\mathbf{r}] - \mathrm{d}\mathbf{r}\cdot\mathrm{d}\mathbf{r}$$

In the theory of *linear* elasticity, we assume that any squares of derivatives of $\mathbf{u}$ are negligible in comparison with linear terms and so finally we obtain

$$(\mathrm{d}s^*)^2 - \mathrm{d}s^2 \approx 2\,\mathrm{d}\mathbf{r}\cdot[(\mathrm{d}\mathbf{r}\cdot\nabla)\mathbf{u}] \tag{3.14}$$

The theory which is developed from this equation has been successfully used to analyse the effects of forces on most structural materials, such as metals and concrete. It cannot, however, be used to describe the larger deformation of highly elastic materials, such as rubber, where the non-linear terms neglected above have to be considered.

Example 3.3

Show that if the displacement vector $\mathbf{u}$ has the form

$$\mathbf{u} = \mathbf{D} + \boldsymbol{\omega} \wedge \mathbf{r}$$

where $\mathbf{D}$ and $\boldsymbol{\omega}$ are constant vectors, then the body undergoes a rigid body displacement.

Solution

We know that

$$(\mathrm{d}s^*)^2 - \mathrm{d}s^2 \approx 2\,\mathrm{d}\mathbf{r}\cdot[(\mathrm{d}\mathbf{r}\cdot\nabla)\mathbf{u}]$$

Now $$(\mathrm{d}\mathbf{r}\cdot\nabla)[\mathbf{D} + \boldsymbol{\omega} \wedge \mathbf{r}] = (\mathrm{d}\mathbf{r}\cdot\nabla)\mathbf{D} + (\mathrm{d}\mathbf{r}\cdot\nabla)(\boldsymbol{\omega} \wedge \mathbf{r})$$

$$= (\mathrm{d}\mathbf{r}\cdot\nabla)(\boldsymbol{\omega} \wedge \mathbf{r})$$

since $\mathbf{D}$ is a constant vector. Now

$$(\mathrm{d}\mathbf{r}\cdot\nabla)(\boldsymbol{\omega} \wedge \mathbf{r}) = \boldsymbol{\omega} \wedge [(\mathrm{d}\mathbf{r}\cdot\nabla)\mathbf{r}]$$

$$= \boldsymbol{\omega} \wedge \mathrm{d}\mathbf{r} \quad \text{from (3.13)}$$

Therefore

$$(\mathrm{d}s^*)^2 - \mathrm{d}s^2 = 2\,\mathrm{d}\mathbf{r}\cdot(\boldsymbol{\omega} \wedge \mathrm{d}\mathbf{r}) = 0$$

since the vectors $\mathrm{d}\mathbf{r}$ and $\boldsymbol{\omega} \wedge \mathrm{d}\mathbf{r}$ are perpendicular.

Hence the final length A^*B^* of each elementary line of particles is the same as the initial length AB; that is, this displacement field corresponds to a rigid body motion. The constant vector $\mathbf{D}$ is a translation of the body as a whole, and $\boldsymbol{\omega} \wedge \mathbf{r}$ is a rotation of the body through an angle $|\boldsymbol{\omega}|$ (in radians).

It is part of the aim of elasticity theory to determine the displacement vector $\mathbf{u}$ for any system of loads applied to a body. To help us in this task we can expand the terms on the right-hand side of (3.14): if we put $\mathbf{u} = u\hat{\mathbf{i}} + v\hat{\mathbf{j}} + w\hat{\mathbf{k}}$, we obtain

$$(\mathrm{d}\mathbf{r}\cdot\nabla)\mathbf{u} = \mathrm{d}x\,\frac{\partial}{\partial x}(\mathbf{u}) + \mathrm{d}y\,\frac{\partial}{\partial y}(\mathbf{u}) + \mathrm{d}z\,\frac{\partial}{\partial z}(\mathbf{u})$$

$$= \hat{\mathbf{i}}\left(\mathrm{d}x\,\frac{\partial u}{\partial x} + \mathrm{d}y\,\frac{\partial u}{\partial y} + \mathrm{d}z\,\frac{\partial u}{\partial z}\right) + \hat{\mathbf{j}}\left(\mathrm{d}x\,\frac{\partial v}{\partial x} + \mathrm{d}y\,\frac{\partial v}{\partial y} + \mathrm{d}z\,\frac{\partial v}{\partial z}\right)$$

$$+ \hat{\mathbf{k}}\left(\mathrm{d}x\,\frac{\partial w}{\partial x} + \mathrm{d}y\,\frac{\partial w}{\partial y} + \mathrm{d}z\,\frac{\partial w}{\partial z}\right)$$

$$\therefore \quad \mathrm{d}\mathbf{r}\cdot[(\mathrm{d}\mathbf{r}\cdot\nabla)\mathbf{u}] = \mathrm{d}x\,\mathrm{d}x\left(\frac{\partial u}{\partial x}\right) + \mathrm{d}x\,\mathrm{d}y\left(\frac{\partial u}{\partial y} + \frac{\partial v}{\partial x}\right) + \mathrm{d}x\,\mathrm{d}z\left(\frac{\partial u}{\partial z} + \frac{\partial w}{\partial x}\right)$$

$$+ \mathrm{d}y\,\mathrm{d}y\left(\frac{\partial v}{\partial y}\right) + \mathrm{d}y\,\mathrm{d}z\left(\frac{\partial v}{\partial z} + \frac{\partial w}{\partial y}\right) + \mathrm{d}z\,\mathrm{d}z\left(\frac{\partial w}{\partial z}\right)$$

The various partial derivatives are grouped together to form three vectors:

$$\boldsymbol{\varepsilon}_x = \left(\frac{\partial u}{\partial x}, \tfrac{1}{2}\left(\frac{\partial u}{\partial y} + \frac{\partial v}{\partial x}\right), \tfrac{1}{2}\left(\frac{\partial u}{\partial z} + \frac{\partial w}{\partial x}\right)\right)$$

$$\boldsymbol{\varepsilon}_y = \left(\tfrac{1}{2}\left(\frac{\partial u}{\partial y} + \frac{\partial v}{\partial x}\right), \frac{\partial v}{\partial y}, \tfrac{1}{2}\left(\frac{\partial v}{\partial z} + \frac{\partial w}{\partial y}\right)\right)$$

$$\boldsymbol{\varepsilon}_z = \left(\tfrac{1}{2}\left(\frac{\partial u}{\partial z} + \frac{\partial w}{\partial x}\right), \tfrac{1}{2}\left(\frac{\partial v}{\partial z} + \frac{\partial w}{\partial y}\right), \frac{\partial w}{\partial z}\right) \qquad \textbf{(3.15)}$$

The component of these three vectors, $(\varepsilon_{xx}, \varepsilon_{xy}, \varepsilon_{xz})$, $(\varepsilon_{yx}, \varepsilon_{yy}, \varepsilon_{yz})$ and $(\varepsilon_{zx}, \varepsilon_{zy}, \varepsilon_{zz})$, respectively, are not all independent; they satisfy, by inspection,

$$\varepsilon_{xy} = \varepsilon_{yx}, \qquad \varepsilon_{xz} = \varepsilon_{zx}, \qquad \varepsilon_{yz} = \varepsilon_{zy} \qquad \textbf{(3.16)}$$

so that only six components are required to define the deformation fully.

If we have knowledge of these three vectors, then precise information about the displacement vector **u** (and, as it turns out, the state of stress in the body) may be obtained, albeit after solving some partial differential equations.

Example 3.4

In the torsion problem of a circular cylinder that is rigidly fixed at (0, 0, 0), first outlined in Example 1.13, it may be shown that the displacement vector is $\mathbf{u} = (-\tau yz, \tau xz, 0)$, where τ is a constant. Determine the strain vectors and show that plane sections originally perpendicular to the axis of the cylinder suffer no distortion apart from a rotation about the axis.

Solution

Here $u = -\tau yz, \qquad v = \tau xz, \qquad w = 0$

and so, from (3.15), the required strain vectors are

$$\boldsymbol{\varepsilon}_x = (0, 0, -\tfrac{1}{2}\tau y), \qquad \boldsymbol{\varepsilon}_y = (0, 0, \tfrac{1}{2}\tau x), \qquad \boldsymbol{\varepsilon}_z = (-\tfrac{1}{2}\tau y, \tfrac{1}{2}\tau x, 0)$$

Note that these strain vectors are directly proportional to the stresses given in Example 1.13, $\boldsymbol{\sigma}_x$, $\boldsymbol{\sigma}_y$ and $\boldsymbol{\sigma}_z$. But we know that **u** is the change in the position vector of a particle, that is,

$$\mathbf{u} = \mathbf{R} - \mathbf{r}$$

Hence, if we consider a particle on a plane perpendicular to the z-axis, then its change in position is

$$\mathbf{R} - \mathbf{r} = -\tau yz\hat{\mathbf{i}} + \tau xz\hat{\mathbf{j}}$$

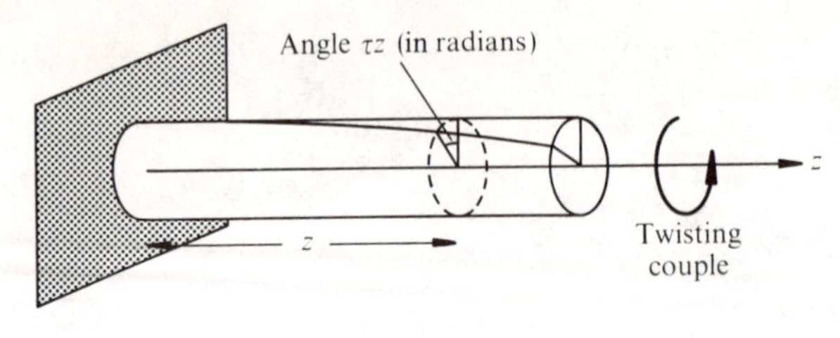

Figure 3.5

But the right-hand side here is of the form $\mathbf{D} + \boldsymbol{\omega} \wedge \mathbf{r}$ where $\mathbf{D} = (0, 0, 0)$ and $\boldsymbol{\omega} = \tau z\hat{\mathbf{k}}$, so that the plane is not translated but simply rotated through an angle τz (see Figure 3.5).

We should note here that cylinders with cross-sections other than circular suffer from **warping** – planes perpendicular to the axis of the beam do not remain planar after deformation (see Supplement 3S.2).

Physical interpretation of the strain vectors

It is a relatively straightforward matter to give a physical interpretation to each component of the strain vectors $\boldsymbol{\varepsilon}_x$, $\boldsymbol{\varepsilon}_y$ and $\boldsymbol{\varepsilon}_z$. We shall first of all consider the interpretation of the so-called **longitudinal** strain components, ε_{xx}, ε_{yy} and ε_{zz}.

Consider an elementary line of particles directed along the x-axis. Then

$$\mathbf{dr} = (\mathrm{d}s, 0, 0)$$

and so, from (3.14),

$$\begin{aligned}(\mathrm{d}s^*)^2 - \mathrm{d}s^2 &= 2\,\mathrm{d}s\,\hat{\mathbf{i}}\cdot\left[\mathrm{d}s\,\frac{\partial}{\partial x}(\mathbf{u})\right] \\ &= 2\,\mathrm{d}s^2\,\hat{\mathbf{i}}\cdot\left[\frac{\partial}{\partial x}(u\hat{\mathbf{i}} + v\hat{\mathbf{j}} + w\hat{\mathbf{k}})\right] \\ &= 2\,\mathrm{d}s^2\,\frac{\partial u}{\partial x} \\ &= 2\,\mathrm{d}s^2\,\varepsilon_{xx}\end{aligned}$$

Therefore

$$\varepsilon_{xx} = \frac{(\mathrm{d}s^*)^2 - \mathrm{d}s^2}{2\,\mathrm{d}s^2} = \frac{(\mathrm{d}s^* - \mathrm{d}s)(\mathrm{d}s^* + \mathrm{d}s)}{2\,\mathrm{d}s^2}$$

But to a very good approximation $ds^* \approx ds$ (only the *difference* between these two small quantities is of significance), and so

$$\varepsilon_{xx} \approx \frac{ds^* - ds}{ds}$$

Hence we can interpret ε_{xx} as the change in length per unit length of a line of particles originally along the x-direction. Note that if $\varepsilon_{xx} > 0$ at a point, this indicates that the region in the close neighbourhood of the point will be stretched, and if $\varepsilon_{xx} < 0$ at a point, then compression is indicated.

Similar interpretations follow for the components ε_{yy} and ε_{zz}.

The physical interpretation of the other strain components – the so-called **shear** components of strain, ε_{xy}, ε_{xz} and ε_{yz}, is less straightforward. Consider two elementary lines of particles which, before deformation, are perpendicular and aligned along the x- and y-axes respectively, so that

$$d\mathbf{r} = (ds, 0, 0) \qquad \text{and} \qquad \delta\mathbf{r} = (0, \delta s, 0)$$

(the notation accounts for the possibility that the elements may be of different lengths). After deformation, they are represented by the vectors $d\mathbf{R}$ and $\delta\mathbf{R}$ respectively, such that $ds^* = |d\mathbf{R}|$ and $\delta s^* = |\delta\mathbf{R}|$, so that the angle between the two lines is θ^*, where

$$\cos\theta^* = \frac{d\mathbf{R}}{ds^*} \cdot \frac{\delta\mathbf{R}}{\delta s^*}$$

(this angle being $\pi/2$ before deformation, of course). But, from (3.11),

$$d\mathbf{R} = (d\mathbf{r} \cdot \nabla)\mathbf{R} \quad \text{and similarly} \quad \delta\mathbf{R} = (\delta\mathbf{r} \cdot \nabla)\mathbf{R}$$

and in this particular case, with $d\mathbf{r} = ds\,\hat{\mathbf{i}}$ and $\delta\mathbf{r} = \delta s\,\hat{\mathbf{j}}$, we have

$$d\mathbf{R} = ds\frac{\partial}{\partial x}(\mathbf{R}) \qquad \text{and} \qquad \delta\mathbf{R} = \delta s\frac{\partial}{\partial y}(\mathbf{R})$$

or

$$d\mathbf{R} = ds\frac{\partial}{\partial x}(\mathbf{u} + \mathbf{r}) \quad \text{and} \quad \delta\mathbf{R} = \delta s\frac{\partial}{\partial y}(\mathbf{u} + \mathbf{r})$$

Therefore

$$d\mathbf{R} = ds\left[\hat{\mathbf{i}}\left(\frac{\partial u}{\partial x} + 1\right) + \hat{\mathbf{j}}\frac{\partial v}{\partial x} + \hat{\mathbf{k}}\frac{\partial w}{\partial x}\right]$$

and

$$\delta\mathbf{R} = \delta s\left[\hat{\mathbf{i}}\frac{\partial u}{\partial y} + \hat{\mathbf{j}}\left(\frac{\partial v}{\partial y} + 1\right) + \hat{\mathbf{k}}\frac{\partial w}{\partial y}\right]$$

Hence

$$\cos\theta^* = \frac{ds}{ds^*}\frac{\delta s}{\delta s^*}\left[\left(\frac{\partial u}{\partial x} + 1\right)\frac{\partial u}{\partial y} + \frac{\partial v}{\partial x}\left(\frac{\partial v}{\partial y} + 1\right) + \frac{\partial w}{\partial x}\frac{\partial w}{\partial y}\right]$$

$$\approx \frac{ds}{ds^*}\frac{\delta s}{\delta s^*}\left[\frac{\partial u}{\partial y} + \frac{\partial v}{\partial x}\right]$$

where we have neglected products of displacement derivatives compared to linear terms. If we take $ds \approx ds^*$ and $\delta s^* \approx \delta s$,

$$\cos\theta^* \approx \left(\frac{\partial u}{\partial y} + \frac{\partial v}{\partial x}\right) = 2\varepsilon_{xy}$$

As a further approximation, if we put

$$\theta^* = \frac{\pi}{2} - \phi^*$$

and note that for small deformations θ^* will be very close to the original value of $\pi/2$ (see Figure 3.6), then

$$\cos\theta^* = \cos\left(\frac{\pi}{2} - \phi^*\right) = \sin\phi^* \approx \phi^*$$

Hence $\varepsilon_{xy} \approx \dfrac{\phi^*}{2}$

so that ε_{xy} directly represents shear at a point. If $\varepsilon_{xy} = 0$ at a point, then elements along the x- and y-axes originally perpendicular will remain perpendicular after the deformation.

Similar interpretations may be deduced for the components ε_{xz} and ε_{yz}.

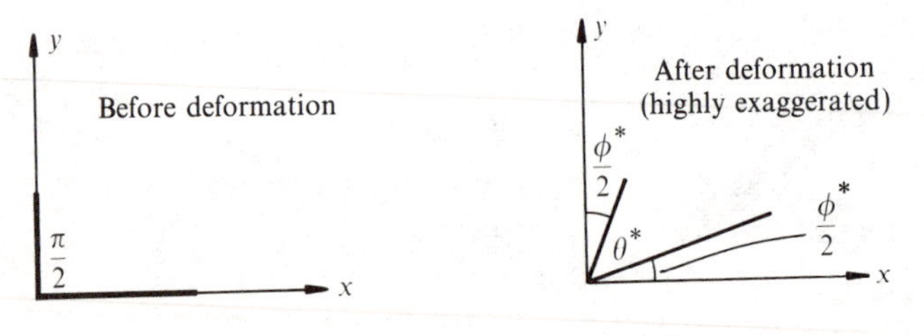

Figure 3.6

3.4 Vector fields

When the aerodynamic properties of a new car design are being tested, the outer surface 'skin' of the car is covered by a large number of small lengths of ribbon and the complete entity placed in a wind tunnel. The effect of the wind orients the ribbons in various directions. The resulting 'wind map' is shown in Figure 3.7. This is an example of a **direction field**.

If the ribbons were slightly more sophisticated (see Figure 3.8a), such that they could extend in length as well as change direction in the wind

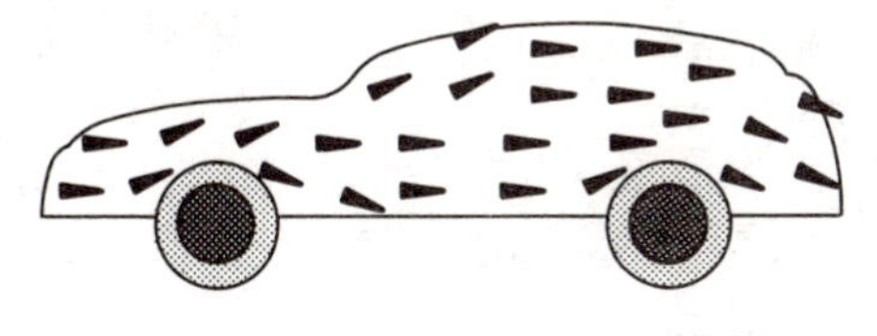

Figure 3.7

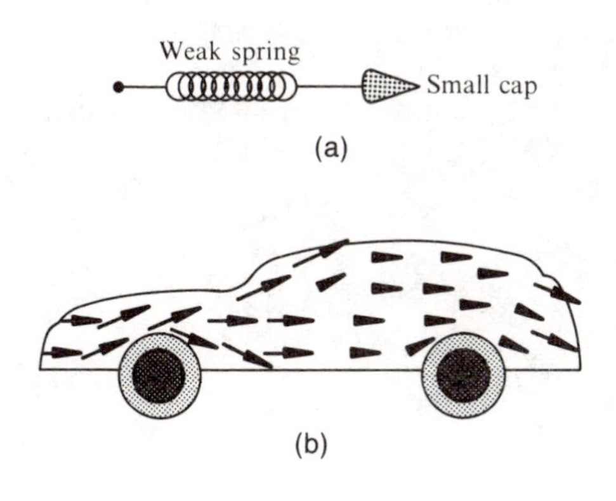

(a)

(b)

Figure 3.8

tunnel, then the picture of Figure 3.8b might be obtained. At each point of the car the ribbon defines a vector – the length of the ribbon is a measure of the wind speed whilst the direction of the ribbon records the direction of the air flow at the point. This collection of vectors (Figure 3.8b) is an example of a **vector field**.

Formally, a vector field is a rule which associates a vector with each point in a region of space. In mathematical terms, a vector field is a vector whose components are functions of the variables x, y and z. (Here we are specifically considering three-dimensional space. In some cases we may want to consider two-dimensional or even four-dimensional space, in which case the components of the vector will be functions of two variables (x and y) or of four variables (x, y, z and t) respectively. Examples of the four-dimensional case will be encountered in Chapter 7 on electromagnetism.)

Vector fields are often denoted by $\mathbf{A}(\mathbf{r})$ or $\mathbf{A}(x, y, z)$:

$$\mathbf{A}(x, y, z) = A_x(x, y, z)\hat{\mathbf{i}} + A_y(x, y, z)\hat{\mathbf{j}} + A_z(x, y, z)\hat{\mathbf{k}} \quad \textbf{(3.17)}$$

One way to obtain a vector field is to take the gradient of a scalar field (although vector fields are not necessarily always gradients of a scalar field). Electric and gravitational fields are common examples of vector fields. For example, the 'source' of the gravitational field is mass. The mere existence of mass creates a gravitational force field (although this field is extremely weak unless the mass is very great). If coordinates are chosen so that the mass is positioned at the origin, then this vector field (like many others) has the inverse square law form:

$$\mathbf{A} = -\frac{\hat{\mathbf{r}}}{|\mathbf{r}|^2} \tag{3.18}$$

where $\mathbf{r} = x\hat{\mathbf{i}} + y\hat{\mathbf{j}} + z\hat{\mathbf{k}}$ and $\hat{\mathbf{r}}$ is the unit vector in the radial direction pointing away from the origin. The negative sign indicates that the force is directed towards the origin, because gravitational forces are attractive. In fact, in this case $\mathbf{A}$ may be shown to be the gradient of the scalar field $1/|\mathbf{r}|$, that is,

$$\mathbf{A} = \nabla\left(\frac{1}{|\mathbf{r}|}\right) \tag{3.19}$$

This is easily verified:

$$\frac{1}{|\mathbf{r}|} = \frac{1}{\sqrt{(x^2+y^2+z^2)}} \quad \text{so} \quad \frac{\partial}{\partial x}\left(\frac{1}{|\mathbf{r}|}\right) = \frac{-x}{(x^2+y^2+z^2)^{3/2}} = \frac{-x}{|\mathbf{r}|^3}$$

Similarly,

$$\frac{\partial}{\partial y}\left(\frac{1}{|\mathbf{r}|}\right) = -\frac{y}{|\mathbf{r}|^3} \quad \text{and} \quad \frac{\partial}{\partial z}\left(\frac{1}{|\mathbf{r}|}\right) = -\frac{z}{|\mathbf{r}|^3}$$

Therefore

$$\nabla\left(\frac{1}{|\mathbf{r}|}\right) = \frac{-x\hat{\mathbf{i}} - y\hat{\mathbf{j}} - z\hat{\mathbf{k}}}{|\mathbf{r}|^3} = -\frac{\mathbf{r}}{|\mathbf{r}|^3} = -\frac{\hat{\mathbf{r}}}{|\mathbf{r}|^2} = \mathbf{A}$$

The scalar $1/|\mathbf{r}|$ is called the **gravitational potential**.

EXERCISES

3.11 Show that if $\mathbf{a}$ is any vector, then $(\mathbf{a}\cdot\nabla)\mathbf{r} = \mathbf{a}$.

3.12 If $\nabla\Phi = 0$ at all points within a region, show that Φ is constant in that region.

3.13 Evaluate the gradient of the following scalar fields ($\mathbf{a}$ is a constant vector):

(a) $\dfrac{1}{|\mathbf{r}-\mathbf{a}|}$ (b) $\dfrac{\mathbf{a}\cdot\mathbf{r}}{r^3}$

3.14 (a) If Φ and ϑ are scalar fields, show that

$$\nabla\left(\frac{\Phi}{\vartheta}\right) = \frac{\vartheta\nabla\Phi - \Phi\nabla\vartheta}{\vartheta^2}$$

if $\vartheta \neq 0$ at the point at which the gradient is taken.

(b) If $\Phi = x^2z + \sin y$ and $\vartheta = xyz$, find the following at the point $(1, \pi/2, -1)$:

(i) $\nabla(\Phi + \vartheta)$ (ii) $\nabla(\Phi\vartheta)$ (iii) $\nabla(\Phi/\vartheta)$

3.15 Sketch the following vector fields:

(a) $\mathbf{A} = x\hat{\mathbf{i}} + y\hat{\mathbf{j}} + z\hat{\mathbf{k}}$ (b) $\mathbf{B} = y\hat{\mathbf{i}} - x\hat{\mathbf{j}}$ (c) $\mathbf{C} = \dfrac{\mathbf{r}}{|\mathbf{r}|^3}$

(d) $\mathbf{D} = \dfrac{y}{\rho^2}\hat{\mathbf{i}} - \dfrac{x}{\rho^2}\hat{\mathbf{j}}$, where $\rho = \sqrt{(x^2 + y^2)}$

3.16 A strain gauge is constructed from a thin wire filament (about 0.0001 cm in diameter) wound so as to orient the major part of the length of the wire in a single direction (see Figure 3.9a). The wire is bonded to thin plastic and the plastic glued to a body at a point of interest. The body strains are transmitted to the gauge, resulting in a change of length of the thin wire and an attendant change in its electrical resistance. This change in resistance can be measured electrically and the value of the strain (longitudinal strain) in the direction of the gauge inferred. If the state of strain at a point is required, then a selection of strain gauges must be used. These clusters of gauges are called ‘rosettes’.

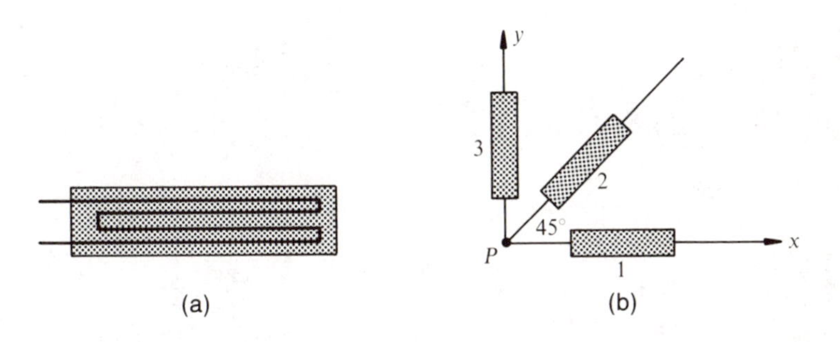

Figure 3.9

Consider a rectangular rosette as shown in Figure 3.9b. The following strain measurements are obtained:

gauge 1 0.002, gauge 2 0.001, gauge 3 −0.004

Show that the shear strain ε_{xy} at the point P is 0.002.

3.17 A small element of magnitude ds is originally in the direction of the unit vector $\hat{\mathbf{l}} = (l_x, l_y, l_z)$. Show that as a result of a deformation defined by the strain vectors $\boldsymbol{\varepsilon}_x$, $\boldsymbol{\varepsilon}_y$ and $\boldsymbol{\varepsilon}_z$, the length becomes

$$ds^* = [1 + (\boldsymbol{\varepsilon}_x \cdot \hat{\mathbf{l}})l_x + (\boldsymbol{\varepsilon}_y \cdot \hat{\mathbf{l}})l_y + (\boldsymbol{\varepsilon}_z \cdot \hat{\mathbf{l}})l_z]\, ds$$

3.5 The divergence and curl of a vector field

We have seen in Chapter 1 that three possible types of product are defined in connection with vectors:

(1) multiplication by a scalar, $\alpha\mathbf{a}$,
(2) scalar or dot product between vectors, $\mathbf{a}\cdot\mathbf{b}$,
(3) vector or cross product between vectors, $\mathbf{a}\wedge\mathbf{b}$.

If we consider ∇ as a vector (even though strictly it is a vector differential operator), then we might also consider three types of product involving ∇ (remembering that order is important):

(1) Multiplication by a scalar field, ∇G. This quantity has already been discussed. It is a vector quantity called the **gradient** of G.

The gradient of a scalar field is a vector field.

(2) Scalar product between ∇ and a vector field, $\nabla\cdot\mathbf{F}$. This is called the **divergence** of $\mathbf{F}$ and is explicitly

$$\left(\hat{\mathbf{i}}\frac{\partial}{\partial x}+\hat{\mathbf{j}}\frac{\partial}{\partial y}+\hat{\mathbf{k}}\frac{\partial}{\partial z}\right)\cdot(F_x\hat{\mathbf{i}}+F_y\hat{\mathbf{j}}+F_z\hat{\mathbf{k}})$$

That is,

$$\nabla\cdot\mathbf{F}=\frac{\partial F_x}{\partial x}+\frac{\partial F_y}{\partial y}+\frac{\partial F_z}{\partial z} \tag{3.20}$$

(using results from Section 1.5 for calculating dot products). Another very common notation for the divergence of a vector field is div $\mathbf{F}$.

The divergence of a vector field is a scalar field.

Note that $\nabla\cdot\mathbf{F}\neq\mathbf{F}\cdot\nabla$ (∇ is not a true vector).

(3) Vector product between ∇ and a vector field, $\nabla\wedge\mathbf{F}$. This is called the **curl** of $\mathbf{F}$ (often written **curl F** or $\nabla\times\mathbf{F}$) and is given by

$$\begin{aligned}&\left(\hat{\mathbf{i}}\frac{\partial}{\partial x}+\hat{\mathbf{j}}\frac{\partial}{\partial y}+\hat{\mathbf{k}}\frac{\partial}{\partial z}\right)\wedge(F_x\hat{\mathbf{i}}+F_y\hat{\mathbf{j}}+F_z\hat{\mathbf{k}})\\&=\hat{\mathbf{i}}\left(\frac{\partial F_z}{\partial y}-\frac{\partial F_y}{\partial z}\right)+\hat{\mathbf{j}}\left(\frac{\partial F_x}{\partial z}-\frac{\partial F_z}{\partial x}\right)+\hat{\mathbf{k}}\left(\frac{\partial F_y}{\partial x}-\frac{\partial F_x}{\partial y}\right)\end{aligned}$$

(using results from Section 1.6 for calculating cross products). In determinant form,

$$\nabla \wedge \mathbf{F} = \begin{vmatrix} \hat{\mathbf{i}} & \hat{\mathbf{j}} & \hat{\mathbf{k}} \\ \dfrac{\partial}{\partial x} & \dfrac{\partial}{\partial y} & \dfrac{\partial}{\partial z} \\ F_x & F_y & F_z \end{vmatrix} \tag{3.21}$$

The curl of a vector field is a vector field.

Example 3.5

Determine the divergence and curl of

(a) $\mathbf{F} = x^2\hat{\mathbf{i}} + y^2\hat{\mathbf{j}} + 3zx\hat{\mathbf{k}}$,

(b) the gravitational field $\mathbf{A} = -\hat{\mathbf{r}}/|\mathbf{r}|^2$, where $\hat{\mathbf{r}}$ is a unit radial vector.

Solution

(a)
$$\nabla \cdot \mathbf{F} = \frac{\partial F_x}{\partial x} + \frac{\partial F_y}{\partial y} + \frac{\partial F_z}{\partial z}$$

$$= \frac{\partial}{\partial x}(x^2) + \frac{\partial}{\partial y}(y^2) + \frac{\partial}{\partial z}(3zx)$$

$$= 2x + 2y + 3x = 5x + 2y$$

which is clearly a scalar field.

$$\nabla \wedge \mathbf{A} = \begin{vmatrix} \hat{\mathbf{i}} & \hat{\mathbf{j}} & \hat{\mathbf{k}} \\ \dfrac{\partial}{\partial x} & \dfrac{\partial}{\partial y} & \dfrac{\partial}{\partial z} \\ x^2 & y^2 & 3zx \end{vmatrix}$$

$$= \hat{\mathbf{i}}(0 - 0) - \hat{\mathbf{j}}(3z - 0) + \hat{\mathbf{k}}(0 - 0) = -3z\hat{\mathbf{j}}$$

which is clearly a vector field, each vector of which points in the y-direction.

(b) $\mathbf{A} = -\hat{\mathbf{r}}/|\mathbf{r}|^2$, where $\mathbf{r} = x\hat{\mathbf{i}} + y\hat{\mathbf{j}} + z\hat{\mathbf{k}}$. Now $\hat{\mathbf{r}} = \mathbf{r}/|\mathbf{r}|$ and $|\mathbf{r}| = \sqrt{(x^2 + y^2 + z^2)}$, therefore

$$\mathbf{A} = -\frac{\mathbf{r}}{|\mathbf{r}|^3} = \frac{-x\hat{\mathbf{i}} - y\hat{\mathbf{j}} - z\hat{\mathbf{k}}}{(x^2 + y^2 + z^2)^{3/2}}$$

Hence
$$\frac{\partial A_x}{\partial x} = -(x^2 + y^2 + z^2)^{-3/2} + \tfrac{3}{2}(x)(2x)(x^2 + y^2 + z^2)^{-5/2}$$

$$= \frac{(2x^2 - y^2 - z^2)}{(x^2 + y^2 + z^2)^{5/2}}$$

In this example, the vector field **A** is symmetric in the variables x, y and z and so we can write immediately

$$\frac{\partial A_y}{\partial y}=\frac{(-x^2+2y^2-z^2)}{(x^2+y^2+z^2)^{5/2}}, \qquad \frac{\partial A_z}{\partial z}=\frac{(-x^2-y^2+2z^2)}{(x^2+y^2+z^2)^{5/2}}$$

giving

$$\nabla\cdot\mathbf{A}=\frac{\partial A_x}{\partial x}+\frac{\partial A_y}{\partial y}+\frac{\partial A_z}{\partial z}=0$$

for all points (x, y, z) (except possibly at the origin, where $x=y=z=0$). Also,

$$\begin{aligned}\nabla\wedge\mathbf{A}&=\begin{vmatrix}\hat{\mathbf{i}} & \hat{\mathbf{j}} & \hat{\mathbf{k}}\\ \dfrac{\partial}{\partial x} & \dfrac{\partial}{\partial y} & \dfrac{\partial}{\partial z}\\ A_x & A_y & A_z\end{vmatrix}\\ &=\hat{\mathbf{i}}(-3zy-3zy)(x^2+y^2+z^2)^{-5/2}\\ &\quad-\hat{\mathbf{j}}(3xz-3zx)(x^2+y^2+z^2)^{-5/2}\\ &\quad+\hat{\mathbf{k}}(3xy-3xy)(x^2+y^2+z^2)^{-5/2}\\ &=0\end{aligned}$$

that is, a vector each of whose components is zero.

The field $\mathbf{A}=-\hat{\mathbf{r}}/|\mathbf{r}|^2$ here is clearly a very special sort of vector field, with its identically zero curl and its zero divergence except at the origin. As we shall see in Chapter 4, the former property is the more important, and the term **conservative** is used to describe such a field. A vector field whose divergence is zero in a region is said to be **solenoidal** in that region.

3.6 Properties of gradient, divergence and curl

If $G(x, y, z)$ is any scalar field and $\mathbf{A}(x, y, z)$ and $\mathbf{B}(x, y, z)$ are any two vector fields, then

(1a) $\nabla\cdot(\mathbf{A}+\mathbf{B})=\nabla\cdot\mathbf{A}+\nabla\cdot\mathbf{B}$

(1b) $\nabla\wedge(\mathbf{A}+\mathbf{B})=\nabla\wedge\mathbf{A}+\nabla\wedge\mathbf{B}$

(2a) $\nabla\cdot(G\mathbf{A})=(\nabla G)\cdot\mathbf{A}+G(\nabla\cdot\mathbf{A})$

(2b) $\nabla\wedge(G\mathbf{A})=(\nabla G)\wedge\mathbf{A}+G(\nabla\wedge\mathbf{A})$

(2c) $\nabla\cdot(\mathbf{A}\wedge\mathbf{B})=\mathbf{B}\cdot(\nabla\wedge\mathbf{A})-\mathbf{A}\cdot(\nabla\wedge\mathbf{B})$

These are the 'obvious' properties to be expected of any derivative operator and compare directly with the formulae developed in Section 2.5

and the properties of the gradient given in equations (3.4) and (3.5). Note that the order of the vectors $(\nabla G) \wedge \mathbf{A}$ is important in (2b).

Other properties of note are the identities:

(3) $\nabla \cdot (\nabla \wedge \mathbf{A}) = 0$, that is, div **curl A** is zero for any vector field **A**.

(4) $\nabla \wedge (\nabla G) = 0$, that is, **curl grad** G is zero for any scalar field G.

The proofs of all of these properties are quite straightforward if one works directly from the definitions. For example, to prove property (4) we note that

$$\nabla G = \hat{\mathbf{i}}\frac{\partial G}{\partial x} + \hat{\mathbf{j}}\frac{\partial G}{\partial y} + \hat{\mathbf{k}}\frac{\partial G}{\partial z}$$

$$\therefore \quad \nabla \wedge (\nabla G) = \begin{vmatrix} \hat{\mathbf{i}} & \hat{\mathbf{j}} & \hat{\mathbf{k}} \\ \dfrac{\partial}{\partial x} & \dfrac{\partial}{\partial y} & \dfrac{\partial}{\partial z} \\ \dfrac{\partial G}{\partial x} & \dfrac{\partial G}{\partial y} & \dfrac{\partial G}{\partial z} \end{vmatrix}$$

$$= \hat{\mathbf{i}}\left(\frac{\partial^2 G}{\partial y\,\partial z} - \frac{\partial^2 G}{\partial z\,\partial y}\right) - \hat{\mathbf{j}}\left(\frac{\partial^2 G}{\partial x\,\partial z} - \frac{\partial^2 G}{\partial z\,\partial x}\right) + \hat{\mathbf{k}}\left(\frac{\partial^2 G}{\partial x\,\partial y} - \frac{\partial^2 G}{\partial y\,\partial x}\right)$$

$$= 0 \quad \text{(zero vector)}$$

owing to the equality of the mixed derivatives. (For equality of mixed second derivatives of G at a point, it is required that the second partial derivatives of G exist and are continuous in a neighbourhood of the point.)

If a vector field **A** has zero curl, that is $\nabla \wedge \mathbf{A} = 0$ in some region of space, then (as one might suspect from property (4)), **A** may be expressed as the gradient of some scalar field. We shall see that vector fields with these properties are of particular importance in connection with line integrals.

The divergence of the vector field obtained by taking the gradient of a scalar field Φ is of considerable importance in applications. It is a scalar quantity called the **Laplacian** of Φ and is denoted by $\nabla^2\Phi$:

$$\nabla \cdot (\nabla \Phi) \equiv \nabla^2 \Phi = \operatorname{div}(\mathbf{grad}\ \Phi) \tag{3.22}$$

Written out explicitly, we have

$$\nabla^2\Phi = \nabla \cdot \left(\hat{\mathbf{i}}\frac{\partial \Phi}{\partial x} + \hat{\mathbf{j}}\frac{\partial \Phi}{\partial y} + \hat{\mathbf{k}}\frac{\partial \Phi}{\partial z}\right)$$

$$= \frac{\partial}{\partial x}\left(\frac{\partial \Phi}{\partial x}\right) + \frac{\partial}{\partial y}\left(\frac{\partial \Phi}{\partial y}\right) + \frac{\partial}{\partial z}\left(\frac{\partial \Phi}{\partial z}\right)$$

$$= \frac{\partial^2 \Phi}{\partial x^2} + \frac{\partial^2 \Phi}{\partial y^2} + \frac{\partial^2 \Phi}{\partial z^2} \tag{3.23}$$

A scalar function is said to be **harmonic** in a region if its Laplacian vanishes in that region; that is, Φ is harmonic if

$$\nabla^2 \Phi = 0 \tag{3.24}$$

This partial differential equation is called **Laplace's equation**. It is a most important equation and arises in many applications. We should note that the potential associated with any vector field which is both solenoidal and conservative satisfies Laplace's equation.

The gradient of a scalar field Φ is, as we have seen, a vector field $\nabla\Phi$. Therefore, we cannot apply the gradient operator twice, since after the first operation we would be trying to take the gradient of a vector field, which is not defined. Similarly, the divergence of a vector field $\mathbf{A}$ gives a scalar field, and we could not take the divergence again. In other words, neither $\nabla\nabla\Phi$ nor $\nabla\cdot(\nabla\cdot\mathbf{A})$ is defined.

On the other hand, the curl of a vector field is itself a vector field and the curl operation could be repeated, so $\nabla \wedge \nabla \wedge \mathbf{A}$ is a meaningful quantity (it is another vector field). It is possible to show, by direct expansion in Cartesian coordinates, that the expansion of $\nabla \wedge \nabla \wedge \mathbf{A}$ is the same as that of the quantity

$$\nabla(\nabla\cdot\mathbf{A}) - \nabla^2\mathbf{A}$$

where $\nabla^2\mathbf{A}$ is *defined* as a vector field whose components are $\nabla^2 A_x$, $\nabla^2 A_y$ and $\nabla^2 A_z$ (see Additional exercise 6). In other words, we have an important vector identity, which is much used in applications:

$$\nabla \wedge \nabla \wedge \mathbf{A} \equiv \nabla(\nabla\cdot\mathbf{A}) - \nabla^2\mathbf{A} \quad \text{or} \quad \textbf{curl curl A} \equiv \textbf{grad}\ \text{div}\ \mathbf{A} - \nabla^2\mathbf{A}$$

(We do not rewrite $\nabla^2\mathbf{A}$ as div **grad A**.)

Example 3.6

Verify the identity for $\nabla \wedge \nabla \wedge \mathbf{A}$ if $\mathbf{A} = xz\hat{\mathbf{i}} + y^2z\hat{\mathbf{j}}$.

Solution

$$\nabla \wedge \mathbf{A} = \begin{vmatrix} \hat{\mathbf{i}} & \hat{\mathbf{j}} & \hat{\mathbf{k}} \\ \dfrac{\partial}{\partial x} & \dfrac{\partial}{\partial y} & \dfrac{\partial}{\partial z} \\ xz & y^2z & 0 \end{vmatrix} = -y^2\hat{\mathbf{i}} + x\hat{\mathbf{j}}$$

$$\therefore \quad \nabla \wedge \nabla \wedge \mathbf{A} = \begin{vmatrix} \hat{\mathbf{i}} & \hat{\mathbf{j}} & \hat{\mathbf{k}} \\ \dfrac{\partial}{\partial x} & \dfrac{\partial}{\partial y} & \dfrac{\partial}{\partial z} \\ -y^2 & x & 0 \end{vmatrix} = (1+2y)\hat{\mathbf{k}}$$

For this vector field, $\nabla\cdot\mathbf{A} = z + 2yz$, so $\nabla(\nabla\cdot\mathbf{A}) = 2z\hat{\mathbf{j}} + (1+2y)\hat{\mathbf{k}}$.
Also,

$$A_x = xz \quad \text{so} \quad \nabla^2 A_x = 0$$

$$A_y = y^2 z \quad \text{so} \quad \nabla^2 A_y = 2z$$

$$A_z = 0 \quad \text{so} \quad \nabla^2 A_z = 0$$

and consequently $\nabla^2\mathbf{A} = 2z\hat{\mathbf{j}}$. Hence

$$\nabla(\nabla\cdot\mathbf{A}) - \nabla^2\mathbf{A} = (1+2y)\hat{\mathbf{k}}$$

which is identical to $\nabla \wedge \nabla \wedge \mathbf{A}$.

It is unfortunate that the terms div (divergence) and curl (rotation) for $\nabla\cdot$ and $\nabla\wedge$ respectively have gained widespread acceptance as descriptors of vector fields. In fact, they only have an unambiguous physical interpretation when used to describe particular kinds of vector fields, of which the steady flow of incompressible fluids is an example. Their *general* application as descriptors of vector fields which exhibit 'divergence' or 'rotation' is not valid, as the following simple examples will show.

Consider the vector fields

$$\mathbf{A} = x\hat{\mathbf{i}} + y\hat{\mathbf{j}} + z\hat{\mathbf{k}} \quad \text{and} \quad \mathbf{B} = y\hat{\mathbf{i}} - x\hat{\mathbf{j}}$$

Diagrammatic views of these fields are shown in Figure 3.10, which suggests that the vector field **A** might be described geometrically as 'diverging' and field **B** as 'curling'. This interpretation is 'confirmed' by calculation:

$$\nabla\cdot\mathbf{A} = \frac{\partial(x)}{\partial x} + \frac{\partial(y)}{\partial y} + \frac{\partial(z)}{\partial z} = 3, \quad \text{hence } \mathbf{A} \text{ has a non-zero divergence.}$$

$$\nabla\wedge\mathbf{A} = 0, \quad \text{hence } \mathbf{A} \text{ does not 'curl'.}$$

$$\nabla\cdot\mathbf{B} = \frac{\partial y}{\partial x} - \frac{\partial x}{\partial y} = 0 - 0 = 0, \quad \text{hence } \mathbf{B} \text{ does not diverge.}$$

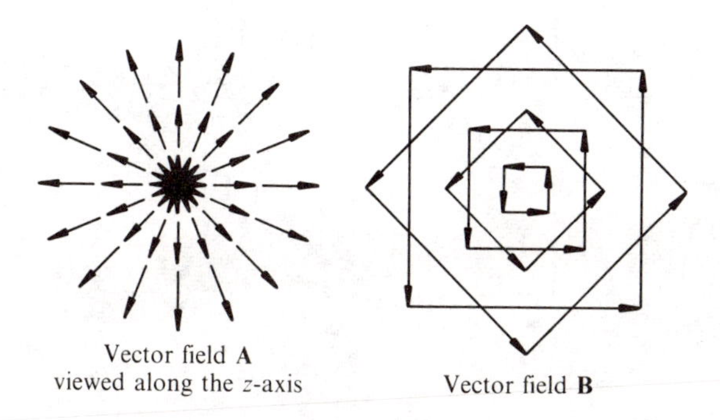

Figure 3.10

$$\nabla \wedge \mathbf{B} = \begin{vmatrix} \hat{\mathbf{i}} & \hat{\mathbf{j}} & \hat{\mathbf{k}} \\ \dfrac{\partial}{\partial x} & \dfrac{\partial}{\partial y} & \dfrac{\partial}{\partial z} \\ y & -x & 0 \end{vmatrix} = -2\hat{\mathbf{k}}, \quad \text{hence } \mathbf{B} \text{ does 'curl'}.$$

However, now consider the closely related vector fields

$$\mathbf{C} = \frac{x}{r^3}\hat{\mathbf{i}} + \frac{y}{r^3}\hat{\mathbf{j}} + \frac{z}{r^3}\hat{\mathbf{k}} \quad \text{where} \quad r = \sqrt{(x^2 + y^2 + z^2)}$$

$$\mathbf{D} = \frac{y}{\rho^2}\hat{\mathbf{i}} - \frac{x}{\rho^2}\hat{\mathbf{j}} \qquad \text{where} \quad \rho = \sqrt{(x^2 + y^2)}$$

Purely from a pictorial point of view, one could describe these vector fields as (**C**) 'diverging' and (**D**) 'curling' (see Figure 3.11). This view is *not* reflected in the calculation of div and curl, however. For vector field **C**,

$$\nabla \cdot \mathbf{C} = \frac{\partial}{\partial x}\left(\frac{x}{r^3}\right) + \frac{\partial}{\partial y}\left(\frac{y}{r^3}\right) + \frac{\partial}{\partial z}\left(\frac{z}{r^3}\right) = 0 \quad \text{(except at the origin)}$$

a result which follows from Example 3.5b, whilst for **D**,

$$\nabla \wedge \mathbf{D} = \begin{vmatrix} \hat{\mathbf{i}} & \hat{\mathbf{j}} & \hat{\mathbf{k}} \\ \dfrac{\partial}{\partial x} & \dfrac{\partial}{\partial y} & \dfrac{\partial}{\partial z} \\ \dfrac{y}{(x^2 + y^2)} & -\dfrac{x}{(x^2 + y^2)} & 0 \end{vmatrix}$$

$$= \hat{\mathbf{k}}\left[-\frac{1}{(x^2 + y^2)} + \frac{2x^2}{(x^2 + y^2)^2} - \frac{1}{(x^2 + y^2)} + \frac{2y^2}{(x^2 + y^2)^2}\right] = 0$$

That is, despite its appearance in Figure 3.11, the vector field **D** has zero curl.

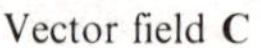

Vector field **C**

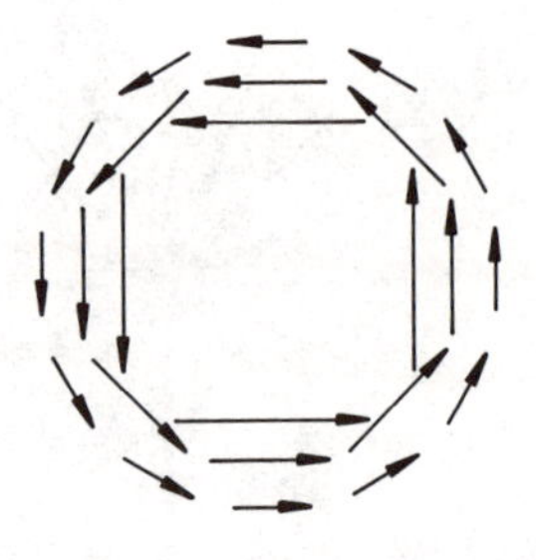

Vector field **D**

Figure 3.11

For general vector fields, then, we should not interpret the terms div and curl too literally as descriptors of the field. None the less, we must emphasize the very important role that the divergence and the curl play in vector field theory, which will become apparent in Chapters 4 and 5. Indeed, even purely notationally, these operators can be used to great advantage considerably to simplify many of the equations that arise in physics and in engineering. A particular example of this occurs in linear isotropic elasticity theory. Certain aspects of this theory have already been described in Sections 1.8 and 3.3, and in the next section we shall continue this vectorial examination, making particular reference to the equations of equilibrium, the stress–strain relations and, finally, to the Navier equation.

EXERCISES

3.18 Is $\nabla \wedge \mathbf{F}$ necessarily perpendicular to $\mathbf{F}$?

3.19 If $\mathbf{A} = 3x^2\hat{\mathbf{i}} + 3y^2\hat{\mathbf{j}} + 3xzy\hat{\mathbf{k}}$, $\mathbf{B} = x^2\hat{\mathbf{i}} + y^2\hat{\mathbf{j}} + z^2\hat{\mathbf{k}}$ and $f = xyz$, compute

(a) ∇f (b) $\nabla \wedge \mathbf{A}$ (c) $\nabla^2 f$ (d) $\nabla \cdot \mathbf{B}$ (e) $\nabla \cdot (\mathbf{A} \wedge \mathbf{B})$

(f) $\nabla \wedge \nabla \wedge \mathbf{A}$

3.20 If $\mathbf{r} = x\hat{\mathbf{i}} + y\hat{\mathbf{j}} + z\hat{\mathbf{k}}$ and $r = |\mathbf{r}|$, prove the following identities (assuming in each that $r \neq 0$):

(a) $\nabla\left(\dfrac{1}{r}\right) = -\dfrac{\mathbf{r}}{r^3}$ (b) $\nabla^2\left(\dfrac{1}{r}\right) = 0$ (c) $\nabla \wedge \left(\dfrac{\hat{\mathbf{i}} \wedge \mathbf{r}}{r^3}\right) = \dfrac{3x\mathbf{r}}{r^5} - \dfrac{\hat{\mathbf{i}}}{r^3}$

(d) $\nabla^2(\ln r) = \dfrac{1}{r^2}$

3.21 If $\Phi = xy/z$ and $\mathbf{A} = \sin x\,\hat{\mathbf{i}} + \cos x\,\hat{\mathbf{j}} + xy\hat{\mathbf{k}}$, verify the following identities:

(a) **curl grad** $\Phi = 0$ (b) div **curl A** $= 0$

(c) **curl curl A** $=$ **grad** div **A** $- \nabla^2\mathbf{A}$

3.22 (a) Prove that $\nabla \wedge [\Phi(r)\mathbf{r}] = 0$.

(b) Using the directional derivative, verify that $\mathrm{d}\Phi/\mathrm{d}r = \nabla\Phi \cdot \hat{\mathbf{r}}$ where $\mathrm{d}\Phi/\mathrm{d}r$ is the rate of change of Φ in the direction of $\hat{\mathbf{r}}$. Hence, or otherwise, show that if $\nabla \cdot [\Phi(r)\mathbf{r}] = 0$ then $\Phi = 1/r^3$.

3.23 Show that the vector field $\mathbf{A} = (y^2 + z^2)yz\hat{\mathbf{i}} + (z^2 + x^2)zx\hat{\mathbf{j}} + (x^2 + y^2)xy\hat{\mathbf{k}}$ is solenoidal and has the same curl as $\mathbf{B} = x^2yz\hat{\mathbf{i}} + y^2zx\hat{\mathbf{j}} + z^2xy\hat{\mathbf{k}}$.

3.24 If **A** and **B** are conservative vector fields, show that $\mathbf{A} \wedge \mathbf{B}$ is solenoidal. Hence show that if Φ and ϑ are scalar fields, then $\nabla\Phi \wedge \nabla\vartheta$ is solenoidal.

3.7 Divergence in elasticity theory (*optional*)

A body may be acted upon by surface forces **T** and by body forces **B**. With respect to the equations of equilibrium, body forces are the more important since each elementary volume of the material (in the interior) must be in equilibrium and it is only the body forces that have any direct effect. The most common body force is gravity but there are others of significance, particularly those which are magnetic in nature.

It may be shown that if the equilibrium of an elemental region of a body, of mass density ρ (dimensions of mass per unit volume), subjected to a body force **B** (dimensions of force per unit mass), is considered, then the equations of equilibrium take the form

$$\begin{aligned} &\nabla \cdot \boldsymbol{\sigma}_z + \rho B_x = 0 \\ &\nabla \cdot \boldsymbol{\sigma}_y + \rho B_y = 0 \\ &\nabla \cdot \boldsymbol{\sigma}_z + \rho B_z = 0 \end{aligned} \tag{3.25}$$

where $\boldsymbol{\sigma}_x$, $\boldsymbol{\sigma}_y$ and $\boldsymbol{\sigma}_z$ are the stress vectors discussed in Section 1.8.

The divergence is also used in the vectorial description of the stress–strain relations which relate the strain vectors $\boldsymbol{\varepsilon}_x$, $\boldsymbol{\varepsilon}_y$ and $\boldsymbol{\varepsilon}_z$ and the displacement vector **u** to the stresses $\boldsymbol{\sigma}_x$, $\boldsymbol{\sigma}_y$ and $\boldsymbol{\sigma}_z$. For linear isotropic materials, these relations take the form

$$\begin{aligned} &\boldsymbol{\sigma}_x = 2\mu\boldsymbol{\varepsilon}_x + \lambda(\nabla \cdot \mathbf{u})\hat{\mathbf{i}} \\ &\boldsymbol{\sigma}_y = 2\mu\boldsymbol{\varepsilon}_y + \lambda(\nabla \cdot \mathbf{u})\hat{\mathbf{j}} \\ &\boldsymbol{\sigma}_z = 2\mu\boldsymbol{\varepsilon}_z + \lambda(\nabla \cdot \mathbf{u})\hat{\mathbf{k}} \end{aligned} \tag{3.26}$$

in which μ and λ are constants, called Lamé constants, which are characteristic of the material. They may be directly related to the more commonly used constants of engineering mechanics, Young's modulus E and the Poisson Ratio ν.

We are now in a position to derive the governing differential equation for the displacement vector **u**, the so-called **Navier equation** of elasticity. Using (3.26) in (3.25) and using the definition of divergence (3.20),

$$\begin{aligned} &2\mu\nabla \cdot \boldsymbol{\varepsilon}_x + \lambda \frac{\partial}{\partial x}(\nabla \cdot \mathbf{u}) + \rho B_x = 0 \\ &2\mu\nabla \cdot \boldsymbol{\varepsilon}_y + \lambda \frac{\partial}{\partial y}(\nabla \cdot \mathbf{u}) + \rho B_y = 0 \\ &2\mu\nabla \cdot \boldsymbol{\varepsilon}_z + \lambda \frac{\partial}{\partial z}(\nabla \cdot \mathbf{u}) + \rho B_z = 0 \end{aligned} \tag{3.27}$$

But utilizing (3.15),

$$2\mu\nabla\cdot\boldsymbol{\varepsilon}_x = 2\mu\left[\frac{\partial^2 u}{\partial x^2} + \frac{1}{2}\left(\frac{\partial^2 u}{\partial y^2} + \frac{\partial^2 v}{\partial y\,\partial x}\right) + \frac{1}{2}\left(\frac{\partial^2 u}{\partial z^2} + \frac{\partial^2 w}{\partial z\,\partial x}\right)\right]$$

$$= 2\mu\left[\frac{1}{2}\left(\frac{\partial^2 u}{\partial x^2} + \frac{\partial^2 u}{\partial y^2} + \frac{\partial^2 u}{\partial z^2}\right) + \frac{1}{2}\frac{\partial}{\partial x}\left(\frac{\partial u}{\partial x} + \frac{\partial v}{\partial y} + \frac{\partial w}{\partial z}\right)\right]$$

$$= \mu\nabla^2 u + \mu\frac{\partial}{\partial x}(\nabla\cdot\mathbf{u}) \tag{3.28a}$$

Similarly,

$$2\mu\nabla\cdot\boldsymbol{\varepsilon}_y = \mu\nabla^2 v + \mu\frac{\partial}{\partial y}(\nabla\cdot\mathbf{u}) \tag{3.28b}$$

$$2\mu\nabla\cdot\boldsymbol{\varepsilon}_z = \mu\nabla^2 w + \mu\frac{\partial}{\partial z}(\nabla\cdot\mathbf{u}) \tag{3.28c}$$

Now multiplying these three equations by $\hat{\mathbf{i}}$, $\hat{\mathbf{j}}$ and $\hat{\mathbf{k}}$ respectively and adding, we find, using (3.27),

$$\mu[\hat{\mathbf{i}}\nabla^2 u + \hat{\mathbf{j}}\nabla^2 v + \hat{\mathbf{k}}\nabla^2 w]$$

$$+ (\lambda + \mu)\left[\hat{\mathbf{i}}\frac{\partial}{\partial x}(\nabla\cdot\mathbf{u}) + \hat{\mathbf{j}}\frac{\partial}{\partial y}(\nabla\cdot\mathbf{u}) + \hat{\mathbf{k}}\frac{\partial}{\partial z}(\nabla\cdot\mathbf{u})\right] + \rho\mathbf{B} = 0$$

therefore

$$\mu\nabla^2\mathbf{u} + (\lambda + \mu)\nabla(\nabla\cdot\mathbf{u}) + \rho\mathbf{B} = 0 \tag{3.29}$$

This fundamental equation which must be satisfied by the displacement vector $\mathbf{u}$ is the Navier equation.

An alternative form may be obtained using the vector identity

$$\nabla\wedge(\nabla\wedge\mathbf{u}) = \nabla(\nabla\cdot\mathbf{u}) - \nabla^2\mathbf{u}$$

to give

$$(\lambda + \mu)(\nabla\wedge(\nabla\wedge\mathbf{u})) + (\lambda + 2\mu)\nabla^2\mathbf{u} + \rho\mathbf{B} = 0 \tag{3.30}$$

ADDITIONAL EXERCISES

1 If $\mathbf{a}$ is a constant vector and $\mathbf{v}$ is defined by $\mathbf{v} = \mathbf{r}(\mathbf{a}\cdot\mathbf{r})$, show that $\nabla\cdot\mathbf{v} = 4(\mathbf{a}\cdot\mathbf{r})$ and $\nabla\wedge\mathbf{v} = \mathbf{a}\wedge\mathbf{r}$.

2 If $\mathbf{G} = 2xye^z\hat{\mathbf{i}} + e^z x^2\hat{\mathbf{j}} + (x^2ye^z + z^2)\hat{\mathbf{k}}$, show that $\nabla\wedge\mathbf{G} = 0$ and find a function $\Phi(x, y, z)$ such that $\mathbf{G} = \nabla\Phi$.

3 If $\mathbf{A}$ is a vector field and Φ is a scalar field, prove that the components of $\mathbf{A}$ normal and tangential to the level surfaces of Φ are, respectively, the vectors

$$\frac{(\mathbf{A}\cdot\nabla\Phi)\nabla\Phi}{|\nabla\Phi|^2} \quad \text{and} \quad \frac{(\nabla\Phi)\wedge(\mathbf{A}\wedge\nabla\Phi)}{|\nabla\Phi|^2}$$

4 If $\mathbf{F} = 2x\hat{\mathbf{i}} + x^2y\hat{\mathbf{j}} + z^2\hat{\mathbf{k}}$ and $\mathbf{G} = x\hat{\mathbf{i}} + y\hat{\mathbf{j}}$, verify the following vector identities:

(a) $\nabla\cdot(\mathbf{F}\wedge\mathbf{G}) = \mathbf{G}\cdot(\nabla\wedge\mathbf{F}) - \mathbf{F}\cdot(\nabla\wedge\mathbf{G})$

(b) $\nabla(\mathbf{F}\cdot\mathbf{G}) = \mathbf{G}\wedge(\nabla\wedge\mathbf{F}) + \mathbf{F}\wedge(\nabla\wedge\mathbf{G}) + (\mathbf{G}\cdot\nabla)\mathbf{F} + (\mathbf{F}\cdot\nabla)\mathbf{G}$

(c) $\nabla\wedge(\mathbf{F}\wedge\mathbf{G}) = (\mathbf{G}\cdot\nabla)\mathbf{F} - (\mathbf{F}\cdot\nabla)\mathbf{G} + \mathbf{F}(\nabla\cdot\mathbf{G}) - \mathbf{G}(\nabla\cdot\mathbf{F})$

Try *proving* identity (a) in general.

5 When a rigid body is rotating, show that the curl of its linear velocity at any point equals twice its angular velocity (see Chapter 1 for a discussion of rigid body motion).

6 Show that for any vector field $\mathbf{F} = F_x\hat{\mathbf{i}} + F_y\hat{\mathbf{j}} + F_z\hat{\mathbf{k}}$,

$$\nabla\wedge(\nabla\wedge\mathbf{F}) = \nabla(\nabla\cdot\mathbf{F}) - \nabla^2\mathbf{F} \quad (\text{or curl curl } \mathbf{F} = \text{grad div } \mathbf{F} - \nabla^2\mathbf{F})$$

where $\nabla^2\mathbf{F} = \nabla^2F_x\hat{\mathbf{i}} + \nabla^2F_y\hat{\mathbf{j}} + \nabla^2F_z\hat{\mathbf{k}}$.

7 In the interior of the sphere $r = R$, there is a field $\mathbf{v}_i = f\hat{\mathbf{a}}$, where f is a constant and $\hat{\mathbf{a}}$ is a unit constant vector. Outside the sphere, there is a field of the form $\mathbf{v}_e = kfR^3[3\mathbf{r}(\hat{\mathbf{a}}\cdot\mathbf{r}) - \hat{\mathbf{a}}r^2]/r^5$.

(a) Show that with different choices for the constant k, either the tangential or the normal components of $\mathbf{v}_i$ and $\mathbf{v}_e$ can be made equal at $r = R$.

(b) Verify that both $\mathbf{v}_i$ and $\mathbf{v}_e$ can be represented as either $-\nabla\Phi$ or $\nabla\wedge\mathbf{A}$, with

$$\Phi_i = -f(\hat{\mathbf{a}}\cdot\mathbf{r}), \qquad \mathbf{A}_i = \tfrac{1}{2}f(\hat{\mathbf{a}}\wedge\mathbf{r})$$

$$\Phi_e = kfR^3\frac{(\hat{\mathbf{a}}\cdot\mathbf{r})}{r^3}, \qquad \mathbf{A}_e = kfR^3\frac{(\hat{\mathbf{a}}\wedge\mathbf{r})}{r^3}$$

SUMMARY

- A **scalar field** is a function $G(x, y, z)$ that defines a scalar at each point of space.
- The **level surfaces** of a scalar field are the surfaces $G(x, y, z) = k$, where k is a constant.
- A **vector field** has the form $A_x(x, y, z)\hat{\mathbf{i}} + A_y(x, y, z)\hat{\mathbf{j}} + A_z(x, y, z)\hat{\mathbf{k}}$ and defines a vector at each point of space.

- The **gradient** of a scalar field G is the vector field $\nabla G = \hat{\mathbf{i}}\dfrac{\partial G}{\partial x} + \hat{\mathbf{j}}\dfrac{\partial G}{\partial y} + \hat{\mathbf{k}}\dfrac{\partial G}{\partial z}$. The gradient operator satisfies the relations

$$\nabla(G_1 + G_2) = \nabla G_1 + \nabla G_2$$

and $$\nabla(G_1 G_2) = G_1 \nabla G_2 + G_2 \nabla G_1$$

- The **rate of change** or **directional derivative** of a scalar field G in the direction of the unit vector $\hat{\mathbf{t}}$ is given by $\dfrac{\mathrm{d}G}{\mathrm{d}s} = \nabla G \cdot \hat{\mathbf{t}}$. $\mathrm{d}G/\mathrm{d}s$ is a maximum in a direction normal to the level surfaces of G.

- If $\mathbf{A}$ is a vector field, then the **divergence** of $\mathbf{A}$ is the scalar field

$$\nabla \cdot \mathbf{A} = \frac{\partial A_x}{\partial x} + \frac{\partial A_y}{\partial y} + \frac{\partial A_z}{\partial z}$$

The divergence satisfies

$$\nabla \cdot (\mathbf{A} + \mathbf{B}) = \nabla \cdot \mathbf{A} + \nabla \cdot \mathbf{B}$$

and $$\nabla \cdot (G\mathbf{A}) = (\nabla G) \cdot \mathbf{A} + G(\nabla \cdot \mathbf{A})$$

- If $\mathbf{A}$ is a vector field, then the **curl** of $\mathbf{A}$ is the vector field

$$\nabla \wedge \mathbf{A} = \hat{\mathbf{i}}\left(\frac{\partial A_z}{\partial y} - \frac{\partial A_y}{\partial z}\right) + \hat{\mathbf{j}}\left(\frac{\partial A_x}{\partial z} - \frac{\partial A_z}{\partial x}\right) + \hat{\mathbf{k}}\left(\frac{\partial A_y}{\partial x} - \frac{\partial A_x}{\partial y}\right)$$

or, in determinant form,

$$\nabla \wedge \mathbf{A} = \begin{vmatrix} \hat{\mathbf{i}} & \hat{\mathbf{j}} & \hat{\mathbf{k}} \\ \dfrac{\partial}{\partial x} & \dfrac{\partial}{\partial y} & \dfrac{\partial}{\partial z} \\ A_x & A_y & A_z \end{vmatrix}$$

The curl satisfies

$$\nabla \wedge (\mathbf{A} + \mathbf{B}) = \nabla \wedge \mathbf{A} + \nabla \wedge \mathbf{B}$$

and $$\nabla \wedge (G\mathbf{A}) = (\nabla G) \wedge \mathbf{A} + G(\nabla \wedge \mathbf{A})$$

- Other important identities are

$$\nabla \cdot (\nabla \wedge \mathbf{A}) \equiv 0 \quad \text{and} \quad \nabla \wedge (\nabla G) \equiv 0$$

- The **Laplacian** of a scalar field G is $\nabla^2 G$ and is defined by

$$\nabla^2 G = \nabla \cdot (\nabla G)$$

- If G satisfies **Laplace's equation** $\nabla^2 G = 0$ in a region, then G is said to be **harmonic** in that region.

SUPPLEMENT

3S.1 Index notation

If G is a scalar function of the coordinates (x_1, x_2, x_3), then the partial derivative of G with respect to variable x_i, $\partial G/\partial x_i$, will be denoted by $G_{,i}$. For example, $G_{,1}$ means $\partial G/\partial x_1$ (or $\partial G/\partial x$), $G_{,2}$ means $\partial G/\partial x_2$ (or $\partial G/\partial y$), and $G_{,3}$ means $\partial G/\partial x_3$ (or $\partial G/\partial z$). Thus the object $G_{,i}$ stands for the collection of derivatives $(\partial G/\partial x_1, \partial G/\partial x_2, \partial G/\partial x_3)$; it is the **indexed version** of the gradient ∇G. Similarly, $G_{,ij}$ means $\partial^2 G/\partial x_i\, \partial x_j$.

A vector field $\mathbf{A}(x, y, z)$ will be denoted by A_i, so the divergence of $\mathbf{A}$ is

$$\nabla \cdot \mathbf{A} = \frac{\partial A_1}{\partial x} + \frac{\partial A_2}{\partial y} + \frac{\partial A_3}{\partial z} = \frac{\partial A_1}{\partial x_1} + \frac{\partial A_2}{\partial x_2} + \frac{\partial A_3}{\partial x_3} = A_{1,1} + A_{2,2} + A_{3,3}$$

If we employ the summation convention, we can write

$$\nabla \cdot \mathbf{A} = A_{i,i}$$

We have seen in Section 3.5 that if a scalar function G satisfies Laplace's equation ($\nabla \cdot (\nabla G) = 0$), then it is called harmonic. In terms of the index notation and again employing the summation convention, G is harmonic if it satisfies

$$G_{,ii} = 0$$

The curl of $\mathbf{A}$ is (notationally) the cross product of ∇ with $\mathbf{A}$, so the ith component of the curl is

$$(\nabla \wedge \mathbf{A})_i = \varepsilon_{ijk} \frac{\partial}{\partial x_j} A_k = \varepsilon_{ijk} A_{k,j}$$

(see (1S.8) in Supplement 1S.1). For example,

$$(\nabla \wedge \mathbf{A})_1 = \varepsilon_{1jk} \frac{\partial}{\partial x_j} A_k = \varepsilon_{123} \frac{\partial A_3}{\partial x_2} + \varepsilon_{132} \frac{\partial A_2}{\partial x_3}$$

on using the summation convention and the properties of ε_{ijk}. But $\varepsilon_{132} = -1$, so

$$(\nabla \wedge \mathbf{A})_1 = \frac{\partial A_3}{\partial x_2} - \frac{\partial A_2}{\partial x_3}, \quad \text{as we know.}$$

To illustrate the use of the index notation we shall again prove property (4) of Section 3.6,

$$\nabla \wedge (\nabla G) = 0$$

The ith component of $\nabla \wedge (\nabla G)$ is $\varepsilon_{ijk}(\nabla G)_{k,j} = \varepsilon_{ijk}(G_{,k})_{,j} = \varepsilon_{ijk} G_{,kj}$

But $\quad G_{,kj} \equiv \dfrac{\partial^2 G}{\partial x_k\, \partial x_j} \equiv \dfrac{\partial^2 G}{\partial x_j\, \partial x_k} \equiv G_{,jk}$

that is, it is symmetric in the indices k and j. But ε_{ijk} is skew-symmetric and therefore, from the result of Example 1S.1 in Supplement 1S.1, $\varepsilon_{ijk}G_{,jk}=0$, which proves the identity.

Example 3S.1

Using the index approach, prove the following identities:

(a) $\nabla \wedge (\mathbf{F} \wedge \mathbf{G}) = (\mathbf{G}\cdot\nabla)\mathbf{F} - (\mathbf{F}\cdot\nabla)\mathbf{G} + \mathbf{F}(\nabla\cdot\mathbf{G}) - \mathbf{G}(\nabla\cdot\mathbf{F})$

(b) $\nabla \wedge (\nabla \wedge \mathbf{A}) = \nabla(\nabla\cdot\mathbf{A}) - \nabla^2\mathbf{A}$

Solution

(a) Let F_i and G_i be the ith components of $\mathbf{F}$ and $\mathbf{G}$ respectively and let H_r denote the rth component of $\mathbf{F} \wedge \mathbf{G}$:

$$H_r = (\mathbf{F} \wedge \mathbf{G})_r = \varepsilon_{rjk} F_j G_k$$

The pth component of $\nabla \wedge \mathbf{H}$ $(= \nabla \wedge (\mathbf{F} \wedge \mathbf{G}))$ is

$$(\nabla \wedge \mathbf{H})_p = \varepsilon_{pqr} H_{r,q}$$

But $$H_{r,q} = (\varepsilon_{rjk} F_j G_k)_{,q} = \varepsilon_{rjk}(F_{j,q} G_k + F_j G_{k,q})$$

$$\therefore \quad (\nabla \wedge \mathbf{H})_p = \varepsilon_{pqr}\varepsilon_{rjk}(F_{j,q} G_k + F_j G_{k,q}) \tag{3S.1}$$

(simply taking the partial derivative of a product and realizing that the terms in ε are constants).

We now employ the basic ε–δ identity first described in Supplement 1S.1:

$$\varepsilon_{pqr}\varepsilon_{rjk} = \varepsilon_{rpq}\varepsilon_{rjk} = \delta_{pj}\delta_{qk} - \delta_{pk}\delta_{qj}$$

$$\therefore \quad \varepsilon_{pqr}\varepsilon_{rjk}(F_{j,q} G_k + F_j G_{k,q}) = (\delta_{pj}\delta_{qk} - \delta_{pk}\delta_{qj})(F_{j,q} G_k + F_j G_{k,q})$$

$$= F_{p,q} G_q + F_p G_{q,q} - F_{q,q} G_p - F_q G_{p,q}$$

Substituting into (3S.1), the pth component of $\nabla \wedge (\mathbf{F} \wedge \mathbf{G})$ may be expressed as

$$[(\mathbf{G}\cdot\nabla)\mathbf{F}]_p + [(\nabla\cdot\mathbf{G})\mathbf{F}]_p - [(\nabla\cdot\mathbf{F})\mathbf{G}]_p - [(\mathbf{F}\cdot\nabla)\mathbf{G}]_p$$

which proves the identity.

(b) This identity is verified using the techniques outlined in (a). Firstly, the mth component of curl $\mathbf{A}$ is

$$(\nabla \wedge \mathbf{A})_m = \varepsilon_{mrq} A_{q,r}$$

$$\therefore \quad [\nabla \wedge (\nabla \wedge \mathbf{A})]_i = \varepsilon_{ipm}(\varepsilon_{mrq} A_{q,r})_{,p}$$

$$= \varepsilon_{ipm}\varepsilon_{mrq} A_{q,rp}$$

$$= (\delta_{ir}\delta_{pq} - \delta_{iq}\delta_{pr}) A_{q,rp}$$

$$= A_{p,ip} - A_{i,rr}$$

But $A_{p,ip} = A_{p,pi}$ (order of differentiation is not important).
Also

$$A_{p,pi} = (A_{p,p})_{,i} = [\nabla(\nabla \cdot \mathbf{A})]_i \quad \text{and} \quad A_{i,pp} = [\nabla^2 \mathbf{A}]_i$$

and so $\nabla \wedge (\nabla \wedge \mathbf{A}) = \nabla(\nabla \cdot \mathbf{A}) - \nabla^2 \mathbf{A}$ **(3S.2)**

The strain tensor

The strain vectors $\boldsymbol{\varepsilon}_x$, $\boldsymbol{\varepsilon}_y$ and $\boldsymbol{\varepsilon}_z$ defined in (3.15) may be grouped together to form the strain tensor η_{ij},

$$\eta_{1j} = (\boldsymbol{\varepsilon}_x)_j \qquad \eta_{2j} = (\boldsymbol{\varepsilon}_y)_j \qquad \eta_{3j} = (\boldsymbol{\varepsilon}_z)_j$$

or directly, in terms of the displacement vector u_i,

$$\eta_{ij} = \tfrac{1}{2}(u_{i,j} + u_{j,i})$$

which is obviously symmetric, that is, $\eta_{ij} = \eta_{ji}$.

The reader will find it instructive to rework Section 3.3 using indices and to verify that the index form of equation (3.14) is

$$(\mathrm{d}s^*)^2 - \mathrm{d}s^2 = 2\eta_{ij}\,\mathrm{d}x_i\,\mathrm{d}x_j, \quad \text{in which } \mathrm{d}x_i = (\mathrm{d}\mathbf{r})_i$$

and also that the relation between the new length $\mathrm{d}s^*$ of a filament originally in the direction of the unit vector l_i (hence $l_i l_i = 1$) of length $\mathrm{d}s$ is

$$\mathrm{d}s^* = (1 + \eta_{ij} l_i l_j)\,\mathrm{d}s$$

Also, using the stress tensor σ_{ij} introduced in Supplement 1S.1, it is easily confirmed that the equilibrium equations (3.25) can be written more concisely as

$$\sigma_{ij,j} + \rho B_i = 0$$

3S.2 Harmonic functions and the torsion problem

Two-dimensional harmonic functions, that is, functions $\Phi(x, y)$ satisfying

$$\frac{\partial^2 \Phi}{\partial x^2} + \frac{\partial^2 \Phi}{\partial y^2} = 0$$

in some region, arise in many areas of application, for example fluid flow, electromagnetism and linear elasticity theory.

In the torsion problem of circular cylinders (see Examples 1.13 and 3.4) we have stated that the displacement $\mathbf{u}$ (actually an example of a vector field) is

$$\mathbf{u} = (-\tau yz, \tau xz, 0)$$

For non-circular cylinders, however, it becomes necessary to assume that the displacement is

$$\mathbf{u} = (-\tau yz, \tau xz, \tau\Phi(x, y)) \tag{3S.3}$$

where $\Phi(x, y)$ is a function of x and y called the **warping function**. It turns out that Φ is a harmonic function and that the stresses corresponding to this displacement field are

$$\boldsymbol{\sigma}_x = \left(0, 0, \mu\tau\left(\frac{\partial\Phi}{\partial x} - y\right)\right)$$

$$\boldsymbol{\sigma}_y = \left(0, 0, \mu\tau\left(\frac{\partial\Phi}{\partial y} + x\right)\right)$$

$$\boldsymbol{\sigma}_z = (0, 0, 0) \tag{3S.4}$$

The stress vector at any constant z-plane is (see (1.50))

$$\mathbf{P}^n = \left(\mu\tau\left(\frac{\partial\Phi}{\partial x} - y\right), \mu\tau\left(\frac{\partial\Phi}{\partial y} + x\right), 0\right)$$

This represents a pure shear of magnitude

$$\mu\tau\sqrt{\left[\left(\frac{\partial\Phi}{\partial x} - y\right)^2 + \left(\frac{\partial\Phi}{\partial y} + x\right)^2\right]}$$

It is a property of a function harmonic within some region D that it cannot have an extremum (maximum or minimum) within D. The proof is achieved by assuming the contrary, namely that the function has such an extremum.

Suppose then that $\Phi(x, y)$ has a maximum at a point P within D. Then it is a maximum along any line passing through P, in particular along lines parallel to the x- and y-axes. For a maximum, this requires that

$$\frac{\partial^2\Phi}{\partial x^2} < 0 \quad \text{and} \quad \frac{\partial^2\Phi}{\partial y^2} < 0 \quad \text{at } P$$

Hence $\dfrac{\partial^2\Phi}{\partial x^2} + \dfrac{\partial^2\Phi}{\partial y^2} < 0$

at P, which contradicts the hypothesis that Φ is a harmonic function. A similar proof shows that Φ cannot have a minimum anywhere in D.

We can use this important theoretical result in a practical situation and show that in the torsion problem of a uniform cylinder the resultant shearing stress achieves its maximum somewhere on the boundary. Again we assume the contrary, namely that there exists a point P in the interior of the cylinder at which the resultant shearing stress is a maximum. We can (at P) orient the coordinate axes so that the x-axis is parallel to the direction of the resultant shearing stress at P. Then, at this point, the magnitude of the shearing stress is

$$\mu\tau\left|\frac{\partial\Phi}{\partial x}-y\right|_P$$

Now $$\frac{\partial^2}{\partial x^2}\left(\frac{\partial\Phi}{\partial x}-y\right)+\frac{\partial^2}{\partial y^2}\left(\frac{\partial\Phi}{\partial x}-y\right)=\frac{\partial}{\partial x}\left(\frac{\partial^2\Phi}{\partial x^2}+\frac{\partial^2\Phi}{\partial y^2}\right)$$

so that if Φ is harmonic then so is $\partial\Phi/\partial x-y$.

Since $\partial\Phi/\partial x-y$ is a harmonic function and is therefore neither a maximum nor a minimum at P, there exists a point Q in the close neighbourhood of P at which

$$\left|\frac{\partial\Phi}{\partial x}-y\right|_Q\geqslant\left|\frac{\partial\Phi}{\partial x}-y\right|_P$$

The resultant shearing stress at Q, say R_Q, is

$$R_Q=\mu\tau\sqrt{\left[\left(\frac{\partial\Phi}{\partial x}-y\right)^2+\left(\frac{\partial\Phi}{\partial y}+x\right)^2\right]}$$

$$\therefore\qquad R_Q\geqslant\mu\tau\left|\frac{\partial\Phi}{\partial x}-y\right|_Q\geqslant\mu\tau\left|\frac{\partial\Phi}{\partial x}-y\right|_P=R_P$$

Hence $R_Q\geqslant R_P$, which contradicts the assumption that the resultant shearing stress is a maximum at P. It follows that if P is a point at which the resultant shearing stress is a maximum it cannot be in the interior – it must be somewhere on the boundary.

Example 3S.2

A cylindrical bar of constant elliptical cross-section with its axis aligned along the z-axis is subjected to equal and opposite twisting moments applied to its plane ends about the z-axis. The equation of the ellipse is

$$\frac{x^2}{a^2}+\frac{y^2}{b^2}=1\qquad(a>b)$$

and the displacement field $\mathbf{u}$ has components

$$u=-\tau yz,\qquad v=\tau xz,\qquad w=-\tau\left(\frac{a^2-b^2}{a^2+b^2}\right)xy$$

Verify that the warping function is harmonic and locate the position of maximum shearing stress.

Solution

The warping function is, using (3S.3),

$$\Phi=-\left(\frac{a^2-b^2}{a^2+b^2}\right)xy$$

Since this is linear in x and in y the second-order partial derivatives must be zero, implying that Φ is indeed harmonic.

Using (3S.4), the stresses are

$$\boldsymbol{\sigma}_x = \left(0, 0, \mu\tau\left[-\left(\frac{a^2-b^2}{a^2+b^2}\right)y - y\right]\right) = \left(0, 0, -2\mu\tau\left(\frac{a^2 y}{a^2+b^2}\right)\right)$$

$$\boldsymbol{\sigma}_y = \left(0, 0, \mu\tau\left[-\left(\frac{a^2-b^2}{a^2+b^2}\right)x + x\right]\right) = \left(0, 0, 2\mu\tau\left(\frac{b^2 x}{a^2+b^2}\right)\right)$$

$$\boldsymbol{\sigma}_z = (0, 0, 0)$$

so the resultant shearing stress is

$$\frac{2\mu\tau}{a^2+b^2}\sqrt{(a^4y^2 + b^4x^2)}$$

We know that the resultant shearing stress is a maximum on the boundary. On the boundary of the elliptical bar, the resultant shearing stress has the form

$$\frac{2\mu\tau b}{a^2+b^2}\sqrt{\left[a^4\left(1-\frac{x^2}{a^2}\right)+b^2x^2\right]} = \frac{2\mu\tau b}{a^2+b^2}\sqrt{[a^4-(a^2-b^2)x^2]}$$

Since $a \geqslant b$, this is clearly a maximum when x assumes its smallest value ($x = 0$). Thus the maximum shearing stress is $2\mu\tau b^2/(a^2+b^2)$ and is located at $(0, \pm b)$ on the cross-section. These sections of the bar would need to be strengthened if it were envisaged that the cylinder would be subjected to very large twisting couples.

Line, Surface and Volume Integrals

PREVIEW

In this chapter we discuss the evaluation of three important types of integral – over curves (line integrals), over surfaces (surface integrals) and through regions of space (volume integrals). Cylindrical and spherical coordinate systems are used where appropriate. The discussion of line integrals gives rise to a convenient classification of vector fields as conservative or non-conservative, a classification which has important consequences in many applications.

4.1 Introduction

Chapter 3 introduced the quantities **grad**, div and **curl**. The approach taken there was very much a first introduction and, although some indication was given of the use of the gradient concept, little was said about the significance of the divergence and **curl** of a vector field. The reason for this approach was two-fold.

In the first place, the notations $\nabla \cdot \mathbf{A}$ and $\nabla \wedge \mathbf{A}$ are clearly far more concise than the expressions they represent; for example, $\nabla \cdot \mathbf{A}$ can simply be considered as a 'shorthand' for a sum of three partial derivatives:

$$\frac{\partial A_x}{\partial x} + \frac{\partial A_y}{\partial y} + \frac{\partial A_z}{\partial z}$$

We have here, in fact, one of the main uses of vector field theory – it enables us to represent mathematical expressions and, indeed, physical laws from a wide variety of applications in a neat and concise form.

In the second place, the definitions used for $\nabla \cdot \mathbf{A}$ and $\nabla \wedge \mathbf{A}$ can actually be derived from more general definitions. These definitions involve vector integrals of a type not yet encountered, namely integrals evaluated along curves in space and over surfaces. That integral definitions exist for divergence and **curl** may seem surprising at the moment, in view of the derivative definitions used in Chapter 3. However, a knowledge of the integral definitions will take us closer to the heart of vector field theory and help us to understand better the physical significance of equations such as div $\mathbf{A} = 0$ or **curl** $\mathbf{H} = \mathbf{J}$.

Thus in this chapter we are going to study **line integrals**, which is the common name for integrals of vector fields along curves. We shall also investigate **surface** or **flux integrals**, both over a closed surface such as a sphere and over a non-closed surface such as a plate. Also, since a closed surface encloses a volume, we shall look briefly at **volume integrals**. The most useful type of volume integral actually involves the integral of a scalar field through a volume, and this is the type we shall concentrate upon. All these types of integral are useful in their own right as well as being used in the general definitions of divergence and **curl**.

4.2 Line integrals

You are probably familiar with the idea of a definite integral of the form $\int_a^b f(x)\,dx$ as the limiting case of a summation. Specifically, if we require the area A under the graph of $y = f(x)$ between $x = a$ and $x = b$, then we can approximate this area by a series of rectangles (see Figure 4.1) so that

$$A \approx \sum_{i=0}^{N-1} f(x_i)\,\Delta x \tag{4.1}$$

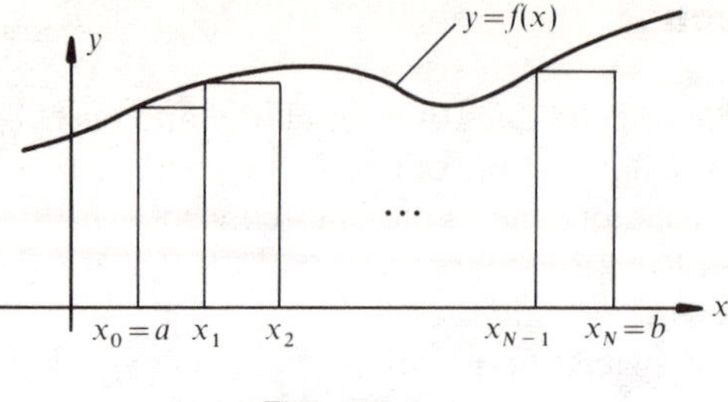

Figure 4.1

The approximation clearly improves if more rectangles are used and, correspondingly, their width $\Delta x = x_i - x_{i-1}$ is reduced. The exact limit of the sum in equation (4.1) as $N \to \infty$ and $\Delta x \to 0$ is referred to as a **definite integral** and gives us an exact expression for the desired area.

In a similar manner, a line integral is obtained as a limiting case of a discrete summation. To take a specific example, consider the problem of moving an object up an inclined plane from point P_1 to P_2 through a distance l (see Figure 4.2a). If a constant force $\mathbf{F}$ is applied to the object, the work done is given by $|\mathbf{F}| \cos \theta\, l$ or, in vector form, $\mathbf{F} \cdot \mathbf{l}$, where $\mathbf{l}$ denotes the vector $\overrightarrow{P_1P_2}$. Now if the force $\mathbf{F}$ varies in magnitude and direction (in which case the angle θ varies), this expression will be incorrect; also, if P_1P_2 is not a straight line some modification of approach is required.

What we can do in all cases is to divide the path P_1P_2 into small segments which can be approximated by straight lines and over which $\mathbf{F}$ can be assumed to be constant, say at a value $\mathbf{F}_i$ for the ith segment (see Figure 4.2b). Then, if $\Delta \mathbf{l}_i$ denotes this ith segment, the work done in moving the object along this segment is approximately $\mathbf{F}_i \cdot \Delta \mathbf{l}_i$, and for the whole path of N segments the total work done is approximately

$$\sum_{i=1}^{N} \mathbf{F}_i \cdot \Delta \mathbf{l}_i$$

The exact work done can be obtained by taking smaller and smaller segments

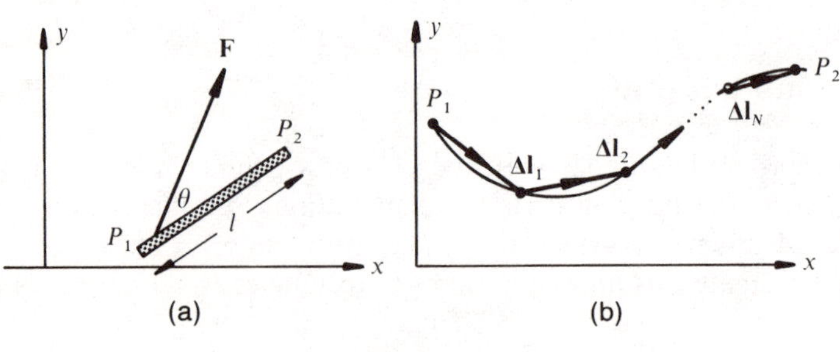

Figure 4.2

and letting N increase. The limit as $N \to \infty$ and all $\Delta \mathbf{l}_i \to 0$ is called the **line integral** of $\mathbf{F}$ along the path C from P_1 to P_2 and is written

$$\int_C \mathbf{F} \cdot \mathrm{d}\mathbf{l}$$

Thus a line integral, like the more familiar definite integral, is defined as the limit of a sum.

Needless to say, the above application of a line integral to finding the work done by a variable force is not the only context in which line integrals arise. They occur, for example, in fluid mechanics where, if $\mathbf{v}$ is the velocity field of the fluid, the line integral

$$\oint_C \mathbf{v} \cdot \mathrm{d}\mathbf{l}$$

around a closed path C has significance and is known as the **circulation** of the fluid. (Note the use here of the symbol $\oint_C$ which is commonly used to denote a line integral over a *closed* path, that is, one where the initial point P_1 and the endpoint P_2 coincide.)

The essential ingredients, then, for a line integral of the form $\int_C \mathbf{F} \cdot \mathrm{d}\mathbf{l}$ are a vector field $\mathbf{F}$ and a path C traversed in a definite sense. Despite the term 'line integral', the path C is not necessarily a straight line. The result of a line integral is a scalar number. If $\mathbf{F}$ denotes a physical quantity, then the units of $\int_C \mathbf{F} \cdot \mathrm{d}\mathbf{l}$ are the units of $\mathbf{F}$ multiplied by the units of distance; for example, if $\mathbf{F}$ is a force measured in newtons and $\mathrm{d}\mathbf{l}$ is a length in metres, then the units of $\int_C \mathbf{F} \cdot \mathrm{d}\mathbf{l}$ are N m or joules (J), the units of *work*.

The notation $\int_C \mathbf{F} \cdot \mathrm{d}\mathbf{l}$ is a common one for a line integral and is clearly neat and concise. Some authors use $\mathrm{d}\mathbf{r}$ instead of $\mathrm{d}\mathbf{l}$ for the 'line element'. In Cartesian coordinates, this quantity has the form

$$\mathrm{d}\mathbf{l} \text{ (or } \mathrm{d}\mathbf{r}) = \mathrm{d}x\,\hat{\mathbf{i}} + \mathrm{d}y\,\hat{\mathbf{j}} + \mathrm{d}z\,\hat{\mathbf{k}}$$

and it can be thought of as a vector between points with coordinates (x, y, z) and $(x + \mathrm{d}x, y + \mathrm{d}y, z + \mathrm{d}z)$ respectively. The z-component of $\mathrm{d}\mathbf{l}$ is not needed if the path C is restricted to the xy-plane.

Two-dimensional line integrals over a path in the xy-plane arise in thermodynamics, often in a 'non-vector form',

$$\int_C \{F_x(x, y)\,\mathrm{d}x + F_y(x, y)\,\mathrm{d}y\}$$

Such a form is readily obtained from the vector form $\int_C \mathbf{F} \cdot \mathrm{d}\mathbf{l}$ by putting

$$\mathbf{F} = F_x\hat{\mathbf{i}} + F_y\hat{\mathbf{j}} \quad \text{and} \quad \mathrm{d}\mathbf{l} = \mathrm{d}x\,\hat{\mathbf{i}} + \mathrm{d}y\,\hat{\mathbf{j}}$$

and using the properties of the dot product.

4.3 Evaluation of line integrals in two dimensions

The general technique for evaluating line integrals is to convert them to definite integrals. We do this by expressing the integration path C in a convenient form; if possible, we obtain the equation of C in parametric form:

$$x = x(t), \quad y = y(t), \quad z = z(t), \qquad t_0 \leqslant t \leqslant t_1$$

The idea, as we saw in Chapter 2, is that as t varies between the two values t_0 and t_1 the corresponding values of x, y and z are such that the path C is traced out. For example, a direct straight-line path from the origin $(0, 0, 0)$ to the point $(2, 2, 2)$ has the parametric form

$$x = 2t, \quad y = 2t, \quad z = 2t, \qquad 0 \leqslant t \leqslant 1$$

For simple two-dimensional examples (whose evaluation we discuss first), we often use the simple parametrization

$$x = t, \quad y = y(x) \quad \text{or} \quad y = t, \quad x = x(y)$$

that is, we convert the given line integral to a definite integral with respect to x (or with respect to y).

If the path C is such that a single parametrization is not possible, then we split the curve C into portions $C_1, C_2, \ldots$ which we can deal with, and use the obvious property

$$\int_C \mathbf{F} \cdot d\mathbf{l} = \int_{C_1} \mathbf{F} \cdot d\mathbf{l} + \int_{C_2} \mathbf{F} \cdot d\mathbf{l} + \cdots$$

Although we are concentrating on techniques for evaluating line integrals in this section, you should not lose sight of the fact that the final answer in each case will be a single number. The significance of this numerical value will depend on that of $\mathbf{F}$.

Example 4.1

Find the value of $\int_C \mathbf{F} \cdot d\mathbf{l}$ where $\mathbf{F} = x^2\hat{\mathbf{i}} + 3xy\hat{\mathbf{j}}$ if

(a) C is the straight-line path from $(0, 0)$ to $(1, 2)$,

(b) C is the parabolic path $y = 2x^2$ from $(0, 0)$ to $(1, 2)$,

(c) C is composed of two straight-line paths – the x-axis from $(0, 0)$ to $(1, 0)$ and then a line parallel to the y-axis from $(1, 0)$ to $(1, 2)$ (see Figure 4.3).

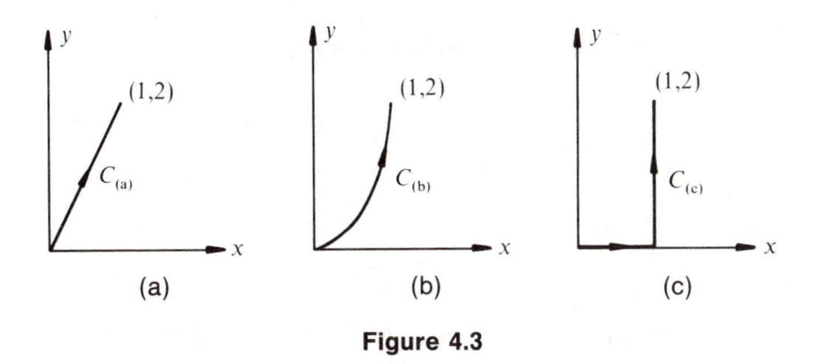

Figure 4.3

Solution

Since the line element $\mathbf{dl}$ lies in the xy-plane, we can express it as $\mathbf{dl} = \mathrm{d}x\,\hat{\mathbf{i}} + \mathrm{d}y\,\hat{\mathbf{j}}$ so that

$$\int_C \mathbf{F}\cdot\mathbf{dl} = \int_C (x^2\hat{\mathbf{i}} + 3xy\hat{\mathbf{j}})\cdot(\mathrm{d}x\,\hat{\mathbf{i}} + \mathrm{d}y\,\hat{\mathbf{j}})$$

$$= \int_C (x^2\,\mathrm{d}x + 3xy\,\mathrm{d}y) \tag{4.2}$$

(a) On this path $y = 2x$, so we can convert (4.2) to a definite integral with respect to x: this is effectively a simple parametrization,

$$x = t, \quad y = 2t, \qquad 0 \leqslant t \leqslant 1$$

Hence $\displaystyle\int_C (x^2\,\mathrm{d}x + 3xy\,\mathrm{d}y) = \int_0^1 (x^2\,\mathrm{d}x + 3x(2x)\,2\,\mathrm{d}x) = \int_0^1 13x^2\,\mathrm{d}x = \tfrac{13}{3}$

Alternatively, of course, we could choose y as the integration variable or parameter and put $x = y/2$, in which case (4.2) becomes

$$\int_0^2 \left\{\left(\frac{y^2}{4}\right)\frac{\mathrm{d}y}{2} + 3\left(\frac{y}{2}\right)y\,\mathrm{d}y\right\} = \int_0^2 \tfrac{13}{8}y^2\,\mathrm{d}y = \tfrac{13}{3}$$

(b) Here we put $y = 2x^2$ and $\mathrm{d}y = 4x\,\mathrm{d}x$ in (4.2), giving the definite integral

$$\int_0^1 \{x^2\,\mathrm{d}x + 3x(2x^2)4x\,\mathrm{d}x\} = \int_0^1 \{x^2\,\mathrm{d}x + 24x^4\,\mathrm{d}x\} = \tfrac{77}{5}$$

which is a different value from that obtained using the direct straight-line path (a).

(c) Referring to Figure 4.3c, we must integrate on the horizontal and vertical portions separately and add the two contributions. On the horizontal section, y is a constant (at zero) so $\mathrm{d}y = 0$ and (4.2) is simply

$$\int_0^1 x^2\,\mathrm{d}x = \tfrac{1}{3}$$

On the vertical section, x is constant at 1 so $dx = 0$ and (4.2) yields

$$\int_0^2 3(1)y\,dy = \int_0^2 3y\,dy = 6$$

Adding the two values, we conclude that the value of the line integral along this third path is $\frac{19}{3}$.

If we reverse the direction of integration in a line integral $\int_C \mathbf{F}\cdot d\mathbf{l}$, then we have to reverse the direction of each of the line elements $d\mathbf{l}$. The net effect of this is to yield the same numerical value for the line integral as before, but with the sign reversed. For example, if we integrate the vector field $\mathbf{F}$ of Example 4.1 from (1, 2) to (0, 0) along the direct straight-line path (a), we obtain a value of $-\frac{13}{3}$.

Example 4.2

Find the value of $\int_C \mathbf{F}\cdot d\mathbf{l}$, where $\mathbf{F} = x^2\hat{\mathbf{i}} + 3xy\hat{\mathbf{j}}$ (as in Example 4.1) and C is the closed path shown in Figure 4.4. (That is, it is made up of the paths (a) and (c) of Example 4.1, but path (a) is reversed. Note that the closed path is being traversed in the positive, or anticlockwise, sense here. We shall adopt this sense for closed path integration unless otherwise stated.)

Solution

Using the results of Example 4.1, parts (a) and (c), we have

$$\int_C \mathbf{F}\cdot d\mathbf{l} = \tfrac{1}{3} + 6 - \tfrac{13}{3} = 2$$

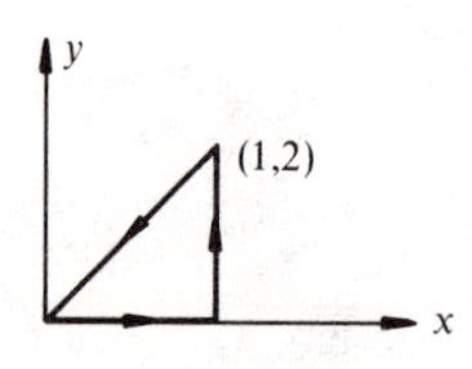

Figure 4.4

Example 4.3

Find the value of $\int_C \mathbf{F}\cdot d\mathbf{l}$, where $\mathbf{F} = (y - 2x)\hat{\mathbf{i}} + (3x + 2y)\hat{\mathbf{j}}$ and C is a circle in the xy-plane with centre the origin and radius 2.

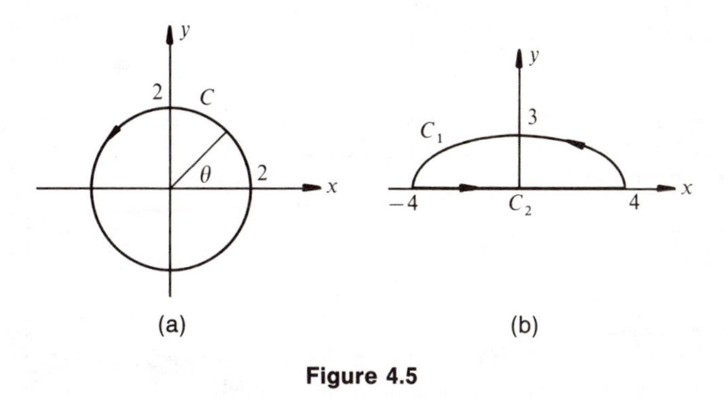

Figure 4.5

Solution

Referring to Figure 4.5a, it is convenient to use the polar angle θ as the integration variable or parameter. Then, for points (x, y) on the path C, we have

$$x = 2\cos\theta \quad \text{so} \quad dx = -2\sin\theta\, d\theta$$

and $$y = 2\sin\theta \quad \text{so} \quad dy = \;\; 2\cos\theta\, d\theta$$

where θ varies from 0 to 2π for the complete traverse of the circle in the sense indicated in Figure 4.5a (anticlockwise). Then

$$\oint_C \mathbf{F}\cdot d\mathbf{l} = \oint_C \{(y-2x)\,dx + (3x+2y)\,dy\}$$

$$= \int_0^{2\pi} \{(2\sin\theta - 4\cos\theta)(-2\sin\theta)\,d\theta$$

$$+ (6\cos\theta + 4\sin\theta)(2\cos\theta)\,d\theta\}$$

$$= \int_0^{2\pi} \{-4\sin^2\theta + 16\sin\theta\cos\theta + 12\cos^2\theta\}\,d\theta \qquad \textbf{(4.3)}$$

The indefinite integrals of $\sin^2\theta$ and $\cos^2\theta$ could readily be found using trigonometric identities. However, for the particular limits 0 and 2π, we simply note that

$$\int_0^{2\pi} (\sin^2\theta + \cos^2\theta)\,d\theta = \int_0^{2\pi} d\theta = 2\pi$$

so that, by symmetry,

$$\int_0^{2\pi} \sin^2\theta\,d\theta = \int_0^{2\pi} \cos^2\theta\,d\theta = \pi$$

Substituting in (4.3),

$$\oint_C \mathbf{F}\cdot d\mathbf{l} = -4\int_0^{2\pi} \sin^2\theta\,d\theta + 8\int_0^{2\pi} \sin 2\theta\,d\theta + 12\int_0^{2\pi} \cos^2\theta\,d\theta$$

$$= -4\pi + 0 + 12\pi = 8\pi$$

Example 4.4

Find the value of $\oint_C \mathbf{F}\cdot d\mathbf{l}$, where $\mathbf{F} = (3x - 4y)\hat{\mathbf{i}} + (4x + 2y)\hat{\mathbf{j}}$ and C is the closed path shown in Figure 4.5b, made up of a semi-ellipse C_1 and a straight-line base C_2.

Solution

For integrating around the elliptical portion, the method is very similar to that used for integrating around the circle in Example 4.3. We parametrize the ellipse in the usual way, namely

$$x = 4\cos\theta, \quad y = 3\sin\theta, \qquad 0 \leqslant \theta \leqslant \pi$$

so that $dx = -4\sin\theta\, d\theta$ and $dy = 3\cos\theta\, d\theta$

Making these substitutions in

$$\int_{C_1} \mathbf{F}\cdot d\mathbf{l} = \int_{C_1} \{(3x - 4y)\, dx + (4x + 2y)\, dy\}$$

gives $\displaystyle\int_0^\pi (48 - 30\sin\theta\cos\theta)\, d\theta = [48\theta + \tfrac{15}{2}\cos 2\theta]_0^\pi = 48\pi$

For the straight-line portion C_2, $y = 0$ and $dy = 0$ so

$$\int_{C_2} \mathbf{F}\cdot d\mathbf{l} = \int_{-4}^{4} 3x\, dx = 0$$

Hence

$$\oint_C \mathbf{F}\cdot d\mathbf{l} = 48\pi + 0 = 48\pi$$

EXERCISES

4.1 If $\mathbf{F} = x\sin y\,\hat{\mathbf{i}} + \cos y\,\hat{\mathbf{j}}$ is a force field, calculate the work done in moving a particle along the following paths:

(a) a straight-line path from $(0, 0)$ to $(0, 1)$ and then another straight-line path to $(1, 1)$,

(b) a straight-line path from $(0, 0)$ to $(1, 0)$ and then to $(1, 1)$,

(c) from $(0, 0)$ to $(1, 1)$ along the parabola $y = x^2$.

4.2 Evaluate $\oint_C \mathbf{F}\cdot d\mathbf{l}$, where $\mathbf{F} = (x - 3y)\hat{\mathbf{i}} + (y - 2x)\hat{\mathbf{j}}$ and C is the ellipse

$$x = 2\cos t, \quad y = 3\sin t, \qquad 0 \leqslant t \leqslant 2\pi$$

4.3 Evaluate $\oint_C \mathbf{A}\cdot d\mathbf{l}$, where $\mathbf{A} = 2x\hat{\mathbf{i}} + (3x-1)\hat{\mathbf{j}}$ and C is the circle in the xy-plane with centre the origin and radius 3.

4.4 Evaluate $\oint_C \mathbf{F}\cdot d\mathbf{l}$, where $\mathbf{F} = y\hat{\mathbf{i}} + (x+z)^2\hat{\mathbf{j}} + (x-z)^2\hat{\mathbf{k}}$ from (0, 0, 0) to (2, 4, 0)

(a) along the parabola $y = x^2$, $z = 0$,

(b) along the straight line $y = 2x$.

4.4 Path dependence of line integrals

A feature common to Examples 4.2, 4.3 and 4.4 is that in each case a non-zero answer is obtained for a line integral evaluated over a closed path. In the specific case of the integral over the triangular path in Example 4.2, the non-zero answer clearly follows from the different answers obtained in Example 4.1 for the values of $\int \mathbf{F}\cdot d\mathbf{l}$ along the two different paths (a) and (c) between the origin and the point (1, 2). However, it sometimes happens that the value of $\int_C \mathbf{F}\cdot d\mathbf{l}$ does *not* depend on the actual path C along which the integral is evaluated, so long as the endpoints remain fixed.

Example 4.5

Find the value of $\int_C \mathbf{F}\cdot d\mathbf{l}$, where $\mathbf{F} = y^2\hat{\mathbf{i}} + 2xy\hat{\mathbf{j}}$ for the two paths

(a) the direct straight line
(b) the parabola

used in Example 4.1 and illustrated in Figure 4.3.

Solution

Along path (a), with $y = 2x$ and $dy = 2\,dx$,

$$\int_{C_{(a)}} \mathbf{F}\cdot d\mathbf{l} = \int_0^1 \{4x^2\,dx + (2x)(2x)2\,dx\} = \int_0^1 12x^2\,dx = 4$$

Along path (b), with $y = 2x^2$ and $dy = 4x\,dx$,

$$\int_{C_{(b)}} \mathbf{F}\cdot d\mathbf{l} = \int_0^1 \{4x^4\,dx + (2x)(2x^2)4x\,dx\} = \int_0^1 20x^4\,dx = 4$$

The equality of the two line integrals in this example could be coincidental. However, if *any* path between the endpoints (0, 0) and (1, 2) is chosen, the value of 4 would still, in fact, be obtained. You could test this statement using, for example, path (c) in Figure 4.3. Since the only difference between Examples 4.1 and 4.5 lies in the different vector fields **F** in the integrand, it is clear that whether a line integral has a value that depends only on the endpoints or whether the actual path also affects the value depends critically on the form of **F**.

It turns out that we can divide vector fields into two classes:

(1) **Conservative vector fields**. These were mentioned briefly in Section 3.5. They have the property that $\int_{P_1}^{P_2} \mathbf{F}\cdot d\mathbf{l}$ does not depend on the path C between points P_1 and P_2 but only on the coordinates of P_1 and P_2. Referring to Figure 4.6a,

$$\int_{C_1} \mathbf{F}\cdot d\mathbf{l} = \int_{C_2} \mathbf{F}\cdot d\mathbf{l}$$

where C_1 and C_2 are any paths joining P_1 to P_2.

Further, for the closed path C made up of paths C_1 and $-C_2$ (see Figure 4.6b),

$$\int_{C} \mathbf{F}\cdot d\mathbf{l} = \int_{C_1} \mathbf{F}\cdot d\mathbf{l} - \int_{C_2} \mathbf{F}\cdot d\mathbf{l} = 0$$

(2) **Non-conservative vector fields**. Clearly, these are vector fields for which the properties of a conservative field do not hold; in particular, $\oint_C \mathbf{F}\cdot d\mathbf{l}$ will not be zero for all closed paths C.

Now, whether a vector field arising in modelling physical phenomena is conservative or not depends on the underlying physics of the situation. A simple force field **F**, such as a gravitational field or an electric field, *does* have the property that

$$\oint_C \mathbf{F}\cdot d\mathbf{l} = 0$$

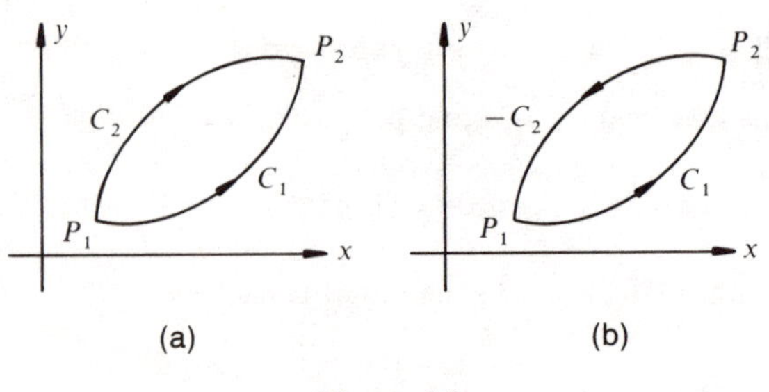

Figure 4.6

for any closed path. This is because the value of the line integral gives the work done in moving an object (or electric charge) from a given position P back to P. It is a simple matter of observation that such a move does not change the energy in the system (in the absence of frictional forces); that is, the overall work done is zero. On the other hand, a magnetic field is non-conservative – magnetic fields are produced by electric currents and one of the basic laws of the subject, consistent with experiment, is that if a magnetic field **H** is integrated around a closed path C which encloses a current I, then

$$\oint_C \mathbf{H} \cdot \mathbf{dl} = I$$

In other words, the line integral of **H** is not zero for every possible choice of closed path C, and **H** is non-conservative.

From a mathematical point of view, we need a simple criterion for testing whether or not a given vector field is conservative. To give us an indication as to how to proceed we consider the two-dimensional fields of Examples 4.1 and 4.5.

From Example 4.1, $\mathbf{F} \equiv \mathbf{F}_1 = x^2\hat{\mathbf{i}} + 3xy\hat{\mathbf{j}}$ is clearly non-conservative, because we have shown that the line integral of $\mathbf{F}_1$ is non-zero for at least one closed path (that shown in Figure 4.4). However, it is clearly impossible to show that the field $\mathbf{F} \equiv \mathbf{F}_2 = y^2\hat{\mathbf{i}} + 2xy\hat{\mathbf{j}}$ is conservative (as we assert) by evaluation of $\oint_C \mathbf{F} \cdot \mathbf{dl}$, because we cannot consider an infinite number of different paths. Fortunately, a criterion can be derived which enables us to deduce whether or not a given vector field is conservative. This test can be readily extended to three-dimensional problems, as we shall see later. The two-dimensional procedure follows from a mathematical theorem known as **Green's theorem in the plane**.

Green's theorem links a double integral over the region R with a line integral around C. (Double integrals are discussed briefly in Appendix 2.) Specifically, if $\mathbf{F} = F_x\hat{\mathbf{i}} + F_y\hat{\mathbf{j}}$ where the components F_x and F_y of the vector field **F** are 'well-behaved' functions of x and y, then Green's theorem tells

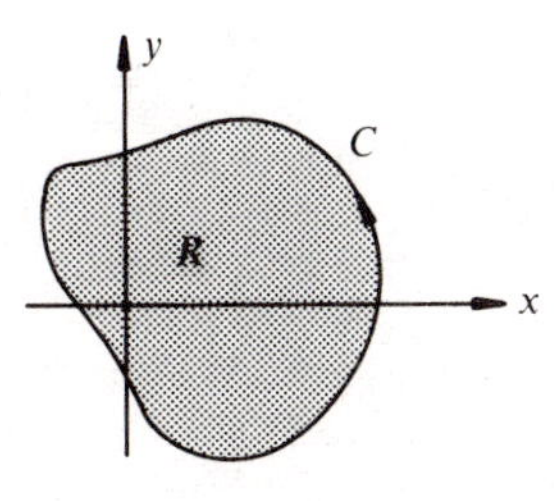

Figure 4.7

us that the line integral

$$\oint_C \mathbf{F}\cdot\mathbf{dl} \quad \text{or equivalently} \quad \oint_C (F_x\,\mathrm{d}x + F_y\,\mathrm{d}y)$$

has the same value as the double integral

$$\iint_R \left(\frac{\partial F_y}{\partial x} - \frac{\partial F_x}{\partial y}\right) \mathrm{d}x\,\mathrm{d}y \tag{4.4}$$

We can use this theorem, for example, to find areas bounded by curves using the indirect method of evaluating a line integral but for our present purposes our interest lies in deducing a test for conservative two-dimensional vector fields. Clearly, if **F** is such that

$$\frac{\partial F_x}{\partial y} = \frac{\partial F_y}{\partial x} \tag{4.5}$$

for all values of x and y, then the double integral (4.4) is zero for all regions R and hence, by Green's theorem, $\oint_C \mathbf{F}\cdot\mathbf{dl}$ must be zero for all simple closed curves C, that is, **F** is conservative.

Example 4.6

(a) Show that $\mathbf{F}_1 = x^2\hat{\mathbf{i}} + 3xy\hat{\mathbf{j}}$ is non-conservative.

(b) Show that $\mathbf{F}_2 = y^2\hat{\mathbf{i}} + 2xy\hat{\mathbf{j}}$ is conservative.

Solution

(a) The components of $\mathbf{F}_1$ are

$$F_x = x^2 \quad \text{so} \quad \frac{\partial F_x}{\partial y} = 0 \quad \text{and} \quad F_y = 3xy \quad \text{so} \quad \frac{\partial F_y}{\partial x} = 3y$$

The non-equality of the partial derivatives tells us what we already knew, namely that $\mathbf{F}_1$ is non-conservative.

(b) More significantly, for the vector field $\mathbf{F}_2$,

$$F_x = y^2 \quad \text{so} \quad \frac{\partial F_x}{\partial y} = 2y \quad \text{and} \quad F_y = 2xy \quad \text{so} \quad \frac{\partial F_y}{\partial x} = 2y$$

The equality of the partial derivatives for all x and y confirms our assertion that $\mathbf{F}_2$ is conservative everywhere, and this field could therefore perhaps be used as a mathematical model for a 'physical' field with the conservative property.

If we can establish, using (4.5), that a given vector field **F** is conservative, then we are in a very powerful position for evaluating line integrals of **F**. We know that for any closed path the line integral will be zero, whilst for a non-closed path between given endpoints P_1 and P_2 we could choose a simple path, for example one or more straight lines, for the evaluation. However, we now show that another method of evaluation is possible, one which will be particularly useful when we consider three-dimensional conservative vector fields.

Example 4.7

Show that the conservative vector field $\mathbf{F}_2 = y^2\hat{\mathbf{i}} + 2xy\hat{\mathbf{j}}$ can be expressed as the gradient of a scalar field, whereas the non-conservative vector field $\mathbf{F}_1 = x^2\hat{\mathbf{i}} + 3xy\hat{\mathbf{j}}$ cannot be so expressed.

Solution

Recall from Chapter 3 that for a scalar field $\Phi(x, y)$,

$$\mathbf{grad}\ \Phi (\equiv \nabla\Phi) = \frac{\partial\Phi}{\partial x}\hat{\mathbf{i}} + \frac{\partial\Phi}{\partial y}\hat{\mathbf{j}}$$

Firstly, if we attempt to put $\mathbf{F}_1 = \nabla\Phi$ then, equating components,

$$\frac{\partial\Phi}{\partial x} = x^2 \quad \text{and} \quad \frac{\partial\Phi}{\partial y} = 3xy$$

Differentiating the first of these equations with respect to y and the second with respect to x gives

$$\frac{\partial^2\Phi}{\partial y\,\partial x} = 0 \quad \text{and} \quad \frac{\partial^2\Phi}{\partial x\,\partial y} = 3y \quad \text{respectively}$$

However, partial differentiation tells us that for well-behaved functions, 'mixed' second derivatives are equal. Their non-equality in this case indicates the incorrectness of the assumption that $\mathbf{F}_1 = \nabla\Phi$.

On the other hand, we *can* express $\mathbf{F}_2$ as a gradient of a scalar Φ. Putting

$$\mathbf{F}_2 = y^2\hat{\mathbf{i}} + 2xy\hat{\mathbf{j}} = \frac{\partial\Phi}{\partial x}\hat{\mathbf{i}} + \frac{\partial\Phi}{\partial y}\hat{\mathbf{j}}$$

and equating components gives

$$\frac{\partial\Phi}{\partial x} = y^2 \quad \text{so} \quad \frac{\partial^2\Phi}{\partial y\,\partial x} = 2y \quad \text{and} \quad \frac{\partial\Phi}{\partial y} = 2xy \quad \text{so} \quad \frac{\partial^2\Phi}{\partial x\,\partial y} = 2y$$

Here the equality of the second derivatives indicates that our assertion that $\mathbf{F}_2 = \nabla\Phi$ was justified.

It is easy to show in general that we can associate a scalar field Φ – usually called a **potential function** – with any conservative vector field. In the two-dimensional case, suppose that $\mathbf{F}$ is conservative and we put

$$\mathbf{F} = \nabla\Phi$$

In component form this becomes

$$F_x = \frac{\partial\Phi}{\partial x} \quad \text{giving} \quad \frac{\partial F_x}{\partial y} = \frac{\partial^2\Phi}{\partial y\,\partial x} \quad \text{and} \quad F_y = \frac{\partial\Phi}{\partial y} \quad \text{giving} \quad \frac{\partial F_y}{\partial x} = \frac{\partial^2\Phi}{\partial x\,\partial y}$$

Equation (4.5) then assures us of the equality of the mixed derivatives and of the validity of the whole procedure. The only remaining problem is to determine the functional form of $\Phi(x, y)$ for a given conservative field $\mathbf{F}$.

Example 4.8

Determine an associated scalar potential Φ for the vector field $\mathbf{F} \equiv \mathbf{F}_2 = y^2\hat{\mathbf{i}} + 2xy\hat{\mathbf{j}}$.

Solution

From Example 4.7, the required function Φ is such that

$$\frac{\partial\Phi}{\partial x} = y^2 \quad \text{and} \quad \frac{\partial\Phi}{\partial y} = 2xy$$

Integrating the first equation with respect to x gives

$$\Phi(x, y) = y^2x + g(y) \tag{4.6}$$

and integrating the second equation with respect to y gives

$$\Phi(x, y) = xy^2 + h(x) \tag{4.7}$$

The functions $g(y)$ and $h(x)$ are arbitrary functions of y and x respectively and arise here, instead of the usual 'constant' of indefinite integration, because we are reversing partial differentiations with respect to x and y respectively. However, to obtain consistency between (4.6) and (4.7) we must have

$$g(y) = h(x) = c, \quad \text{a constant}$$

Hence $\Phi(x, y) = xy^2 + c$ **(4.8)**

The existence of an arbitrary constant c need not concern us, as we shall see in the next example.

An alternative approach for finding the potential Φ in this sort of problem is as follows. Integrate $\partial\Phi/\partial x = y^2$ with respect to x to give $\Phi = y^2x + g(y)$ as in equation (4.6), then differentiate this expression for

Φ with respect to y, giving

$$\frac{\partial \Phi}{\partial y} = 2yx + \frac{\mathrm{d}g}{\mathrm{d}y}$$

Comparing this last expression with the known result that $\partial\Phi/\partial y = 2yx$, we see that $g(y)$ must be such that its derivative $g'(y) = 0$, that is, $g(y)$ must be a constant, c.

Example 4.9

Evaluate $\int_C \mathbf{F}_2 \cdot \mathrm{d}\mathbf{l}$ between points P_1 (0, 0) and P_2 (1, 2) using the associated scalar potential Φ.

Solution

We know that, since $\mathbf{F}_2$ is conservative, the actual path C is immaterial.

$$\begin{aligned}
\int_{P_1}^{P_2} \mathbf{F}_2 \cdot \mathrm{d}\mathbf{l} &= \int_{P_1}^{P_2} (y^2\,\mathrm{d}x + 2xy\,\mathrm{d}y) \\
&= \int_{P_1}^{P_2} \left(\frac{\partial \Phi}{\partial x}\mathrm{d}x + \frac{\partial \Phi}{\partial y}\mathrm{d}y \right) \\
&= \int_{P_1}^{P_2} \mathrm{d}\Phi = \Phi(P_2) - \Phi(P_1) \\
&= \Big[xy^2 + c \Big]_{(0,0)}^{(1,2)}, \quad \text{using (4.8)} \\
&= 1(2)^2 + c - 0 - c = 4
\end{aligned}$$

This confirms the result of Example 4.5. Note the cancellation of the constant c.

Clearly, using Φ we can readily evaluate any other line integral of $\mathbf{F}_2$. Suppose P_1 is the point (1, 1) and P_2 is the point $(-1, 2)$, for example. Then

$$\int_{P_1}^{P_2} \mathbf{F}_2 \cdot \mathrm{d}\mathbf{l} = \Big[xy^2 + c \Big]_{(1,1)}^{(-1,2)} = (-1)(4) + c - (1)(1) - c = -5$$

You could of course confirm this by taking any simple path between P_1 and P_2, for example the one in Figure 4.8, and carrying out direct evaluation of the integral.

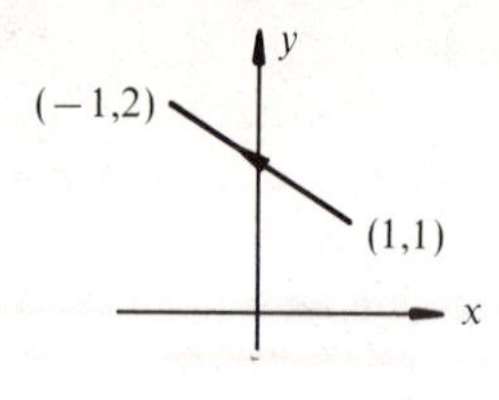

Figure 4.8

We have now concluded our study of two-dimensional line integrals. However, we shall see that the powerful result for conservative fields,

$$\int_{P_1}^{P_2} \mathbf{F}\cdot \mathrm{d}\mathbf{l} = \Phi(P_2) - \Phi(P_1), \quad \text{where } \mathbf{F} = \nabla\Phi$$

still holds in the three-dimensional case.

4.5 Line integrals in three dimensions

For an integration contour C which moves outside the xy-plane, we have three components in the line element $\mathbf{dl}$:

$$\mathrm{d}\mathbf{l} = \mathrm{d}x\,\hat{\mathbf{i}} + \mathrm{d}y\,\hat{\mathbf{j}} + \mathrm{d}z\,\hat{\mathbf{k}}$$

Hence for a vector field $\mathbf{F} = F_x\hat{\mathbf{i}} + F_y\hat{\mathbf{j}} + F_z\hat{\mathbf{k}}$, the form of the line integral is

$$\int_C \mathbf{F}\cdot \mathrm{d}\mathbf{l} = \int_C (F_x\,\mathrm{d}x + F_y\,\mathrm{d}y + F_z\,\mathrm{d}z)$$

which is a straightforward extension of the two-dimensional representation. However, the actual evaluation of three-dimensional line integrals can be slightly more involved than in the two-dimensional case. Natural first questions to ask when confronted with the need for such an evaluation are whether the vector field $\mathbf{F}$ in question is conservative and, if so, whether we can avoid actual integration by the use of a potential function $\Phi(x, y, z)$.

It turns out that we can readily test whether $\mathbf{F}$ is conservative by using the concept of the **curl** of $\mathbf{F}$, which we first met in Section 3.5. Let us first express the two-dimensional criterion for a conservative field, equation (4.5), in terms of **curl F**. If $\mathbf{F} = F_x\hat{\mathbf{i}} + F_y\hat{\mathbf{j}}$, where $\mathbf{F}$ and its components are only functions of x and y, then, using the determinant form,

$$\mathbf{curl\,F} \equiv \nabla \wedge \mathbf{F} = \begin{vmatrix} \hat{\mathbf{i}} & \hat{\mathbf{j}} & \hat{\mathbf{k}} \\ \dfrac{\partial}{\partial x} & \dfrac{\partial}{\partial y} & \dfrac{\partial}{\partial z} \\ F_x & F_y & F_z \end{vmatrix} = \left(\frac{\partial F_y}{\partial x} - \frac{\partial F_x}{\partial y}\right)\hat{\mathbf{k}}$$

Thus (4.5) is equivalent to the equation

$$\nabla \wedge \mathbf{F} \equiv 0$$

It turns out, although fuller proof must wait until Chapter 5, that the vanishing of the curl at all points in a region is a sufficient criterion for any vector field (in two or three dimensions) to be classified as conservative in that region. In the three-dimensional case, this implies the simultaneous holding of three conditions:

- The x-component of **curl F** is identically zero, that is

$$\frac{\partial F_z}{\partial y} = \frac{\partial F_y}{\partial z}$$

- The y-component of **curl F** is identically zero, that is

$$\frac{\partial F_z}{\partial x} = \frac{\partial F_x}{\partial z}$$

- The z-component of **curl F** is identically zero, that is

$$\frac{\partial F_y}{\partial x} = \frac{\partial F_x}{\partial y}$$

If we can establish that a vector field **F** is conservative, then we can express it as the gradient of a scalar field $\Phi(x, y, z)$. This follows from the result, shown in Section 3.6, that

$$\nabla \wedge \nabla\Phi \equiv 0 \quad (\text{or } \mathbf{curl\ grad}\ \Phi \equiv 0) \quad \text{for any scalar field } \Phi$$

Hence if $\nabla \wedge \mathbf{F} \equiv 0$ we can, with complete consistency, find a potential function Φ such that $\mathbf{F} = \nabla\Phi$.

Example 4.10

Show that the vector field

$$\mathbf{F} = (y^2 \cos x + z^3)\hat{\mathbf{i}} + (2y \sin x - 4)\hat{\mathbf{j}} + (3xz^2 + 2)\hat{\mathbf{k}}$$

is conservative, and deduce the corresponding scalar potential Φ.

Solution

We simply find $\nabla \wedge \mathbf{F}$:

$$\nabla \wedge \mathbf{F} = \begin{vmatrix} \hat{\mathbf{i}} & \hat{\mathbf{j}} & \hat{\mathbf{k}} \\ \dfrac{\partial}{\partial x} & \dfrac{\partial}{\partial y} & \dfrac{\partial}{\partial z} \\ (y^2 \cos x + z^3) & (2y \sin x - 4) & (3xz^2 + 2) \end{vmatrix}$$

x-component

$$(\nabla \wedge \mathbf{F})_x = \frac{\partial}{\partial y}(3xz^2 + 2) - \frac{\partial}{\partial z}(2y \sin x - 4) = 0 - 0 = 0$$

y-component

$$(\nabla \wedge \mathbf{F})_y = \frac{\partial}{\partial z}(y^2 \cos x + z^3) - \frac{\partial}{\partial x}(3xz^2 + 2) = 3z^2 - 3z^2 = 0$$

z-component

$$(\nabla \wedge \mathbf{F})_z = \frac{\partial}{\partial x}(2y \sin x - 4) - \frac{\partial}{\partial y}(y^2 \cos x + z^3)$$

$$= 2y \cos x - 2y \cos x = 0$$

Since all three components vanish identically, we have that $\nabla \wedge \mathbf{F} \equiv 0$, proving that $\mathbf{F}$ is conservative.

To obtain Φ such that $\mathbf{F} = \nabla\Phi$, we proceed by comparing components of $\mathbf{F}$ and $\nabla\Phi$ and integrating (as in the two-dimensional Example 4.8). Comparing x-components,

$$\frac{\partial \Phi}{\partial x} = y^2 \cos x + z^3$$

Integrating with respect to x and remembering that y and z are constants for a partial differentiation with respect to x,

$$\Phi = y^2 \sin x + xz^3 + f_1(y, z) \tag{4.9}$$

Comparing y-components,

$$\frac{\partial \Phi}{\partial y} = 2y \sin x - 4$$

Integrating with respect to y gives

$$\Phi = y^2 \sin x - 4y + f_2(x, z) \tag{4.10}$$

Comparing z-components,

$$\frac{\partial \Phi}{\partial z} = 3xz^2 + 2$$

Integrating with respect to z gives

$$\Phi = xz^3 + 2z + f_3(x, y) \tag{4.11}$$

The functions f_1, f_2 and f_3 can be readily determined by requiring that these three equations give a consistent expression for Φ. Hence

$$f_1(y, z) = 2z - 4y, \qquad f_2(x, z) = xz^3, \qquad f_3(x, y) = y^2 \sin x$$

so that the required function is

$$\Phi = y^2 \sin x + xz^3 + 2z - 4y$$

plus a possible constant.

Having found Φ, we can readily integrate a conservative vector field $\mathbf{F}$ between given endpoints P_1 with coordinates (x_1, y_1, z_1) and P_2 with coordinates (x_2, y_2, z_2). We have, for any path C between P_1 and P_2,

$$\begin{aligned}\int_{P_1}^{P_2} \mathbf{F}\cdot\mathrm{d}\mathbf{l} &= \int_{P_1}^{P_2} \nabla\Phi\cdot\mathrm{d}\mathbf{l}\\ &= \int_{P_1}^{P_2}\left(\frac{\partial\Phi}{\partial x}\,\mathrm{d}x + \frac{\partial\Phi}{\partial y}\,\mathrm{d}y + \frac{\partial\Phi}{\partial z}\,\mathrm{d}z\right)\\ &= \int_{P_1}^{P_2}\mathrm{d}\Phi\\ &= \Phi(x_2, y_2, z_2) - \Phi(x_1, y_1, z_1)\end{aligned}$$

Example 4.11

Show that the vector field $\mathbf{F} = (yz + 2y)\hat{\mathbf{i}} + (xz + 2x)\hat{\mathbf{j}} + (xy + 3)\hat{\mathbf{k}}$ is conservative. If P_1 is the origin $(0, 0, 0)$ and P_2 is the point $(1, 1, 1)$, find $\int_{P_1}^{P_2} \mathbf{F}\cdot\mathrm{d}\mathbf{l}$

(a) by using the associated scalar potential Φ,

(b) by evaluation of the line integral along the straight-line path between P_1 and P_2,

(c) by evaluation of the line integral along a path made up of the three straight-line segments C_1, C_2 and C_3 shown in Figure 4.9.

Solution

The conservative property of the field $\mathbf{F}$ is, of course, established by showing that $\nabla \wedge \mathbf{F}$ vanishes identically. We omit the details.

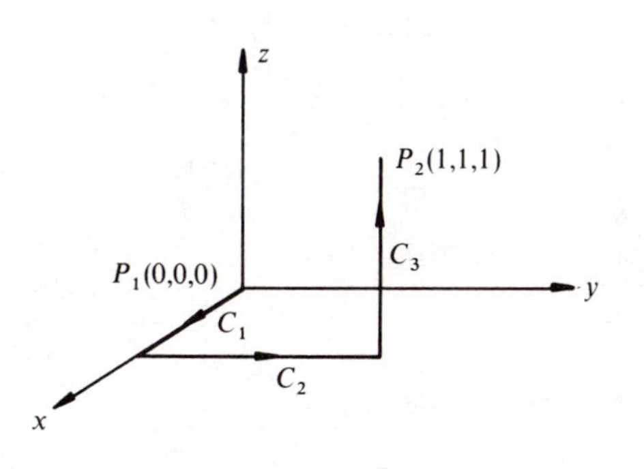

Figure 4.9

(a) We find Φ by equating components of $\mathbf{F}$ and $\nabla\Phi$.

$$\frac{\partial \Phi}{\partial x} = yz + 2y, \quad \text{giving} \quad \Phi = xyz + 2xy + f_1(y, z)$$

$$\frac{\partial \Phi}{\partial y} = xz + 2x, \quad \text{giving} \quad \Phi = xyz + 2xy + f_2(x, z)$$

$$\frac{\partial \Phi}{\partial z} = xy + 3, \quad \text{giving} \quad \Phi = xyz + 3z + f_3(x, y)$$

hence $\Phi = xyz + 2xy + 3z$ (apart from a constant). At the origin,

$$\Phi(0, 0, 0) = 0 + 0 + 0 = 0$$

whilst at the endpoint P_2,

$$\Phi(1, 1, 1) = 1 + 2 + 3 = 6$$

so
$$\int_{P_1}^{P_2} \mathbf{F} \cdot \mathrm{d}\mathbf{l} = 6 - 0 = 6$$

(b) We can represent the direct straight-line path in a simple parametric form:

$$x = t, \quad y = t, \quad z = t, \qquad 0 \leqslant t \leqslant 1$$

Along this path,

$$\mathrm{d}\mathbf{l} = \mathrm{d}t\, \hat{\mathbf{i}} + \mathrm{d}t\, \hat{\mathbf{j}} + \mathrm{d}t\, \hat{\mathbf{k}}$$

Thus
$$\int_{P_1}^{P_2} \mathbf{F} \cdot \mathrm{d}\mathbf{l} = \int_{P_1}^{P_2} \{(yz + 2y)\, \mathrm{d}x + (xz + 2x)\, \mathrm{d}y + (xy + 3)\, \mathrm{d}z\}$$

$$= \int_0^1 \{(t^2 + 2t)\, \mathrm{d}t + (t^2 + 2t)\, \mathrm{d}t + (t^2 + 3)\, \mathrm{d}t\} = 6$$

(c) On the portion C_1, $y = z = 0$, $\mathrm{d}\mathbf{l} = \mathrm{d}x\, \hat{\mathbf{i}}$ and $\mathbf{F} = 2x\hat{\mathbf{j}} + 3\hat{\mathbf{k}}$, so $\mathbf{F} \cdot \mathrm{d}\mathbf{l} = 0$ and the line integral is zero.

On C_2, $x = 1$, $z = 0$ and $\mathrm{d}\mathbf{l} = \mathrm{d}y\, \hat{\mathbf{j}}$, so

$$\mathbf{F} \cdot \mathrm{d}\mathbf{l} = \{2y\hat{\mathbf{i}} + 2\hat{\mathbf{j}} + (y + 3)\hat{\mathbf{k}}\} \cdot \mathrm{d}y\, \hat{\mathbf{j}} = 2\, \mathrm{d}y$$

giving

$$\int_{C_2} \mathbf{F} \cdot \mathrm{d}\mathbf{l} = \int_0^1 2\, \mathrm{d}y = 2$$

On C_3, $x = y = 1$ and $\mathrm{d}\mathbf{l} = \mathrm{d}z\, \hat{\mathbf{k}}$, so

$$\mathbf{F} \cdot \mathrm{d}\mathbf{l} = \{(z + 2)\hat{\mathbf{i}} + (z + 2)\hat{\mathbf{j}} + 4\hat{\mathbf{k}}\} \cdot \mathrm{d}z\, \hat{\mathbf{k}} = 4\, \mathrm{d}z$$

giving

$$\int_{C_3} \mathbf{F} \cdot \mathrm{d}\mathbf{l} = \int_0^1 4\, \mathrm{d}z = 4$$

Adding the results for C_1, C_2 and C_3 gives $0 + 2 + 4 = 6$.

Clearly, the evaluation of the line integral of a non-conservative vector field in three dimensions, which would have to be evaluated over a specified path, could be a difficult process. Fortunately, sufficient symmetry is present in most elementary applications to reduce the task to manageable proportions.

We conclude this section with an engineering application of the procedure for showing that a vector field is conservative and deducing a corresponding scalar potential.

So far we have not discussed how vector field lines originate or terminate but clearly in many physical situations they do so. For example, we can have fluid emanating from a 'source' and disappearing into a 'sink'. The simplest situation is probably a 'point source' (see Figure 4.10a), where it is evident by symmetry that a vector field $\mathbf{v}$, called the velocity field of the fluid, exists at each point and is directed radially. The magnitude of $\mathbf{v}$ falls off as the inverse square of the distance r from the source. This is because the amount of fluid crossing the surface of a sphere of radius r (and area $4\pi r^2$) must be independent of r and this will clearly only occur if $|\mathbf{v}|$ falls off as $1/r^2$.

An analogous situation occurs in electrostatics, where it is assumed that a vector field called an electric field $\mathbf{E}$ is produced by a point electric charge and, by the same reasoning as that used in the fluid case,

$$\mathbf{E} = \frac{K}{r^2}\hat{\mathbf{r}} \qquad (r^2 = x^2 + y^2 + z^2)$$

where $\hat{\mathbf{r}}$ is a unit radial vector and the constant K depends on the system of units employed.

Yet a third application of a 'point source', or rather 'point sink', is the gravitational field of a point mass, as discussed in Section 3.4.

(a)

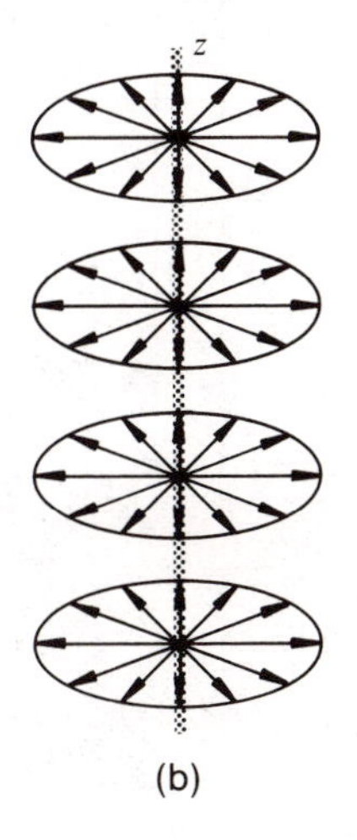

(b)

Figure 4.10

A slightly more complicated case is illustrated in Figure 4.10b, which shows an infinite line source of a vector field along the z-direction; the source could be a fluid source or an array of electric charges producing an electric field. Clearly, by symmetry, the direction of the resulting vector field ($\mathbf{v}$ or $\mathbf{E}$) is radially outwards from the source: there is no z-component. The magnitude falls off as $1/\rho$, where ρ is the distance from the line source. (Can you suggest why this is so, using an argument similar to that used above for deducing the inverse-square-law dependence of the field of a point source?) Thus the field due to the infinite line source has the form

$$\mathbf{F} = \frac{1}{\rho}\hat{\boldsymbol{\rho}} \qquad (\rho^2 = x^2 + y^2)$$

where $\hat{\boldsymbol{\rho}}$ is a unit vector directed radially from the source. Constants have been omitted from this expression so that it is independent of the precise application.

Example 4.12

Show that the vector field $\mathbf{F} = \hat{\boldsymbol{\rho}}/\rho$ is conservative and determine a corresponding scalar potential Φ. Show that Φ satisfies the Laplace equation, $\nabla^2\Phi = 0$.

Solution

In Cartesian coordinates,

$$\hat{\boldsymbol{\rho}} = \frac{x\hat{\mathbf{i}} + y\hat{\mathbf{j}}}{\sqrt{(x^2 + y^2)}} \quad \text{so} \quad \mathbf{F} = \frac{x\hat{\mathbf{i}} + y\hat{\mathbf{j}}}{x^2 + y^2}$$

Thus

$$\nabla \wedge \mathbf{F} = \begin{vmatrix} \hat{\mathbf{i}} & \hat{\mathbf{j}} & \hat{\mathbf{k}} \\ \dfrac{\partial}{\partial x} & \dfrac{\partial}{\partial y} & \dfrac{\partial}{\partial z} \\ \dfrac{x}{x^2 + y^2} & \dfrac{y}{x^2 + y^2} & 0 \end{vmatrix}$$

$$= \hat{\mathbf{k}}\left[\frac{\partial}{\partial x}\left(\frac{y}{x^2 + y^2}\right) - \frac{\partial}{\partial y}\left(\frac{x}{x^2 + y^2}\right)\right] \tag{4.12}$$

Now

$$\frac{\partial}{\partial x}\left(\frac{y}{x^2 + y^2}\right) = \frac{\partial}{\partial \rho}\left(\frac{y}{\rho^2}\right)\frac{\partial \rho}{\partial x} = \left(-\frac{2y}{\rho^3}\right)\left(\frac{x}{\rho}\right) = \frac{-2yx}{\rho^4}$$

Similarly, noting the symmetry,

$$\frac{\partial}{\partial y}\left(\frac{x}{x^2 + y^2}\right) = -\frac{2xy}{\rho^4}$$

so, from (4.12),

$$\nabla \wedge \mathbf{F} \equiv 0$$

that is, **F** is indeed conservative.

Putting $\mathbf{F} = \nabla \Phi$, therefore, and equating components, we obtain

$$\frac{\partial \Phi}{\partial x} = \frac{x}{x^2 + y^2} \quad \text{and} \quad \frac{\partial \Phi}{\partial y} = \frac{y}{x^2 + y^2} \tag{4.13}$$

Integrating the first of these equations with respect to x,

$$\Phi = \int \frac{x\,dx}{x^2 + y^2} = \frac{1}{2}\int \frac{2x\,dx}{x^2 + y^2} = \tfrac{1}{2}\ln(x^2 + y^2) + g(y)$$

Differentiating with respect to y,

$$\frac{\partial \Phi}{\partial y} = \frac{y}{x^2 + y^2} + g'(y)$$

so, by comparison with the second of equations (4.13), $g'(y) = 0$. Therefore, $g(y)$ is a constant c, and the required potential is

$$\Phi = \tfrac{1}{2}\ln(x^2 + y^2) + c = \tfrac{1}{2}\ln \rho^2 + c = \ln \rho + c$$

Now consider $\nabla^2 \Phi$ (or div grad Φ, as discussed in Section 3.6). In two dimensions,

$$\nabla^2 \Phi = \frac{\partial^2 \Phi}{\partial x^2} + \frac{\partial^2 \Phi}{\partial y^2}$$

Here, $\dfrac{\partial^2 \Phi}{\partial x^2} = \dfrac{\partial}{\partial x}\left(\dfrac{x}{x^2 + y^2}\right)$, using the first of equations (4.13)

$$= x\frac{\partial}{\partial x}\left(\frac{1}{x^2 + y^2}\right) + \frac{1}{x^2 + y^2}$$

$$= \frac{-2x^2}{(x^2 + y^2)^2} + \frac{1}{x^2 + y^2}$$

$$= \frac{y^2 - x^2}{(x^2 + y^2)^2} \tag{4.14}$$

Similarly, again noting the symmetry,

$$\frac{\partial^2 \Phi}{\partial y^2} = \frac{x^2 - y^2}{(y^2 + x^2)^2} \tag{4.15}$$

Adding equations (4.14) and (4.15) tells us that $\nabla^2 \Phi = 0$ here, so that the potential Φ is a solution of the two-dimensional Laplace equation. This is just one example of a general result of vector field theory that was first mentioned in Section 3.6, namely that if **F** is a conservative vector field ($\nabla \wedge \mathbf{F} \equiv 0$) and, in addition, is a **solenoidal** vector field ($\nabla \cdot \mathbf{F} \equiv 0$), then the associated potential Φ is such that

$$\nabla \cdot \mathbf{F} = \nabla \cdot \nabla \Phi = 0, \quad \text{that is} \quad \nabla^2 \Phi = 0$$

(The reader can readily check that the vector field in this example, namely $\mathbf{F} = (x\hat{\mathbf{i}} + y\hat{\mathbf{j}})/(x^2 + y^2)$, has zero divergence.)

Clearly, if $\mathbf{F}$ is conservative but non-solenoidal ($\nabla \cdot \mathbf{F} \neq 0$), then the scalar potential Φ would *not* satisfy $\nabla^2 \Phi = 0$. To be precise, if $\nabla \cdot \mathbf{F} = f(x, y, z)$ in general, then

$$\nabla^2 \Phi = f(x, y, z)$$

This partial differential equation is known as **Poisson's equation.**

EXERCISES

4.5 Determine which of the following vector fields are conservative:

(a) $\mathbf{F} = (1 + 4x - y)\hat{\mathbf{i}} + (e^y - 2y - x)\hat{\mathbf{j}}$ (b) $\mathbf{F} = (-x^2 + y^2)\hat{\mathbf{i}} + (x^2 - y)\hat{\mathbf{j}}$

(c) $\mathbf{F} = e^{xy}(y\hat{\mathbf{i}} + x\hat{\mathbf{j}})$ (d) $\mathbf{F} = z\hat{\mathbf{i}} + \hat{\mathbf{j}} + x\hat{\mathbf{k}}$

(e) $\mathbf{F} = (3x^2 - y^2)\hat{\mathbf{i}} - 2xy\hat{\mathbf{j}} - \hat{\mathbf{k}}$ (f) $\mathbf{F} = 3x^2yz^2\hat{\mathbf{i}} + x^3z^2\hat{\mathbf{j}} + x^3yz\hat{\mathbf{k}}$

For those fields which are conservative, find an associated scalar potential Φ.

4.6 A force field is given by $\mathbf{F} = [(\frac{1}{2}x + 2y)\hat{\mathbf{i}} + 2x\hat{\mathbf{j}}]$N. Find by integration the work done in moving an object

(a) from the origin to the point B (6, 0, 0)m,

(b) from the point B to the point C (6, 3, 0)m,

(c) around the complete triangular path $OBCO$.

Is the vector field conservative?

4.7 (a) If $\mathbf{F}_1$ and $\mathbf{F}_2$ are conservative vector fields with associated scalar potentials Φ_1 and Φ_2 respectively, show that the vector field $\mathbf{F} = \mathbf{F}_1 + \mathbf{F}_2$ is also conservative. Determine a scalar potential for $\mathbf{F}$.

(b) Verify your answers to (a) for the particular fields

$$\mathbf{F}_1 = x\hat{\mathbf{i}} + y\hat{\mathbf{j}} \quad \text{and} \quad \mathbf{F}_2 = x^2y\hat{\mathbf{i}} + \tfrac{1}{3}x^3\hat{\mathbf{j}}$$

4.8 An electric field varies as

$$\mathbf{E} = (y^2\hat{\mathbf{i}} - x^2\hat{\mathbf{j}} + xyz\hat{\mathbf{k}})\,\mathrm{V\,m^{-1}}$$

Find the work done in moving a charge of two coulombs

(a) along the x-axis from -1 mm to $+1$ mm,

(b) along the parabola $z = x^2$ from (0, 2, 0)m to (1, 2, 1)m.

Check your answers by finding the corresponding scalar potential.

4.9 Find the work done against the force field

$$\mathbf{F} = [(yz + 2z)\hat{\mathbf{i}} + xz\hat{\mathbf{j}} + (xy + 2x)\hat{\mathbf{k}}]\text{N}$$

in moving a particle from the point A $(1, 0, 1)$m to B $(0, 1, 1)$m along the following paths:

(a) the circle $x^2 + y^2 = 1$, $z = 1$,

(b) a path consisting of two straight-line segments AC and CB, where C is the point $(0, 0, 1)$m.

Show that this vector field **F** is conservative and find an associated scalar potential. Check your answers to (a) and (b) using this potential.

4.6 Occurrence of surface integrals

As mentioned in Section 4.1, part of our motivation for studying surface integrals lies in the need to obtain a more general definition of the divergence concept. Rather more directly, however, the need to sum (or, in the limiting case, integrate) a physical quantity across a surface arises in a number of areas of engineering and science.

Consider conduction of heat, for example. Heat will be conducted through a medium if a temperature gradient exists. In a simple one-dimensional case, such as heat flowing in a thin rod whose surfaces are insulated and which is lying along the x-direction, the basic physical law is that the rate Q at which heat flows across a unit area is proportional to the temperature gradient $\mathrm{d}T/\mathrm{d}x$. Mathematically, we can express this law by the simple scalar equation

$$Q = -\kappa \frac{\mathrm{d}T}{\mathrm{d}x} \tag{4.16}$$

where the constant of proportionality κ is called the **thermal conductivity** of the medium.

In the more general three-dimensional case, we can represent the heat flow by a vector field **Q** whose direction at any point gives the direction of the heat flow and whose magnitude (as in the one-dimensional case) gives the rate at which heat crosses a surface of unit area normal to the flow. The three-dimensional equivalent to (4.16) is, assuming that κ is a constant and independent of direction,

$$\mathbf{Q} = -\kappa \nabla T \tag{4.17}$$

where ∇T is the gradient of the scalar field T.

Now if **Q** is constant in direction and magnitude then, by its definition, the rate at which heat crosses a flat surface of area A normal to the direction

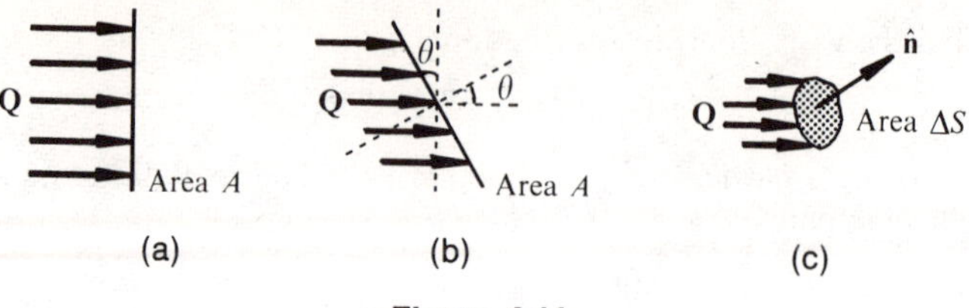

Figure 4.11

of flow is $f = |\mathbf{Q}|\, A$ (see Figure 4.11a). If the area A is inclined at an angle θ (see Figure 4.11b), then the rate of heat flow is $f = |\mathbf{Q}| \cos\theta\, A$, that is, we take the component of $\mathbf{Q}$ normal to A.

If $\mathbf{Q}$ is varying across A, or if A is not flat, these simple results will not be valid. However, if we take a sufficiently small surface element of area ΔS, say (a more common notation than ΔA), then the surface element can be assumed approximately flat, and any variation of $\mathbf{Q}$ across ΔS will be small. Hence $\Delta f = |\mathbf{Q}| \cos\theta\, \Delta S$ will give us the approximate rate of heat flow across ΔS.

It is convenient to rewrite this equation in vector form. To do this, we associate with the area ΔS a direction specified by a unit vector $\hat{\mathbf{n}}$ taken perpendicular or normal to ΔS (see Figure 4.11c) and define a **vector area**

$$\Delta \mathbf{S} = \Delta S\, \hat{\mathbf{n}}$$

The expression $|\mathbf{Q}| \cos\theta\, \Delta S$ then becomes $\mathbf{Q}\cdot\hat{\mathbf{n}}\, \Delta S$ or $\mathbf{Q}\cdot\Delta\mathbf{S}$. (Note that there are two possible unit normals $\hat{\mathbf{n}}_1$ and $\hat{\mathbf{n}}_2$ to a given surface element ΔS. However, $\hat{\mathbf{n}}_1 = -\hat{\mathbf{n}}_2$, so the only possible ambiguity is one of sign rather than of magnitude.)

For a general surface S (see Figure 4.12a), the total heat flow crossing S is approximated by adding the flows across the N elements ΔS_i $(i = 1, 2, \dots, N)$ which make up S. For the ith such element, ΔS_i,

$$\text{rate of heat flow} = \Delta f_i \approx \mathbf{Q}_i\cdot\hat{\mathbf{n}}_i\, \Delta S_i \quad \text{or} \quad \mathbf{Q}_i\cdot\Delta\mathbf{S}_i$$

so that the total rate of heat flow across S is

$$f \approx \sum_{i=1}^{N} \mathbf{Q}_i\cdot\hat{\mathbf{n}}_i\, \Delta S_i \quad \text{or} \quad \sum_{i=1}^{N} \mathbf{Q}_i\cdot\Delta\mathbf{S}_i$$

If we now increase N and correspondingly let the elements ΔS_i become smaller and smaller, the sum will tend to a limit and the rate of heat flow will be given by

$$f = \lim_{N\to\infty} \sum_{i=1}^{N} \mathbf{Q}_i\cdot\hat{\mathbf{n}}_i\, \Delta S_i \quad \text{which is written as} \quad \iint_S \mathbf{Q}\cdot\hat{\mathbf{n}}\, \mathrm{d}S$$

This integral is called the **flux** or **surface integral** of $\mathbf{Q}$ over S. The alternative notation $\iint_S \mathbf{Q}\cdot\mathrm{d}\mathbf{S}$ is also used.

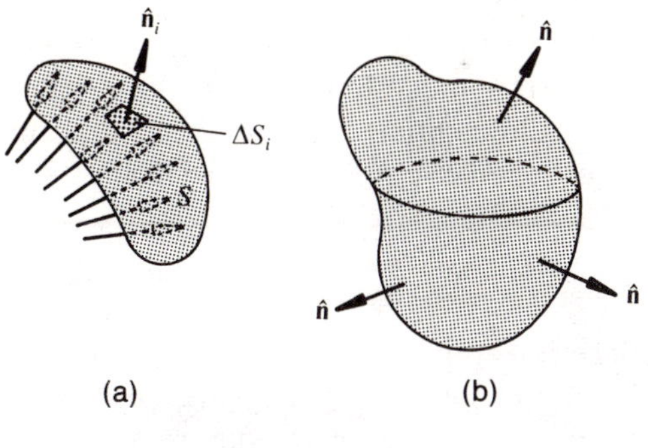

Figure 4.12

If the surface S is *closed* then, by convention, we use for $\hat{\mathbf{n}}$ the unit normal which is directed *outwards* from the volume enclosed by S (see Figure 4.12b), and the flux integral is written as

$$\oiint_S \mathbf{Q}\cdot\hat{\mathbf{n}}\,\mathrm{d}S$$

In general, for any vector field $\mathbf{F}$ we can denote the flux of $\mathbf{F}$ across a surface S by either of the representations

$$\iint_S \mathbf{F}\cdot\hat{\mathbf{n}}\,\mathrm{d}S \quad \text{or} \quad \iint_S \mathbf{F}\cdot\mathrm{d}\mathbf{S}$$

where the integrals are limiting cases of summations over S as already discussed. If $\mathbf{F}$ represents a vector field in a dynamic context, such as the heat flow vector $\mathbf{Q}$ above or a velocity field $\mathbf{v}$ in a fluid, then the flux integral has a direct physical significance. In the fluids context, if ρ_V denotes the density of the fluid (a scalar field), then the flux integral

$$\iint_S \rho_V\,\mathbf{v}\cdot\hat{\mathbf{n}}\,\mathrm{d}S$$

gives us the rate of flow of fluid across S as a mass flow per unit time.

Even in a static context where nothing is 'flowing' however, the flux concept can still arise. For example, the flux integral of an electrostatic field $\mathbf{E}$ across a closed surface S,

$$\oiint_S \mathbf{E}\cdot\hat{\mathbf{n}}\,\mathrm{d}S \quad \text{or} \quad \oiint_S \mathbf{E}\cdot\mathrm{d}\mathbf{S}$$

is a valid and indeed useful concept, as is the magnetic flux

$$\iint_S \mathbf{B}\cdot\hat{\mathbf{n}}\,\mathrm{d}S$$

where $\mathbf{B}$ is a magnetic field vector; indeed $\mathbf{B}$ is often referred to as the magnetic **flux density** vector, the name signifying, of course, that the total flux across S is obtained by integration over S of the normal component of $\mathbf{B}$.

Finally we should mention that surface integrals that are not strictly flux integrals also arise. If $\Phi(x, y, z)$ is a scalar field, then

$$\iint_S \Phi \, \mathrm{d}S$$

is called the surface integral of Φ across S. For example, if Φ denotes an electric charge density (in coulombs per square metre, say), then this surface integral gives us the total charge on the surface S: yet again the integral is a limiting case of a summation process. If the integrand $\Phi(x, y, z)$ is identically 1 at all points on a surface S, then

$$\iint_S \Phi \, \mathrm{d}S \equiv \iint_S \mathrm{d}S$$

and in this case (and only in this case), the surface integral over S gives us the magnitude of the area of S.

4.7 Evaluation of elementary surface integrals

The simplest case of a flux integral is where the vector field $\mathbf{F}$ is everywhere normal to the surface S and where $|\mathbf{F}|$ is constant over S.

Example 4.13

For the inverse square law field,

$$\mathbf{F} = \frac{K}{r^2}\hat{\mathbf{r}} \tag{4.18}$$

deduce the flux of $\mathbf{F}$ across a sphere with centre the origin and radius R.

Solution

The flux $\oiint_S \mathbf{F}\cdot\hat{\mathbf{n}} \, \mathrm{d}S$ can be expressed as $\oiint_S |\mathbf{F}| \, |\hat{\mathbf{n}}| \, \mathrm{d}S$ in this case because $\mathbf{F}$ and $\hat{\mathbf{n}}$ are everywhere parallel, both being radial vectors (see Figure 4.13a). Hence

$$\oiint_S \mathbf{F}\cdot\hat{\mathbf{n}} \, \mathrm{d}S = \oiint_S \frac{K}{R^2} \mathrm{d}S = \frac{K}{R^2}\oiint_S \mathrm{d}S = \frac{K}{R^2} 4\pi R^2 = 4\pi K$$

because, as mentioned above, $\oiint_S \mathrm{d}S$ (that is, an integrand of 1) just yields the area of S.

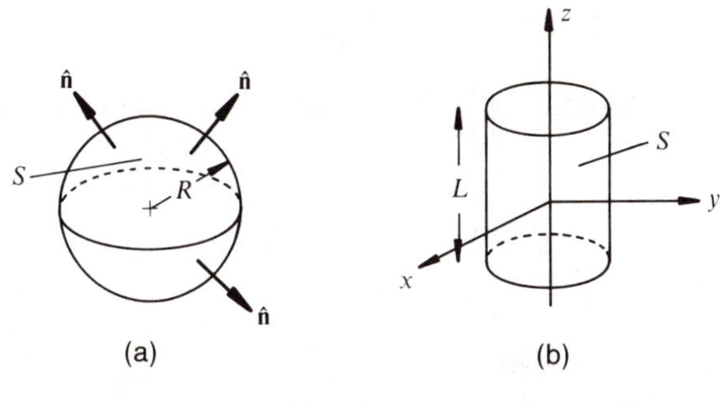

Figure 4.13

One particular case of the inverse square law is the electric field **E** due to a point charge q at the origin. For this case, (4.18) becomes

$$\mathbf{E} = \frac{q}{4\pi\varepsilon_0 r^2}\hat{\mathbf{r}} \qquad (\varepsilon_0 \text{ denotes an electrical constant})$$

so, from the result of Example 4.13,

$$\oiint_S \mathbf{E}\cdot \mathrm{d}\mathbf{S} = \frac{q}{\varepsilon_0}$$

that is, the electric flux across a spherical surface S equals the charge inside S (divided by the constant ε_0). It can be shown in elementary electrostatics that this result holds for *any* closed surface S where the total charge enclosed by S is denoted by $\sum Q$. This result, known as **Gauss' law**, is used for finding the magnitude of the electric field **E** due to a highly symmetrical charge arrangement. This is done by carefully choosing a 'Gaussian surface' S such that

- the vectors **E** and $\hat{\mathbf{n}}$ are always parallel (that is, **E** is normal to S everywhere),
- $\mathbf{E}\cdot\hat{\mathbf{n}}$ is constant over S.

For example, if we consider an infinite line charge along the z-direction (see Figure 4.13b), we reason, by symmetry and as in Section 4.5, that the electric field lines are radial, with no z-component. Also, again using symmetry and the fact that the charge is of infinite length, $|\mathbf{E}|$ depends only on the distance ρ from the line charge. Hence, if we consider the flux of **E** over the closed cylindrical surface S of radius ρ shown in Figure 4.13b:

(1) For the top and bottom surfaces, $\iint_S \mathbf{E}\cdot\hat{\mathbf{n}}\,\mathrm{d}S$ is zero as **E** is parallel to these surfaces, that is normal to $\hat{\mathbf{n}}$.

(2) For the curved surface of length L,

$$\iint_S \mathbf{E}\cdot\hat{\mathbf{n}}\,\mathrm{d}S = \iint_S |\mathbf{E}|\,\mathrm{d}S \quad \text{(as } \mathbf{E} \text{ is normal to this surface, that is parallel to } \hat{\mathbf{n}})$$

$$= |\mathbf{E}|\iint_S \mathrm{d}S \quad \text{(as } |\mathbf{E}| \text{ is constant over this surface)}$$

$$= |\mathbf{E}|2\pi\rho L$$

Hence, by Gauss' law,

$$|\mathbf{E}|2\pi\rho L = \frac{\sigma L}{\varepsilon_0}$$

where σ is the charge per unit length of the line. Therefore

$$|\mathbf{E}| = \frac{\sigma}{2\pi\varepsilon_0\rho}$$

We now consider the evaluation of flux integrals over some regular geometrical shapes, but where the quantity $\mathbf{F}\cdot\hat{\mathbf{n}}$ may vary over the surface S so that actual integration is necessary.

Example 4.14

Evaluate the flux of the vector field $\mathbf{F} = 2x^2\hat{\mathbf{i}} + xyz\hat{\mathbf{j}} - xy^2\hat{\mathbf{k}}$ over the six surfaces of the rectangular solid bounded by the planes $x=0$, $x=2$, $y=0$, $y=1$, $z=0$ and $z=\frac{3}{2}$ (see Figure 4.14a).

Solution

This is quite a long example but it is not difficult. We evaluate the flux integral

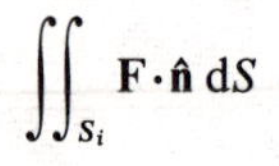

$$\iint_{S_i} \mathbf{F}\cdot\hat{\mathbf{n}}\,\mathrm{d}S$$

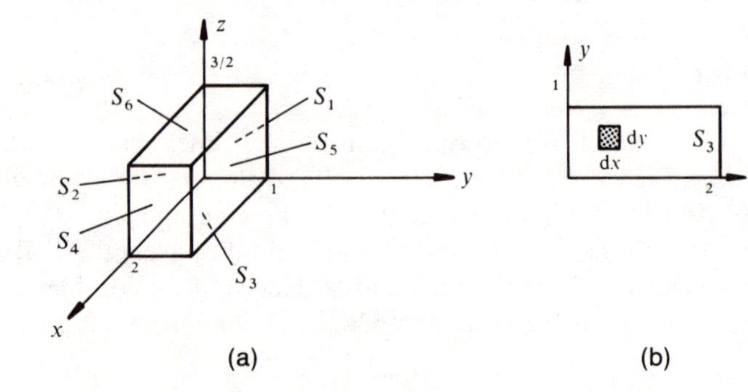

Figure 4.14

over each of the six surfaces S_i ($i = 1, 2, \ldots, 6$) in turn, and finally add the six contributions. The outward unit normal $\hat{\mathbf{n}}$ to each surface is obvious.

S_1: over this back face $x = 0$, $\hat{\mathbf{n}} = -\hat{\mathbf{i}}$ (remember it is the *outward* normal),

$$\therefore \quad \mathbf{F}\cdot\hat{\mathbf{n}} = (2x^2\hat{\mathbf{i}} + xyz\hat{\mathbf{j}} - xy^2\hat{\mathbf{k}})\cdot(-\hat{\mathbf{i}}) = -2x^2 = 0 \quad \text{on } S_1$$

$$\therefore \quad \iint_{S_1} \mathbf{F}\cdot\hat{\mathbf{n}}\,\mathrm{d}S = 0$$

S_2: over this left-hand-side face $y = 0$, $\hat{\mathbf{n}} = -\hat{\mathbf{j}}$,

$$\therefore \quad \mathbf{F}\cdot\hat{\mathbf{n}} = -xyz = 0$$

$$\therefore \quad \iint_{S_2} \mathbf{F}\cdot\hat{\mathbf{n}}\,\mathrm{d}S = 0$$

S_3: over this face, the base $z = 0$, $\hat{\mathbf{n}} = -\hat{\mathbf{k}}$,

$$\therefore \quad \mathbf{F}\cdot\hat{\mathbf{n}} = xy^2$$

$$\therefore \quad \iint_{S_3} \mathbf{F}\cdot\hat{\mathbf{n}}\,\mathrm{d}S = \iint_{S_3} xy^2\,\mathrm{d}S$$

To evaluate this integral over the flat rectangular surface S_3, we simply put $\mathrm{d}S = \mathrm{d}x\,\mathrm{d}y$ (a rectangular element, see Figure 4.14b). The surface integral $\iint_{S_3} xy^2\,\mathrm{d}S$ can be expressed as a **double integral** $\iint_{S_3} xy^2\,\mathrm{d}x\,\mathrm{d}y$, where we insert limits such that the surface of integration is covered. Here, clearly, x varies from 0 to 2 and y varies from 0 to 1. If we choose (arbitrarily) to integrate, with respect to x first, we can lay the integral out as

$$\int_0^1 \mathrm{d}y \int_0^2 xy^2\,\mathrm{d}x = \int_0^1 y^2\left[\frac{x^2}{2}\right]_0^2 \mathrm{d}y = 2\int_0^1 y^2\,\mathrm{d}y = \tfrac{2}{3}$$

The reverse order of integration is, of course, equally valid. (This sort of double integral, incidentally arises in a number of other contexts, some of them quite remote from vector field theory. Evaluation of double integrals is covered more fully in Appendix 2.)

S_4: over this front face $x = 2$, $\hat{\mathbf{n}} = +\hat{\mathbf{i}}$,

$$\therefore \quad \iint_{S_4} \mathbf{F}\cdot\hat{\mathbf{n}}\,\mathrm{d}S = \iint_{S_4} 2x^2\,\mathrm{d}S = 8\iint_{S_3} \mathrm{d}S = 8(\text{area of } S_4) = 12$$

S_5: over this right-hand-side face $y = 1$, $\hat{\mathbf{n}} = +\hat{\mathbf{j}}$,

$$\therefore \quad \iint_{S_5} \mathbf{F}\cdot\hat{\mathbf{n}}\,\mathrm{d}S = \iint_{S_5} xyz\,\mathrm{d}S = \iint_{S_5} xz\,\mathrm{d}x\,\mathrm{d}z = \int_0^{3/2} z\,\mathrm{d}z \int_0^2 x\,\mathrm{d}x$$

$$= \tfrac{9}{8} \times 2 = \tfrac{9}{4}$$

S_6: over this top face $z = \frac{3}{2}$, $\hat{\mathbf{n}} = +\hat{\mathbf{k}}$,

$$\therefore \quad \iint_{S_6} \mathbf{F} \cdot \hat{\mathbf{n}} \, dS = -\iint_{S_6} xy^2 \, dS = -\iint_{S_6} xy^2 \, dx \, dy$$

$$= -\int_0^2 x \, dx \int_0^1 y^2 \, dy = -2 \times \tfrac{1}{3} = -\tfrac{2}{3}$$

Adding these six contributions gives $\oiint_S \mathbf{F} \cdot \hat{\mathbf{n}} \, dS = \frac{57}{4}$ for the total flux of $\mathbf{F}$ over all six surfaces.

We note that evaluating a flux integral gives us a number. The significance of the number depends on the context, that is, upon the physical quantity, if any, represented by $\mathbf{F}$. (You may recall that we made similar remarks about evaluating a line integral.)

In Example 4.14 there was no difficulty in finding the unit normal $\hat{\mathbf{n}}$ to the surfaces because they were flat and, in each case, perpendicular to a coordinate axis. In other situations, finding $\hat{\mathbf{n}}$ may not be so easy; indeed, for a curved surface S the direction of $\hat{\mathbf{n}}$ will vary from point to point. Fortunately, we have already shown in Chapter 3 how to obtain normals to surfaces via the use of the gradient concept.

Recall, from Section 3.1, that at any point P in a scalar field $\Phi(x, y, z)$, the vector quantity $\nabla\Phi$ (or **grad** Φ) is perpendicular to the particular level surface $\Phi = \text{constant}$ passing through P. Hence, to obtain a *unit* vector $\hat{\mathbf{n}}$ normal to a surface $\Phi = \text{constant}$, we simply divide $\nabla\Phi$ by its magnitude (see Figure 4.15):

$$\hat{\mathbf{n}} = \frac{\nabla\Phi}{|\nabla\Phi|} \qquad \textbf{(4.19)}$$

For example, for the cylindrical surface

$x^2 + y^2 = 16$

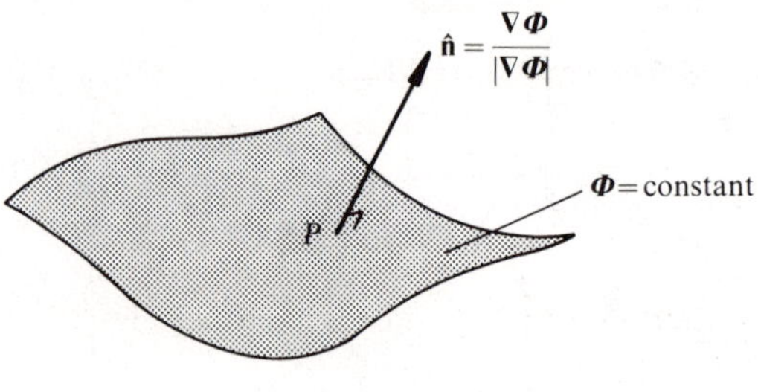

$\hat{\mathbf{n}} = \dfrac{\nabla\Phi}{|\nabla\Phi|}$

P

$\Phi = \text{constant}$

Figure 4.15

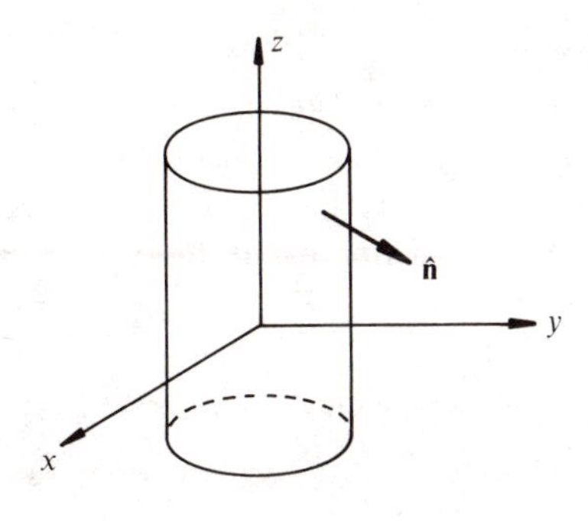

Figure 4.16

(see Figure 4.16), we can consider the cylinder as the level surface $\Phi = 16$, where $\Phi(x, y, z) = x^2 + y^2$. Hence, $\nabla\Phi = 2x\hat{\mathbf{i}} + 2y\hat{\mathbf{j}}$ is a vector normal to S, and a unit normal is

$$\hat{\mathbf{n}} = \frac{\nabla\Phi}{|\nabla\Phi|} = \frac{2x\hat{\mathbf{i}} + 2y\hat{\mathbf{j}}}{2=\sqrt{(x^2 + y^2)}} = \frac{x\hat{\mathbf{i}} + y\hat{\mathbf{j}}}{4}$$

This is clearly a vector directed radially outwards from the z-axis, as we would expect.

4.8 Cylindrical and spherical coordinate systems

Cylindrical polar coordinates

It is possible to evaluate flux integrals over cylindrical surfaces using Cartesian coordinates, as we shall see in Section 4.9. However, it is perhaps more natural to use a cylindrical coordinate system. In this system, shown in Figure 4.17a, any point P is considered as having a position on a cylinder whose axis is the z-axis. If Q is the projection of P onto the xy-plane, then

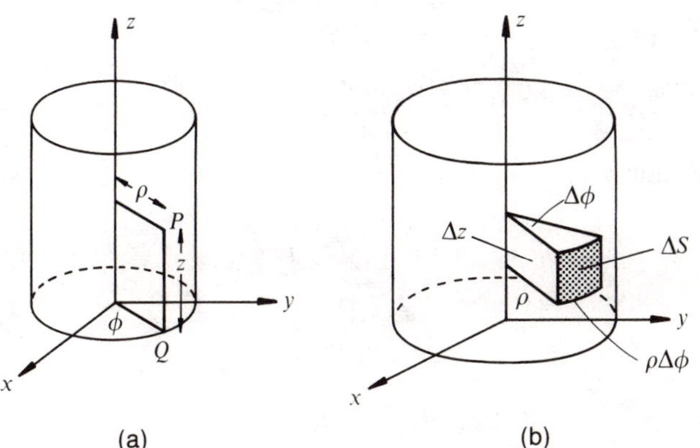

Figure 4.17

the cylindrical coordinates of P are the usual polar coordinates of Q with the addition of the z-coordinate. We use ρ and ϕ to denote the polar coordinates of Q (avoiding the usual r and θ because we use these letters later as spherical coordinates). We have the obvious relationships between (ρ, ϕ, z) and (x, y, z), the coordinates of P in cylindrical and Cartesian coordinates respectively:

$$x = \rho \cos \phi, \qquad y = \rho \sin \phi, \qquad z = z$$

$$\rho = \sqrt{(x^2 + y^2)}, \qquad \phi = \tan^{-1}\left(\frac{y}{x}\right), \qquad z = z$$

The surface $\rho = R$ (a constant) passing through P has Cartesian representation $x^2 + y^2 = R^2$, so it is indeed a cylinder of radius R with axis along the z-axis. On such a cylinder we can readily construct a surface element ΔS by incrementing z by Δz and ϕ by $\Delta\phi$ (see Figure 4.17b). This gives an element which is approximately rectangular and has area

$$\Delta S \approx \rho\,\Delta\phi\,\Delta z = R\,\Delta\phi\,\Delta z$$

Clearly, the curved surface area of a cylinder radius R and length l can be obtained by simple integration of ΔS:

$$\text{area} = \iint dS = \int_{z=0}^{l} \int_{\phi=0}^{2\pi} R\,d\phi\,dz = R \int_0^l dz \int_0^{2\pi} d\phi = 2\pi R l$$

Example 4.15

Deduce the flux of the vector field $\mathbf{F} = 4x\hat{\mathbf{i}} - 2y^2\hat{\mathbf{j}} + z^2x^2\hat{\mathbf{k}}$ over the curved surface of the cylinder $x^2 + y^2 = 4$, $0 \leqslant z \leqslant 3$ (see Figure 4.18a).

Solution

We require $\iint_S \mathbf{F}\cdot\hat{\mathbf{n}}\,dS$, where

$$\hat{\mathbf{n}} = \frac{\nabla(x^2 + y^2)}{|\nabla(x^2 + y^2)|} = \frac{2x\hat{\mathbf{i}} + 2y\hat{\mathbf{j}}}{2\sqrt{(x^2 + y^2)}} = \frac{x\hat{\mathbf{i}} + y\hat{\mathbf{j}}}{2}$$

$$\therefore \qquad \mathbf{F}\cdot\hat{\mathbf{n}} = 2x^2 - y^3$$

The surface is a cylinder of radius 2, so we convert to cylindrical coordinates:

$$x = 2\cos\phi, \quad y = 2\sin\phi \quad \text{and} \quad dS = 2\,d\phi\,dz$$

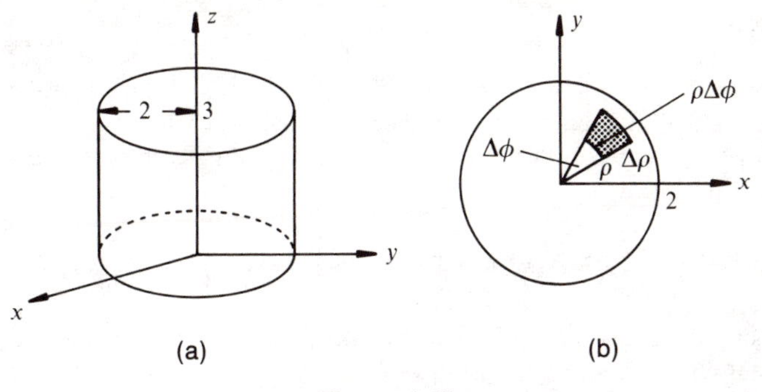

Figure 4.18

Therefore

$$\iint_S \mathbf{F}\cdot\hat{\mathbf{n}}\,dS = \int_0^3 \int_0^{2\pi} (8\cos^2\phi - 8\sin^3\phi)2\,d\phi\,dz$$

$$= 16\int_0^{2\pi} (\cos^2\phi - \sin^3\phi)\,d\phi \int_0^3 dz$$

$$= 48\int_0^{2\pi} (\cos^2\phi - \sin^3\phi)\,d\phi$$

$$= 48\pi$$

since $\int_0^{2\pi} \cos^2\phi\,d\phi = \pi$ and (by symmetry) $\int_0^{2\pi} \sin^3\phi\,d\phi = 0$.

We can also use cylindrical coordinates to find the flux of a vector field over the flat circular end of a cylinder. Viewing from the positive z-direction, we can construct a surface element as shown in Figure 4.18b where, for small enough $\Delta\phi$, the element can be approximated as a rectangle of area

$$\Delta S = \rho\,\Delta\phi\,\Delta\rho$$

Clearly, the area of the circle radius 2 is given by the integral

$$\iint dS = \int_{\rho=0}^{2} \int_{\phi=0}^{2\pi} \rho\,d\phi\,d\rho = \int_0^2 \rho\,d\rho \int_0^{2\pi} d\phi = 2\pi\frac{2^2}{2}$$

which is $\pi(\text{radius})^2$, as we would expect.

Example 4.16

Find the flux of $\mathbf{F} = 4x\hat{\mathbf{i}} - 2y^2\hat{\mathbf{j}} + z^2x^2\hat{\mathbf{k}}$ over the end faces of the cylinder shown in Figure 4.18a.

Solution

On the base $z=0$, $\hat{\mathbf{n}}=-\hat{\mathbf{k}}$, so $\mathbf{F}\cdot\hat{\mathbf{n}}=-x^2z^2=0$ and there is no flux across the base. On the top surface $z=3$, $\hat{\mathbf{n}}=\hat{\mathbf{k}}$, so $\mathbf{F}\cdot\hat{\mathbf{n}}=z^2x^2=9x^2=9\rho^2\cos^2\phi$. The flux is then given by the integral over the circle:

$$\int_0^2\int_0^{2\pi} 9\rho^2\cos^2\phi\,\rho\,\mathrm{d}\phi\,\mathrm{d}\rho=9\int_0^2\rho^3\,\mathrm{d}\rho\int_0^{2\pi}\cos^2\phi\,\mathrm{d}\phi=36\pi$$

Using these answers and the answer to Example 4.15, we see that the total flux over all three surfaces of the cylinder is 84π.

Spherical polar coordinates

Just as we normally use cylindrical coordinates when cylindrical geometry is present in a problem, so, correspondingly, we can tackle situations involving spherical geometry using a spherical coordinate system. In spherical coordinates, any point P is considered as lying on a sphere with centre the origin, and the distance of P from the origin is denoted by r. (Clearly, a surface over which r is constant is a sphere.) The other two 'spherical' coordinates of P are the angles θ and ϕ shown in Figure 4.19a. Note that the angle ϕ is the same 'polar' angle as used in the cylindrical coordinate system.

A careful study of the expanded version, Figure 4.19b, should convince you that the relations between the spherical coordinates (r, θ, ϕ) of P and its Cartesian coordinates (x, y, z) are

$$x=r\sin\theta\cos\phi,\qquad y=r\sin\theta\sin\phi,\qquad z=r\cos\theta$$

$$r=\sqrt{(x^2+y^2+z^2)},\qquad \theta=\tan^{-1}\left(\frac{\sqrt{(x^2+y^2)}}{z}\right),\qquad \phi=\tan^{-1}\left(\frac{y}{x}\right)$$

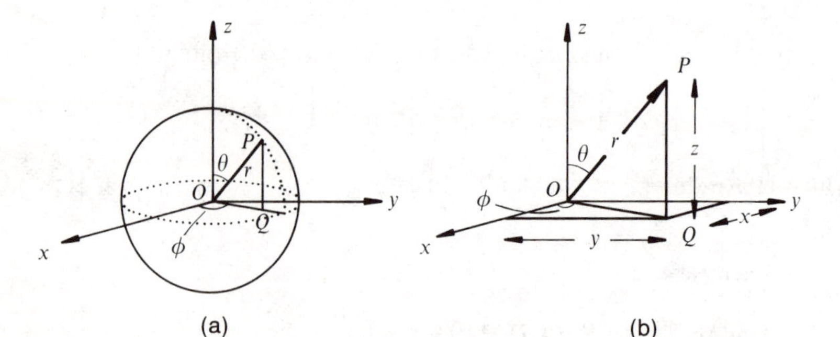

Figure 4.19

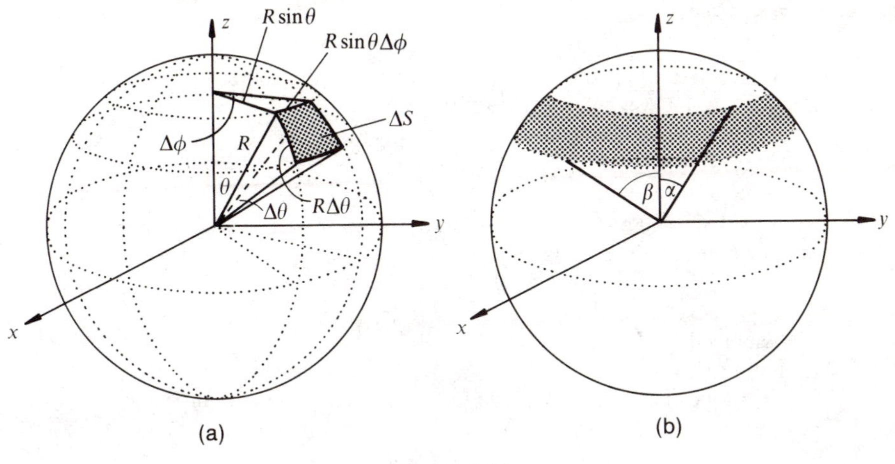

Figure 4.20

(The ρ of cylindrical coordinates (the distance OQ) is related to the r coordinate by $\rho = r \sin \theta$.)

Note that, for spherical coordinates,

- r must be positive,
- θ has a minimum of zero (for a point on the positive z-axis) and a maximum of π radians (for a point on the negative z-axis),
- the angle ϕ can vary from zero (for a point on the positive x-axis) to 2π radians.

To enable us to evaluate the flux of a vector field over a spherical surface $r = R$ we have already (in Example 3.1) deduced the unit normal $\hat{\mathbf{n}}$ so we must now deduce a surface element on this sphere. Figure 4.20a shows increments $\Delta\theta$ and $\Delta\phi$ producing arcs of length $R\,\Delta\theta$ and $R \sin \theta\,\Delta\phi$ respectively on the sphere $r = R$. For small increments, these arcs can be approximated as straight lines, so the element ΔS is approximately rectangular, with area

$$\Delta S \approx R^2 \sin \theta\, \Delta\theta\, \Delta\phi$$

Example 4.17

Using the surface element described above, deduce

(a) the surface area of a sphere of radius R,

(b) the area of the strip on the surface of the sphere $x^2 + y^2 + z^2 = R^2$ lying between the angles $\theta = \alpha$ and $\theta = \beta$ (see Figure 4.20b).

Solution

(a) $$\text{Area} = \iint_S \mathrm{d}S = R^2 \int_{\phi=0}^{2\pi} \int_{\theta=0}^{\pi} \sin\theta \, \mathrm{d}\theta \, \mathrm{d}\phi = 4\pi R^2$$

(b) $$\text{Area} = R^2 \int_{\phi=0}^{2\pi} \int_{\theta=\alpha}^{\beta} \sin\theta \, \mathrm{d}\theta \, \mathrm{d}\phi = 2\pi R^2(\cos\alpha - \cos\beta)$$

Clearly, if $\alpha = 0$ and $\beta = \pi$, we reproduce the answer to (a), because then the whole sphere is covered.

Example 4.18

If $\mathbf{r}$ is a position vector, deduce $\iint_S \mathbf{r} \cdot \hat{\mathbf{n}} \, \mathrm{d}S$, where S is the surface of a hemisphere with centre the origin and radius a (see Figure 4.21).

Solution

From Example 3.1,

$$\hat{\mathbf{n}} = \left(\frac{x}{a}\right)\hat{\mathbf{i}} + \left(\frac{y}{a}\right)\hat{\mathbf{j}} + \left(\frac{z}{a}\right)\hat{\mathbf{k}}$$

is the unit outward normal at any point on this sphere. Since $\mathbf{r} = x\hat{\mathbf{i}} + y\hat{\mathbf{j}} + z\hat{\mathbf{k}}$, we have

$$\mathbf{r} \cdot \hat{\mathbf{n}} = \frac{x^2 + y^2 + z^2}{a} = a$$

Using the fact that the surface element on the sphere $r = a$ is $a^2 \sin\theta \, \mathrm{d}\theta \, \mathrm{d}\phi$, we obtain

$$\iint_S \mathbf{r} \cdot \hat{\mathbf{n}} \, \mathrm{d}S = \int_{\phi=0}^{2\pi} \int_{\theta=0}^{\pi/2} a^3 \sin\theta \, \mathrm{d}\theta \, \mathrm{d}\phi = a^3 \int_0^{2\pi} \mathrm{d}\phi \int_0^{\pi/2} \sin\theta \, \mathrm{d}\theta$$
$$= 2\pi a^3$$

(Note the limits 0 and $\pi/2$ on the angle θ, which are needed to cover the hemispherical surface.)

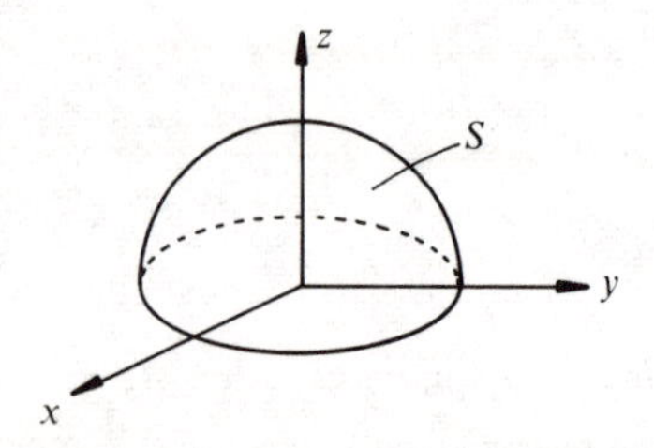

Figure 4.21

One of the obvious applications of a flux integral over a spherical surface occurs when we have a point source of a vector field. The highly symmetric case of a field that depends only on the distance r from the source has already been mentioned and is relatively trivial. More interesting applications arise when directional effects are present – when the field depends on one or both of the angles θ and ϕ as well as upon r. One such situation arises in antenna theory (an antenna is a device for radiating, or indeed receiving, electrical energy for communication purposes).

Example 4.19

A certain antenna at the origin produces electric and magnetic fields whose magnitudes are given by

$$|\mathbf{E}| = \left(\frac{E_0}{r}\right) \sin\theta \cos(\omega t - \beta r) \quad \text{(electric)}$$

$$|\mathbf{H}| = \left(\frac{H_0}{r}\right) \sin\theta \cos(\omega t - \beta r) \quad \text{(magnetic)}$$

and whose respective directions are along lines of latitude and longitude on the surface of a sphere (see Figure 4.22). Calculate (a) the instantaneous power, and (b) the average power radiated by the antenna.

Solution

We will show in Chapter 7 that the instantaneous power is given by $\mathbf{E} \wedge \mathbf{H}$ (known to electrical engineers as the Poynting vector, $\mathbf{P}$). Here, $\mathbf{P}$ is radial (see Figure 4.22), so

$$\mathbf{P} = \left(\frac{E_0 H_0}{r^2}\right) \sin^2\theta \cos^2(\omega t - \beta r)\hat{\mathbf{r}}$$

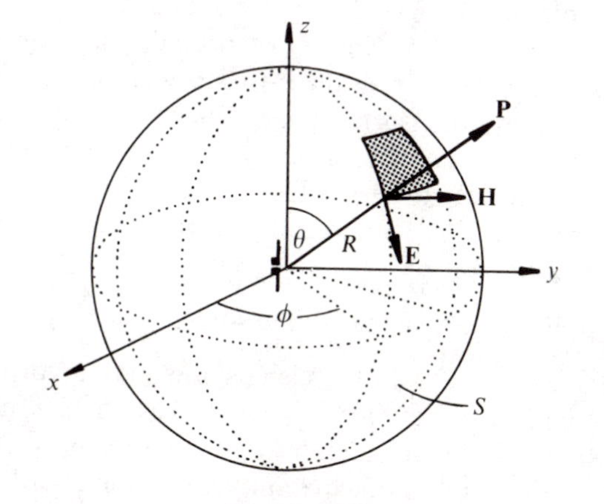

Figure 4.22

Note the directional effect due to the $\sin^2\theta$ factor, with no power being transmitted along the z-direction.

The total instantaneous power W that is radiated is the flux integral of $\mathbf{P}$ over a spherical surface of radius R:

$$W = \iint_S \mathbf{P}\cdot\hat{\mathbf{n}}\,\mathrm{d}S = \iint_S |\mathbf{P}|\,\mathrm{d}S$$

Using the expression for the surface element on such a sphere:

$$W = E_0H_0 \int_0^{2\pi}\int_0^{\pi}\left(\frac{\sin^2\theta}{R^2}\right)\cos^2(\omega t - \beta R)\,R^2\sin\theta\,\mathrm{d}\theta\,\mathrm{d}\phi$$

$$= E_0H_0\cos^2(\omega t - \beta R)\int_0^{2\pi}\mathrm{d}\phi\int_0^{\pi}\sin^3\theta\,\mathrm{d}\theta$$

Now $$\int_0^{\pi}\sin^3\theta\,\mathrm{d}\theta = \int_0^{\pi}(1-\cos^2\theta)\sin\theta\,\mathrm{d}\theta = \int_0^{\pi}(-1+\cos^2\theta)\,\mathrm{d}(\cos\theta)$$

$$= \left[-\cos\theta + \frac{\cos^3\theta}{3}\right]_0^{\pi} = \tfrac{4}{3}$$

so $$W = E_0H_0\cos^2(\omega t - \beta R)\,2\pi(\tfrac{4}{3}) = \tfrac{8}{3}\pi E_0H_0\cos^2(\omega t - \beta R)$$

The average power flow taken over one period of the cosine term is

$$W_{\mathrm{av}} = \tfrac{8}{3}\pi E_0H_0[\cos^2(\omega t - \beta R)]_{\mathrm{av}} = \tfrac{4}{3}\pi E_0H_0$$

4.9 Evaluation of surface integrals by projection

So far we have considered flux integrals over surfaces where only two coordinates are varying, for example ϕ and z over a cylindrical surface or θ and ϕ over a spherical surface. In principle, of course, any surface integral will involve two integrations. This is because, although the integrand $\mathbf{F}\cdot\hat{\mathbf{n}}$ is, in general, a scalar function of three variables, for example x, y and z in Cartesian coordinates, the surface over which the integration is being performed is usually known in a form, say $z = f(x, y)$, such that $\mathbf{F}\cdot\hat{\mathbf{n}}$ can be expressed in terms of only two variables x and y, with z being replaced by $f(x, y)$. In other words, in a surface integral the integrand can be replaced by $g(x, y)\,\mathrm{d}S$, where

$$g(x, y) = (\mathbf{F}\cdot\hat{\mathbf{n}}) \quad \text{evaluated on the surface } z = f(x, y)$$

However, in some cases the element ΔS may involve increments in all three coordinates. For example, if we had to evaluate a flux integral over the triangle S shown in Figure 4.23, how would we express ΔS? The answer is that we *project* the whole surface S onto one of the coordinate planes, for example the xy-plane, and convert the surface integral into an equivalent

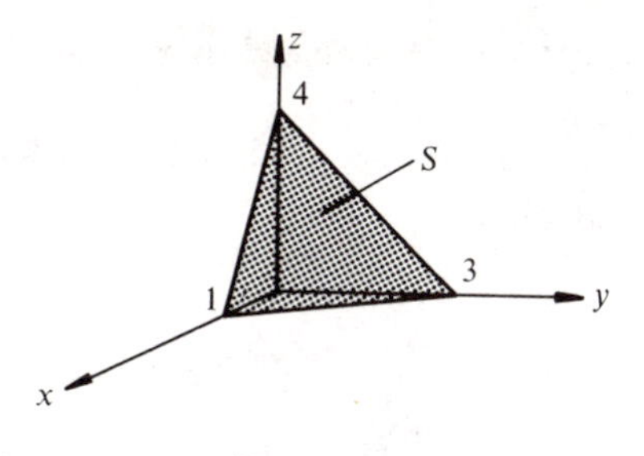

Figure 4.23

double integral over the projected region. To understand the procedure fully, we recall that any vector **F** can be expressed in terms of its three Cartesian components, F_x, F_y and F_z, where these quantities are given (see Figure 4.24a) by

$$F_x = \mathbf{F}\cdot\hat{\mathbf{i}} = \text{projection of } \mathbf{F} \text{ along the } x\text{-direction}$$

$$F_y = \mathbf{F}\cdot\hat{\mathbf{j}} = \text{projection of } \mathbf{F} \text{ along the } y\text{-direction}$$

$$F_z = \mathbf{F}\cdot\hat{\mathbf{k}} = \text{projection of } \mathbf{F} \text{ along the } z\text{-direction}$$

Similarly, a planar surface element $\Delta\mathbf{S}$ which, as we have seen, can be regarded as a vector

$$\Delta\mathbf{S} = \Delta S\,\hat{\mathbf{n}}$$

can be projected into each of the three coordinate planes, giving

$$\Delta\mathbf{S} = (\Delta D_1)\hat{\mathbf{i}} + (\Delta D_2)\hat{\mathbf{j}} + (\Delta D_3)\hat{\mathbf{k}}$$

where (see Figure 4.24b)

$$\Delta D_1 = |\Delta\mathbf{S}\cdot\hat{\mathbf{i}}| = \Delta S\,|\hat{\mathbf{n}}\cdot\hat{\mathbf{i}}| = \text{projection of } \Delta S \text{ on the } yz\text{-plane}$$

$$\Delta D_2 = |\Delta\mathbf{S}\cdot\hat{\mathbf{j}}| = \Delta S\,|\hat{\mathbf{n}}\cdot\hat{\mathbf{j}}| = \text{projection of } \Delta S \text{ on the } xz\text{-plane}$$

$$\Delta D_3 = |\Delta\mathbf{S}\cdot\hat{\mathbf{k}}| = \Delta S\,|\hat{\mathbf{n}}\cdot\hat{\mathbf{k}}| = \text{projection of } \Delta S \text{ on the } xy\text{-plane}$$

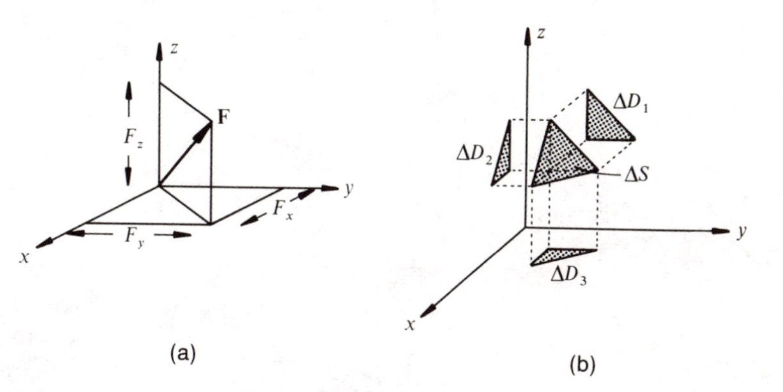

Figure 4.24

Thus an integrand $\mathbf{F}\cdot\hat{\mathbf{n}}\,\Delta S$ may be replaced by $\mathbf{F}\cdot\hat{\mathbf{n}}\,\Delta D_3/|\hat{\mathbf{n}}\cdot\hat{\mathbf{k}}|$ if a projection is made onto the xy-plane, for example.

Figure 4.25 shows a surface S projected onto the xy-plane, giving a region D. We know that the flux integral of a vector field $\mathbf{F}$ across S is the limit of the sum

$$\sum_{i=1}^{N} \mathbf{F}_i\cdot\hat{\mathbf{n}}_i\,\Delta S_i$$

so after projection this becomes

$$\sum_{i=1}^{N} \mathbf{F}_i\cdot\hat{\mathbf{n}}_i\,\frac{\Delta D_3}{|\hat{\mathbf{n}}_i\cdot\hat{\mathbf{k}}|}$$

which as $N\to\infty$ becomes a double integral, expressible in Cartesian coordinates as

$$\iint_D \mathbf{F}\cdot\hat{\mathbf{n}}\,\frac{\mathrm{d}x\,\mathrm{d}y}{|\hat{\mathbf{n}}\cdot\hat{\mathbf{k}}|}$$

It is possible to project a surface S onto a coordinate plane if any line perpendicular to that plane meets the surface S in no more than one point. This restriction does not usually cause difficulty. For example, suppose we wish to use the projection method to evaluate the flux integral of Example 4.15, where S is the curved surface of the cylinder shown in Figure 4.18a. We cannot project S onto the xy-plane here, but projection onto the xz- or yz-plane is possible. Choosing the former, we clearly do *not* have a one-to-one correspondence between points on S and points in the xz-plane. However, we can easily proceed by considering S as made up of two separate surfaces S_1 and S_2 (see Figure 4.26), so that the rectangular region D is the projection of both S_1 and S_2.

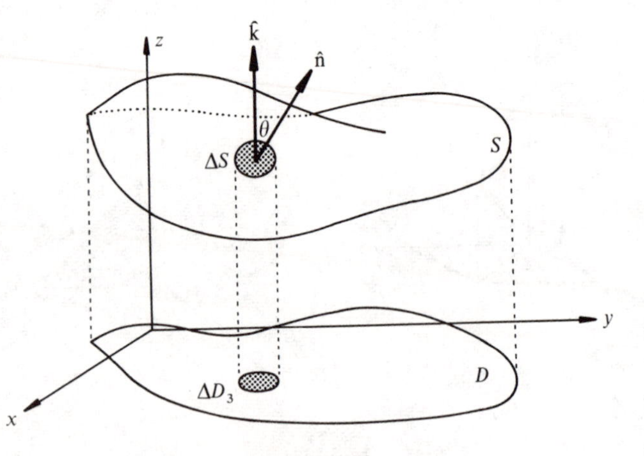

Figure 4.25

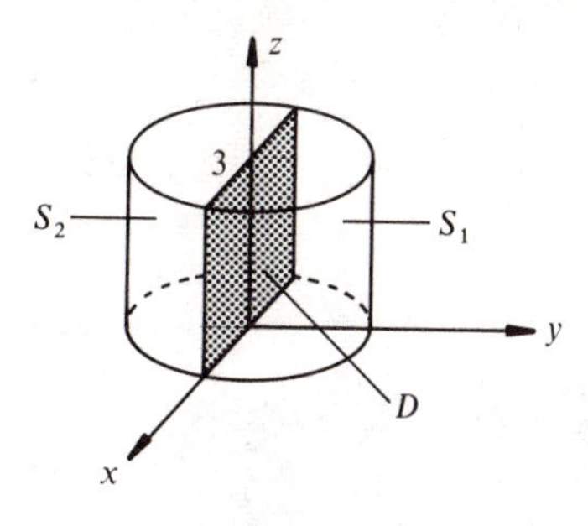

Figure 4.26

Example 4.20

Rework Example 4.15 using only Cartesian coordinates and using projection onto the xz-plane.

Solution

Segmenting S into S_1 and S_2 as explained and initially proceeding as in Example 4.15,

$$\iint_{S_1} \mathbf{F}\cdot\hat{\mathbf{n}}\,\mathrm{d}S = \iint_{S_1} (2x^2 - y^3)\,\mathrm{d}S = \iint_D (2x^2 - y^3)\frac{\mathrm{d}x\,\mathrm{d}z}{|\hat{\mathbf{n}}\cdot\hat{\mathbf{j}}|}$$

Now $\hat{\mathbf{n}}\cdot\hat{\mathbf{j}} = (\frac{1}{2}x\hat{\mathbf{i}} + \frac{1}{2}y\hat{\mathbf{j}})\cdot\hat{\mathbf{j}} = \frac{1}{2}y$, so on S_1, where $y > 0$, $|\hat{\mathbf{n}}\cdot\hat{\mathbf{j}}| = |\frac{1}{2}y| = \frac{1}{2}y$. Since the double integral over D must be expressed in terms of x and z, we eliminate y using the equation of S_1, namely $y = \sqrt{(4 - x^2)}$. Hence

$$\iint_{S_1} \mathbf{F}\cdot\hat{\mathbf{n}}\,\mathrm{d}S = \iint_D \left(\frac{4x^2}{y} - 2y^2\right)\mathrm{d}x\,\mathrm{d}z$$

$$= \iint_D \left\{\frac{4x^2}{\sqrt{(4 - x^2)}} - 2(4 - x^2)\right\}\mathrm{d}x\,\mathrm{d}z \qquad \textbf{(4.20)}$$

Similarly, projecting S_2 onto D, $|\hat{\mathbf{n}}\cdot\hat{\mathbf{j}}| = |\frac{1}{2}y| = -\frac{1}{2}y$ as y is negative on S_2. Also, $y = -\sqrt{(4 - x^2)}$ on S_2, so

$$\iint_{S_2} \mathbf{F}\cdot\hat{\mathbf{n}}\,\mathrm{d}S = \iint_D \left(-\frac{4x^2}{y} + 2y^2\right)\mathrm{d}x\,\mathrm{d}z$$

$$= \iint_D \left\{\frac{4x^2}{\sqrt{(4 - x^2)}} + 2(4 - x^2)\right\}\mathrm{d}x\,\mathrm{d}z \qquad \textbf{(4.21)}$$

Great care is needed with signs here. Adding (4.20) and (4.21) gives

$$\oiint_S \mathbf{F}\cdot\hat{\mathbf{n}}\,\mathrm{d}S = \iint_D \frac{8x^2}{\sqrt{(4 - x^2)}}\,\mathrm{d}x\,\mathrm{d}z = 8\int_0^3 \mathrm{d}z \int_{-2}^{2} \frac{x^2}{\sqrt{(4 - x^2)}}\,\mathrm{d}x$$

(the limits being appropriate to the rectangle D shown in Figure 4.26).

The z-integral is trivial, so

$$\iint_S \mathbf{F}\cdot\hat{\mathbf{n}}\,\mathrm{d}S = 48\int_0^2 \frac{x^2}{\sqrt{(4-x^2)}}\,\mathrm{d}x = 48\pi$$

which is, of course, the same answer as obtained by direct integration using cylindrical coordinates.

Example 4.21

If $\mathbf{F} = z^2\hat{\mathbf{k}}$, find the flux of $\mathbf{F}$ over that part of the surface $z = 2 - (x^2 + y^2)$ which is above the xy-plane (see Figure 4.27).

Solution

The surface S is a level surface of the function $\Phi(x, y, z) = z - 2 + (x^2 + y^2)$, so we can find $\hat{\mathbf{n}}$ using (4.19):

$$\hat{\mathbf{n}} = \frac{\nabla\Phi}{|\nabla\Phi|} = \frac{2x\hat{\mathbf{i}} + 2y\hat{\mathbf{j}} + \hat{\mathbf{k}}}{\sqrt{(4x^2 + 4y^2 + 1)}} \quad \text{so} \quad \mathbf{F}\cdot\hat{\mathbf{n}} = \frac{z^2}{\sqrt{(4x^2 + 4y^2 + 1)}}$$

Projecting onto the xy-plane, $\mathrm{d}S$ is replaced by

$$\frac{\mathrm{d}D}{|\hat{\mathbf{n}}\cdot\hat{\mathbf{k}}|} = \sqrt{(4x^2 + 4y^2 + 1)}\,\mathrm{d}D$$

Hence $\iint_S \mathbf{F}\cdot\hat{\mathbf{n}}\,\mathrm{d}S = \iint_D z^2\,\mathrm{d}D$

Putting $z = 2 - x^2 - y^2$ (over S) we finally obtain

$$\iint_S \mathbf{F}\cdot\hat{\mathbf{n}}\,\mathrm{d}S = \iint_D (2 - x^2 - y^2)\,\mathrm{d}D$$

As D is the circle $x^2 + y^2 = 2$, we use polar coordinates for the double integral:

$$x = \rho\cos\phi, \qquad y = \rho\sin\phi, \qquad \mathrm{d}D = \rho\,\mathrm{d}\phi\,\mathrm{d}\rho$$

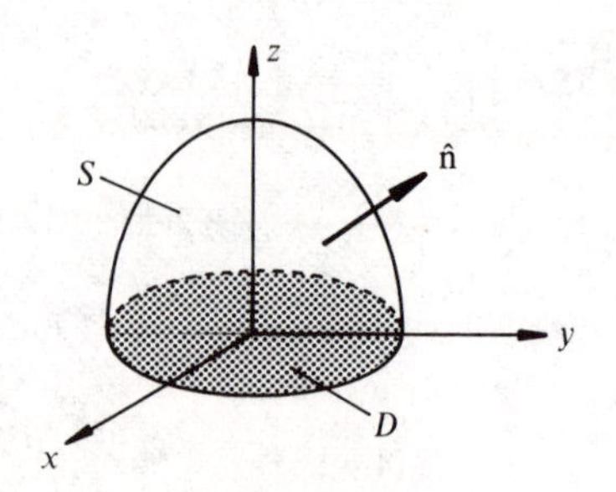

Figure 4.27

Therefore

$$\iint_S \mathbf{F}\cdot\hat{\mathbf{n}}\, dS = \int_0^{2\pi}\int_0^{\sqrt{2}} (2-\rho^2)\rho\, d\rho\, d\phi = 2\pi \int_0^{\sqrt{2}} (2\rho-\rho^3)\, d\rho = 2\pi$$

EXERCISES

4.10 Distinguish between the area of a surface and a surface integral.

4.11 Find the flux of the vector field $\mathbf{F} = x\hat{\mathbf{i}} + y\hat{\mathbf{j}}$ over the complete closed surface bounded by the planes $x = \pm 1, y = \pm 1, z = \pm 1$.

4.12 Find the total electric flux crossing the cylinder whose boundary surfaces are

$$\rho = 2, \qquad z = 0, \qquad z = 10 \quad \text{(distances are in metres)}$$

if the electric flux density is

$$\mathbf{D} = 30\exp(-\tfrac{1}{2}\rho)\cos^2\phi\,\hat{\boldsymbol{\rho}} - z\hat{\mathbf{k}} \quad \mathrm{C\,m^{-2}} \quad \text{(in cylindrical coordinates)}$$

4.13 If $\mathbf{F} = x\hat{\mathbf{i}} + y\hat{\mathbf{j}} + (z^2-1)\hat{\mathbf{k}}$, calculate $\oiint_S \mathbf{F}\cdot\hat{\mathbf{n}}\, dS$ over the entire closed surface bounded by the cylinder $x^2+y^2=a^2$, the plane $z=0$ and the plane $z=b$.

4.14 A vector field $\mathbf{F}$ is given by $\mathbf{F} = \frac{1}{2}y^2\hat{\mathbf{i}} + x\hat{\mathbf{j}}$.

(a) Find the flux of $\nabla \wedge \mathbf{F}$ over the hemisphere $z = +\sqrt{(4-x^2-y^2)}$

(i) using spherical coordinates, (ii) by the method of projections.

(b) Evaluate the line integral $\oint_C \mathbf{F}\cdot d\mathbf{l}$ over the circle $x^2+y^2=4$, $z=0$. (You should obtain the same answer as for the flux integral.)

4.15 If $\mathbf{F} = y\hat{\mathbf{i}} + 2\hat{\mathbf{j}} + \hat{\mathbf{k}}$ and S is that portion of the spherical surface $x^2+y^2+z^2=9$ in the first octant (that is, the region bounded by planes $x=0$, $y=0$ and $z=0$), show that the flux integral $\iint_S \mathbf{F}\cdot\hat{\mathbf{n}}\, dS$ may be transformed using spherical coordinates into

$$9\int_0^{\pi/2}\int_0^{\pi/2} (3\sin^3\theta\sin\phi\cos\phi + 2\sin^2\theta\sin\phi + \sin\theta\cos\theta)\, d\theta\, d\phi$$

Hence evaluate the flux integral.

4.16 Given the vector field $\mathbf{A} = (x+y^2)\hat{\mathbf{i}} - 2x\hat{\mathbf{j}} + 2yz\hat{\mathbf{k}}$, find the flux of $\mathbf{A}$ across the surface of the plane $2x+y+2z=6$ in the first octant.

4.17 If $\mathbf{F} = y^2\hat{\mathbf{i}} + 2yz\hat{\mathbf{j}} + xy\hat{\mathbf{k}}$, calculate $\iint_S \nabla \wedge \mathbf{F} \cdot \mathrm{d}\mathbf{S}$ where S is the surface defined by $2x + 2y + z = 2$ and bounded by $x = 0$, $y = 0$ and $z = 0$ in the first octant.

4.10 Volume integrals

We have now studied line integrals along curves in space, and integrals which are taken over surfaces. Therefore it will not surprise you to learn that a volume integral is one taken over a closed region or volume in space. Like their line and surface counterparts, volume integrals are also limiting cases of summations.

To begin with a physical example, we learn in elementary science that the mass of a body occupying a volume V is obtained by multiplying V by the mass density of the body, ρ_V. However, this is only valid if ρ_V is constant throughout the body. If ρ_V varies from point to point, that is, $\rho_V = f(x, y, z)$, then to find the total mass we first divide the volume V into N small elements, $\Delta V_1, \Delta V_2, \ldots, \Delta V_N$, where V is the sum of all these small volumes, that is, $V = \sum_{i=1}^{N} \Delta V_i$. We can then say that if (x_i, y_i, z_i) are the coordinates of a point within the volume element ΔV_i, the mass of the element ΔV_i is

$$\Delta M_i \approx \rho_V(x_i, y_i, z_i)\, \Delta V_i$$

and the total mass of the whole body is given approximately by the summation

$$M \approx \sum_{i=1}^{N} \rho_V(x_i, y_i, z_i)\, \Delta V_i \qquad \textbf{(4.22)}$$

To improve the approximation, we take smaller and smaller elements and correspondingly let N tend to ∞. Assuming that ρ_V is a well-behaved function of the coordinates x, y and z, it can be shown that the sum in (4.22) tends to a finite limit, which is called the **volume integral** of ρ_V over the volume V. The mass M is then given exactly by

$$M = \iiint_V \rho_V \,\mathrm{d}V \qquad \textbf{(4.23)}$$

(the reason for using three integral signs will emerge shortly).

The considerations above are particularly relevant to the case of a compressible fluid. However, we can define the volume integral of f over a volume V in the same way for any scalar field $f(x, y, z)$:

$$\iiint_V f(x, y, z)\,\mathrm{d}V = \lim_{N\to\infty} \sum_{i=1}^{N} f(x_i, y_i, z_i)\, \Delta V_i$$

A further example of a volume integral is the calculation of the total electric charge Q in a region V from the charge density (in coulombs per metre^3) $\sigma(x, y, z)$:

$$Q = \iiint_V \sigma(x, y, z)\, \mathrm{d}V$$

As a volume integral involves integrating a scalar field over a volume (to give, of course, a scalar answer), you might be wondering at this stage what volume integrals have to do with vector fields. A full discussion of the link must wait until Chapter 5, but we already know that the divergence $\nabla \cdot \mathbf{F}$ of a vector field is a scalar field and so could be integrated over a volume. Indeed, volume integrals of the type

$$\iiint_V \nabla \cdot \mathbf{F}\, \mathrm{d}V$$

turn out to be highly significant.

The evaluation of volume integrals is a fairly straightforward process. We select a volume element ΔV in an appropriate coordinate system and decide on suitable limits for the integral, the limits being such that we cover the correct volume, no more and no less. Finding the limits is the main source of difficulty, the actual integration process usually being relatively routine.

In Cartesian coordinates, the volume element is a rectangular parallelepiped $\Delta V = \Delta x\, \Delta y\, \Delta z$ (see Figure 4.28a) obtained by incrementing each of the three coordinates in turn. The form of the volume integral in Cartesian coordinates is, therefore,

$$\iiint_V f(x, y, z)\, \mathrm{d}x\, \mathrm{d}y\, \mathrm{d}z \qquad \textbf{(4.24)}$$

which suggests that a volume integral is a **triple integral** (that is, just one step beyond a double integral) and is evaluated by successive integration with respect to x, then y, then z (or, in principle, in any order we choose).

In cylindrical coordinates (ρ, ϕ, z), the volume element is obtained by incrementing the coordinates ρ, ϕ and z by $\Delta\rho$, $\Delta\phi$ and Δz. The resulting volume element is as shown in Figure 4.28b where, assuming that the increments are sufficiently small for the element to be approximated as a rectangular parallelepiped,

$$\Delta V \approx \Delta\rho\, (\rho\, \Delta\phi)\, \Delta z = \rho\, \Delta\rho\, \Delta\phi\, \Delta z$$

Thus, a volume integral in cylindricai coordinates has the form

$$\iiint_V f(\rho, \phi, z) \rho\, \mathrm{d}\rho\, \mathrm{d}\phi\, \mathrm{d}z \qquad \textbf{(4.25)}$$

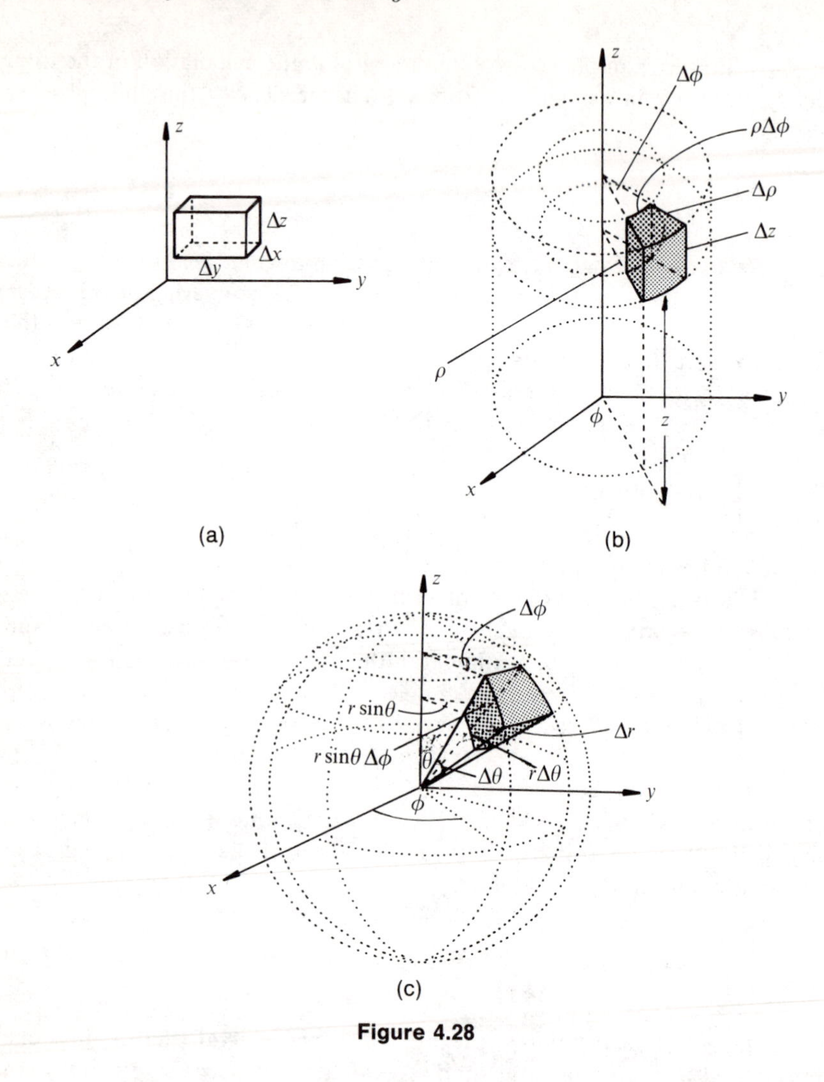

Figure 4.28

The substitutions $x = \rho \cos \phi$, $y = \rho \sin \phi$ are made, if necessary, to express the integrand in cylindrical coordinates.

In spherical coordinates (r, θ, ϕ), the increments Δr, $\Delta\theta$ and $\Delta\phi$ produce, as shown in Figure 4.28c, a solid based on sides of lengths Δr, $r\,\Delta\theta$ and $r \sin\theta\,\Delta\phi$. Again, for small increments, the solid is approximately cuboid, with volume

$$\Delta V \approx \Delta r\,(r\,\Delta\theta)(r \sin\theta\,\Delta\phi) = r^2 \sin\theta\,\Delta r\,\Delta\theta\,\Delta\phi$$

Hence, a volume integral in spherical coordinates has the form

$$\iiint_V f(r, \theta, \phi) r^2 \sin\theta \,\mathrm{d}r\,\mathrm{d}\theta\,\mathrm{d}\phi \tag{4.26}$$

The substitutions already discussed in Section 4.8,

$$x = r \sin\theta \cos\phi, \qquad y = r \sin\theta \, \sin\phi, \qquad z = r\cos\theta$$

are made, if necessary, to express the integrand in spherical coordinates.

4.11 Worked examples on volume integrals

Example 4.22

If $f = 4x + xz$, evaluate $\iiint_V f \, \mathrm{d}V$ over the rectangular solid bounded by $x = 0$, $x = 2$, $y = 0$, $y = 1$, $z = 0$ and $z = \frac{3}{2}$ (see Figure 4.14a and Example 4.14).

Solution

It is natural to use Cartesian coordinates here, and the limits are obvious from Figure 4.14a. Hence

$$\iiint_V f \, \mathrm{d}V = \iiint_V (4x + xz) \, \mathrm{d}x \, \mathrm{d}y \, \mathrm{d}z$$

If we choose to integrate with respect to x, then y, then z (although any other order is possible), the integral can be laid out as

$$\begin{aligned}
\int_0^{3/2} \mathrm{d}z \int_0^1 \mathrm{d}y \int_0^2 (4x + xz) \, \mathrm{d}x &= \int_0^{3/2} \mathrm{d}z \int_0^1 \left[2x^2 + \tfrac{1}{2}x^2 z \right]_0^2 \mathrm{d}y \\
&= \int_0^{3/2} \mathrm{d}z \int_0^1 (8 + 2z) \, \mathrm{d}y \\
&= \int_0^{3/2} \left[8y + 2zy \right]_0^1 \mathrm{d}z \\
&= \int_0^{3/2} (8 + 2z) \, \mathrm{d}z = \tfrac{57}{4}
\end{aligned}$$

This answer is the same as that obtained in Example 4.14 for the *flux* integral of

$$\mathbf{F} = 2x^2 \hat{\mathbf{i}} + xyz \hat{\mathbf{j}} - xy^2 \hat{\mathbf{k}}$$

over the surface of this volume V. The scalar function $f = 4x + xz$ is clearly the divergence $\nabla \cdot \mathrm{F}$, so we have equality between a *flux* integral across a surface and a *volume* integral over the volume bounded by that surface. Specifically,

$$\oiint_S \mathbf{F} \cdot \hat{\mathbf{n}} \, \mathrm{d}S = \iiint_V \nabla \cdot \mathbf{F} \, \mathrm{d}V$$

We shall see in Chapter 5 that this is, in fact, a general and most powerful result, known as the divergence theorem.

Example 4.23

Evaluate $\iiint_V (2x + y)\,\mathrm{d}V$, where V is the region bounded by the surface $z = 4 - x^2$ and the planes $x = 0$, $y = 0$, $y = 3$ and $z = 0$.

Solution

Referring to Figure 4.29, if we choose to integrate with respect to z first, we obtain

$$I = \int_0^3 \mathrm{d}y \int_0^2 \mathrm{d}x \int_0^{4-x^2} (2x + y)\,\mathrm{d}z$$

The first integral corresponds to a summation over a column on a rectangular base $\Delta x\,\Delta y$ in the $z = 0$ plane, the top of the column being where it intersects the surface $z = 4 - x^2$. Carrying out the first integration (with respect to z) gives

$$\left[2xz + yz\right]_0^{4-x^2} = 2x(4 - x^2) + y(4 - x^2)$$

hence $$I = \int_0^3 \mathrm{d}y \int_0^2 (8x - 2x^3 + 4y - yx^2)\,\mathrm{d}x$$

which is just an ordinary double integral over the base rectangle $0 \leqslant x \leqslant 2$, $0 \leqslant y \leqslant 3$. On evaluation, this gives an answer of 48.

As was the case with double integrals, the order of integration in a triple integral is, in principle, immaterial although sometimes one order of integration may be more difficult than others. Here, for example, we could choose to integrate with respect to x first, and lay the integral out as

$$I \equiv \int_0^3 \mathrm{d}y \int_0^4 \mathrm{d}z \int_0^{\sqrt{(4-z)}} (2x + y)\,\mathrm{d}x$$

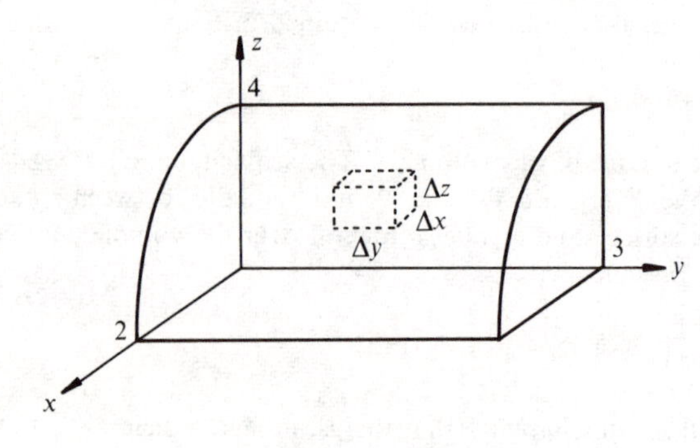

Figure 4.29

The first integral gives

$$\left[x^2 + yx\right]_0^{\sqrt{(4-z)}} = 4 - z + y\sqrt{(4-z)}$$

What remains is the double integral

$$I = \int_0^3 \mathrm{d}y \int_0^4 \{4 - z + y\sqrt{(4-z)}\}\,\mathrm{d}z$$

which is taken over the back rectangle $0 \leqslant y \leqslant 3$, $0 \leqslant z \leqslant 4$ in Figure 4.29. The reader is encouraged to complete the integration to show that the same answer, 48, is obtained. Yet a third formulation, the first integration being with respect to y, is also possible.

A volume integral of the type $\iiint_V f\,\mathrm{d}V$ is, as we have seen, interpreted as a summation of a continuously varying quantity f over a volume V. In the particular case $f(x, y, z) = 1$,

$$\iiint_V f\,\mathrm{d}V = \iiint_V \mathrm{d}V$$

which is just the volume of V.

Example 4.24

Deduce the volumes of (a) a cylinder of length l and radius R, and (b) a sphere of radius R.

Solution

(a) Using (4.25) with $f(\rho, \phi, z) \equiv 1$ and using limits appropriate to the cylinder ($0 \leqslant z \leqslant l$, $0 \leqslant \rho \leqslant R$, $0 \leqslant \phi \leqslant 2\pi$), we have

$$\text{volume} = \int_0^l \int_0^{2\pi} \int_0^R \rho\,\mathrm{d}\rho\,\mathrm{d}\phi\,\mathrm{d}z$$

$$= \int_0^l \mathrm{d}z \int_0^{2\pi} \mathrm{d}\phi \int_0^R \rho\,\mathrm{d}\rho \quad \text{(a product of three independent integrals)}$$

$$= \left[z\right]_0^l \left[\phi\right]_0^{2\pi} \left[\tfrac{1}{2}\rho^2\right]_0^R = \pi R^2 l$$

(b) Similarly, from (4.26) with $f(r, \theta, \phi) \equiv 1$ and defining the sphere by $0 \leqslant r \leqslant R$, $0 \leqslant \theta \leqslant \pi$, $0 \leqslant \phi \leqslant 2\pi$,

$$\text{volume} = \int_0^{2\pi} \int_0^{\pi} \int_0^{R} r^2 \sin\theta \, dr \, d\theta \, d\phi = \int_0^{2\pi} d\phi \int_0^{\pi} \sin\theta \, d\theta \int_0^{R} r^2 \, dr$$

$$= 2\pi \Big[-\cos\theta \Big]_0^{\pi} \Big[\tfrac{1}{3} r^3 \Big]_0^{R} = \tfrac{4}{3}\pi R^3$$

The examples above were comparatively easy and were designed to give confidence in the use of less familiar coordinate systems. The following examples are slightly more substantial.

Example 4.25

Evaluate $\iiint_V (x^2 + y^2 + z^2) \, dV$, where V is the interior of the sphere with centre the origin and radius a.

Solution

The equation of the *surface* of this sphere is

$$x^2 + y^2 + z^2 = a^2$$

However, since the integral is taken over the whole *volume* of the sphere, we cannot replace the integrand by a^2. Instead, we switch to spherical coordinates and put $f(r, \theta, \phi) = r^2$ $(= x^2 + y^2 + z^2)$, so the integral becomes, using (4.26),

$$\iiint_V r^2 r^2 \sin\theta \, dr \, d\theta \, d\phi = \int_0^{2\pi} d\phi \int_0^{\pi} \sin\theta \, d\theta \int_0^{a} r^4 \, dr = \tfrac{4}{5}\pi a^5$$

Example 4.26

The density of the material of a cylindrical body of radius R is directly proportional to the distance from the axis of the cylinder. Deduce (a) the total mass of the cylinder, and (b) the moment of inertia about the axis.

Solution

(a) We assume, as usual, that the cylinder has axis along the z-axis and is of length l. The density can be expressed in cylindrical coordinates as $\rho_V = \kappa\rho$, $0 \leqslant \rho \leqslant R$, where κ is a constant of proportionality. The total mass is therefore

$$M = \iiint_V \rho_V \, dV = \int_0^{l} \int_0^{2\pi} \int_0^{R} \kappa\rho \, \rho \, d\rho \, d\phi \, dz$$

In order to illustrate a particular point we integrate first with respect to ϕ, then z, then ρ:

$$M = \kappa \int_0^R \rho^2\, d\rho \int_0^l dz \int_0^{2\pi} d\phi = \kappa \int_0^R \rho 2\pi\rho l\, d\rho$$

Note that $2\pi\rho l\,\Delta\rho$ is the volume of the annulus between cylinders of radii ρ and $\rho + \Delta\rho$; for a problem with **cylindrical symmetry**, such as this, we could begin with $\Delta V = 2\pi\rho l\,\Delta\rho$ as the volume element and just evaluate a single integral with respect to ρ.

Continuing the problem,

$$M = 2\pi\kappa l \int_0^R \rho^2\, d\rho = \tfrac{2}{3}\pi\kappa l R^3 \tag{4.27}$$

(b) The moment of inertia of a point mass about an axis is

mass × (distance of mass from the axis)^2

so for the basic volume element at a distance ρ from the axis of the cylinder,

$$\Delta I \approx (\Delta V\,\kappa\rho)(\rho^2) = (\text{mass})(\text{distance})^2$$

Hence, the total moment of inertia is

$$I = \iiint_V \kappa\rho^3\, dV = \kappa \iiint_V \rho^3 \rho\, d\rho\, d\phi\, dz = \kappa \int_0^R \rho^4\, d\rho \int_0^{2\pi} d\phi \int_0^l dz$$

$$= \kappa \int_0^R \rho^4 2\pi l\, d\rho \quad \text{(which could have been our starting point)}$$

$$= \tfrac{2}{5}\pi\kappa l R^5$$

or, using (4.27),

$$I = \tfrac{3}{5} M R^2$$

Example 4.27

If $\mathbf{F}$ is the vector field $\mathbf{F} = 4x\hat{\mathbf{i}} - 2y^2\hat{\mathbf{j}} + z^2x^2\hat{\mathbf{k}}$ (as in Example 4.15), find the volume integral

$$\iiint_V \nabla\cdot\mathbf{F}\, dV$$

over the region inside the cylinder defined by surfaces $x^2 + y^2 = 4$, $z = 0$ and $z = 3$ (see Figure 4.18a).

Solution

We have

$$\nabla\cdot\mathbf{F} = 4 - 4y + 2zx^2$$

Converting to cylindrical coordinates,

$$\nabla\cdot\mathbf{F} = 4 - 4\rho\sin\phi + 2z\rho^2\cos^2\phi$$

Hence $$\iiint_V \nabla\cdot\mathbf{F}\,\mathrm{d}V = \iiint_V \nabla\cdot\mathbf{F}\rho\,\mathrm{d}\rho\,\mathrm{d}\phi\,\mathrm{d}z$$

$$= \int_0^3\int_0^{2\pi}\int_0^2 (4\rho - 4\rho^2\sin\phi + 2z\rho^3\cos^2\phi)\,\mathrm{d}\rho\,\mathrm{d}\phi\,\mathrm{d}z$$

$$= 84\pi \quad \text{by straightforward integration (in any order)}$$

Note that this value is the same as the total flux integral of $\mathbf{F}$ over the whole surface (curved and flat ends) of the cylinder, a similar result to that obtained in Example 4.22.

Example 4.28

Find the moment of inertia and the radius of gyration of a thick hollow sphere about a diameter. Assume constant density.

Solution

Let a be the inner radius of the sphere, b the outer radius and ρ_V the constant density. Because of symmetry, we can consider the eighth of the sphere in the first octant (see Figure 4.30). The volume element ΔV of mass $\rho_V\,\Delta V$ has a moment of inertia about the z-axis (a diameter) of

$$\Delta I = (\rho_V\,\Delta V)\rho^2 = (\rho_V\,\Delta V)r^2\sin^2\theta$$

so for the whole octant, using the normal volume element in spherical coordinates,

$$I = \int_0^{\pi/2}\int_0^{\pi/2}\int_a^b \rho_V r^2\sin^2\theta\, r^2\sin\theta\,\mathrm{d}r\,\mathrm{d}\theta\,\mathrm{d}\phi$$

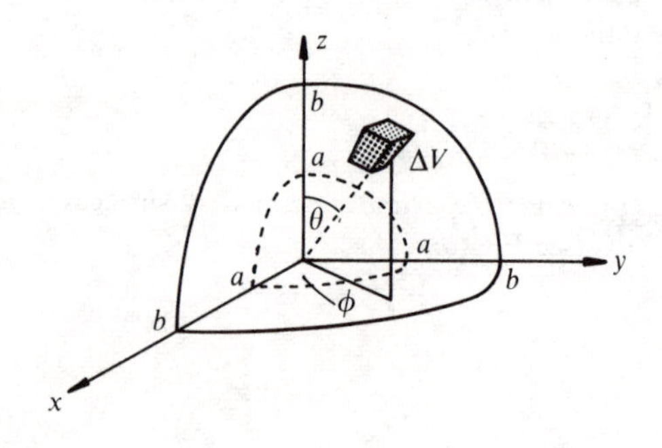

Figure 4.30

(note the limits carefully). Thus

$$I = \rho_V \int_0^{\pi/2} d\phi \int_0^{\pi/2} \sin^3\theta \, d\theta \int_a^b r^4 \, dr = \rho_V (b^5 - a^5) \frac{\pi}{10} \int_0^{\pi/2} \sin^3\theta \, d\theta$$

(leaving the only difficult integral until last!). Now (see Example 4.19 for a similar integral),

$$\int_0^{\pi/2} \sin^3\theta \, d\theta = \int_0^{\pi/2} (1 - \cos^2\theta) \, d(-\cos\theta)$$

$$= \left[-\cos\theta + \tfrac{1}{3}\cos^3\theta \right]_0^{\pi/2} = \tfrac{2}{3}$$

so that $I = \frac{1}{15}\pi\rho_V(b^5 - a^5)$

or, for the whole sphere, eight times this value.

The radius of gyration k is linked to the moment of inertia by $I = Mk^2$, where M is the total mass of the body. Here $M = \frac{4}{3}\pi(b^3 - a^3)\rho_V$, for which we scarcely need to do a volume integration! Hence

$$k = \sqrt{\frac{I}{M}} = \sqrt{\frac{2(b^5 - a^5)}{5(b^3 - a^3)}}$$

EXERCISES

4.18 Distinguish carefully between a volume integral and the volume enclosed by a closed surface.

4.19 Given the vector field $\mathbf{A} = (2x^2 - 3z)\hat{\mathbf{i}} - 2xy\hat{\mathbf{j}} - 4x\hat{\mathbf{k}}$, evaluate the volume integral $\iiint_V \nabla \cdot \mathbf{A} \, dV$, where V is the region bounded by the planes $x = 0$, $y = 0$, $z = 0$ and $2x + 2y + z = 4$.

4.20 Sketch the region bounded by the surfaces

$$z = -1, \qquad z = 2, \qquad \phi = 30°, \qquad \phi = 60°, \qquad \rho = 3$$

in cylindrical coordinates (distances are in metres). If the electric charge density in the region is

$$\sigma = \rho \sin 2\phi \quad \mathrm{C\,m^{-3}}$$

calculate the total charge.

4.21 Given the vector field $\mathbf{F} = \frac{1}{2}x^2\hat{\mathbf{i}}$, evaluate the volume integral $\iiint_V \nabla \cdot \mathbf{F} \, dV$, where V is the interior of the hemisphere with centre the origin and radius 3 and over which $x > 0$. What are the units of the numerical answer (in terms of the units of $\mathbf{F}$)?

4.22 The elastic energy per unit volume in a material is given by the expression $q^2/2EI$, where E and I are constants and q is the stress. In a certain cylinder of radius R and length l, the stress is directly proportional to the distance from the axis, being zero on the axis and having a value q_0 on the outer surface. Calculate the total elastic energy in the cylinder.

4.23 (a) Evaluate the volume integral $\iiint_V \nabla \cdot \mathbf{F}\, dV$, where $\mathbf{F} = 4yz\hat{\mathbf{j}}$ and V is the region enclosed by a sphere with centre the origin and unit radius.

(b) Find the flux integral $\oiint_S \mathbf{F} \cdot \hat{\mathbf{n}}\, dS$ over the surface S of the above sphere. (You should obtain the same answer as for (a).)

ADDITIONAL EXERCISES

1 A vector force field is given, in cylindrical coordinates, by $\mathbf{F} = -(\kappa/\rho)\hat{\boldsymbol{\rho}}$ where κ is a constant and $\hat{\boldsymbol{\rho}}$ is a unit vector directed radially outwards from the z-axis. Show that the work needed to move an object from any radial distance ρ_1 to a point at twice that radial distance is independent of ρ_1.

2 Show that if $\mathbf{r}$ is a position vector then the vector field $\mathbf{F} = r^2\mathbf{r}$ is conservative. Deduce the corresponding scalar potential.

3 If $\mathbf{F} = 3x^2\hat{\mathbf{i}} + 4xy\hat{\mathbf{j}}$, evaluate the line integral $\int_C \mathbf{F} \cdot d\mathbf{l}$ along the curve whose parametric equations are

$$x = 2\cos t, \quad y = 3\sin t, \qquad 0 \leqslant t \leqslant \pi$$

Sketch the path of integration.

4 For each of the following paths, calculate the work done in moving a charge of 5 C from the point (4, 2, 0) to the point (1, 1, 0) in the electric field $\mathbf{E} = (2y\hat{\mathbf{i}} + 2x\hat{\mathbf{j}})\,\mathrm{V\,m^{-1}}$:

(a) along the straight line $3y = x + 2$,

(b) along the parabola $y^2 = x$,

(c) along the hyperbola $x(7 - 3y) = 4$.

Would you expect the results to be the same in each case?

5 If $\mathbf{A} = (x + 2y + \alpha z)\hat{\mathbf{i}} + (\beta x - 3y - z)\hat{\mathbf{j}} + (4x + \gamma y + 2z)\hat{\mathbf{k}}$, find the constants α, β and γ if $\mathbf{A}$ is to be conservative. Hence deduce the corresponding scalar potential. Write down the value of $\int_C \mathbf{A} \cdot d\mathbf{l}$, where C is any path from (1, 0, 0) to (1, 2, 1). Confirm your answer by direct integration along a straight-line path.

6 If $\mathbf{F} = z\hat{\mathbf{i}} + x\hat{\mathbf{j}} - 3y^2z\hat{\mathbf{k}}$, calculate the flux of $\mathbf{F}$ over that portion of the curved surface of the cylinder $x^2 + y^2 = 16$ in the positive octant between $z = 0$ and $z = 5$.

7 If $\mathbf{r}$ is the position vector of a point (x, y, z), calculate $\oint\!\!\oint_S \mathbf{r}\cdot\hat{\mathbf{n}}\,\mathrm{d}S$ where

(a) S is the surface of the sphere $x^2 + y^2 + z^2 = a^2$,

(b) S is the surface of the unit cube bounded by the planes $x = 0$, $x = 2$, $y = 0$, $y = 2$, $z = 0$ and $z = 2$.

Calculate the volume integral $\iiint_V \nabla\cdot\mathbf{r}\,\mathrm{d}V$ over the enclosed volume in each case, and show that the same answers are obtained.

8 If $\mathbf{F} = (x + y)\hat{\mathbf{i}} - 2z\hat{\mathbf{j}} + y\hat{\mathbf{k}}$, evaluate $\oint\!\!\oint_S \mathbf{F}\cdot\hat{\mathbf{n}}\,\mathrm{d}S$ over the hemisphere defined by $x^2 + y^2 + z^2 = 25$ with $z \geqslant 0$.

9 If $\mathbf{F} = y\hat{\mathbf{i}} - x\hat{\mathbf{j}}$, evaluate the flux integral $\iint_S \nabla\wedge\mathbf{F}\cdot\mathrm{d}\mathbf{S}$ over the hemisphere $x^2 + y^2 + z^2 = 4$, $z \geqslant 0$. Also evaluate the line integral $\oint_C \mathbf{F}\cdot\mathrm{d}\mathbf{l}$, where C is the circle $x^2 + y^2 = 4$, and show that the same answer is obtained.

10 If $\mathbf{F} = x\hat{\mathbf{i}} - 2y\hat{\mathbf{j}} - z\hat{\mathbf{k}}$, evaluate $\iint_S \mathbf{F}\cdot\hat{\mathbf{n}}\,\mathrm{d}S$ where S is that part of the plane $x + 2y + 3z = 6$ which is located in the positive octant ($x \geqslant 0$, $y \geqslant 0$ and $z \geqslant 0$).

11 Given the vector field $\mathbf{F} = (2x^2 - 3z)\hat{\mathbf{i}} - 2xy\hat{\mathbf{j}} - 4x\hat{\mathbf{k}}$, evaluate the volume integral $\iiint_V \nabla\cdot\mathbf{F}\,\mathrm{d}V$, where V is the closed region bounded by the planes $x = 0$, $y = 0$, $z = 0$ and $2x + 2y + z = 4$.

12 Calculate the volume of the region bounded by the surface $z = \mathrm{e}^{-(x^2+y^2)}$, the cylinder $x^2 + y^2 = 1$ and the plane $z = 0$.

13 A gas holder has the form of a vertical cylinder of radius a and height h with a hemispherical top also of radius a. The density of the gas inside varies with height z above the base according to the relation $\sigma = C\,\mathrm{e}^{-z}$, where C is a constant. Calculate the total mass of the gas in the holder and deduce its mean density.

14 Calculate the volume of the two regions bounded by the sphere $x^2 + y^2 + z^2 = 16$ and the cone $z^2 = x^2 + y^2$.

15 A solid sphere with centre at the origin and radius a has a variable mass density σ given by $\sigma = k(a - z)$, where k is a constant. Find the total mass and the centre of gravity of the sphere.

SUMMARY

- Line integrals, flux integrals and volume integrals are all limiting cases of summations – along çurves, over surfaces and through volumes respectively. All three types of integral yield scalar answers.

- The form of a **line integral** is $\int_C \mathbf{F}\cdot d\mathbf{l}$, where $\mathbf{F}$ is a given vector field and C is a given path between points P_1 and P_2. If P_1 and P_2 coincide, that is, if C is a **closed path**, we write $\oint_C \mathbf{F}\cdot d\mathbf{l}$. A line integral can be evaluated by using the equation of the path to convert it to a definite integral.

- If $\oint_C \mathbf{F}\cdot d\mathbf{l} = 0$ for all paths in a region R, we say that $\mathbf{F}$ is a **conservative** vector field in R. A necessary and sufficient condition for $\mathbf{F}$ to be conservative in R is the vanishing of its **curl** everywhere in that region, that is $\nabla \wedge \mathbf{F} = 0$. In a two-dimensional region, this condition reduces to

$$\frac{\partial F_x}{\partial y} = \frac{\partial F_y}{\partial x}$$

- A conservative vector field $\mathbf{F}$ can be expressed as the gradient of a scalar function Φ, called the **potential**. In this case, $\int_C \mathbf{F}\cdot d\mathbf{l} = \Phi(P_2) - \Phi(P_1)$, where C is any path from P_1 to P_2.

- The form of a **flux integral** of a vector field across a surface S is $\iint_S \mathbf{F}\cdot\hat{\mathbf{n}}\, dS$ or, equivalently, $\iint_S \mathbf{F}\cdot d\mathbf{S}$. For a **closed surface** S, that is, one which encloses a volume, we write $\oiint_S \mathbf{F}\cdot\hat{\mathbf{n}}\, dS$, where $\hat{\mathbf{n}}$ is an outwardly-directed unit normal vector at a point on S.

- The **unit normal** $\hat{\mathbf{n}}$ to a surface, if not obvious by inspection, can be calculated using $\hat{\mathbf{n}} = \dfrac{\nabla\Phi}{|\nabla\Phi|}$ when the equation of the surface is written in the form $\Phi = \text{constant}$.

- Flux integrals over cylinders may be evaluated using **cylindrical coordinates** (ρ, ϕ, z). The surface elements are

$$\Delta S = R\,\Delta\phi\,\Delta z \quad \text{on the cylindrical surface } \rho = R$$

$$\Delta S = \rho\,\Delta\phi\,\Delta\rho \quad \text{on an end surface } z = \text{constant}$$

- Flux integrals over spherical surfaces may be evaluated using **spherical coordinates** (r, θ, ϕ). The surface element on a sphere $r = a$ is

$$\Delta S = a^2 \sin\theta \, \Delta\theta \, \Delta\phi$$

- Projecting a surface onto a coordinate plane can be a useful method for evaluating surface integrals.

- The form of the **volume integral** of a scalar field is $\iiint_V f \, dV$, where V is a given volume.

- The basic volume elements are

$$\Delta V = \Delta x \, \Delta y \, \Delta z \quad \text{in Cartesian coordinates}$$

$$\Delta V = \rho \, \Delta\rho \, \Delta\phi \, \Delta z \quad \text{in cylindrical coordinates}$$

$$\Delta V = r^2 \sin\theta \, \Delta r \, \Delta\theta \, \Delta\phi \quad \text{in spherical coordinates}$$

  Volume integrals are evaluated as triple integrals using limits, and a coordinate system, that are appropriate to the volume of integration.

SUPPLEMENT

4S.1 Surface integrals in elasticity theory

In Section 1.7 we described the effect on a body of forces $\mathbf{F}_1, \mathbf{F}_2, \ldots, \mathbf{F}_n$ with lines of action passing through points whose position vectors relative to an origin O are $\mathbf{r}_1, \mathbf{r}_2, \ldots, \mathbf{r}_n$ respectively. We deduced that such a system of forces is equivalent to a single force $\mathbf{R}$ through O,

$$\mathbf{R} = \sum_{i=1}^{n} \mathbf{F}_i$$

together with a single couple $\mathbf{G}$, whose torque is equal to the vector sum of the torques of the separate forces about O:

$$\mathbf{G} = \sum_{i=1}^{n} \mathbf{r}_i \wedge \mathbf{F}_i$$

The resultant force $\mathbf{R}$ is the same for all choices of origin O, but this is not the case for the couple $\mathbf{G}$. For if we take as origin a point O', whose

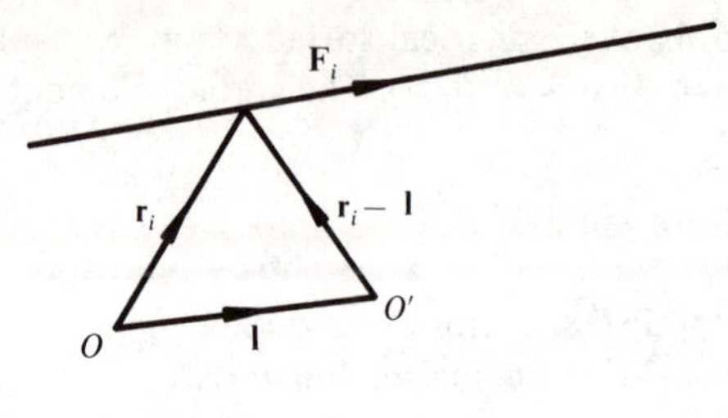

Figure 4S.1

position vector relative to O is $\mathbf{l}$ (see Figure 4S.1), we find that the system is equivalent to a force $\mathbf{R}$ (as before) through O', together with a couple whose torque is

$$\begin{aligned}\mathbf{G}' &= \sum_{i=1}^{n} \mathbf{r}_i' \wedge \mathbf{F}_i = \sum_{i=1}^{n} (\mathbf{r}_i - \mathbf{l}) \wedge \mathbf{F}_i \\ &= \sum_{i=1}^{n} \mathbf{r}_i \wedge \mathbf{F}_i - \mathbf{l} \wedge \sum_{i=1}^{n} \mathbf{F}_i \\ &= \mathbf{G} - \mathbf{l} \wedge \mathbf{R}\end{aligned}$$

This relation is, of course, obvious from the fact that a force $\mathbf{R}$ at O' is equivalent to an equal force at O together with a couple of torque $\mathbf{l} \wedge \mathbf{R}$. However, if $\mathbf{R}$ is zero then the resultant couple due to those forces is independent of the origin, and in this case any convenient origin may be chosen to evaluate the resultant couple.

In linear elasticity theory, we may express the resultant forces and couples in terms of the stresses $\boldsymbol{\sigma}_x$, $\boldsymbol{\sigma}_y$ and $\boldsymbol{\sigma}_z$ (see Section 1.8) using surface integrals. Consider a surface S within a body, an origin O and a small surface element at Q which is approximately planar, with vector area $\hat{\mathbf{n}}\,\mathrm{d}S$ (see Figure 4S.2). The force developed across this area is $\mathbf{P}^n\,\mathrm{d}S$, where $\mathbf{P}^n$ is the stress vector at Q.

If O' is any point of interest, then the total forces on S are equivalent to a single force through O',

$$\iint_S \mathbf{P}^n\,\mathrm{d}S = \iint_S \{(\boldsymbol{\sigma}_x \cdot \hat{\mathbf{n}})\hat{\mathbf{i}} + (\boldsymbol{\sigma}_y \cdot \hat{\mathbf{n}})\hat{\mathbf{j}} + (\boldsymbol{\sigma}_z \cdot \hat{\mathbf{n}})\hat{\mathbf{k}}\}\,\mathrm{d}S$$

and a single couple about O',

$$\iint_S (\mathbf{r} - \mathbf{l}) \wedge \mathbf{P}^n\,\mathrm{d}S = \iint_S \mathbf{r} \wedge \mathbf{P}^n\,\mathrm{d}S - \mathbf{l} \wedge \iint_S \mathbf{P}^n\,\mathrm{d}S$$

since $\mathbf{l}$ is a fixed vector.

If the resultant force on S is zero, then $\iint_S \mathbf{P}^n\,\mathrm{d}S = 0$ and the resultant

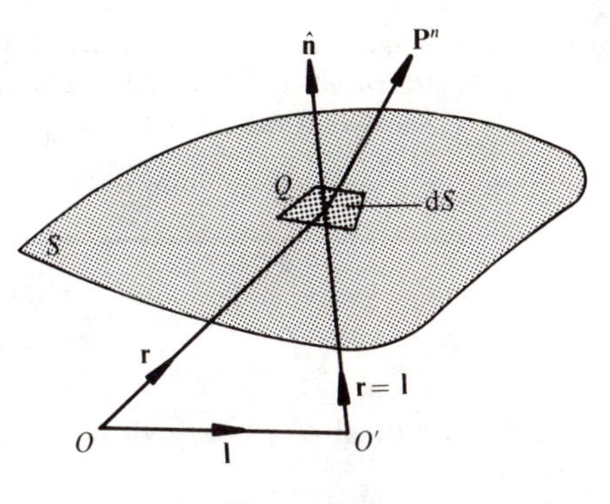

Figure 4S.2

couple is $\mathbf{G} = \iint_S \mathbf{r} \wedge \mathbf{P}^n \, dS$, which is the couple obtained at O so that, in this case, $\mathbf{G}$ is independent of origin. In component form,

$$G_x = \iint_S \{y(\boldsymbol{\sigma}_z \cdot \hat{\mathbf{n}}) - z(\boldsymbol{\sigma}_y \cdot \hat{\mathbf{n}})\} \, dS$$

$$G_y = \iint_S \{z(\boldsymbol{\sigma}_x \cdot \hat{\mathbf{n}}) - x(\boldsymbol{\sigma}_z \cdot \hat{\mathbf{n}})\} \, dS$$

$$G_z = \iint_S \{x(\boldsymbol{\sigma}_y \cdot \hat{\mathbf{n}}) - y(\boldsymbol{\sigma}_x \cdot \hat{\mathbf{n}})\} \, dS$$

Such integrals often have to be evaluated in elasticity theory, particularly when it is necessary to check that a solution for the stresses is consistent with the forces applied to the body. Note that these are examples of the more general form of surface integral, not the flux integrals considered in depth in this chapter.

Example 4S.1

The stress distribution in a beam of length l of circular cross-section and radius a, with its axis aligned along the x-axis, is given by

$$\sigma_{xx} = \frac{4W}{\pi a^4}(x - l)z, \qquad \sigma_{xy} = \frac{-2W(\frac{1}{2} + \nu)}{\pi a^4(1 + \nu)}(yz),$$

$$\sigma_{xz} = \frac{2W}{\pi a^4(1 + \nu)}[(\tfrac{3}{4} + \tfrac{1}{2}\nu)(a^2 - z^2) - (\tfrac{1}{4} - \tfrac{1}{2}\nu)y^2]$$

$$\sigma_{yx} = \sigma_{xy}, \qquad \sigma_{zx} = \sigma_{xz}, \qquad \sigma_{yy} = \sigma_{yz} = \sigma_{zz} = 0$$

where W and ν are constants. Verify that the lateral surface of the cylinder $y^2+z^2=a^2$ is stress-free, and that the resultant of the forces on the end $x=l$ is $W\hat{\mathbf{k}}$. Show that the resultant couple about the origin is $(0, -lW, 0)$.

Solution

On $y^2+z^2=a^2$, a unit normal is $y\hat{\mathbf{j}}+z\hat{\mathbf{k}}/a$, therefore on the lateral surface of the cylinder,

$$P_x^n=\boldsymbol{\sigma}_x\cdot\hat{\mathbf{n}}=\sigma_{xy}\left(\frac{y}{a}\right)+\sigma_{xz}\left(\frac{z}{a}\right)$$

$$=\frac{2W}{\pi a^4(1+\nu)}\left(\frac{z}{a}\right)[-(\tfrac{1}{2}+\nu)y^2+(\tfrac{3}{4}+\tfrac{1}{2}\nu)(a^2-z^2)-(\tfrac{1}{4}-\tfrac{1}{2}\nu)y^2]$$

$$=\frac{2W}{\pi a^4(1+\nu)}\left(\frac{z}{a}\right)[-(\tfrac{1}{2}+\nu)y^2+(\tfrac{3}{4}+\tfrac{1}{2}\nu)y^2-(\tfrac{1}{4}-\tfrac{1}{2}\nu)y^2]$$

$$=0$$

Also, $P_y^n=\boldsymbol{\sigma}_y\cdot\hat{\mathbf{n}}=0$ and $P_z^n=\boldsymbol{\sigma}_z\cdot\hat{\mathbf{n}}=0$

Hence the curved surface is indeed stress-free.

On the end $x=l$, $\hat{\mathbf{n}}=\hat{\mathbf{i}}$, therefore

$$P_x^n=\boldsymbol{\sigma}_x\cdot\hat{\mathbf{n}}=\sigma_{xx}=\frac{4W}{\pi a^4}(x-l)z=0 \quad \text{on } x=l$$

$$P_y^n=\boldsymbol{\sigma}_y\cdot\hat{\mathbf{n}}=\sigma_{yx}=-\frac{2W(\tfrac{1}{2}+\nu)}{\pi a^4(1+\nu)}(yz)$$

$$P_z^n=\boldsymbol{\sigma}_z\cdot\hat{\mathbf{n}}=\sigma_{zx}=\frac{2W}{\pi a^4(1+\nu)}[(\tfrac{3}{4}+\tfrac{1}{2}\nu)(a^2-z^2)-(\tfrac{1}{4}-\tfrac{1}{2}\nu)y^2]$$

Thus the total force in the y-direction is given by integrating over the surface S, that is,

$$\iint_S P_y^n\,\mathrm{d}S=-\frac{2W(\tfrac{1}{2}+\nu)}{\pi a^4(1+\nu)}\iint_S yz\,\mathrm{d}S$$

If we change to polar coordinates, putting $y=\rho\cos\phi$ and $z=\rho\sin\phi$,

$$\iint_S yz\,\mathrm{d}S=\int_0^{2\pi}\int_0^a\rho^3\left(\frac{\sin 2\phi}{2}\right)\mathrm{d}\rho\,\mathrm{d}\phi=0$$

(since $\int_0^{2\pi}\sin 2\phi\,\mathrm{d}\phi=0$). Hence the resultant force on S has zero y-component.

The total force in the z-direction is

$$\iint_S P_z^n \, \mathrm{d}S = K \iint_S \{(\tfrac{3}{4} + \tfrac{1}{2}\nu)(a^2 - z^2) - (\tfrac{1}{4} - \tfrac{1}{2}\nu)y^2\} \, \mathrm{d}S$$

where $K = \dfrac{2W}{\pi a^4(1+\nu)}$

Again using polar coordinates, the integral becomes

$$K \int_0^{2\pi} \int_0^a \{(\tfrac{3}{4} + \tfrac{1}{2}\nu)(a^2 - \rho^2 \sin^2 \phi) - (\tfrac{1}{4} - \tfrac{1}{2}\nu)\rho^2 \cos^2 \phi\} \rho \, \mathrm{d}\rho \, \mathrm{d}\phi$$

$$= K \int_0^{2\pi} \int_0^a \{(\tfrac{3}{4} + \tfrac{1}{2}\nu)a^2\rho - (\tfrac{3}{4} + \tfrac{1}{2}\nu)\rho^3 \sin^2 \phi - (\tfrac{1}{4} - \tfrac{1}{2}\nu)\rho^3 \cos^2 \phi\} \, \mathrm{d}\rho \, \mathrm{d}\phi$$

$$= K \int_0^{2\pi} \left\{\frac{a^4}{2}(\tfrac{3}{4} + \tfrac{1}{2}\nu) - \frac{a^4}{4}(\tfrac{3}{4} + \tfrac{1}{2}\nu) \sin^2 \phi - \frac{a^4}{4}(\tfrac{1}{4} - \tfrac{1}{2}\nu) \cos^2 \phi\right\} \mathrm{d}\phi$$

$$= \frac{2W}{\pi a^4(1+\nu)} \left\{\frac{a^4}{2}(\tfrac{3}{4} + \tfrac{1}{2}\nu) - \frac{a^4}{4}(\tfrac{3}{4} + \tfrac{1}{2}\nu)\tfrac{1}{2} - \frac{a^4}{4}(\tfrac{1}{4} - \tfrac{1}{2}\nu)\right\} 2\pi$$

$$= W$$

Since the resultant force on S is not zero, the resultant couple will be dependent on the choice of origin. If we choose an origin at (0, 0, 0), then the resultant couple has components (here with $\hat{\mathbf{n}} = \hat{\mathbf{i}}$)

$$G_x = \iint_S (y\sigma_{xz} - z\sigma_{xy}) \, \mathrm{d}S$$

$$G_y = \iint_S (z\sigma_{xx} - x\sigma_{zx}) \, \mathrm{d}S = -l \iint_S \sigma_{zx} \, \mathrm{d}S = -lW$$

$$G_z = \iint_S (x\sigma_{yx} - y\sigma_{xx}) \, \mathrm{d}S = l \iint_S \sigma_{yx} \, \mathrm{d}S = 0$$

The results for G_y and G_z follow easily from the surface integrals already

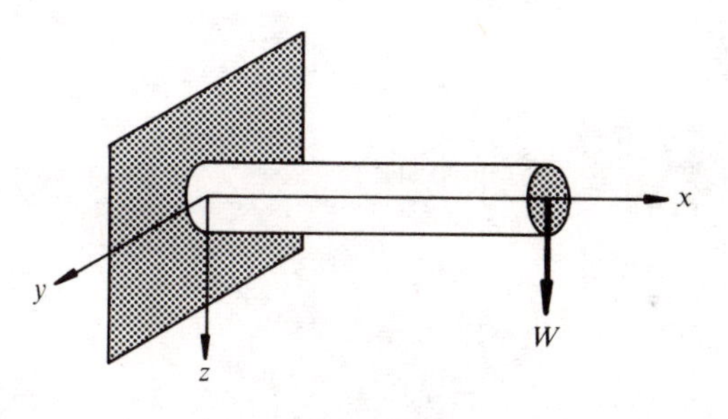

Figure 4S.3

calculated. Also, it is relatively easy to show that

$$\iint_S y\sigma_{xz}\,\mathrm{d}S = 0 \quad \text{and} \quad \iint_S z\sigma_{xy}\,\mathrm{d}S = 0$$

so that $G_x = 0$. Therefore the resultant couple about $(0, 0, 0)$ is $(0, -lW, 0)$.

In conclusion, the forces on the end $x = l$ are equivalent to a single force $W\hat{\mathbf{k}}$ together with an attendant couple $-lW\hat{\mathbf{j}}$ about the origin. This implies that this stress system may be used to describe a cantilever deformed by an end load W, as shown in Figure 4S.3.

Integral Theorems of Vector Analysis

PREVIEW

In this chapter, which is central to the theory of vector fields, we link the various types of integral using two fundamental theorems – the divergence theorem and the theorem due to Stokes. Applications to fluid mechanics and heat transfer are discussed and, perhaps more importantly, we obtain quite general definitions of divergence and curl, independent of any particular coordinate system.

5.1 Introduction

In this chapter we reach the heart of the theory of vector fields by introducing two powerful theorems. These are the **divergence theorem** and **Stokes' theorem** and they are a major unifying influence in the subject in that they provide connections between the various types of integral – line, surface and volume – that we studied in Chapter 4. A knowledge of these theorems will also enable us to delve more deeply into the significance of the divergence $\nabla \cdot \mathbf{F}$ and the curl $\nabla \wedge \mathbf{F}$ of a vector field. The deeper understanding that we obtain will enable us, in Chapter 6, to develop specific expressions for these quantities in coordinate systems other than Cartesian. Perhaps even more importantly, we shall, hopefully, learn to apply vector field concepts with confidence in appropriate areas of science and engineering. Examples of such applications will be the subject of Chapter 7.

Our approach to both theorems will be relatively informal, in that we shall demonstrate only simplified proofs using Cartesian coordinates and will then proceed to illustrate the theorems using examples.

5.2 Outline derivation of the divergence theorem

We recall from Section 4.6 that the flux of a vector field $\mathbf{F}$ across a closed surface S is the integral $\oiint_S \mathbf{F} \cdot \hat{\mathbf{n}}\, dS$, where $\hat{\mathbf{n}}$ is the outwardly-directed unit normal to S. We now consider the evaluation of this integral over a particularly simple closed surface, namely the six bounding faces of the small rectangular parallelepiped shown in Figure 5.1. The edges of this rectangular 'box' are parallel to the three coordinate axes and are of lengths Δx, Δy and Δz respectively.

If the centre of the box is a point P with coordinates (x_0, y_0, z_0), then the centre of the back face $ABCD$ has coordinates $(x_0 - \frac{1}{2}\Delta x, y_0, z_0)$. Hence if $F_x(x_0, y_0, z_0)$ is the x-component of the vector field $\mathbf{F}$ at P, then at

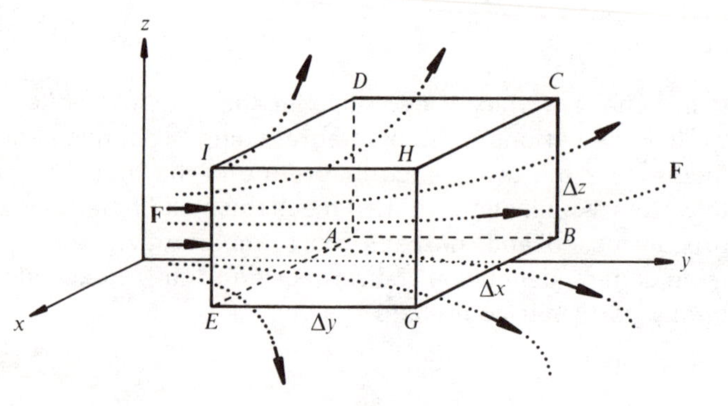

Figure 5.1

$(x_0 - \frac{1}{2}\Delta x, y_0, z_0)$ this component will be approximately

$$F_x(x_0, y_0, z_0) - \frac{\partial F_x}{\partial x}\frac{\Delta x}{2}$$

Hence the flux of **F** over the face *ABCD* which has area $\Delta y\, \Delta z$ will be

$$\iint_{ABCD} \mathbf{F}\cdot\hat{\mathbf{n}}\, \mathrm{d}S = \iint_{ABCD} \mathbf{F}\cdot(-\hat{\mathbf{i}})\, \mathrm{d}S$$

$$\approx -\left(F_x(x_0, y_0, z_0) - \frac{\partial F_x}{\partial x}\frac{\Delta x}{2}\right)\Delta y\, \Delta z$$

where we have assumed that the face is small enough for the value of **F** to be treated as a constant and we have approximated the total flux as the value of **F** at the centre of the face multiplied by the area.

By similar reasoning, we have for the flux over the front face *EGHI*,

$$\iint_{EGHI} \mathbf{F}\cdot\hat{\mathbf{n}}\, \mathrm{d}S = \iint_{EGHI} \mathbf{F}\cdot(+\hat{\mathbf{i}})\, \mathrm{d}S$$

$$\approx +\left(F_x(x_0, y_0, z_0) + \frac{\partial F_x}{\partial x}\frac{\Delta x}{2}\right)\Delta y\, \Delta z$$

Adding these results, we obtain

$$\left(\frac{\partial F_x}{\partial x}\Delta x\right)\Delta y\, \Delta z$$

as an approximate value for the net flux over these two faces.

By similar, symmetric, reasoning,

$$\text{net flux over } ADIE \text{ and } BCHG \approx \left(\frac{\partial F_y}{\partial y}\Delta y\right)\Delta x\, \Delta z$$

$$\text{net flux over } AEGB \text{ and } CDIH \approx \left(\frac{\partial F_z}{\partial z}\Delta z\right)\Delta x\, \Delta y$$

Adding the three results above gives, as an approximation to the total flux,

$$\oiint_S \mathbf{F}\cdot\hat{\mathbf{n}}\, \mathrm{d}S \approx \left(\frac{\partial F_x}{\partial x} + \frac{\partial F_y}{\partial y} + \frac{\partial F_z}{\partial z}\right)\Delta x\, \Delta y\, \Delta z$$

The bracketed quantity is easily recognizable as the divergence $\nabla\cdot\mathbf{F}$, while the product $\Delta x\, \Delta y\, \Delta z$ is of course the volume ΔV of the rectangular parallelepiped. Hence, for this very simple surface,

$$\oiint_S \mathbf{F}\cdot\hat{\mathbf{n}}\, \mathrm{d}S \approx \nabla\cdot\mathbf{F}\, \Delta V \tag{5.1}$$

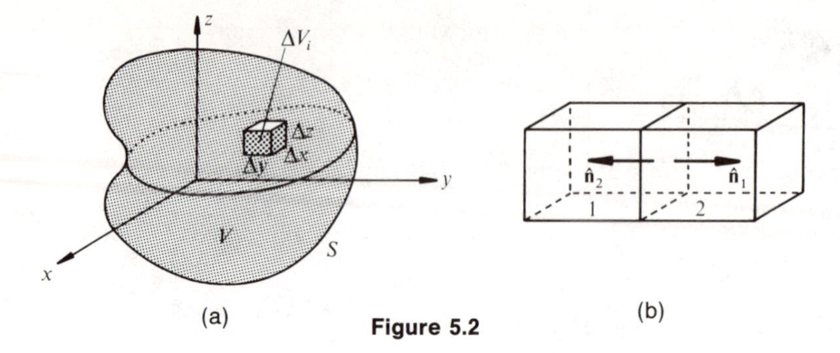

Figure 5.2

We now consider this elementary parallelepiped ΔV as just one small element ΔV_i of a volume V bounded by a closed surface S (see Figure 5.2a). We can apply (5.1) to each of the N such elements ΔV_i making up the whole volume V. We then add both sides of (5.1) over all these elements.

From the right-hand side of (5.1) we obtain $\sum_{i=1}^{N} \nabla \cdot \mathbf{F}\, \Delta V_i$. Taking smaller and smaller elements, that is letting N tend to ∞, this sum becomes the volume integral $\iiint_V \nabla \cdot \mathbf{F}\, \mathrm{d}V$. From the left-hand side of (5.1), we are summing all the flux across all the surfaces of N parallelepipeds. However, recalling that $\hat{\mathbf{n}}$ always denotes the outward normal, the outward flux across one surface will be 'cancelled' by the outward flux from the adjacent parallelepiped over the face that the two have in common (see Figure 5.2b). The only 'non-cancelling' contributions will come from the faces on the *bounding* surface S, and these will add to give the net outward flux across S. Hence, we finally obtain

$$\oiint_S \mathbf{F} \cdot \hat{\mathbf{n}}\, \mathrm{d}S = \iiint_V \nabla \cdot \mathbf{F}\, \mathrm{d}V \qquad \textbf{(5.2)}$$

which is the basic statement of the **divergence theorem**.

Using the vector area concept, the flux integral on the left-hand side of (5.2) can be expressed in the alternative form $\oiint_S \mathbf{F} \cdot \mathrm{d}\mathbf{S}$.

Points to note in the divergence theorem

The above 'derivation' of the divergence theorem made a number of implicit assumptions, which would need to be highlighted and considered in detail in a more rigorous treatment of vector analysis. In particular, we have implicitly assumed that the vector field $\mathbf{F}$ has components which are sufficiently smooth throughout V (that is, that each component has partial derivatives which are continuous) in order that the divergence may be taken at each point.

Also, the surface S is required to be sufficiently 'smooth' in order that the surface integral $\iint_S \mathbf{F} \cdot \hat{\mathbf{n}}\, \mathrm{d}S$ and the volume integral $\iiint_V \nabla \cdot \mathbf{F}\, \mathrm{d}V$ can be assigned unambiguous values. The usual restriction placed on S is that any line perpendicular to a coordinate plane should meet the surface in no more than two points. (The surface may then be split into two parts, each of which may be defined by a single-valued function of two variables.)

Also, using a cuboid for the infinitesimal region in the 'derivation' is over simplistic and, in a rigorous treatment, would lead to problems on the bounding surface S. It would be more sensible to segment the region V by infinitesimal cylinders with sides parallel to the z-axis (or the x-axis or the y-axis) of cross-sectional area $\Delta x\ \Delta y$ which may be bounded on one face (for those cylinders impinging on S) by part of S itself. The derivation in this case would proceed along similar lines to that for the cuboid.

The divergence theorem can be extended to surfaces for which lines perpendicular to the coordinate planes meet the surface in more than two points. The extension is obtained by subdividing S into smaller surfaces which do satisfy this criterion. This process of subdivision has already been used in the evaluation, by the projection method, of surface integrals over closed surfaces.

The theorem can also be extended to situations in which S is formed from two or more disjoint pieces. The volume V is then the region bounded by the disjoint surfaces. A situation where this extended form of the divergence theorem may be used is as follows. We have seen earlier, equation (3.18), that the gravitational vector field $\mathbf{F}$ of a 'point' mass situated at the origin is

$$\mathbf{F} = -\frac{m\mathbf{r}}{|\mathbf{r}|^3}$$

We demonstrate that the field satisfies Gauss' law for gravitational fields, namely that, for any closed surface S,

$$\oiint_S \mathbf{F}\cdot\hat{\mathbf{n}}\,\mathrm{d}S = \begin{cases} -4\pi m & \text{if the mass is inside } S \\ 0 & \text{if the mass is outside } S \end{cases}$$

We first suppose that the mass is outside S. The vector field $\mathbf{F}$ is differentiable everywhere within S since its only 'singularity' is at the position of m – at the origin, $\mathbf{r} = 0$. We can thus determine the divergence of $\mathbf{F}$ at each point of V (bounded by S), and we obtain (see Example 3.5) $\nabla\cdot\mathbf{F} = 0$. Therefore, by the divergence theorem,

$$\oiint_S \mathbf{F}\cdot\hat{\mathbf{n}}\,\mathrm{d}S = \iiint_V \nabla\cdot\mathbf{F}\,\mathrm{d}V = 0$$

Considering the second possibility, in which the mass is within the region bounded by S, we can no longer use the divergence theorem directly since $\mathbf{F}$ is not sufficiently smooth throughout V, and $\nabla\cdot\mathbf{F}$ cannot be determined at *all* points of V (the calculation fails at the origin). However, the surface integral is meaningful since $\mathbf{F}$ is well defined at each point on S. To evaluate the surface integral, we first convert it into an easier form by appealing to the extended form of the divergence theorem. We surround the

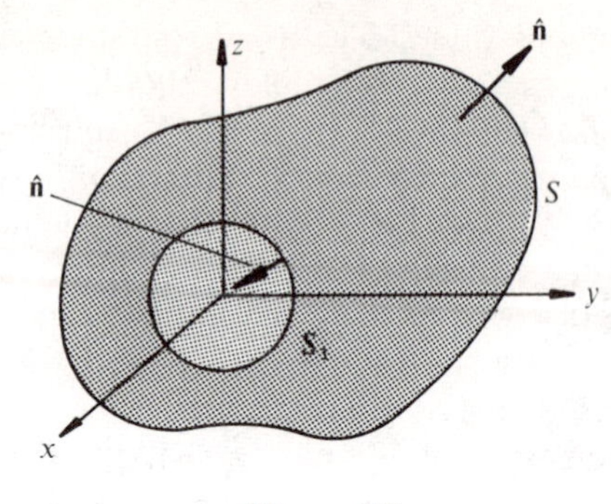

Figure 5.3

origin by a sphere S_1 of radius a, this radius being sufficiently small that the sphere is contained entirely within S (see Figure 5.3).

We consider the application of the divergence theorem to the region U bounded by S_1 and S. We note that on both S and S_1 the normal is pointing out of U (again see Figure 5.3). In the region U, $\nabla\cdot\mathbf{F}=0$ everywhere, and so the flux of $\mathbf{F}$ over the combined surfaces is zero, that is,

$$\oiint_S \mathbf{F}\cdot\hat{\mathbf{n}}\,\mathrm{d}S+\oiint_{S_1}\mathbf{F}\cdot\hat{\mathbf{n}}\,\mathrm{d}S=0$$

Now on S_1, which is a sphere, a unit normal pointing towards the origin is $\hat{\mathbf{n}}=-\hat{\mathbf{r}}$ and so, on this surface,

$$\mathbf{F}\cdot\hat{\mathbf{n}}=\frac{m\mathbf{r}\cdot\hat{\mathbf{r}}}{|\mathbf{r}|^3}=\frac{m\hat{\mathbf{r}}\cdot\hat{\mathbf{r}}}{|\mathbf{r}|^2}=\frac{m}{a^2}$$

Also, on this surface the element of surface area is $\mathrm{d}S=a^2\sin\theta\,\mathrm{d}\theta\,\mathrm{d}\phi$, so

$$\begin{aligned}\oiint_S \mathbf{F}\cdot\hat{\mathbf{n}}\,\mathrm{d}S=-\oiint_{S_1}\mathbf{F}\cdot\hat{\mathbf{n}}\,\mathrm{d}S&=-\oiint_{S_1}\left(\frac{m}{a^2}\right)a^2\sin\theta\,\mathrm{d}\theta\,\mathrm{d}\phi\\&=-\int_0^{2\pi}\int_0^{\pi}m\sin\theta\,\mathrm{d}\theta\,\mathrm{d}\phi\\&=+\int_0^{2\pi}\Big[m\cos\theta\Big]_0^{\pi}\mathrm{d}\phi\\&=-4\pi m\end{aligned}$$

which validates Gauss' law.

5.3 Worked examples using the divergence theorem

The divergence theorem (5.2) provides a link between a flux integral over a closed surface S and a volume integral over the volume V bounded by S.

We have, in fact, already encountered two specific examples of the theorem in Chapter 4. In Examples 4.15 and 4.16, we showed that the flux of the vector field

$$\mathbf{F} = 4x\hat{\mathbf{i}} - 2y^2\hat{\mathbf{j}} + x^2z^2\hat{\mathbf{k}}$$

over the three surfaces of the cylinder shown in Figure 4.18a was 84π. (Note that all three surfaces of the cylinder had to be considered because the divergence theorem only applies to *closed* surfaces.) Then, in Example 4.27, we evaluated the volume integral of

$$\nabla \cdot \mathbf{F} = 4 - 4y + 2x^2z$$

over the region inside this cylinder. Again the answer was 84π, which of course verifies the divergence theorem for this particular vector field and particular geometry.

Similarly, we demonstrated, in Examples 4.14 and 4.22, the equality of the flux of a vector field $\mathbf{F}$ over the six surfaces of a rectangular solid and the integral of $\nabla \cdot \mathbf{F}$ over the volume enclosed.

Example 5.1

Verify the divergence theorem for the radial vector field $\mathbf{F} = z^2\hat{\mathbf{r}}$ using a sphere centre the origin and radius 3. ($\hat{\mathbf{r}}$ denotes a unit vector directed radially outwards from the origin.)

Solution

We need to find the flux integral $\oiint_S \mathbf{F} \cdot \hat{\mathbf{n}}\, \mathrm{d}S$ over the surface of the given sphere and to show that this is equal to the volume integral $\iiint_V \nabla \cdot \mathbf{F}\, \mathrm{d}V$ through the volume of the sphere. We shall use spherical coordinates for the final evaluations, but at this stage we only know how to evaluate $\nabla \cdot \mathbf{F}$ in Cartesian coordinates. Hence we convert to spherical coordinates after taking the divergence.

We have $\mathbf{F} = z^2\hat{\mathbf{r}}$ (see Figure 5.4). If $\mathbf{r} = x\hat{\mathbf{i}} + y\hat{\mathbf{j}} + z\hat{\mathbf{k}}$ is the position vector of any point P,

$$\hat{\mathbf{r}} = \frac{\mathbf{r}}{|\mathbf{r}|} = \frac{1}{\sqrt{(x^2 + y^2 + z^2)}}(x\hat{\mathbf{i}} + y\hat{\mathbf{j}} + z\hat{\mathbf{k}})$$

so our given vector field is, in Cartesian coordinates,

$$\mathbf{F} = \frac{1}{\sqrt{(x^2 + y^2 + z^2)}}(xz^2\hat{\mathbf{i}} + yz^2\hat{\mathbf{j}} + z^3\hat{\mathbf{k}})$$

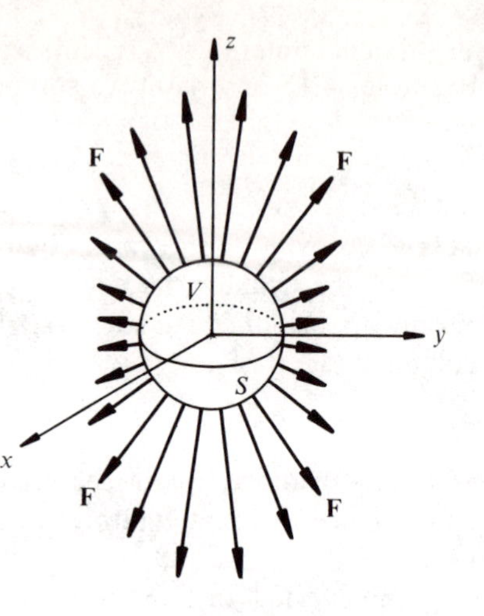

Figure 5.4

A straightforward calculation yields

$$\nabla \cdot \mathbf{F} = \frac{4z^2}{r}, \quad \text{where } r = |\mathbf{r}|$$

Converting to spherical coordinates (see Section 4.8),

$$\nabla \cdot \mathbf{F} = 4r \cos^2 \theta$$

so we obtain the volume integral

$$\iiint_V \nabla \cdot \mathbf{F} \, \mathrm{d}V = \iiint_V (4r \cos^2 \theta) r^2 \sin \theta \, \mathrm{d}r \, \mathrm{d}\theta \, \mathrm{d}\phi$$

where we have used the normal volume element of spherical coordinates. Integrating over the volume enclosed by the sphere,

$$\iiint_V \nabla \cdot \mathbf{F} \, \mathrm{d}V = 4 \int_0^{2\pi} \mathrm{d}\phi \int_0^{\pi} \cos^2 \theta \sin \theta \, \mathrm{d}\theta \int_0^3 r^3 \, \mathrm{d}r = 108\pi$$

For the flux integral, we have

$$\mathbf{F} \cdot \hat{\mathbf{n}} = \mathbf{F} \cdot \hat{\mathbf{r}} = z^2 \quad \text{so} \quad \iint_S \mathbf{F} \cdot \hat{\mathbf{n}} \, \mathrm{d}S = \iint_S (9 \cos^2 \theta) 9 \sin \theta \, \mathrm{d}\theta \, \mathrm{d}\phi$$

where we have replaced z by $r \cos \theta$ with $r = 3$, and used the normal surface element $r^2 \sin \theta \, \mathrm{d}\theta \, \mathrm{d}\phi$ on the spherical surface $r = 3$. Hence

$$\iint_S \mathbf{F} \cdot \hat{\mathbf{n}} \, \mathrm{d}S = 81 \int_0^{2\pi} \mathrm{d}\phi \int_0^{\pi} \cos^2 \theta \sin \theta \, \mathrm{d}\theta = 108\pi$$

The equality of the volume and surface integrals means that we have indeed verified the divergence theorem for this situation.

Example 5.2

Use the divergence theorem to show that, if u and v are scalar fields, then

(a) $$\oint\!\!\oint_S u\,\nabla v\cdot d\mathbf{S} = \iiint_V (u\,\nabla^2 v + \nabla u\cdot\nabla v)\,dV$$

(b) $$\oint\!\!\oint_S (u\,\nabla v - v\,\nabla u)\cdot d\mathbf{S} = \iiint_V (u\,\nabla^2 v - v\,\nabla^2 u)\,dV$$

where V is the volume bounded by surface S. (These results are known as **Green's identities**; they are used, for example, in variational calculus which is the basis of the finite element method for solving differential equations. They may be regarded as generalizations of 'integration by parts' in one dimension.)

Solution

Since we are required to prove the equality of a flux integral and a volume integral, it should be clear that the divergence theorem is required. Hence, putting $\mathbf{F} = u\,\nabla v$ (a vector quantity, of course) in (5.2),

$$\oint\!\!\oint_S u\,\nabla v\cdot d\mathbf{S} = \iiint_V (\nabla\cdot(u\,\nabla v))\,dV$$

The integrand in the volume integral has the expansion

$$\nabla\cdot(u\,\nabla v) = u\,\nabla\cdot(\nabla v) + \nabla v\cdot\nabla u = u\,\nabla^2 v + \nabla v\cdot\nabla u$$

or $$\text{div}(u\,\mathbf{grad}\,v) = u\,\text{div}(\mathbf{grad}\,v) + \mathbf{grad}\,v\cdot\mathbf{grad}\,u$$

(see Section 3.6). This proves identity (a). If we now simply interchange the scalars u and v, we obtain

$$\oint\!\!\oint_S v\,\nabla u\cdot d\mathbf{S} = \iiint_V (v\,\nabla^2 u + \nabla v\cdot\nabla u)\,dV$$

Subtraction from the identity in (a) then yields identity (b).

5.4 A general definition of divergence

From the divergence theorem (5.2), we can readily obtain an alternative and more general definition of the divergence of a vector field than the Cartesian definition

$$\nabla\cdot\mathbf{F} = \frac{\partial F_x}{\partial x} + \frac{\partial F_y}{\partial y} + \frac{\partial F_z}{\partial z}$$

that we have already encountered. Applying (5.2) to the case of a volume V

small enough for $\nabla\cdot\mathbf{F}$ to be approximately constant over V, we obtain (compare with result (5.1) for a rectangular volume)

$$\oiint_S \mathbf{F}\cdot\hat{\mathbf{n}}\,\mathrm{d}S = \iiint_V \nabla\cdot\mathbf{F}\,\mathrm{d}V \approx (\nabla\cdot\mathbf{F})V$$

or
$$\nabla\cdot\mathbf{F} \approx \frac{1}{V}\oiint_S \mathbf{F}\cdot\hat{\mathbf{n}}\,\mathrm{d}S \tag{5.3}$$

Then, as the volume V shrinks to a 'point', we have a point definition of divergence:

$$\nabla\cdot\mathbf{F} = \lim_{V\to 0}\left(\frac{1}{V}\oiint_S \mathbf{F}\cdot\hat{\mathbf{n}}\,\mathrm{d}S\right) \tag{5.4}$$

In words, (5.4) defines the divergence at a point P in a vector field as the flux of $\mathbf{F}$ per unit volume over a vanishingly small surface S enclosing P. The dimensions of $\nabla\cdot\mathbf{F}$ are, therefore, the dimensions of $\mathbf{F}$ divided by length. This is a 'coordinate-free' definition of divergence. Indeed, it is perfectly possible to take (5.4) as the starting point for defining $\nabla\cdot\mathbf{F}$ and use it to obtain all the previous results, including the divergence theorem and the Cartesian representation of divergence.

In Chapter 6, we shall use (5.4) to obtain expressions for the divergence in coordinate systems other than Cartesian. More immediately, we can use it to give us a more physical interpretation of the divergence concept. Consider the quantity

$$I = \frac{1}{V}\oiint_S \mathbf{F}\cdot\hat{\mathbf{n}}\,\mathrm{d}S$$

where S encloses the volume V. If, at each point on S, $\mathbf{F}$ is directed outwards from the enclosed region, then $\mathbf{F}\cdot\hat{\mathbf{n}} > 0$ at each point of S, leading to the result that $I > 0$. Similarly, if $\mathbf{F}$ is directed inwards at every point on S, then $\mathbf{F}\cdot\hat{\mathbf{n}} < 0$ at every point on S and so $I < 0$ (see Figure 5.5). But to obtain $\nabla\cdot\mathbf{F}$ at a point P inside S, we must, by (5.4), let S shrink to the point P. If the situation depicted in Figure 5.5a, namely $I > 0$, holds even in this limiting case, then $\nabla\cdot\mathbf{F}$ will certainly be positive at P, and we say P is a **source** of the vector field $\mathbf{F}$. Conversely, if the situation shown in Figure 5.5b holds in the limiting case, then $\nabla\cdot\mathbf{F}$ will be negative at P, and P is called a **sink** of the vector field $\mathbf{F}$.

The reader should not conclude that the geometrical picture of the vector fields in the neighbourhood of a source or a sink is necessarily as described in Figure 5.5. Rather, these diagrams show particular examples of sources and sinks that often arise in applications, and it is useful to keep these images in mind when thinking about 'real' vector fields. In the particular examples shown in Figure 5.5, the field lines appear to terminate (sink) or originate (source) at P. Although, again, this is not universally typical of

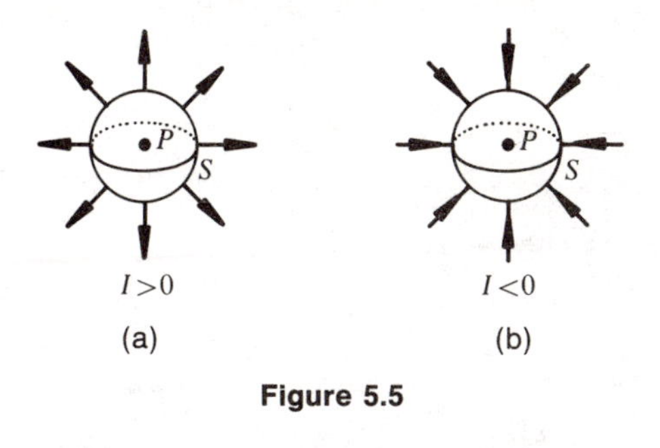

Figure 5.5

vector fields $\mathbf{F}$ for which $\nabla \cdot \mathbf{F} \neq 0$, it is sufficiently common in applications for it to be a useful characteristic to bear in mind.

A zero value for the divergence at a point will imply that the point is neither a source nor a sink, and the net flux over a small surface S enclosing P will be zero (see Figure 5.6a for a 'typical' example of such a vector field). Indeed, there are some important vector fields whose divergence is zero at *all* points, so that a characteristic of such fields is that the field lines have no beginning and no end. An example of such a vector field is the magnetic field $\mathbf{B}$ produced by an electric current. Two examples of magnetic fields due to different current arrangements are shown in Figures 5.6b and 5.6c. Note that the field lines either form closed loops or start and finish 'at infinity' in all cases.

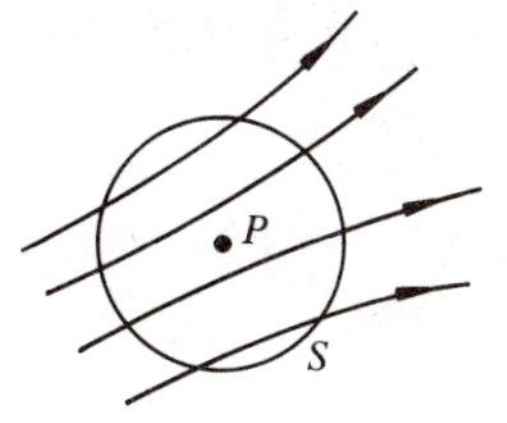

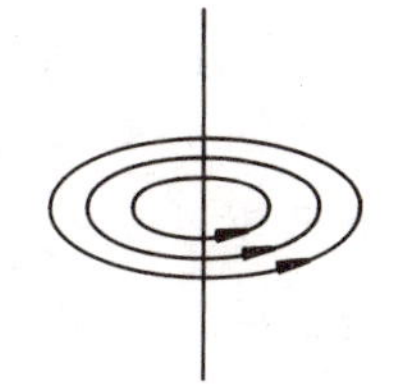

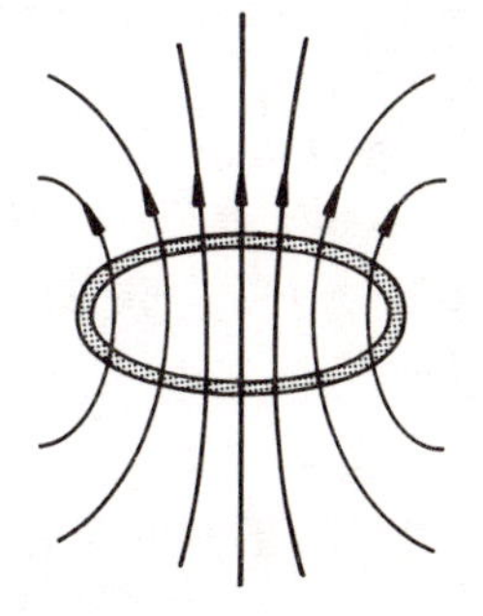

Magnetic field due to an infinite current-carrying wire

Magnetic field due to a toroid

(a) (b) (c)

Figure 5.6

EXERCISES

5.1 If $\mathbf{r}$ is the position vector $x\hat{\mathbf{i}} + y\hat{\mathbf{j}} + z\hat{\mathbf{k}}$, use the divergence theorem to evaluate the flux integral $\oiint_S \mathbf{r}\cdot\hat{\mathbf{n}}\,\mathrm{d}S$

(a) over the surface $x^2 + y^2 + z^2 = a^2$,

(b) over the surface of the unit cube bounded by the planes $x = 0$, $x = 1$, $y = 0$, $y = 1$, $z = 0$ and $z = 1$.

5.2 If $\mathbf{F}$ is the vector field $\mathbf{F} = xy\hat{\mathbf{i}} - yz\hat{\mathbf{j}} + 3\hat{\mathbf{k}}$ and S is the surface (curved surface and ends) of the cylinder shown in Figure 5.7, express the flux of $\mathbf{F}$ across S in terms of a volume integral and hence evaluate it.

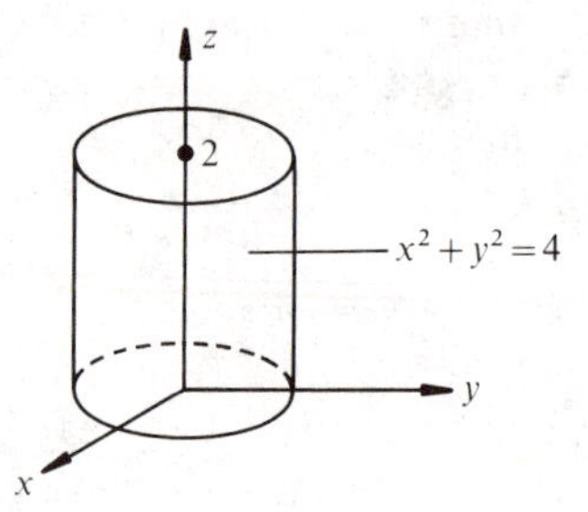

Figure 5.7

5.3 If f is the scalar field $f = x^4 + y^4 + z^4$, use the divergence theorem to calculate the flux of ∇f over the surface $x^2 + y^2 + z^2 = 9$.

5.4 Verify the divergence theorem for each of the following situations:

(a) $\mathbf{F} = x^2\hat{\mathbf{i}} + z\hat{\mathbf{j}} + yz\hat{\mathbf{k}}$ and the cube whose vertices are (0, 0, 0), (1, 0, 0), (1, 1, 0), (0, 1, 0), (0, 0, 1), (1, 0, 1), (1, 1, 1) and (0, 1, 1),

(b) $\mathbf{F} = 2xz\hat{\mathbf{i}} + yz\hat{\mathbf{j}} + z^2\hat{\mathbf{k}}$ and the upper half of the sphere $x^2 + y^2 + z^2 = a^2$,

(c) $\mathbf{F} = -xyz\hat{\mathbf{k}}$ and the cylinder $x^2 + y^2 = 4$, $0 \leqslant z \leqslant 2$.

5.5 Two vector fields $\mathbf{A}$ and $\mathbf{B}$ are linked by

$$\nabla \wedge \mathbf{A} = -\frac{\partial \mathbf{B}}{\partial t} \quad \text{and} \quad \nabla \wedge \mathbf{B} = \frac{\partial \mathbf{A}}{\partial t}$$

Prove that

$$\tfrac{1}{2}\frac{\mathrm{d}}{\mathrm{d}t}\iiint_V (|\mathbf{A}|^2 + |\mathbf{B}|^2)\,\mathrm{d}V = -\oiint_S (\mathbf{A}\wedge\mathbf{B})\cdot\hat{\mathbf{n}}\,\mathrm{d}S$$

(*Hint:* $\mathrm{d}(\mathbf{F}\cdot\mathbf{G})/\mathrm{d}t = \mathbf{F}\cdot\mathrm{d}\mathbf{G}/\mathrm{d}t + \mathbf{G}\cdot\mathrm{d}\mathbf{F}/\mathrm{d}t$.)

5.6 The heat conduction equation (see Section 5.5) is

$$\nabla^2 u = \frac{c\rho_V}{\kappa}\frac{\partial u}{\partial t}$$

where c, ρ_V and κ are constants. If S is a closed surface over which the temperature u is zero, apply the divergence theorem to the vector $\kappa u\,\nabla u$ to show that

$$\iiint_V c\rho_V u\frac{\partial u}{\partial t}\,\mathrm{d}V = -\iiint_V \kappa|\nabla u|^2\,\mathrm{d}V$$

where V is the volume enclosed by S.

5.5 Continuity equations and heat conduction

Continuity equations arise in a number of areas of science and engineering. Their mathematical description provides a useful illustration of many of the concepts of vector field theory that we have been discussing. We shall illustrate continuity equations using elementary ideas in fluid mechanics and heat flow, but much the same ideas apply to flow of electric current. This latter application will be dealt with in Chapter 7.

Specifically then, let $\rho_V(x, y, z, t)$ be the density of a fluid at a point with coordinates (x, y, z) at a time t. Then the mass of a volume V at instant t is

$$M = \iiint_V \rho_V\,\mathrm{d}V$$

(see Section 4.10). Also, the rate at which mass crosses the surface S enclosing V is given by

$$\oiint_S \rho_V\mathbf{v}\cdot\hat{\mathbf{n}}\,\mathrm{d}S$$

where $\mathbf{v}(x, y, z, t)$ is the fluid velocity, a vector field (see Section 4.6).

We can equate the rate of mass flow across S with the rate of decrease of mass within V. This is 'continuity of mass' and assumes that there are no sources or sinks of fluid within V. In mathematical terms,

$$\oiint_S \rho_V\mathbf{v}\cdot\hat{\mathbf{n}}\,\mathrm{d}S = -\frac{\mathrm{d}}{\mathrm{d}t}\iiint_V \rho_V\,\mathrm{d}V \tag{5.5}$$

the negative sign being inserted because a net outflow of fluid across S makes the left-hand side of (5.5) positive ($\hat{\mathbf{n}}$ is an outward normal) and the rate of change on the right-hand side negative.

Equation (5.5) is the integral form of the continuity equation. It can readily be transformed into differential or point form by applying the divergence theorem to the flux integral and interchanging the order of the time differentiation and volume integration on the right-hand side. We obtain

$$\iiint_V \nabla\cdot(\rho_V \mathbf{v})\,\mathrm{d}V = -\iiint_V \frac{\partial \rho_V}{\partial t}\,\mathrm{d}V$$

or, on transposing,

$$\iiint_V \left\{\nabla\cdot(\rho_V \mathbf{v}) + \frac{\partial \rho_V}{\partial t}\right\}\mathrm{d}V = 0$$

Since this result must hold for any volume V, the integrand must vanish identically, which gives the continuity equation at a point:

$$\nabla\cdot(\rho_V \mathbf{v}) = -\frac{\partial \rho_V}{\partial t}$$

For the special case of an incompressible fluid, $\partial \rho_V/\partial t = 0$ and so

$$\nabla\cdot(\rho_V \mathbf{v}) = \rho_V\,\nabla\cdot\mathbf{v} = 0 \quad \text{or} \quad \nabla\cdot\mathbf{v} = 0$$

In other words, the velocity vector of an incompressible fluid, like the magnetic field vector discussed earlier, is one whose divergence vanishes everywhere within a region. The technical term for such vector fields is **solenoidal**.

These ideas are readily extended to the analysis of heat conduction. As we saw in Section 4.6, the rate at which heat crosses a unit area normal to the direction of flow depends upon the thermal conductivity κ of the material present and the temperature gradient ∇u. Thus, by comparison with the left-hand side of (5.5),

$$\oint\!\!\oint_S (-\kappa\,\nabla u)\cdot\hat{\mathbf{n}}\,\mathrm{d}S$$

gives us the rate of heat flow across a closed surface S. But if ρ_V and c denote the density and specific heat capacity respectively, then

$$\frac{\mathrm{d}}{\mathrm{d}t}\iiint_V c\rho_V u\,\mathrm{d}V$$

must be the rate of increase of the heat contained in the volume V enclosed by S. Hence, by continuity, we must have

$$\oint\!\!\oint_S (\kappa\,\nabla u)\cdot\hat{\mathbf{n}}\,\mathrm{d}S = \frac{\mathrm{d}}{\mathrm{d}t}\iiint_V c\rho_V u\,\mathrm{d}V = \iiint_V c\rho_V \frac{\partial u}{\partial t}\,\mathrm{d}V$$

assuming that c and ρ_V are constants. Transforming the left-hand side by the divergence theorem, we obtain

$$\iiint_V \nabla\cdot(\kappa\,\nabla u)\,\mathrm{d}V = \iiint_V c\rho_V \frac{\partial u}{\partial t}\,\mathrm{d}V$$

or
$$\iiint_V \left\{ \nabla \cdot (\kappa \nabla u) - c\rho_V \frac{\partial u}{\partial t} \right\} \mathrm{d}V = 0$$

Since V is arbitrary we find, making the further assumption that κ is constant, that

$$\nabla \cdot \nabla u = \nabla^2 u = \frac{c\rho_V}{\kappa} \frac{\partial u}{\partial t}$$

which is the basic differential equation of heat conduction.

5.6 The divergence theorem and strain energy (*optional*)

The previous section showed that the divergence theorem may be usefully applied in the construction of continuity equations and in deriving the heat conduction equation. In this section we shall show that the theorem is also invaluable in solid mechanics.

When an elastic body is deformed by forces, then those forces do work. As a result, the body, in its deformed state, gains energy known as **strain energy**. This strain energy is used to return the body to its original configuration when the deforming forces are removed.

We begin by considering perhaps the simplest elastic system – a spring of stiffness k. The force F developed in the spring is linearly related to the measured extension u, so that a spring of relaxed length l^* extended to a length x sustains a force $k(x - l^*)$. If the spring is extended by a further amount $\mathrm{d}x$, then the extra work performed is $k(x - l^*)\,\mathrm{d}x$. The total work required to extend a spring from l^* to l is

$$\begin{aligned} \int_{l^*}^{l} k(x - l^*)\,\mathrm{d}x &= \left[\tfrac{1}{2}k(x - l^*)^2 \right]_{l^*}^{l} \\ &= \tfrac{1}{2}k(l - l^*)^2 \\ &= \tfrac{1}{2}[k(l - l^*)][l - l^*] \end{aligned}$$

In words,

work done, $W = \tfrac{1}{2}$(final force)(final extension)

or, in vector terms,

$$W = \tfrac{1}{2}\mathbf{F} \cdot \mathbf{u}$$

where, using the obvious choice of axes (with the x-axis along the axis of the spring), $\mathbf{u} = u\hat{\mathbf{i}}$ and $\mathbf{F} = F\hat{\mathbf{i}}$.

If we now consider deforming an elastic body such that its displacement vector

$$\mathbf{u} = u\hat{\mathbf{i}} + v\hat{\mathbf{j}} + w\hat{\mathbf{k}}$$

is produced by increasing the surface forces and body forces from zero to final values $\mathbf{P}^n$ and $\mathbf{B}$ respectively, then the total work can be shown to be

$$W = \tfrac{1}{2} \oiint_S \mathbf{P}^n \cdot \mathbf{u} \, dS + \tfrac{1}{2} \iiint_V \rho \mathbf{B} \cdot \mathbf{u} \, dV \tag{5.6}$$

where ρ is the density of the material, which has a volume V and is bounded by a surface S. This result is obtained by considering the work done by the surface forces on an element of the boundary and by the body forces on an element of the interior of the body as they act through the displacement. The expression can be directly related to the stresses and strains developed within the body.

Now, by (1.50),

$$\tfrac{1}{2} \oiint_S \mathbf{P}^n \cdot \mathbf{u} \, dS = \tfrac{1}{2} \oiint_S (\boldsymbol{\sigma}_x u + \boldsymbol{\sigma}_y v + \boldsymbol{\sigma}_z w) \cdot \hat{\mathbf{n}} \, dS$$

The right-hand side of this equation may be transformed into a volume integral by the divergence theorem, giving

$$\tfrac{1}{2} \iiint_V \{\nabla \cdot (\boldsymbol{\sigma}_x u + \boldsymbol{\sigma}_y v + \boldsymbol{\sigma}_z w)\} \, dV$$

Also, $\rho \mathbf{B} \cdot \mathbf{u} = \rho(B_x u + B_y v + B_z w) = -(u \, \nabla \cdot \boldsymbol{\sigma}_x + v \, \nabla \cdot \boldsymbol{\sigma}_y + w \, \nabla \cdot \boldsymbol{\sigma}_z)$

using (3.25). Hence, substituting into (5.6),

$$\begin{aligned} W &= \tfrac{1}{2} \iiint_V \{\nabla \cdot (\boldsymbol{\sigma}_x u + \boldsymbol{\sigma}_y v + \boldsymbol{\sigma}_z w) - u \, \nabla \cdot \boldsymbol{\sigma}_x - v \, \nabla \cdot \boldsymbol{\sigma}_y - w \, \nabla \cdot \boldsymbol{\sigma}_z\} \, dV \\ &= \tfrac{1}{2} \iiint_V (\boldsymbol{\sigma}_x \cdot \nabla u + \boldsymbol{\sigma}_y \cdot \nabla v + \boldsymbol{\sigma}_z \cdot \nabla w) \, dV \end{aligned}$$

using identity (2a) of Section 3.6. But it is easily verified using (1.52) and (3.15) that

$$\boldsymbol{\sigma}_x \cdot \boldsymbol{\varepsilon}_x + \boldsymbol{\sigma}_y \cdot \boldsymbol{\varepsilon}_y + \boldsymbol{\sigma}_z \cdot \boldsymbol{\varepsilon}_z = \boldsymbol{\sigma}_x \cdot \nabla u + \boldsymbol{\sigma}_y \cdot \nabla v + \boldsymbol{\sigma}_z \cdot \nabla w$$

and so finally

$$W = \tfrac{1}{2} \iiint_V (\boldsymbol{\sigma}_x \cdot \boldsymbol{\varepsilon}_x + \boldsymbol{\sigma}_y \cdot \boldsymbol{\varepsilon}_y + \boldsymbol{\sigma}_z \cdot \boldsymbol{\varepsilon}_z) \, dV$$

W is called the strain energy of the body.

As an example of the central importance that strain energy enjoys in solid mechanics, we simply quote one of the Castigliano theorems: Suppose an elastic body is rigidly supported (that is, the supports are not springs) and is acted on by a system of forces including a point force $\mathbf{F}$ on the surface, to produce a strain energy W: if $\mathbf{u}$ is the displacement measured at the point of application of **F**, then

$$\mathbf{F} = \hat{\mathbf{i}}\frac{\partial W}{\partial u} + \hat{\mathbf{j}}\frac{\partial W}{\partial v} + \hat{\mathbf{k}}\frac{\partial W}{\partial w}$$

This result is fundamental in the theoretical development of structural mechanics. Structures are formed by connecting together members (elements) at points called 'nodes'. The nodes would be given an arbitrary (virtual) displacement consistent with the boundary conditions. From this, an expression for the strain energy can be developed. The Castigliano theorem quoted above is then used to relate the known applied forces to the unknown displacements. The resulting equations are linear and may be easily solved to obtain the unknown displacements.

5.7 Outline derivation of Stokes' theorem

The divergence theorem links a volume integral with a flux integral over a closed surface. Our approach to demonstrating the theorem began by finding the flux over the surface of an elemental volume. Stokes' theorem is 'one dimension down', so to speak, on the divergence theorem and links a flux integral over a non-closed surface with a line integral over the closed curve bounding that surface. Accordingly, we approach Stokes' theorem by first considering a line integral around an elementary closed path. The path we choose is the parallelogram $PQRS$ shown in Figure 5.8a. The sides of the parallelogram are of lengths Δu and Δv, and hence the vector area is, by (1.37),

$$\Delta\mathbf{S} = \Delta u\,\hat{\mathbf{u}} \wedge \Delta v\,\hat{\mathbf{v}} = \Delta u\,\Delta v\,(\hat{\mathbf{u}} \wedge \hat{\mathbf{v}})$$

where $\hat{\mathbf{u}}$ and $\hat{\mathbf{v}}$ are unit vectors parallel to $\overrightarrow{PQ}$ and $\overrightarrow{PS}$ respectively (see Figure 5.8b).

If **F** is a vector field with the Cartesian representation

$$\mathbf{F} = F_x\hat{\mathbf{i}} + F_y\hat{\mathbf{j}} + F_z\hat{\mathbf{k}}$$

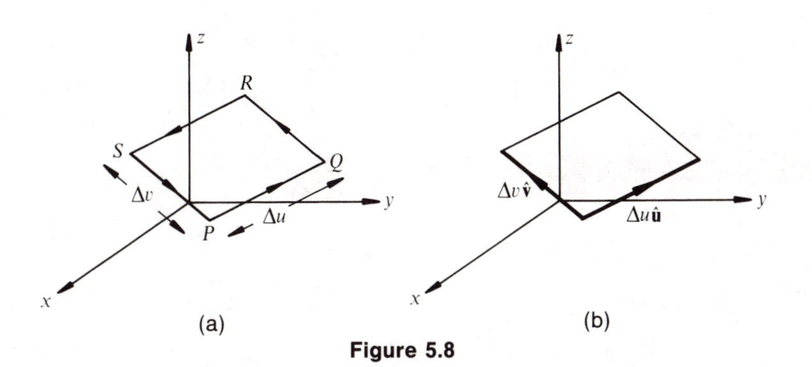

Figure 5.8

we calculate the line integral of $\mathbf{F}$ around the boundary $PQRS$ (in that order) of the parallelogram. We assume that Δu and Δv are sufficiently small for us to treat $\mathbf{F}$ as a constant vector along each boundary.

For the section PQ,

$$\int_{PQ} \mathbf{F}\cdot d\mathbf{l} \approx \mathbf{F}_P\cdot \Delta u\, \hat{\mathbf{u}}$$

where $\mathbf{F}_P$ denotes the vector field $\mathbf{F}$ at the point P.

Similarly, for the section RS,

$$\int_{RS} \mathbf{F}\cdot d\mathbf{l} = -\int_{SR} \mathbf{F}\cdot d\mathbf{l} \approx -\mathbf{F}_S\cdot \Delta u\, \hat{\mathbf{u}} \tag{5.7}$$

where $\mathbf{F}_S$ is the value of $\mathbf{F}$ at the point S. We can obtain the components of $\mathbf{F}_S$ in terms of the components of $\mathbf{F}_P$ as follows:

$$(F_x)_S \approx (F_x)_P + \frac{\partial F_x}{\partial v}\Delta v, \qquad (F_y)_S \approx (F_y)_P + \frac{\partial F_y}{\partial v}\Delta v,$$

$$(F_z)_S \approx (F_z)_P + \frac{\partial F_z}{\partial v}\Delta v$$

since Δv is the distance of the point S from P, and where we have neglected higher-order terms. But, from our discussion of the gradient concept in Section 3.2, we know that

$$\frac{\partial F_x}{\partial v} = \nabla F_x\cdot \hat{\mathbf{v}}$$

and similarly for the other partial derivatives. Hence,

$$\begin{aligned}\mathbf{F}_S &= (F_x)_S\hat{\mathbf{i}} + (F_y)_S\hat{\mathbf{j}} + (F_z)_S\hat{\mathbf{k}} \\ &= \mathbf{F}_P + [(\nabla F_x\cdot\hat{\mathbf{v}})\hat{\mathbf{i}} + (\nabla F_y\cdot\hat{\mathbf{v}})\hat{\mathbf{j}} + (\nabla F_z\cdot\hat{\mathbf{v}})\hat{\mathbf{k}}]\,\Delta v\end{aligned}$$

Substituting in (5.7),

$$\int_{RS} \mathbf{F}\cdot d\mathbf{l} \approx -\mathbf{F}_P\cdot\Delta u\,\hat{\mathbf{u}} - [(\nabla F_x\cdot\hat{\mathbf{v}})\hat{\mathbf{i}} + (\nabla F_y\cdot\hat{\mathbf{v}})\hat{\mathbf{j}} + (\nabla F_z\cdot\hat{\mathbf{v}})\hat{\mathbf{k}}]\,\Delta v\cdot\Delta u\,\hat{\mathbf{u}}$$

By similar reasoning,

$$\int_{SP} \mathbf{F}\cdot d\mathbf{l} \approx +\mathbf{F}_P\cdot(-\Delta v\,\hat{\mathbf{v}}) = -\mathbf{F}_P\cdot\Delta v\,\hat{\mathbf{v}}$$

and $$\int_{QR} \mathbf{F}\cdot d\mathbf{l} \approx +\mathbf{F}_P\cdot\Delta v\,\hat{\mathbf{v}} + [(\nabla F_x\cdot\hat{\mathbf{u}})\hat{\mathbf{i}} + (\nabla F_y\cdot\hat{\mathbf{u}})\hat{\mathbf{j}} + (\nabla F_z\cdot\hat{\mathbf{u}})\hat{\mathbf{k}}]\,\Delta u\cdot\Delta v\,\hat{\mathbf{v}}$$

Adding the contributions from the four sides,

$$\oint_C \mathbf{F}\cdot d\mathbf{l} \approx [(\nabla F_x\cdot\hat{\mathbf{u}})\hat{\mathbf{i}} + (\nabla F_y\cdot\hat{\mathbf{u}})\hat{\mathbf{j}} + (\nabla F_z\cdot\hat{\mathbf{u}})\hat{\mathbf{k}}]\cdot\hat{\mathbf{v}}\,\Delta u\,\Delta v$$
$$- [(\nabla F_x\cdot\hat{\mathbf{v}})\hat{\mathbf{i}} + (\nabla F_y\cdot\hat{\mathbf{v}})\hat{\mathbf{j}} + (\nabla F_z\cdot\hat{\mathbf{v}})\hat{\mathbf{k}}]\cdot\hat{\mathbf{u}}\,\Delta u\,\Delta v \qquad \textbf{(5.8)}$$

Although the right-hand side of (5.8) looks formidable, it is readily simplified by considering the terms in pairs. If the unit vectors $\hat{\mathbf{u}}$ and $\hat{\mathbf{v}}$ have components (u_x, u_y, u_z) and (v_x, v_y, v_z) respectively,

$$\nabla F_x\cdot\hat{\mathbf{u}} = \frac{\partial F_x}{\partial x}u_x + \frac{\partial F_x}{\partial y}u_y + \frac{\partial F_x}{\partial z}u_z$$

so $$(\nabla F_x\cdot\hat{\mathbf{u}})\hat{\mathbf{i}}\cdot\hat{\mathbf{v}} = \left(\frac{\partial F_x}{\partial x}u_x + \frac{\partial F_x}{\partial y}u_y + \frac{\partial F_x}{\partial z}u_z\right)v_x$$

and, similarly,

$$(\nabla F_x\cdot\hat{\mathbf{v}})\hat{\mathbf{i}}\cdot\hat{\mathbf{u}} = \left(\frac{\partial F_x}{\partial x}v_x + \frac{\partial F_x}{\partial y}v_y + \frac{\partial F_x}{\partial z}v_z\right)u_x$$

On subtraction of this pair of terms in (5.8), we obtain

$$\frac{\partial F_x}{\partial y}(u_y v_x - u_x v_y) + \frac{\partial F_x}{\partial z}(u_z v_x - u_x v_z)$$

Similarly, for the second terms in the brackets of (5.8),

$$(\nabla F_y\cdot\hat{\mathbf{u}})\hat{\mathbf{j}}\cdot\hat{\mathbf{v}} - (\nabla F_y\cdot\hat{\mathbf{v}})\hat{\mathbf{j}}\cdot\hat{\mathbf{u}} = \frac{\partial F_y}{\partial x}(u_x v_y - u_y v_x) + \frac{\partial F_y}{\partial z}(u_z v_y - u_y v_z)$$

and $$(\nabla F_z\cdot\hat{\mathbf{u}})\hat{\mathbf{k}}\cdot\hat{\mathbf{v}} - (\nabla F_z\cdot\hat{\mathbf{v}})\hat{\mathbf{k}}\cdot\hat{\mathbf{u}} = \frac{\partial F_z}{\partial x}(u_x v_z - u_z v_x) + \frac{\partial F_z}{\partial y}(u_y v_z - u_z v_y)$$

Hence, substituting into (5.8) and collecting terms,

$$\frac{1}{\Delta u\,\Delta v}\oint_C \mathbf{F}\cdot d\mathbf{l} \approx (u_z v_y - u_y v_z)\left(\frac{\partial F_y}{\partial z} - \frac{\partial F_z}{\partial y}\right) + (u_z v_x - u_x v_z)\left(\frac{\partial F_x}{\partial z} - \frac{\partial F_z}{\partial x}\right)$$
$$+ (u_y v_x - u_x v_y)\left(\frac{\partial F_x}{\partial y} - \frac{\partial F_y}{\partial x}\right) \qquad \textbf{(5.9)}$$

The right-hand side of (5.9) is a sum of three products, which suggests that it is the scalar product of two vectors, for example **A** and **B**, where

$$A_x \doteq u_z v_y - u_y v_z \text{ or its negative} \quad \text{and} \quad B_x = \frac{\partial F_y}{\partial z} - \frac{\partial F_z}{\partial y} \text{ or its negative}$$

In fact, recalling the Cartesian form of the cross product (1.45), we can put $\mathbf{A} = \mathbf{u} \wedge \mathbf{v}$. Similarly, from the Cartesian form of the curl of a vector

(3.21), we can express **B** as $\nabla \wedge \mathbf{F}$. Hence finally we have, from (5.9),

$$\oint_C \mathbf{F} \cdot d\mathbf{l} \approx (\mathbf{u} \wedge \mathbf{v}) \cdot (\nabla \wedge \mathbf{F})\, \Delta u\, \Delta v = \Delta \mathbf{S} \cdot (\nabla \wedge \mathbf{F}) = \nabla \wedge \mathbf{F} \cdot \hat{\mathbf{n}}\, \Delta S \tag{5.10}$$

where $\hat{\mathbf{n}}$ is the unit normal to the parallelogram of area ΔS. Since $\hat{\mathbf{n}}$ is parallel to $\mathbf{u} \wedge \mathbf{v}$ we see from Figure 5.8b that the sense of $\hat{\mathbf{n}}$ is linked to the sense of travel around C by a right-hand screw rule.

We can now use (5.10) to give a simplified proof of Stokes' theorem. Consider a surface S bounded by a closed curve C (see Figure 5.9a). We use a similar technique to that used for demonstrating the divergence theorem in Section 5.2. We divide the surface S into N small elements of areas $\Delta S_1, \Delta S_2, \ldots, \Delta S_N$, where each element is approximately flat (see Figure 5.9b). For a typical element ΔS_i with boundary C_i, we apply (5.10) to give

$$\oint_{C_i} \mathbf{F} \cdot d\mathbf{l} \approx \nabla \wedge \mathbf{F} \cdot \hat{\mathbf{n}}_i\, \Delta S_i$$

where $\hat{\mathbf{n}}_i$ is the unit normal to ΔS_i. Repeating the process for all the N elements and adding, we obtain

$$\sum_{i=1}^{N} \oint_{C_i} \mathbf{F} \cdot d\mathbf{l} \approx \sum_{i=1}^{N} \nabla \wedge \mathbf{F} \cdot \hat{\mathbf{n}}_i\, \Delta S_i \tag{5.11}$$

If we take smaller and smaller elements (and, correspondingly, let N tend to ∞), the sum on the right-hand side of (5.11) becomes the flux integral

$$\iint_S \nabla \wedge \mathbf{F} \cdot \hat{\mathbf{n}}\, dS$$

The left-hand side of (5.11) involves the sum of all the line integrals over all the boundaries of the small elements into which S has been divided. However, as we can see from Figure 5.9b, contributions to the line integrals are in opposite senses on the internal boundaries, and such contributions will

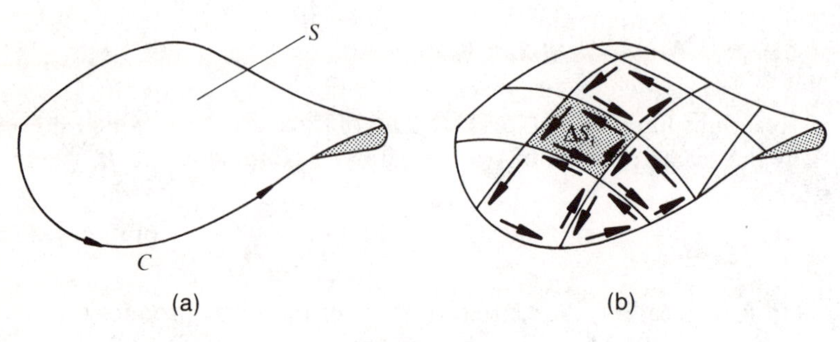

Figure 5.9

cancel. No such cancellation will occur on the boundary C, so that the left-hand side of (5.11) will, as N tends to ∞, tend to the line integral of $\mathbf{F}$ around the boundary. Finally, we obtain

$$\oint_C \mathbf{F}\cdot d\mathbf{l} = \iint_S \nabla \wedge \mathbf{F}\cdot\hat{\mathbf{n}}\, dS \qquad \textbf{(5.12)}$$

Equation (5.12) is the mathematical statement of **Stokes' theorem**. More informally, the theorem states that the **circulation** of $\mathbf{F}$ around C equals the **flux** of curl $\mathbf{F}$ across S. Note also from Figure 5.9b that the sense of the unit normal $\hat{\mathbf{n}}$ in the flux integral is linked to the sense of travel around C by a right-hand screw rule; thus, if we reversed the sense of the line integral we would have to reverse $\hat{\mathbf{n}}$.

Points to note in Stokes' theorem

The above 'derivation' of Stokes' theorem does not stand up well to a rigorous examination. As with the divergence theorem derivation, we have made a number of implicit assumptions which in some important applications may need to be carefully considered.

Firstly, the vector field $\mathbf{F}$ must be sufficiently smooth (each component of $\mathbf{F}$ should have continuous partial derivatives) in order that the curl of $\mathbf{F}$ may be taken at each point of S. Secondly, we have implicitly assumed that both the line integral $\oint_C \mathbf{F}\cdot d\mathbf{l}$ and the surface integral $\iint_S \nabla \wedge \mathbf{F}\cdot\hat{\mathbf{n}}\, dS$ may be assigned unambiguous values. This will certainly be the case if the Cartesian form of the equation of S, $z = f(x, y)$ (or $x = g(y, z)$ or $y = h(x, z)$), is such that f (or g or h) is single valued and differentiable. These conditions imply that lines parallel to the z-axis intersect the surface in just one point (in the case when $z = f(x, y)$). If Stokes' theorem did only apply when these conditions on S held strictly, then it would be of limited use. In fact, however, it may be applied to more general surfaces S^* (bounded by a curve C^*) which do not satisfy these criteria but which may be split into surfaces $S_1, S_2, \ldots, S_n$ with boundaries $C_1, C_2, \ldots, C_n$ which do satisfy them, so that Stokes' theorem holds for each subsurface. On adding the surface integrals together, the surface integral over S^* is obtained. On adding the line integrals (involving various cancellations), only the line integral over C^* remains.

This extended form of Stokes' theorem may be used, for example, in the derivation of Ampere's law from the vector field – in this case a magnetic field $\mathbf{H}$ – due to an infinite line current I along the z-axis,

$$\mathbf{H} = \frac{I}{2\pi\rho}\hat{\boldsymbol{\phi}} \qquad \text{in cylindrical coordinates}$$

$$= \frac{I}{2\pi}\left(\frac{y\hat{\mathbf{i}} - x\hat{\mathbf{j}}}{x^2 + y^2}\right) \qquad \text{in Cartesian coordinates}$$

Ampere's law states that

$$\oint_C \mathbf{H}\cdot d\mathbf{l} = \begin{cases} I & \text{if } C \text{ contains } I \\ 0 & \text{if } C \text{ does not contain } I \end{cases}$$

We first suppose that C does not surround the z-axis (the path of the current). Then we can always find a surface S, bounded by C, which is not cut by the z-axis. $\mathbf{H}$ is continuously differentiable everywhere on S (it is singular only on the z-axis), and we find that $\nabla \wedge \mathbf{H} = 0$. Hence, using Stokes' theorem for this case,

$$\oint_C \mathbf{H}\cdot d\mathbf{l} = \iint_S (\nabla \wedge \mathbf{H})\cdot\hat{\mathbf{n}}\, dS = 0$$

Considering the second possibility, in which C does surround the z-axis, there is now no surface S, bounded by C, which is not intersected by the z-axis. Thus there will always be at least one point of S for which $\nabla \wedge \mathbf{H}$ does not exist and so Stokes' theorem may not be used directly. However, the line integral $\oint_C \mathbf{H}\cdot d\mathbf{l}$ is meaningful and may be assigned an unambiguous value. To determine this value, we convert the line integral to an easier form by surrounding the z-axis by a circle C_1 of radius a in a plane parallel to the xy-plane. A surface S may now be found, bounded by C and C_1, which is not intersected by the z-axis and to which the extended form of Stokes' theorem may be applied. Everywhere on this surface $\nabla \wedge \mathbf{H} = 0$, so the circulation of $\mathbf{H}$ over the boundary of S (namely C *and* C_1) is zero, that is,

$$\oint_C \mathbf{H}\cdot d\mathbf{l} + \oint_{C_1} \mathbf{H}\cdot d\mathbf{l} = 0$$

Note carefully that, having chosen a particular orientation in which C is described, we are necessarily led to choose the *opposite* orientation on C_1 because the tangential direction on C_1, the normal to S, and the tangent direction to the surface at C_1 pointing out of S form a right-handed system

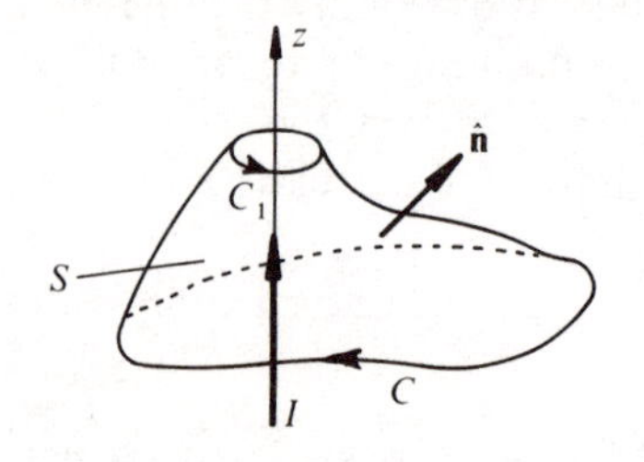

Figure 5.10

(see Figure 5.10). Now, on C_1,

$$\mathbf{H} = \frac{I}{2\pi}\frac{\hat{\boldsymbol{\phi}}}{a} \quad \text{and} \quad \mathbf{dl} = a\,\mathrm{d}\phi\,\hat{\boldsymbol{\phi}}$$

So $$\oint_{C_1} \mathbf{H}\cdot\mathbf{dl} = \int_{2\pi}^{0} \frac{I}{2\pi a}\,a\,\mathrm{d}\phi = \int_{2\pi}^{0} \frac{I}{2\pi}\,\mathrm{d}\phi = -I$$

(note the limits, which are due to the reverse orientation on C_1). Here we have performed the calculation in polar coordinates; the reader should rework the problem in Cartesian coordinates.

Finally then, when C contains the z-axis we have

$$\oint_{C} \mathbf{H}\cdot\mathbf{dl} = -\oint_{C_1} \mathbf{H}\cdot\mathbf{dl} = I$$

which validates Ampere's law.

5.8 Worked examples using Stokes' theorem

Example 5.3

Verify Stokes' theorem for the vector field $\mathbf{F} = x\hat{\mathbf{i}} + x\hat{\mathbf{j}} + 2xy\hat{\mathbf{k}}$ using the hemisphere $x^2 + y^2 + z^2 = 4$, $z \leqslant 0$.

Solution

We need to evaluate both the line integral of $\mathbf{F}$ around the bounding curve C in the sense shown in Figure 5.11a and the flux of $\nabla \wedge \mathbf{F}$ over the hemispherical surface S.

The line integral is straightforward:

$$\oint_{C} \mathbf{F}\cdot\mathbf{dl} = \oint_{C} (x\,\mathrm{d}x + x\,\mathrm{d}y)$$

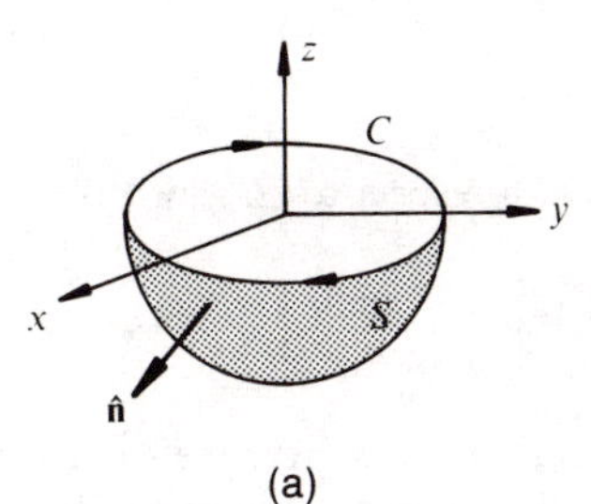

(a)

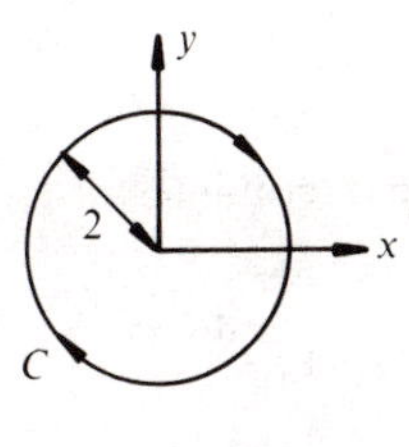

(b)

Figure 5.11

(the z-component of $\mathbf{F}$ does not contribute). Using polar coordinates with $x = 2\cos\phi$ and $y = 2\sin\phi$ because C is a circle of radius 2, we obtain

$$\oint_C \mathbf{F}\cdot d\mathbf{l} = \int_{2\pi}^{0} (2\cos\phi)(-2\sin\phi)\, d\phi + \int_{2\pi}^{0} (2\cos\phi)(2\cos\phi)\, d\phi$$

$$= 0 + 4\int_{2\pi}^{0} \cos^2\phi\, d\phi = -4\pi$$

Note that the limits on the line integral are from 2π to 0, rather than from 0 to 2π. This is because the sense of travel around C and the sense of the unit normal $\hat{\mathbf{n}}$ are linked by the right-hand screw rule (see Figure 5.11).

For the flux integral we require

$$\nabla \wedge \mathbf{F} = \begin{vmatrix} \hat{\mathbf{i}} & \hat{\mathbf{j}} & \hat{\mathbf{k}} \\ \dfrac{\partial}{\partial x} & \dfrac{\partial}{\partial y} & \dfrac{\partial}{\partial z} \\ x & x & 2xy \end{vmatrix} = 2x\hat{\mathbf{i}} - 2y\hat{\mathbf{j}} + \hat{\mathbf{k}}$$

The unit outward normal to the hemisphere is

$$\hat{\mathbf{n}} = \frac{\mathbf{r}}{|\mathbf{r}|} = \tfrac{1}{2}(x\hat{\mathbf{i}} + y\hat{\mathbf{j}} + z\hat{\mathbf{k}})$$

so $$(\nabla \wedge \mathbf{F})\cdot\hat{\mathbf{n}} = x^2 - y^2 + \tfrac{1}{2}z$$

and the relevant flux integral is

$$\iint_S \nabla \wedge \mathbf{F}\cdot\hat{\mathbf{n}}\, dS = \iint_S (x^2 - y^2 + \tfrac{1}{2}z)\, dS$$

To integrate over the hemispherical surface, we convert the integrand to spherical coordinates with $r = 2$; we obtain

$$\int_0^{2\pi}\int_{\pi/2}^{\pi} (4\sin^2\theta\cos^2\phi - 4\sin^2\theta\sin^2\phi + \cos\theta)(4\sin\theta)\, d\theta\, d\phi$$

$$= \int_0^{2\pi}\int_{\pi/2}^{\pi} (16\sin^3\theta\cos 2\phi + 2\sin 2\theta)\, d\theta\, d\phi$$

The first term integrates to zero with respect to ϕ, whilst the second yields

$$2\int_0^{2\pi} d\phi \int_{\pi/2}^{\pi} \sin 2\theta\, d\theta = 4\pi\left[\frac{-\cos 2\theta}{2}\right]_{\pi/2}^{\pi} = -4\pi$$

We see that the flux integral of $\nabla \wedge \mathbf{F}$ over the surface has the same value as the line integral of $\mathbf{F}$ around the boundary. Hence Stokes' theorem has been verified for this situation.

Example 5.4

Verify Stokes' theorem for the vector field $\mathbf{F} = (x - y)\hat{\mathbf{i}} + 2z\hat{\mathbf{j}} + x^2\hat{\mathbf{k}}$ using the cone $z = +\sqrt{(x^2 + y^2)}$, $0 < z < 2$.

Solution

The line integral is

$$\oint_C \mathbf{F} \cdot \mathrm{d}\mathbf{l} = \oint_C \{(x - y)\,\mathrm{d}x + 4\,\mathrm{d}y\}$$

where C is the circle $x^2 + y^2 = 4$, $z = 2$ (see Figure 5.12). This can be readily evaluated to give a value 4π.

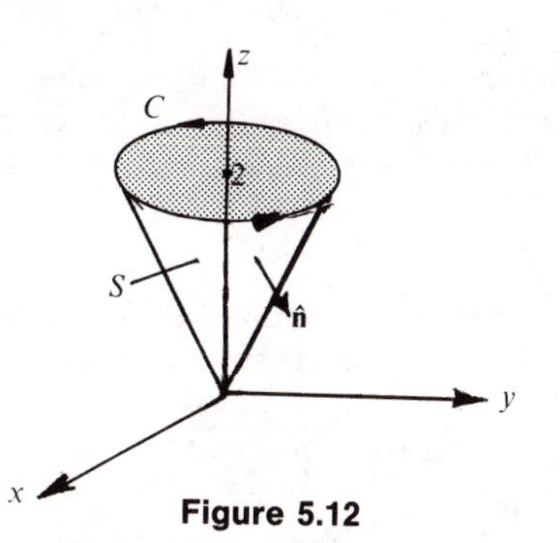

Figure 5.12

The flux integral is more difficult. The unit normal $\hat{\mathbf{n}}$ to the conical surface is given by

$$\hat{\mathbf{n}} = \frac{\nabla f}{|\nabla f|} \quad \text{where} \quad f = z - \sqrt{(x^2 + y^2)}$$

But $$\nabla f = -\frac{x}{\sqrt{(x^2 + y^2)}}\hat{\mathbf{i}} - \frac{y}{\sqrt{(x^2 + y^2)}}\hat{\mathbf{j}} + \hat{\mathbf{k}} = -\left(\frac{x}{z}\right)\hat{\mathbf{i}} - \left(\frac{y}{z}\right)\hat{\mathbf{j}} + \hat{\mathbf{k}}$$

Hence $$|\nabla f| = \sqrt{\left(\frac{x^2 + y^2}{z^2} + 1\right)} = \sqrt{2} \quad \text{on } S$$

so $$\hat{\mathbf{n}} = \frac{1}{\sqrt{2}}\left[-\left(\frac{x}{z}\right)\hat{\mathbf{i}} - \left(\frac{y}{z}\right)\hat{\mathbf{j}} + \hat{\mathbf{k}}\right]$$

Also, $$\nabla \wedge \mathbf{F} = \begin{vmatrix} \hat{\mathbf{i}} & \hat{\mathbf{j}} & \hat{\mathbf{k}} \\ \dfrac{\partial}{\partial x} & \dfrac{\partial}{\partial y} & \dfrac{\partial}{\partial z} \\ x - y & 2z & x^2 \end{vmatrix} = -2\hat{\mathbf{i}} - 2x\hat{\mathbf{j}} + \hat{\mathbf{k}}$$

so $$\nabla \wedge \mathbf{F} \cdot \hat{\mathbf{n}} = \frac{1}{\sqrt{2}}\left[2\left(\frac{x}{z}\right) + 2\left(\frac{xy}{z}\right) + 1\right]$$

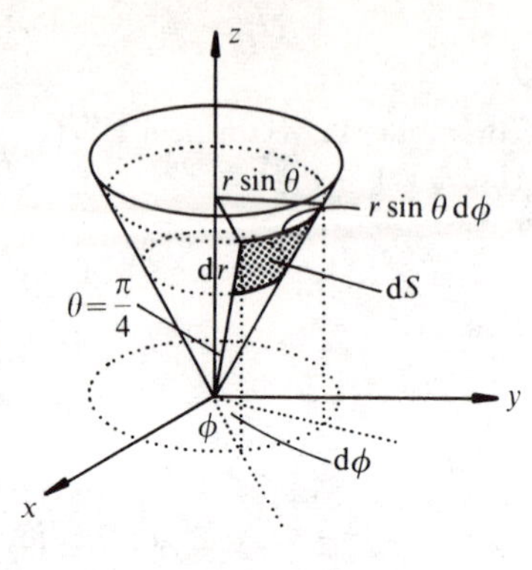

Figure 5.13

On S, the spherical coordinate θ has value $\pi/4$, so the coordinate r $(=z/\cos\theta)$ varies from 0 to $2\sqrt{2}$ as z varies from 0 to 2. Also, referring to Figure 5.13, the surface element on the cone is

$$dS = r\sin\theta \, dr \, d\phi = \frac{r}{\sqrt{2}} \, dr \, d\phi$$

Hence the flux of $\nabla \wedge \mathbf{F}$ across the surface of the cone is

$$\iint_S \nabla \wedge \mathbf{F} \cdot \hat{\mathbf{n}} \, dS = \tfrac{1}{2} \int_0^{2\pi} \int_0^{2\sqrt{2}} \left\{ 2\left(\frac{x}{z}\right) + 2\left(\frac{xy}{z}\right) + 1 \right\} r \, dr \, d\phi$$

Finally, we replace x, y and z by their equivalents in spherical coordinates,

$$x = r\sin\theta\cos\phi = \frac{r}{\sqrt{2}}\cos\phi, \qquad y = r\sin\theta\sin\phi = \frac{r}{\sqrt{2}}\sin\phi,$$

$$z = r\cos\theta = \frac{r}{\sqrt{2}}$$

so
$$\iint_S \nabla \wedge \mathbf{F} \cdot \hat{\mathbf{n}} \, dS = \int_0^{2\pi} \int_0^{2\sqrt{2}} \left(\cos\phi + \frac{r}{\sqrt{2}} \cos\phi \sin\phi + \tfrac{1}{2} \right) r \, dr \, d\phi$$

$$= \tfrac{1}{2} 2\pi \left[\frac{r^2}{2} \right]_0^{2\sqrt{2}} = 4\pi$$

Again, this has the same value as the line integral, thus completing the verification of Stokes' theorem.

5.9 A general definition of curl

So far, we have only encountered the curl of a vector field in Cartesian coordinates. However, just as we used the divergence theorem to obtain an alternative and more general definition of the divergence concept, so we can use Stokes' theorem (5.12) to carry out an analogous procedure for the curl.

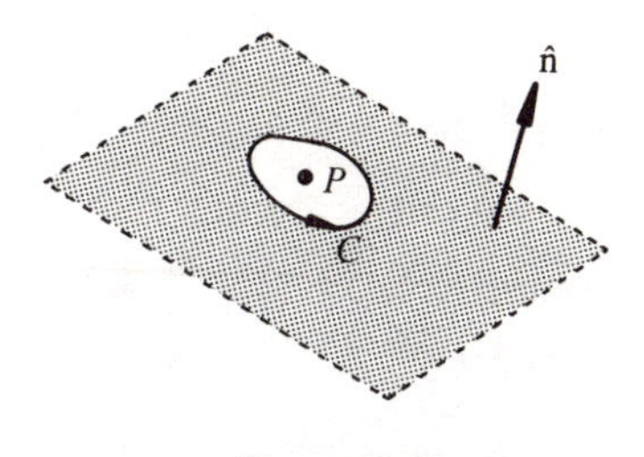

Figure 5.14

Consider a point P in a vector field **F** and define a direction by means of a unit vector $\hat{\mathbf{n}}$ (see Figure 5.14). If we consider an elementary closed path C lying in a plane through P with normal $\hat{\mathbf{n}}$, we can apply Stokes' theorem,

$$\oint_C \mathbf{F}\cdot d\mathbf{l} = \iint_{\Delta S} \nabla \wedge \mathbf{F}\cdot\hat{\mathbf{n}}\, dS$$

where ΔS is any elementary surface bounded by C. However, if ΔS is small enough for $\nabla \wedge \mathbf{F}$ not to vary significantly over it, the flux of $\nabla \wedge \mathbf{F}$ can be approximated by using the value of $\nabla \wedge \mathbf{F}$ at P, that is,

$$\oint_C \mathbf{F}\cdot d\mathbf{l} \approx (\nabla \wedge \mathbf{F})_P\cdot\hat{\mathbf{n}}\,\Delta S$$

or
$$(\nabla \wedge \mathbf{F})_P\cdot\hat{\mathbf{n}} \approx \frac{1}{\Delta S}\oint_C \mathbf{F}\cdot d\mathbf{l} \tag{5.13}$$

If we now take limits in such a way that ΔS shrinks 'to the point P', we obtain a point definition of curl:

$$(\nabla \wedge \mathbf{F})_P\cdot\hat{\mathbf{n}} = \lim_{\Delta S\to 0}\left(\frac{1}{\Delta S}\oint_C \mathbf{F}\cdot d\mathbf{l}\right) \tag{5.14}$$

That is, the component of curl **F** along the direction of $\hat{\mathbf{n}}$ is defined as the limit of a circulation per unit area. The dimensions of $\nabla \wedge \mathbf{F}$ are therefore the dimensions of **F** divided by length.

Equation (5.14) is a coordinate-free definition of (one component of) curl **F** at a point. We could, had we so chosen, have begun our discussion with the definition (5.14) and readily obtained Stokes' theorem and the Cartesian representation of curl. We shall, in fact, use (5.14) in Chapter 6 to obtain expressions for $\nabla \wedge \mathbf{F}$ in non-Cartesian coordinate systems.

5.10 Conservative vector fields revisited

In Chapter 4 we learnt that vector fields within a region of space, R_C say, could be divided into two classes, conservative and non-conservative. For

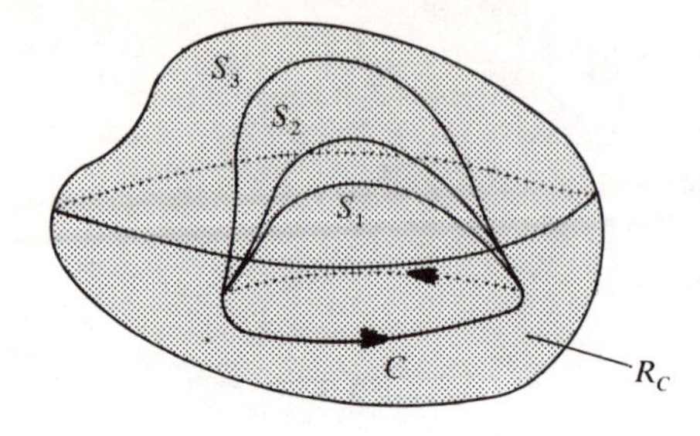

Figure 5.15

conservative fields, the line integral $\oint_C \mathbf{F}\cdot d\mathbf{l}$ is identically zero for any closed path C lying entirely within R_C, whilst for non-conservative fields there must be at least one closed path yielding a non-zero value for the line integral.

Now any closed path is the boundary of an infinite number of 'capping' surfaces $S_1, S_2, \ldots$ all lying within R_C (see Figure 5.15). Hence if $\oint_C \mathbf{F}\cdot d\mathbf{l} = 0$ then, by Stokes' theorem,

$$\iint_{S_1} \nabla \wedge \mathbf{F}\cdot\hat{\mathbf{n}}\, dS = \iint_{S_2} \nabla \wedge \mathbf{F}\cdot\hat{\mathbf{n}}\, dS = \cdots = 0$$

Thus, for a field to be conservative within R_C, we must have

$$\iint_S \nabla \wedge \mathbf{F}\cdot\hat{\mathbf{n}}\, dS = 0$$

for *any* arbitrary non-closed surface S within R_C. This can only be the case if $\nabla \wedge \mathbf{F}$ is identically zero everywhere in the region. In other words,

$$\nabla \wedge \mathbf{F} = 0 \quad \text{in } R_C$$

is a *necessary* condition for a vector field to be conservative within the region R_C.

Conversely, of course, if we know that $\nabla \wedge \mathbf{F} = 0$ everywhere R_C, then $\iint_S \nabla \wedge \mathbf{F}\cdot\hat{\mathbf{n}}\, dS = 0$ for any surface S lying entirely in R_C and, again using Stokes' theorem, $\oint_C \mathbf{F}\cdot d\mathbf{l}$ must be zero for any closed path C lying in this region. Thus $\nabla \wedge \mathbf{F} = 0$ is also a *sufficient* condition for $\mathbf{F}$ to be conservative. Indeed, we have already used this result in Chapter 4.

In Cartesian coordinates, the vector condition $\nabla \wedge \mathbf{F} = 0$ is, of course, equivalent to the three scalar conditions

$$\frac{\partial F_y}{\partial z} = \frac{\partial F_z}{\partial y}, \qquad \frac{\partial F_z}{\partial x} = \frac{\partial F_x}{\partial z}, \qquad \frac{\partial F_x}{\partial y} = \frac{\partial F_y}{\partial x}$$

and for the particular case of a vector field $\mathbf{F}$ of the form

$$\mathbf{F} = F_x(x, y)\hat{\mathbf{i}} + F_y(x, y)\hat{\mathbf{j}}$$

only the single condition $\partial F_y/\partial x = \partial F_x/\partial y$ arises.

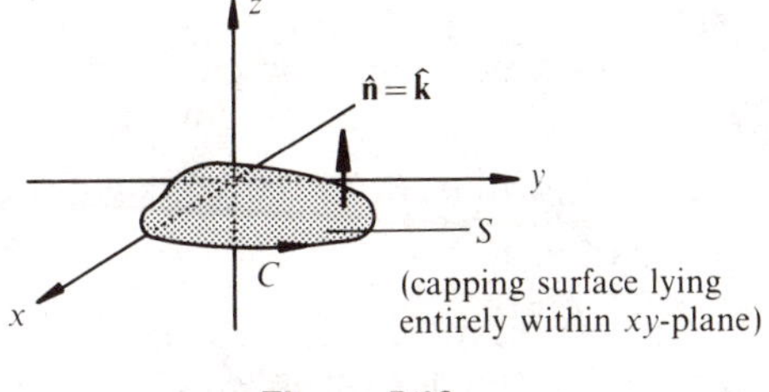

Figure 5.16

To see fully why this is so, let us write out Stokes' theorem for the two-dimensional situation (see Figure 5.16). Using Cartesian coordinates, the 'line integral side' of Stokes' theorem (5.12) is

$$\oint_C \{F_x(x, y)\,\mathrm{d}x + F_y(x, y)\,\mathrm{d}y\}$$

while $$\nabla \wedge \mathbf{F} = \begin{vmatrix} \hat{\mathbf{i}} & \hat{\mathbf{j}} & \hat{\mathbf{k}} \\ \dfrac{\partial}{\partial x} & \dfrac{\partial}{\partial y} & \dfrac{\partial}{\partial z} \\ F_x & F_y & 0 \end{vmatrix} = \left(\frac{\partial F_y}{\partial x} - \frac{\partial F_x}{\partial y}\right)\hat{\mathbf{k}}$$

Also, if we choose the capping surface S to lie entirely in the xy-plane, then the unit vector $\hat{\mathbf{n}}$ will be directed along the positive z-direction. Hence the 'flux integral' $\iint_S \nabla \wedge \mathbf{F}\cdot\hat{\mathbf{n}}\,\mathrm{d}S$ will be

$$\iint_S (\nabla \wedge \mathbf{F})\cdot\hat{\mathbf{k}}\,\mathrm{d}S = \iint_S \left(\frac{\partial F_y}{\partial x} - \frac{\partial F_x}{\partial y}\right)\mathrm{d}x\,\mathrm{d}y$$

and Stokes' theorem has the form

$$\oint_C (F_x\,\mathrm{d}x + F_y\,\mathrm{d}y) = \iint_S \left(\frac{\partial F_y}{\partial x} - \frac{\partial F_x}{\partial y}\right)\mathrm{d}x\,\mathrm{d}y \qquad \textbf{(5.15)}$$

This result, which we have already encountered and used in Chapter 4 and which is dignified by the name of **Green's theorem in the plane**, is clearly seen as just a special case of Stokes' theorem. The condition $\partial F_y/\partial x = \partial F_x/\partial y$ for a 'two-dimensional' conservative vector field clearly follows from (5.15).

It might prove useful at this point to summarize the properties of a conservative vector field. A conservative vector field is a continuously differentiable vector field $\mathbf{F}$, defined in a region R, and which possesses the following properties:

(1) Its line integral $\oint_C \mathbf{F}\cdot\mathrm{d}\mathbf{l}$ around any simple closed curve within R is zero.

(2) Its line integral between given endpoints P and Q is independent of the actual path of integration between P and Q, provided the path lies within R.

(3) $\mathbf{F}$ can be expressed, within R, as the gradient of a scalar field.

(4) Within R, the curl of $\mathbf{F}$, $\nabla \wedge \mathbf{F}$, is zero.

Although we have not proved it fully, it can be shown that if any one of these properties is known to hold, then so must the other three.

EXERCISES

5.7 If $\mathbf{A} = (x^2 + y^2)\hat{\mathbf{i}} - 2xy\hat{\mathbf{j}} + xyz\hat{\mathbf{k}}$, find, by two separate methods, the flux of $\nabla \wedge \mathbf{A}$ over the rectangular surface in the xy-plane bounded by $x = 0$, $x = a$, $y = 0$ and $y = b$.

5.8 Verify Stokes' theorem for the vector field $\mathbf{F} = z\hat{\mathbf{i}} + x\hat{\mathbf{j}} + y\hat{\mathbf{k}}$ and the surface S which is that portion of the plane $2x + y + 2z = 3$ lying in the first octant $(x \geqslant 0,\ y \geqslant 0,\ z \geqslant 0)$.

5.9 If $\mathbf{A} = -y\hat{\mathbf{i}} + x\hat{\mathbf{j}}$ and S is the surface of the hemisphere $x^2 + y^2 + z^2 = a^2$, $z \geqslant 0$, evaluate

(a) $\oint_C \mathbf{F} \cdot \mathrm{d}\mathbf{l}$ where C is the bounding surface of S,

(b) $\iint_S (\nabla \wedge \mathbf{F}) \cdot \hat{\mathbf{n}}\, \mathrm{d}S$.

Hence verify Stokes' theorem for this case.

5.10 If $\mathbf{F}$ is a vector field and u and v are scalar fields, use Stokes' theorem to deduce the following results:

(a) $\iint_S u \nabla \wedge \mathbf{F} \cdot \mathrm{d}\mathbf{S} = \oint_C u\mathbf{F} \cdot \mathrm{d}\mathbf{l} - \iint_S (\nabla u \wedge \mathbf{F}) \cdot \mathrm{d}\mathbf{S}$

(b) $\oint_C u\, \mathbf{grad}\, v \cdot \mathrm{d}\mathbf{l} = -\oint_C v\, \mathbf{grad}\, u \cdot \mathrm{d}\mathbf{l}$

(In each case, S is a surface bounded by a closed curve C.)

5.11 Evaluate the line integral $\oint_C \mathbf{F}\cdot d\mathbf{l}$ around the closed path C shown in Figure 5.17 where $\mathbf{F} = x\hat{\mathbf{i}} - z\hat{\mathbf{j}} + y\hat{\mathbf{k}}$. Perform the calculation (a) directly, as a line integral, and (b) indirectly, using Stokes' theorem.

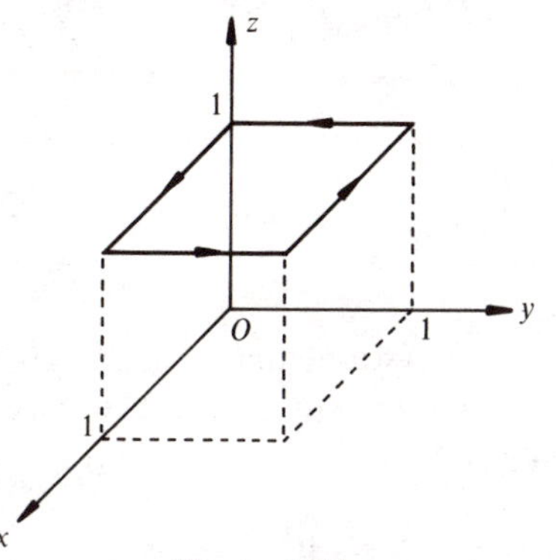

Figure 5.17

5.12 Verify Stokes' theorem for each of the following cases:

(a) $\mathbf{F} = -y\hat{\mathbf{i}} + 2yz\hat{\mathbf{j}} + y^2\hat{\mathbf{k}}$ for the surface S of the hemisphere $x^2 + y^2 + z^2 = 1$, $z \geqslant 0$, and its boundary,

(b) $\mathbf{F} = 2y\hat{\mathbf{i}} - x\hat{\mathbf{j}} + xz\hat{\mathbf{k}}$ for the surface S of the hemisphere $x^2 + y^2 + z^2 = 4$, $z \geqslant 0$, and its boundary.

5.13 A surface S consists of that part of the cylinder $x^2 + y^2 = 9$ between $z = 0$ and $z = 4$ for $y \geqslant 0$ and two semicircles of radius 3 in the planes $z = 0$ and $z = 4$ (see Figure 5.18). If $\mathbf{F} = z\hat{\mathbf{i}} + xy\hat{\mathbf{j}} + xz\hat{\mathbf{k}}$, find the flux of $\nabla \wedge \mathbf{F}$ over S using a line integral.

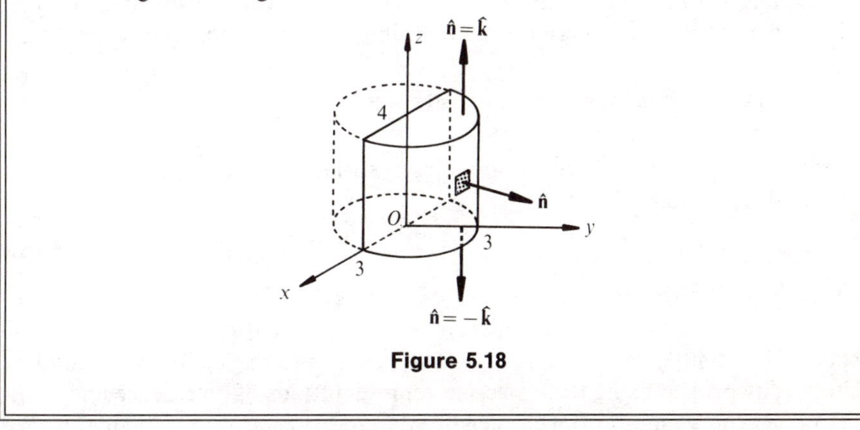

Figure 5.18

5.11 Further consequences of the integral theorems

Stokes' theorem involves the flux of $\nabla \wedge \mathbf{F}$ across a *non-closed* surface. The flux of $\nabla \wedge \mathbf{F}$ across a *closed* surface is also of interest and is, in fact, readily

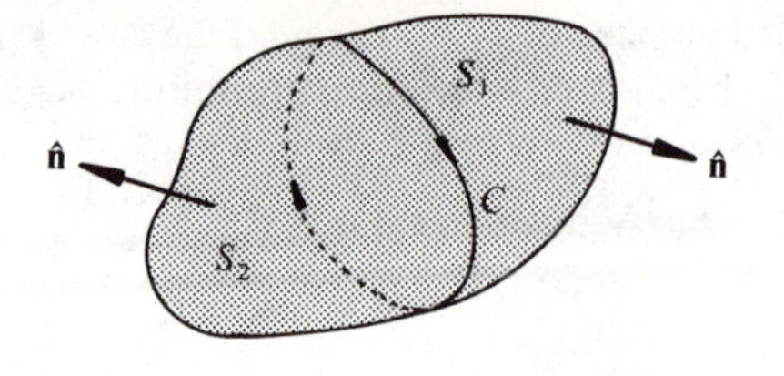

Figure 5.19

shown always to be zero. Assume that a closed surface S is divided into two 'subsurfaces' S_1 and S_2 by a closed curve C (see Figure 5.19). Applying Stokes' theorem separately to both S_1 and S_2, we have

$$\iint_{S_1} \nabla \wedge \mathbf{F} \cdot \hat{\mathbf{n}} \, \mathrm{d}S = \oint_C \mathbf{F} \cdot \mathrm{d}\mathbf{l}$$

and
$$\iint_{S_2} \nabla \wedge \mathbf{F} \cdot \hat{\mathbf{n}} \, \mathrm{d}S = \oint_{-C} \mathbf{F} \cdot \mathrm{d}\mathbf{l} = -\oint_C \mathbf{F} \cdot \mathrm{d}\mathbf{l}$$

Adding these results

$$\oiint_S \nabla \wedge \mathbf{F} \cdot \hat{\mathbf{n}} \, \mathrm{d}S = \iint_{S_1} \nabla \wedge \mathbf{F} \cdot \hat{\mathbf{n}} \, \mathrm{d}S + \iint_{S_2} \nabla \wedge \mathbf{F} \cdot \hat{\mathbf{n}} \, \mathrm{d}S$$

$$= \oint_C \mathbf{F} \cdot \mathrm{d}\mathbf{l} - \oint_C \mathbf{F} \cdot \mathrm{d}\mathbf{l} = 0$$

In dealing with a surface integral over a *closed* surface S enclosing a volume V, we can bring the divergence theorem into action. In this particular case,

$$\oiint_S \nabla \wedge \mathbf{F} \cdot \hat{\mathbf{n}} \, \mathrm{d}S = 0 = \iiint_V \nabla \cdot (\nabla \wedge \mathbf{F}) \, \mathrm{d}V$$

Since this result must hold for any closed surface S, and hence any volume V, we deduce that

$$\nabla \cdot (\nabla \wedge \mathbf{F}) = 0 \qquad \textbf{(5.16)}$$

That is, the divergence of the curl of any vector field is identically zero. Hence a vector field $\nabla \wedge \mathbf{F}$ is always a divergenceless or solenoidal vector field. The identity (5.16), which was first referred to in Section 3.6, can also be readily proved using the Cartesian representations of divergence and curl. However our general proof emphasizes its validity in any coordinate system. (Note that the result (5.16) assumes that the divergence of $\nabla \wedge \mathbf{F}$ exists at *each* point of V.)

It follows from (5.16) that any vector field which is *known* to be solenoidal can be expressed as the curl of another vector field, say $\mathbf{A}$. This second vector field is usually referred to as a **vector potential**. In particular, the magnetic vector potential is a valid and useful concept in electromagnetics, as we shall see in Chapter 7. At this point, however, we shall merely note that there is not a unique vector potential $\mathbf{A}$ for a given solenoidal field $\mathbf{F}$. This follows because if we put

$$\mathbf{A}' = \mathbf{A} + \nabla\Phi$$

where Φ is an arbitrary scalar field, then

$$\nabla \wedge \mathbf{A}' = \nabla \wedge (\mathbf{A} + \nabla\Phi) = \nabla \wedge \mathbf{A} + \nabla \wedge \nabla\Phi$$

But for any scalar field Φ, $\nabla \wedge \nabla\Phi$ is identically zero (see Section 3.6), so $\mathbf{A}$ and $\mathbf{A}'$ have the same curl, and hence $\mathbf{F}$ has an infinite number of associated vector potentials. In practice, to ensure a unique potential $\mathbf{A}$, we must place an additional constraint on $\mathbf{A}$ such as, for example, $\nabla \cdot \mathbf{A} = 0$.

The reader may wonder why specifying the divergence of $\mathbf{A}$, as well as its curl, ensures uniqueness. To show this, we shall give a proof of what is known as **Helmholtz's theorem**. This theorem states that we can uniquely determine a vector field $\mathbf{A}$ in a volume V enclosed by a closed surface S if both the divergence, $\nabla \cdot \mathbf{A}$, and the curl, $\nabla \wedge \mathbf{A}$, are specified throughout V as well as the normal component of $\mathbf{A}$ on S. The derivation of Helmholtz's theorem uses many of the basic results of vector field theory and the reader is advised to work through its proof in the following example.

Example 5.5

Prove Helmholtz's theorem.

Solution

The proof is by contradiction, in that we first assume that $\mathbf{A}$ and $\mathbf{B}$ are distinct vectors satisfying the given conditions then show that $\mathbf{C} = \mathbf{A} - \mathbf{B}$ is a vector which is identically zero throughout V.

Thus we assume

$$\nabla \cdot \mathbf{A} = \nabla \cdot \mathbf{B} \quad \text{and} \quad \nabla \wedge \mathbf{A} = \nabla \wedge \mathbf{B}$$

throughout the volume V and that

$$\mathbf{A} \cdot \hat{\mathbf{n}} = \mathbf{B} \cdot \hat{\mathbf{n}}$$

on the bounding surface S. Consider the divergence, the curl and the normal component of $\mathbf{C}$. We must have

$$\nabla \cdot \mathbf{C} = \nabla \cdot (\mathbf{A} - \mathbf{B}) = \nabla \cdot \mathbf{A} - \nabla \cdot \mathbf{B} = 0$$

and $\quad \nabla \wedge \mathbf{C} = \nabla \wedge (\mathbf{A} - \mathbf{B}) = \nabla \wedge \mathbf{A} - \nabla \wedge \mathbf{B} = 0$

throughout V and

$$\mathbf{C} \cdot \hat{\mathbf{n}} = (\mathbf{A} - \mathbf{B}) \cdot \hat{\mathbf{n}} = \mathbf{A} \cdot \hat{\mathbf{n}} - \mathbf{B} \cdot \hat{\mathbf{n}} = 0$$

on S. But if $\nabla \wedge \mathbf{C} = 0$, then $\mathbf{C}$ is conservative throughout V and we can put $\mathbf{C} = \nabla\Phi$, where Φ is some scalar potential. Hence

$$\nabla \cdot \mathbf{C} = \nabla \cdot \nabla\Phi = \nabla^2\Phi = 0 \tag{5.17}$$

throughout V and

$$\mathbf{C}\cdot\hat{\mathbf{n}} = \nabla\Phi\cdot\hat{\mathbf{n}} = \frac{\partial\Phi}{\partial n}\hat{\mathbf{n}}\cdot\hat{\mathbf{n}} = \frac{\partial\Phi}{\partial n} = 0 \tag{5.18}$$

on S, where we have used a property of $\nabla\Phi$ discussed in Section 3.2.

We now invoke the first of Green's identities proved in Example 5.2 but with both u and v replaced by Φ. Hence we put $\nabla v\cdot\hat{\mathbf{n}} = \partial\Phi/\partial n$, and we obtain

$$\oint\!\!\oint_S \Phi\frac{\partial\Phi}{\partial n}\,\mathrm{d}S = \iiint_V (\Phi\,\nabla^2\Phi + \nabla\Phi\cdot\nabla\Phi)\,\mathrm{d}V$$

which, on using (5.17) and (5.18), becomes

$$\iiint_V \nabla\Phi\cdot\nabla\Phi\,\mathrm{d}V = \iiint_V |\nabla\Phi|^2\,\mathrm{d}V = 0$$

But the integrand $|\nabla\Phi|^2$ is clearly non-negative, so we must have that $\nabla\Phi$, and hence $\mathbf{C}$, is zero throughout V. Thus $\mathbf{A} = \mathbf{B}$ throughout V, and the uniqueness of $\mathbf{A}$ is established.

ADDITIONAL EXERCISES

1 If $\mathbf{r}$ denotes a position vector from the origin to a point with coordinates (x, y, z), show that

(a) $\oint_C \mathbf{r}\cdot\mathrm{d}\mathbf{l} = 0$ for any closed surface C,

(b) $\oint\!\!\oint_S \mathbf{r}\cdot\hat{\mathbf{n}}\,\mathrm{d}S = 3$ (volume enclosed by S).

(c) If S is the surface bounded by the paraboloid $z = 4 - (x^2 + y^2)$ and the xy-plane, show that $\oint\!\!\oint_S \mathbf{r}\cdot\hat{\mathbf{n}}\,\mathrm{d}S = 24\pi$.

2 (a) Show that the volume enclosed by any closed surface S is $V = \frac{1}{6}\oint\!\!\oint_S \nabla r^2\cdot\hat{\mathbf{n}}\,\mathrm{d}S$.

(b) If there exists a scalar field $h(x, y, z)$ such that $\nabla^2\Phi = h\Phi$ and $\nabla^2\vartheta = h\vartheta$ in a region R bounded by a closed surface S, prove that

$$\oint\!\!\oint_S (\Phi\,\nabla\vartheta - \vartheta\,\nabla\Phi)\cdot\hat{\mathbf{n}}\,\mathrm{d}S = 0$$

(c) If Φ is harmonic in a region R, show that

$$\oiint_S \Phi \nabla\Phi \cdot \hat{\mathbf{n}}\, dS = \iiint_R |\nabla\Phi|^2\, dV$$

(d) Show that if $\mathbf{A} = \nabla\Phi$ and Φ is harmonic, then

$$\iiint_R |\mathbf{A}|^2\, dV = \oiint_S \Phi\mathbf{A} \cdot \hat{\mathbf{n}}\, dS$$

3 Verify the divergence theorem for the vector field $\mathbf{F} = xy\hat{\mathbf{i}} - yz\hat{\mathbf{j}} + 3\hat{\mathbf{k}}$ and the hemisphere $x^2 + y^2 + z^2 = 1$, $z \geqslant 0$.

4 Verify the divergence theorem for the vector field $\mathbf{F} = 2x^2y\hat{\mathbf{i}} - y^2\hat{\mathbf{j}} + 4xz^2\hat{\mathbf{k}}$ taken over the region in the first octant ($x \geqslant 0$, $y \geqslant 0$, $z \geqslant 0$) bounded by $y^2 + z^2 = 9$ and $x = 2$.

5 If $\oint_C \mathbf{E} \cdot d\mathbf{l} = -\frac{\partial}{\partial t}\iint_S \mathbf{B} \cdot \hat{\mathbf{n}}\, dS$, where S is any surface bounded by the curve C, show that $\nabla \wedge \mathbf{E} = -\partial\mathbf{B}/\partial t$.

6 Verify Stokes' theorem for the vector field $\mathbf{F} = x\hat{\mathbf{i}} + (2z - x)\hat{\mathbf{j}} + y^2\hat{\mathbf{k}}$ using two surfaces each bounded by the circle $x^2 + y^2 = 4$, $z = 1$.

SUMMARY – An overview of integration

We now discuss the various types of integral that we have encountered, and the associated theorems, and try to extract common threads.

Definite integrals

The basic result here is familiar enough:

$$\int_a^b \frac{df}{dx}\, dx = f(b) - f(a) \tag{5.19}$$

This is the fundamental theorem of calculus.

Line integrals

For a scalar line integral of a conservative vector field $\mathbf{F}$ we have, for given endpoints,

$$\int_{P_1}^{P_2} \mathbf{F} \cdot d\mathbf{l} = \Phi(P_2) - \Phi(P_1), \quad \text{where } \mathbf{F} = \nabla\Phi$$

It will prove convenient to write the left-hand side of this result in terms of the 'intrinsic' parameter on the curve – the arc length s. In Cartesian coordinates,

$$\int_{P_1}^{P_2} \mathbf{F}\cdot\mathrm{d}\mathbf{l} = \int_{P_1}^{P_2} \left(\frac{\partial \Phi}{\partial x}\hat{\mathbf{i}} + \frac{\partial \Phi}{\partial y}\hat{\mathbf{j}} + \frac{\partial \Phi}{\partial z}\hat{\mathbf{k}}\right)\cdot(\mathrm{d}x\,\hat{\mathbf{i}} + \mathrm{d}y\,\hat{\mathbf{j}} + \mathrm{d}z\,\hat{\mathbf{k}})$$

$$= \int_{P_1}^{P_2} \left(\frac{\partial \Phi}{\partial x}\mathrm{d}x + \frac{\partial \Phi}{\partial y}\mathrm{d}y + \frac{\partial \Phi}{\partial z}\mathrm{d}z\right)$$

Hence parametrizing with the arc-length parameter s, we obtain

$$\int_{P_1}^{P_2} \left(\frac{\partial \Phi}{\partial x}\frac{\mathrm{d}x}{\mathrm{d}s} + \frac{\partial \Phi}{\partial y}\frac{\mathrm{d}y}{\mathrm{d}s} + \frac{\partial \Phi}{\partial z}\frac{\mathrm{d}z}{\mathrm{d}s}\right)\mathrm{d}s$$

or
$$\int_{P_1}^{P_2} \frac{\mathrm{d}\Phi}{\mathrm{d}s}\,\mathrm{d}s = \Phi(P_2) - \Phi(P_1) \tag{5.20}$$

The result can readily be seen as a generalization of (5.19). Indeed, we can extend (5.20) to the case of a **vector line integral**, defined by

$$\int_{P_1}^{P_2} \frac{\mathrm{d}\mathbf{F}}{\mathrm{d}s}\,\mathrm{d}s = \hat{\mathbf{i}}\int_{P_1}^{P_2} \frac{\mathrm{d}F_x}{\mathrm{d}s}\,\mathrm{d}s + \hat{\mathbf{j}}\int_{P_1}^{P_2} \frac{\mathrm{d}F_y}{\mathrm{d}s}\,\mathrm{d}s + \hat{\mathbf{k}}\int_{P_1}^{P_2} \frac{\mathrm{d}F_z}{\mathrm{d}s}\,\mathrm{d}s$$

Using (5.20) for each component,

$$\int_{P_1}^{P_2} \frac{\mathrm{d}\mathbf{F}}{\mathrm{d}s}\,\mathrm{d}s = \hat{\mathbf{i}}[F_x(P_2) - F_x(P_1)] + \hat{\mathbf{j}}[F_y(P_2) - F_y(P_1)]$$
$$+ \hat{\mathbf{k}}[F_z(P_2) - F_z(P_1)]$$
$$= \mathbf{F}(P_2) - \mathbf{F}(P_1) \tag{5.21}$$

If we imagine the path of integration C from P_1 to P_2 to be a part of a longer path (see Figure 5.20), then the points P_1 and P_2 separate the set of points on this longer path into two regions – points on C and points not on C: we can say that P_1 and P_2 constitute the **boundary** of C. With respect to the longer path, the boundary points of C allow a natural definition of the terms *inside* and *outside*. We can also define, again quite naturally, an *intrinsic* positive direction at the boundary points to be that direction which points from *inside* to *outside* the path C. The positive direction for the arc-length parameter is that shown in Figure 5.20; it is clearly opposite to the intrinsic positive direction at P_1. The result (5.21) then tells us that the integral over a region of a differential operator $\mathrm{d}/\mathrm{d}s$ acting on $\mathbf{F}$ is equal to the sum $-\mathbf{F}(P_1) + \mathbf{F}(P_2)$ evaluated

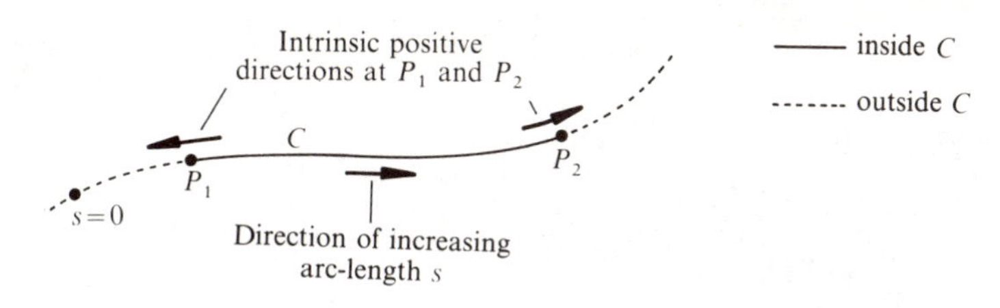

Figure 5.20

at the boundary of the region. (The negative sign is included in this 'sum' since at P_1 the intrinsic positive direction is in the opposite direction to the measurement of arc length.)

Stokes' theorem

This theorem can be thought of as similar in form to (5.21) but of dimension one higher:

$$\iint_S \nabla \wedge \mathbf{F} \cdot \hat{\mathbf{n}} \, dS = \oint_C \mathbf{F} \cdot d\mathbf{l} \tag{5.22}$$

Here the 'region' of integration is a surface S and the boundary is the closed curve C. The surface S may be regarded as part of a larger surface (see Figure 5.21a). Again there is a naturally defined 'inside' and 'outside' and a naturally defined positive direction (on S) pointing outwards (see Figure 5.21b). The positive outward direction (perpendicular to C) together with the tangent direction to C naturally defines a positive normal to S according to the right-hand rule (see Figure 5.21b at the point P). The similarity of (5.21) to (5.22) is clear: the operator d/ds is now the curl, $\nabla \wedge$. The integral of $\nabla \wedge \mathbf{F}$ over a region S is equal to the 'sum' evaluated on the boundary (not surprisingly, this 'sum' is a line integral as the boundary now comprises the infinite number of points on C). Of course, the direction chosen for positive $\hat{\mathbf{n}}$ is intimately connected to the 'positive' direction on C: if C is traversed in the opposite direction, then $\hat{\mathbf{n}}$ also must be reversed.

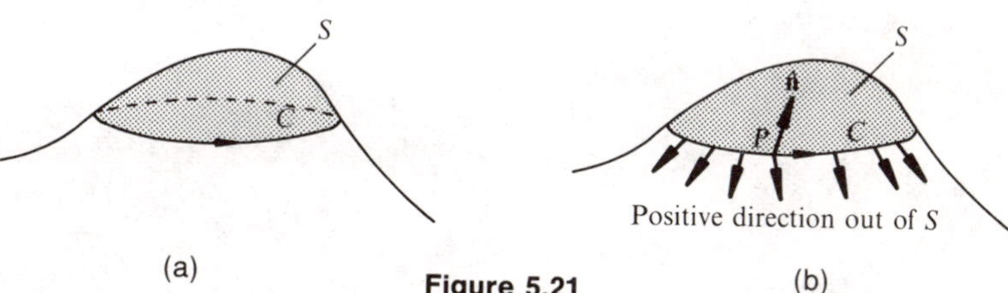

Figure 5.21

Divergence theorem

This, too, can be thought of as similar in form to (5.21) but of dimension two higher:

$$\iiint_V \nabla \cdot \mathbf{F} \, \mathrm{d}V = \oiint_S \mathbf{F} \cdot \hat{\mathbf{n}} \, \mathrm{d}S \tag{5.23}$$

Here the 'region' of integration is a volume V and the boundary is the closed surface S. Here V may be regarded as part of a larger volume allowing a natural definition for the 'positive' normal to S (pointing from the 'inside' to the 'outside', as shown in Figure 5.22). The similarity with (5.21) and (5.22) is now obvious: the integral of $\nabla \cdot \mathbf{F}$ over a region V is equal to the 'sum' evaluated on the boundary, and in this case the 'sum' is the surface integral of $\mathbf{F}$ over S.

If various terms are interpreted appropriately, then the results (5.19) to (5.23) can be summarized in the form:

> The **integral** over a **region** of a **differential operator** acting on a **vector** is equal to the **sum** of the vector evaluated on the **boundary** of the region.

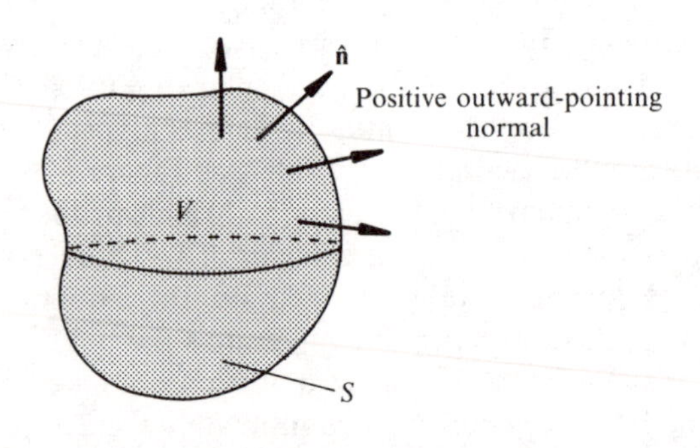

Figure 5.22

6 Orthogonal Coordinate Systems

PREVIEW

In this chapter we discuss, in a systematic way, how we can represent scalar fields, vector fields and, in particular, the gradient, divergence, curl and Laplacian operators in any convenient coordinate system in which the base vectors are orthogonal. Knowledge of these representations enables us readily to apply vector field concepts to situations where Cartesian coordinates would be inappropriate.

6.1 Curvilinear coordinates

In this text we have chosen, more often than not, to describe quantities of interest in terms of Cartesian coordinates. But, when the situation warranted, we have used two other coordinate systems – spherical and cylindrical polars. In particular, in Chapter 4 we showed how to evaluate volume integrals and flux integrals using these two coordinate systems.

The usefulness of any particular coordinate system is highly dependent on the application we have in mind. For example, if we wished to determine the equation of a sphere of radius a centred on the origin we would prefer to work in spherical polar coordinates, because in this system the equation takes its simplest form,

$$r = a \tag{6.1}$$

whereas in Cartesians it takes the (relatively) complicated form

$$x^2 + y^2 + z^2 = a^2 \tag{6.2}$$

In cylindrical polars the corresponding equation is

$$\rho^2 + z^2 = a^2 \tag{6.3}$$

We may draw the general conclusion that other applications that involve a high degree of spherical symmetry will also be simplified if spherical polar coordinates are used. Similarly, if the problem of interest displays cylindrical symmetry, then we would normally work in cylindrical polar coordinates.

In the instances where we have already used coordinate systems other than Cartesians, their application has been to simplify certain scalar quantities such as integrands in volume integrals. It is our aim in this chapter to define what is meant by the components of a *vector* in non-Cartesian coordinate systems – also called **curvilinear** systems. Further, we shall develop expansions for **grad**, div and **curl** in these curvilinear systems as well as expressions for the basic 'elements of space' – the line element, the surface element and the volume element.

We shall treat the subject in some generality but will again make particular reference to spherical and cylindrical polars, as these are the two most commonly used non-Cartesian coordinate systems.

We begin by examining the replacement of the (x, y, z) coordinates by some general set of coordinates (u_1, u_2, u_3) through a **transformation of coordinates** of the form

$$x = x(u_1, u_2, u_3), \qquad y = y(u_1, u_2, u_3), \qquad z = z(u_1, u_2, u_3) \tag{6.4}$$

in which the right-hand sides are *given* functions of u_1, u_2 and u_3. We suppose that the system (6.4) may be 'inverted' so that we can determine single-valued expressions for (u_1, u_2, u_3) in terms of (x, y, z):

$$u_1 = u_1(x, y, z), \qquad u_2 = u_2(x, y, z), \qquad u_3 = u_3(x, y, z) \tag{6.5}$$

Then, given a point P in space with Cartesian coordinates (x, y, z), we can determine a unique set of coordinates (u_1, u_2, u_3), called the curvilinear coordinates of P, from (6.5).

If we take u_1 to be a constant in (6.4), then each of x, y and z are functions of just two variables (u_2 and u_3) and hence describe a surface. This surface is called a u_1-constant **coordinate surface**. Similarly, taking u_2 as a constant and u_3 as a constant in turn, defines the u_2-constant and u_3-constant coordinate surfaces respectively. Each pair of coordinate surfaces intersects in a curve called a **coordinate curve**; for example, the u_1-coordinate curve is

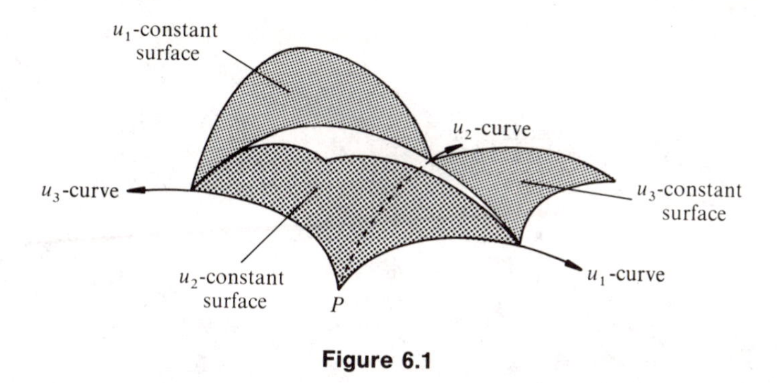

Figure 6.1

the intersection of the u_2-constant and u_3-constant coordinate surfaces. We conclude, generally, that three coordinate curves pass through every point P in space (see Figure 6.1). If the coordinate surfaces intersect at right angles at every point, then the curvilinear coordinate system is called **orthogonal**. Although a complete theory of general curvilinear coordinates may be developed, we shall consider only orthogonal coordinate systems, of which, as we shall see, Cartesians, spherical polars and cylindrical polars are particular cases.

Example 6.1

Determine the coordinate curves and coordinate surfaces for (a) Cartesian, (b) spherical polar, and (c) cylindrical polar coordinate systems.

Solution

(a) If (u_1, u_2, u_3) represents a Cartesian system, then equations (6.5) take the trivial form

$$u_1 = x, \qquad u_2 = y, \qquad u_3 = z$$

and the transformation (6.4) is simply

$$x = u_1, \qquad y = u_2, \qquad z = u_3$$

The variables x, y and z are each allowed to vary from $-\infty$ to $+\infty$.

The coordinate surfaces $u_1 = \text{constant}$, $u_2 = \text{constant}$ and $u_3 = \text{constant}$ are the surfaces $x = \text{constant}$, $y = \text{constant}$ and $z = \text{constant}$ respectively (see Figure 6.2). Clearly, these 'surfaces' are planes parallel to the Cartesian coordinate planes. They intersect in straight lines parallel to the x-, y- and z-coordinate axes. These intersections are the 'coordinate curves' for this case.

(b) If (u_1, u_2, u_3) represents a spherical polar system, then it is usual to relabel u_1, u_2 and u_3 as r, θ and ϕ respectively. The transformation

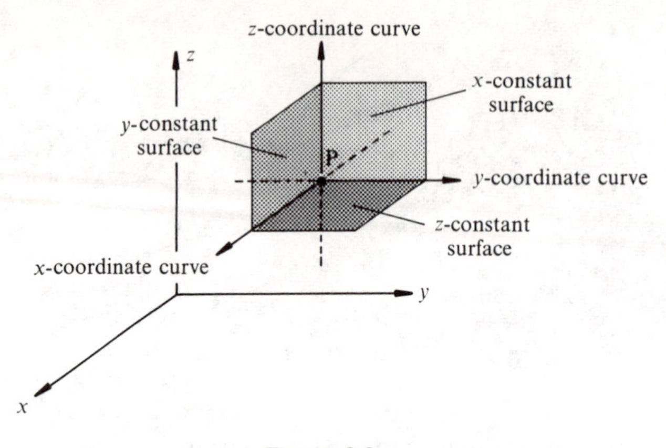

Figure 6.2

equations (6.4) for this case are

$$x = u_1 \sin u_2 \cos u_3, \qquad y = u_1 \sin u_2 \sin u_3, \qquad z = u_1 \cos u_2$$

or, equivalently,

$$x = r \sin\theta \cos\phi, \qquad y = r \sin\theta \sin\phi, \qquad z = r\cos\theta \tag{6.6}$$

while the inverse relations (6.5) become

$$r = +\sqrt{(x^2 + y^2 + z^2)}, \qquad \theta = \tan^{-1}\frac{\sqrt{(x^2 + y^2)}}{z}, \qquad \phi = \tan^{-1}\left(\frac{y}{x}\right) \tag{6.7}$$

(see Section 4.8 and Figure 4.19). The ranges allowed for r, θ and ϕ are

$$0 \leqslant r < \infty, \qquad 0 \leqslant \theta \leqslant \pi, \qquad 0 \leqslant \phi \leqslant 2\pi \tag{6.8}$$

The coordinate surfaces for this system are

- $r =$ constant: a sphere centred on the origin,
- $\phi =$ constant: a half-plane with the z-axis as one edge,
- $\theta =$ constant: a cone with apex at the origin.

The coordinate curves for this system are

- the ϕ-coordinate curves: latitude lines on a sphere, formed by the intersection of cones of constant θ with spheres of constant r,
- the r-coordinate curves: half-lines with one end at the origin, formed by the intersection of cones of constant θ with planes of constant ϕ,
- the θ-coordinate curves: great circles on a sphere, formed by the intersection of spheres of constant r and planes of constant ϕ.

See Figure 6.3.

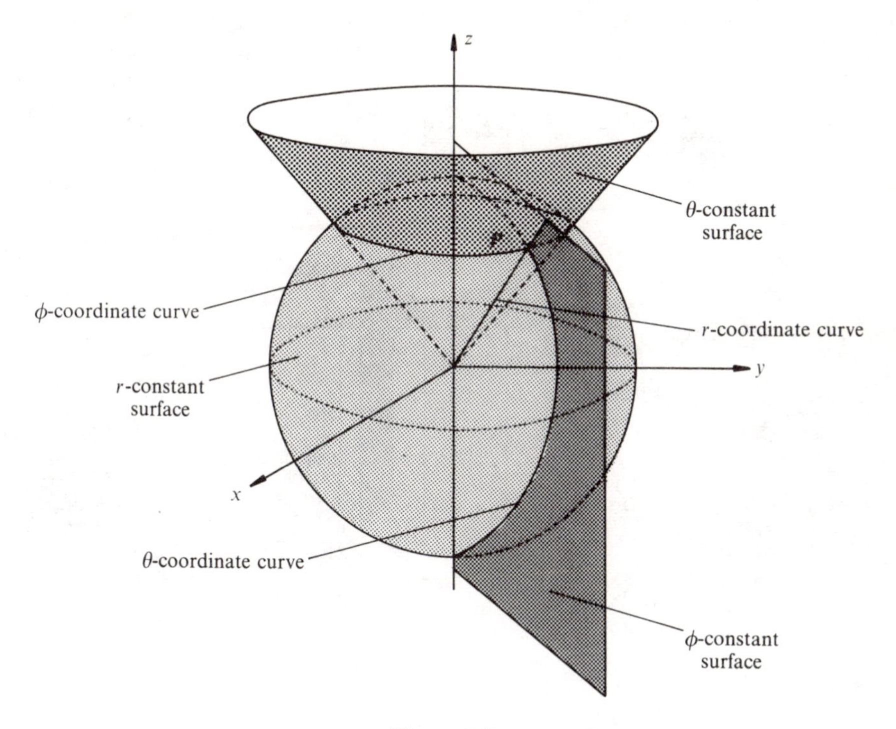

Figure 6.3

(c) If (u_1, u_2, u_3) represents a cylindrical polar coordinate system, then we relabel u_1, u_2 and u_3 as ρ, ϕ and z respectively. The transformation equations (6.4) are now

$$x = u_1 \cos u_2, \qquad y = u_1 \sin u_2, \qquad z = u_3 \tag{6.9}$$

or, equivalently,

$$x = \rho \cos \phi, \qquad y = \rho \sin \phi, \qquad z = z \tag{6.10}$$

The inverse relations are

$$\rho = +\sqrt{(x^2 + y^2)}, \qquad \phi = \tan^{-1}\left(\frac{y}{x}\right), \qquad z = z \tag{6.11}$$

(see Section 4.8 and Figure 4.17).

As we have anticipated in the notation, the coordinate u_3 is identical to the Cartesian coordinate z and the coordinate ϕ is identical to the corresponding spherical coordinate. The allowed ranges are

$$0 \leqslant \rho < \infty, \qquad 0 \leqslant \phi \leqslant 2\pi, \qquad -\infty < z < \infty \tag{6.12}$$

The coordinate surfaces $\phi =$ constant and $z =$ constant have already been defined above in parts (b) and (a) respectively. The coordinate

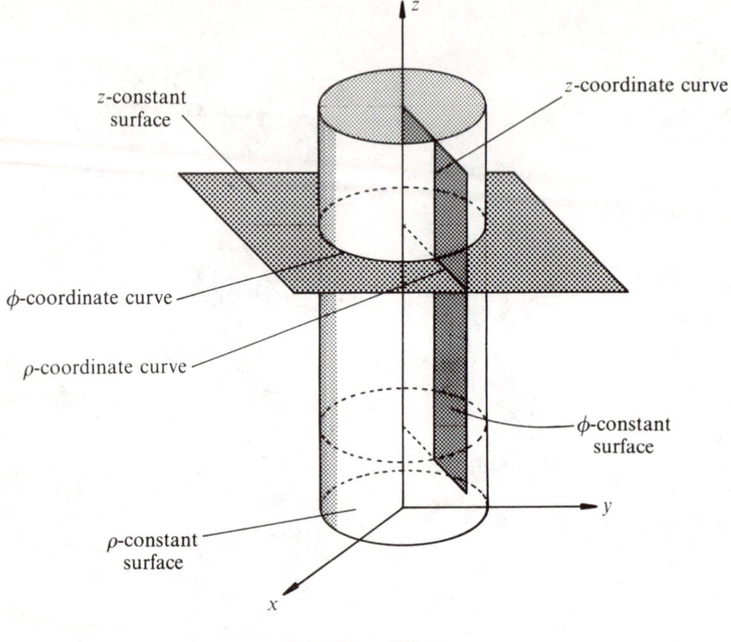

Figure 6.4

surface $\rho=$ constant is a circular cylinder with axis coincident with the z-axis (see Figure 6.4).

The coordinate curves are

- the z-coordinate curves: straight lines parallel to the z-axis, formed by the intersection of planes of constant ϕ and cylinders of constant ρ,
- the ϕ-coordinate curves: circles on a cylinder, formed by the intersection of the cylinders of constant ρ with planes of constant z,
- the ρ-coordinate curves: half-lines originating on the z-axis and parallel to the xy-plane, formed by the intersection of planes of constant ϕ and planes of constant z.

Returning to the analysis of general orthogonal curvilinear coordinates, we shall now consider the representation of vectors. We remind the reader that the concept of a vector is independent of the coordinate system in which we work: it is only when we want to develop an expression for the components of a vector that we have to choose a particular coordinate system.

Let us consider the position $\mathbf{r}$ of a point P with curvilinear coordinates (u_1, u_2, u_3). We should first note that we *cannot* write

$$\mathbf{r}=u_1\hat{\mathbf{i}}+u_2\hat{\mathbf{j}}+u_3\hat{\mathbf{k}}$$

since this is only true if (u_1, u_2, u_3) are the Cartesian coordinates of the endpoint of the vector **r**.

We have seen in Section 2.4 that, if **r** denotes a position vector at any point on a curve that is described in terms of a parameter t, then $d\mathbf{r}/dt$ gives us, at any point on that curve, a tangent vector. On a u_1-coordinate curve, where u_2 and u_3 are held constant, we can regard u_1 as a parameter, so that $\partial\mathbf{r}/\partial u_1$ must be a tangent vector. Similarly, $\partial\mathbf{r}/\partial u_2$ and $\partial\mathbf{r}/\partial u_3$ are tangent vectors to the u_2- and u_3-coordinate curves respectively. Hence we can obtain three unit vectors, say $\hat{\mathbf{e}}_1$, $\hat{\mathbf{e}}_2$ and $\hat{\mathbf{e}}_3$, using the relations

$$\hat{\mathbf{e}}_i = \frac{\partial\mathbf{r}}{\partial u_i}\bigg/\left|\frac{\partial\mathbf{r}}{\partial u_i}\right|, \qquad i = 1, 2, 3 \tag{6.13}$$

The quantities $|\partial\mathbf{r}/\partial u_1|$, $|\partial\mathbf{r}/\partial u_2|$ and $|\partial\mathbf{r}/\partial u_3|$ are called **scale factors** and are denoted by h_1, h_2 and h_3 respectively. Thus, from (6.13),

$$\frac{\partial\mathbf{r}}{\partial u_i} = h_i\hat{\mathbf{e}}_i, \qquad i = 1, 2, 3 \tag{6.14a}$$

where $$h_i = \left|\frac{\partial\mathbf{r}}{\partial u_i}\right|, \qquad i = 1, 2, 3 \tag{6.14b}$$

At each point of space, then, we can define a triad of unit vectors $\hat{\mathbf{e}}_1$, $\hat{\mathbf{e}}_2$ and $\hat{\mathbf{e}}_3$ (see Figure 6.5), which corresponds directly to the introduction of $\hat{\mathbf{i}}$, $\hat{\mathbf{j}}$ and $\hat{\mathbf{k}}$ in the Cartesian description of vectors.

It is clear that the curvilinear system of coordinates (u_1, u_2, u_3) will be an *orthogonal* system if, and only if, $\hat{\mathbf{e}}_1$, $\hat{\mathbf{e}}_2$ and $\hat{\mathbf{e}}_3$ are *mutually perpendicular* at every point. It is also clear that an arbitrary vector **A** located at a particular point P in space may (by the parallelogram law of combination) be written in curvilinear component form as

$$\mathbf{A} = A_1\hat{\mathbf{e}}_1 + A_2\hat{\mathbf{e}}_2 + A_3\hat{\mathbf{e}}_3 \tag{6.15a}$$

where, using the orthogonality of the unit vectors,

$$|\mathbf{A}|^2 = \mathbf{A}\cdot\mathbf{A} = A_1^2 + A_2^2 + A_3^2 \tag{6.15b}$$

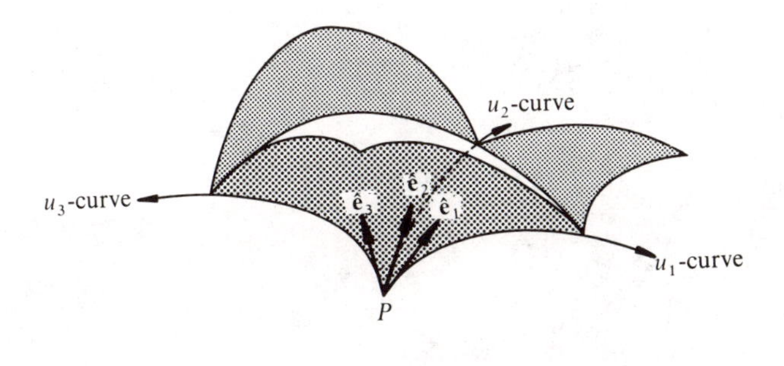

Figure 6.5

Example 6.2

Determine the scale factors for the spherical coordinate system and express the unit vectors in this system in terms of $\hat{\mathbf{i}}$, $\hat{\mathbf{j}}$ and $\hat{\mathbf{k}}$.

Solution

We must use (6.14) and (6.15), and we begin with the usual expression for the position vector $\mathbf{r}$ in Cartesian coordinates:

$$\mathbf{r} = x\hat{\mathbf{i}} + y\hat{\mathbf{j}} + z\hat{\mathbf{k}}, \quad \text{where } |\mathbf{r}| = \sqrt{(x^2 + y^2 + z^2)} = r$$

Then, using (6.6),

$$\mathbf{r} = r \sin\theta \cos\phi\,\hat{\mathbf{i}} + r \sin\theta \sin\phi\,\hat{\mathbf{j}} + r\cos\theta\,\hat{\mathbf{k}}$$

$$\therefore \qquad \frac{\partial \mathbf{r}}{\partial r} = \sin\theta\cos\phi\,\hat{\mathbf{i}} + \sin\theta\sin\phi\,\hat{\mathbf{j}} + \cos\theta\,\hat{\mathbf{k}}$$

and $|\partial\mathbf{r}/\partial r| = 1$ by a simple calculation. Hence the scale factor associated with the r-coordinate is, by (6.14b),

$$h_r = \left|\frac{\partial \mathbf{r}}{\partial r}\right| = 1$$

where we have relabelled h_1 as h_r. Then, by (6.14a) and using $\hat{\mathbf{r}}$ (instead of $\hat{\mathbf{e}}_1$) to denote the unit vector,

$$\hat{\mathbf{r}} = \frac{1}{h_r}\frac{\partial \mathbf{r}}{\partial r} = \sin\theta\cos\phi\,\hat{\mathbf{i}} + \sin\theta\sin\phi\,\hat{\mathbf{j}} + \cos\theta\,\hat{\mathbf{k}} \qquad \textbf{(6.16)}$$

Similar calculations give us the other two scale factors, which we denote by h_θ and h_ϕ (instead of h_2 and h_3), and also the corresponding unit vectors, which we denote by $\hat{\boldsymbol{\theta}}$ and $\hat{\boldsymbol{\phi}}$ (instead of $\hat{\mathbf{e}}_2$ and $\hat{\mathbf{e}}_3$):

$$\frac{\partial \mathbf{r}}{\partial \theta} = r\cos\theta\cos\phi\,\hat{\mathbf{i}} + r\cos\theta\sin\phi\,\hat{\mathbf{j}} - r\sin\theta\,\hat{\mathbf{k}}$$

$$h_\theta = \left|\frac{\partial \mathbf{r}}{\partial \theta}\right| = r$$

$$\hat{\boldsymbol{\theta}} = \frac{1}{h_\theta}\frac{\partial \mathbf{r}}{\partial \theta} = \cos\theta\cos\phi\,\hat{\mathbf{i}} + \cos\theta\sin\phi\,\hat{\mathbf{j}} - \sin\theta\,\hat{\mathbf{k}} \qquad \textbf{(6.17)}$$

Similarly,

$$\frac{\partial \mathbf{r}}{\partial \phi} = -r\sin\theta\sin\phi\,\hat{\mathbf{i}} + r\sin\theta\cos\phi\,\hat{\mathbf{j}}$$

$$h_\phi = \left|\frac{\partial \mathbf{r}}{\partial \phi}\right| = r\sin\theta$$

$$\hat{\boldsymbol{\phi}} = \frac{1}{h_\phi}\frac{\partial \mathbf{r}}{\partial \phi} = -\sin\phi\,\hat{\mathbf{i}} + \cos\phi\,\hat{\mathbf{j}} \qquad \textbf{(6.18)}$$

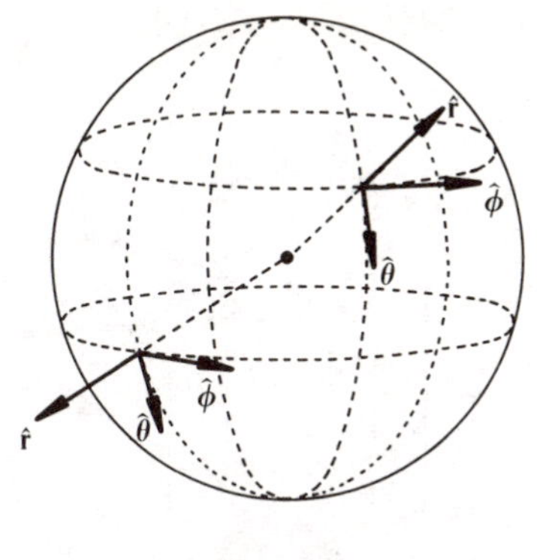

Figure 6.6

It is easily checked that $\hat{\mathbf{r}}\cdot\hat{\boldsymbol{\phi}}=\hat{\mathbf{r}}\cdot\hat{\boldsymbol{\theta}}=\hat{\boldsymbol{\phi}}\cdot\hat{\boldsymbol{\theta}}=0$, that is, the three unit vectors are mutually perpendicular and the spherical polar coordinate system is indeed orthogonal (see Figure 6.6).

Using (6.14) and (6.15), it is easy to show that for Cartesian coordinates, where

$$\mathbf{r}=x\hat{\mathbf{i}}+y\hat{\mathbf{j}}+z\hat{\mathbf{k}}$$

the unit vectors are

$$\hat{\mathbf{e}}_1=\hat{\mathbf{i}}, \qquad \hat{\mathbf{e}}_2=\hat{\mathbf{j}}, \qquad \hat{\mathbf{e}}_3=\hat{\mathbf{k}}$$

and the scale factors are

$$h_1=\left|\frac{\partial\mathbf{r}}{\partial x}\right|=|\hat{\mathbf{i}}|=1, \qquad h_2=|\hat{\mathbf{j}}|=1, \qquad h_3=|\hat{\mathbf{k}}|=1 \tag{6.19}$$

In cylindrical coordinates,

$$\mathbf{r}=\rho\cos\phi\,\hat{\mathbf{i}}+\rho\sin\phi\,\hat{\mathbf{j}}+z\hat{\mathbf{k}}$$

so $$\frac{\partial\mathbf{r}}{\partial\rho}=\cos\phi\,\hat{\mathbf{i}}+\sin\phi\,\hat{\mathbf{j}} \quad\text{and hence}\quad \left|\frac{\partial\mathbf{r}}{\partial\rho}\right|=1$$

$$\therefore \qquad h_1\equiv h_\rho=1 \tag{6.20a}$$

and the corresponding unit vector is

$$\hat{\mathbf{e}}_1\equiv\hat{\boldsymbol{\rho}}=\cos\phi\,\hat{\mathbf{i}}+\sin\phi\,\hat{\mathbf{j}}$$

The reader should easily be able to show by similar calculations that

$$h_2\equiv h_\phi=\rho \quad\text{and}\quad h_3\equiv h_z=1 \tag{6.20b}$$

and hence that

$$\hat{\mathbf{e}}_2\equiv\hat{\boldsymbol{\phi}}=-\sin\phi\,\hat{\mathbf{i}}+\cos\phi\,\hat{\mathbf{j}} \quad\text{and}\quad \hat{\mathbf{e}}_3\equiv\hat{\mathbf{z}}=\hat{\mathbf{k}}$$

(see Figure 6.7).

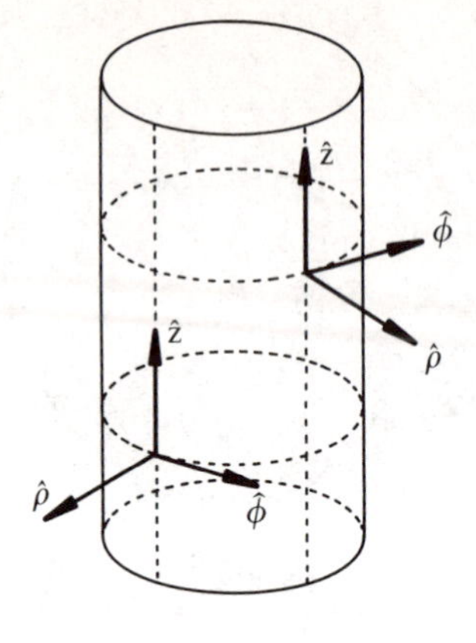

Figure 6.7

EXERCISES

6.1 Verify the relations $\hat{\mathbf{r}} \wedge \hat{\boldsymbol{\theta}} = \hat{\boldsymbol{\phi}}$, $\hat{\boldsymbol{\theta}} \wedge \hat{\boldsymbol{\phi}} = \hat{\mathbf{r}}$ and $\hat{\boldsymbol{\phi}} \wedge \hat{\mathbf{r}} = \hat{\boldsymbol{\theta}}$ in spherical polars, and the relations $\hat{\boldsymbol{\rho}} \wedge \hat{\boldsymbol{\phi}} = \hat{\mathbf{k}}$, $\hat{\boldsymbol{\phi}} \wedge \hat{\mathbf{k}} = \hat{\boldsymbol{\rho}}$ and $\hat{\mathbf{k}} \wedge \hat{\boldsymbol{\rho}} = \hat{\boldsymbol{\phi}}$ in cylindrical polars.

6.2 (a) Express $\hat{\mathbf{i}}$, $\hat{\mathbf{j}}$ and $\hat{\mathbf{k}}$ in terms of the corresponding unit vectors in spherical polar coordinates, and hence express the following in spherical coordinates:

(i) the vector $\mathbf{A} = 3\hat{\mathbf{i}} + 5\hat{\mathbf{j}} - \hat{\mathbf{k}}$,

(ii) the position vector $\mathbf{r}$.

(b) Repeat part (a) using cylindrical polar coordinates.

6.2 The gradient in generalized coordinates

We have already defined the gradient of a scalar field G in Cartesian coordinates in Chapter 3:

$$\nabla G = \frac{\partial G}{\partial x}\hat{\mathbf{i}} + \frac{\partial G}{\partial y}\hat{\mathbf{j}} + \frac{\partial G}{\partial z}\hat{\mathbf{k}} \qquad \textbf{(6.21)}$$

In order to extend this definition to other orthogonal coordinate systems, we express G in generalized coordinates so that any level surface of G can be described by an equation of the form

$$G(u_1, u_2, u_3) = \text{constant}$$

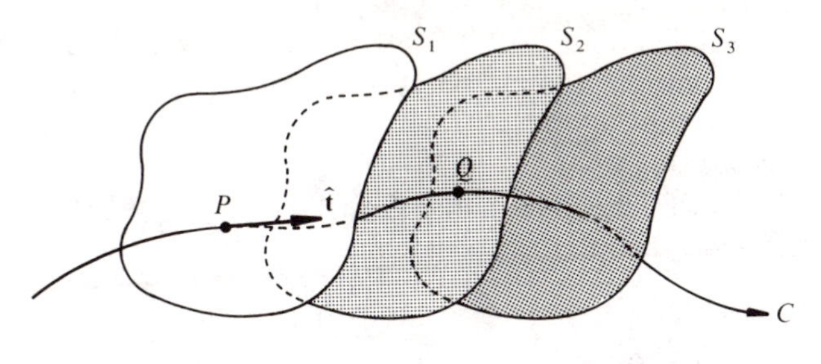

Figure 6.8

Suppose that P is a point in the scalar field and that S_1 is the level surface at P. Suppose C is an arbitrary curve passing through P and that we move from P to a neighbouring point Q such that the generalized coordinates change by $\mathrm{d}u_1$, $\mathrm{d}u_2$ and $\mathrm{d}u_3$ (see Figure 6.8). The change in G as we move from P to Q is

$$\mathrm{d}G \approx \frac{\partial G}{\partial u_1}\mathrm{d}u_1 + \frac{\partial G}{\partial u_2}\mathrm{d}u_2 + \frac{\partial G}{\partial u_3}\mathrm{d}u_3 \tag{6.22}$$

If C is parametrized using the arc-length parameter s, then the rate of change of G along C is

$$\frac{\mathrm{d}G}{\mathrm{d}s} = \frac{\partial G}{\partial u_1}\frac{\mathrm{d}u_1}{\mathrm{d}s} + \frac{\partial G}{\partial u_2}\frac{\mathrm{d}u_2}{\mathrm{d}s} + \frac{\partial G}{\partial u_3}\frac{\mathrm{d}u_3}{\mathrm{d}s} \tag{6.23}$$

But if $\mathbf{r}$ is the position vector of the point P, then we know from Section 2.6 that the unit tangent vector at P is $\hat{\mathbf{t}} = \mathrm{d}\mathbf{r}/\mathrm{d}s$ or, using generalized coordinates,

$$\begin{aligned}\hat{\mathbf{t}} &= \frac{\partial \mathbf{r}}{\partial u_1}\frac{\mathrm{d}u_1}{\mathrm{d}s} + \frac{\partial \mathbf{r}}{\partial u_2}\frac{\mathrm{d}u_2}{\mathrm{d}s} + \frac{\partial \mathbf{r}}{\partial u_3}\frac{\mathrm{d}u_3}{\mathrm{d}s} \\ &= h_1\hat{\mathbf{e}}_1\frac{\mathrm{d}u_1}{\mathrm{d}s} + h_2\hat{\mathbf{e}}_2\frac{\mathrm{d}u_2}{\mathrm{d}s} + h_3\hat{\mathbf{e}}_3\frac{\mathrm{d}u_3}{\mathrm{d}s}\end{aligned} \tag{6.24}$$

For an orthogonal coordinate system, we can express (6.23) in the form of a scalar product:

$$\frac{\mathrm{d}G}{\mathrm{d}s} = \left(\frac{1}{h_1}\frac{\partial G}{\partial u_1}\hat{\mathbf{e}}_1 + \frac{1}{h_2}\frac{\partial G}{\partial u_2}\hat{\mathbf{e}}_2 + \frac{1}{h_3}\frac{\partial G}{\partial u_3}\hat{\mathbf{e}}_3\right)\cdot\left(h_1\hat{\mathbf{e}}_1\frac{\mathrm{d}u_1}{\mathrm{d}s} + h_2\hat{\mathbf{e}}_2\frac{\mathrm{d}u_2}{\mathrm{d}s} + h_3\hat{\mathbf{e}}_3\frac{\mathrm{d}u_3}{\mathrm{d}s}\right)$$

The second vector here is $\hat{\mathbf{t}}$ and we define the first vector as the gradient of G, that is,

$$\nabla G = \frac{1}{h_1}\frac{\partial G}{\partial u_1}\hat{\mathbf{e}}_1 + \frac{1}{h_2}\frac{\partial G}{\partial u_2}\hat{\mathbf{e}}_2 + \frac{1}{h_3}\frac{\partial G}{\partial u_3}\hat{\mathbf{e}}_3 = \sum_{i=1}^{3}\frac{1}{h_i}\frac{\partial G}{\partial u_i}\hat{\mathbf{e}}_i \tag{6.25}$$

$$\therefore \quad \frac{\mathrm{d}G}{\mathrm{d}s} = \nabla G\cdot\hat{\mathbf{t}} \tag{6.26}$$

The general definition (6.25) is of course completely consistent with the Cartesian definition (6.21), because in that case $h_1 = h_2 = h_3 = 1$, $(\hat{\mathbf{e}}_1, \hat{\mathbf{e}}_2, \hat{\mathbf{e}}_3) \equiv (\hat{\mathbf{i}}, \hat{\mathbf{j}}, \hat{\mathbf{k}})$ and $(u_1, u_2, u_3) \equiv (x, y, z)$.

We have already obtained (6.26) using only Cartesian coordinates in Section 3.2. Clearly, if C lies in the surface $G(u_1, u_2, u_3) = k$, then $\hat{\mathbf{t}}$ is tangential to the surface and $\mathrm{d}G/\mathrm{d}s = 0$ (G is unchanging on C). In this case, therefore,

$$\nabla G \cdot \hat{\mathbf{t}} = 0$$

that is, ∇G is normal to the surface $G(u_1, u_2, u_3) = k$, as we know.

We can now use (6.25) very readily to obtain the form of ∇G in the three common coordinate systems using the scale factors and unit vectors already calculated:

- Cartesians

$$\nabla G = \hat{\mathbf{i}} \frac{\partial G}{\partial x} + \hat{\mathbf{j}} \frac{\partial G}{\partial y} + \hat{\mathbf{k}} \frac{\partial G}{\partial z} \tag{6.27}$$

- Cylindrical polars

$$\nabla G = \hat{\boldsymbol{\rho}} \frac{\partial G}{\partial \rho} + \frac{1}{\rho} \hat{\boldsymbol{\phi}} \frac{\partial G}{\partial \phi} + \hat{\mathbf{k}} \frac{\partial G}{\partial z} \tag{6.28}$$

- Spherical polars

$$\nabla G = \hat{\mathbf{r}} \frac{\partial G}{\partial r} + \frac{1}{r} \hat{\boldsymbol{\theta}} \frac{\partial G}{\partial \theta} + \frac{1}{r \sin \theta} \hat{\boldsymbol{\phi}} \frac{\partial G}{\partial \phi} \tag{6.29}$$

Example 6.3

Determine, in spherical polars, a vector normal to the surface $x^2 + y^2 + z = 0$.

Solution

We express the equation of the surface in spherical polars:

$$r^2 \sin^2 \theta \cos^2 \phi + r^2 \sin^2 \theta \sin^2 \phi + r \cos \theta = 0$$

or

$$r^2 \sin^2 \theta + r \cos \theta = 0$$

The equation of the surface is therefore $G = 0$, where

$$G \equiv r^2 \sin^2 \theta + r \cos \theta$$

Using (6.29), a normal vector is (see Figure 6.9)

$$\begin{aligned} \mathbf{n} = \nabla G &= \hat{\mathbf{r}}(2r \sin^2 \theta + \cos \theta) + \frac{1}{r} \hat{\boldsymbol{\theta}}(2r^2 \sin \theta \cos \theta - r \sin \theta) \\ &= \hat{\mathbf{r}}(2r \sin^2 \theta + \cos \theta) + \hat{\boldsymbol{\theta}}(r \sin 2\theta - \sin \theta) \end{aligned}$$

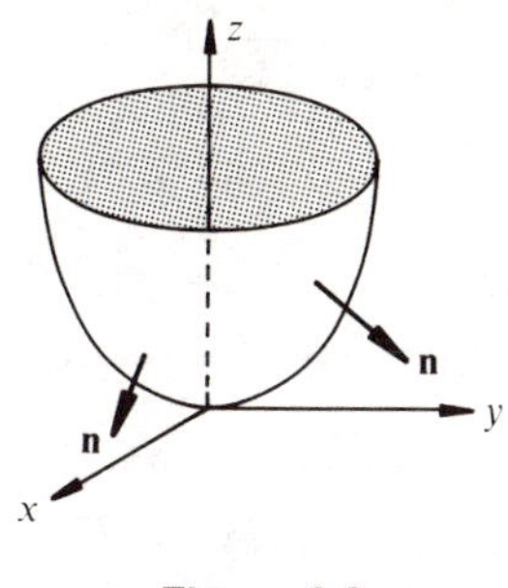

Figure 6.9

A *unit* normal $\hat{n}$ may be obtained easily, since $\hat{\mathbf{r}}$ and $\hat{\boldsymbol{\theta}}$ are mutually orthogonal and so

$$|\mathbf{n}|^2 = \mathbf{n}\cdot\mathbf{n} = (2r\sin^2\theta + \cos\theta)^2 + (r\sin 2\theta - \sin\theta)^2$$

and then $\hat{\mathbf{n}} = \mathbf{n}/|\mathbf{n}|$.

EXERCISES

6.3 Show that $\partial(\nabla\Phi)/\partial x = \nabla(\partial\Phi/\partial x)$, where Φ is any scalar field. If (u_1, u_2, u_3) are orthogonal curvilinear coordinates, is the equation $\partial(\nabla\Phi)/\partial u_1 = \nabla(\partial\Phi/\partial u_1)$ always satisfied?

6.4 Determine the gradients of the following scalar fields ($\mathbf{a}$ is a constant vector):

(a) r^2 (b) $\dfrac{1}{r}$ (c) $\mathbf{a}\cdot\mathbf{r}$ (d) $\dfrac{\mathbf{a}\cdot\mathbf{r}}{r^3}$

6.5 Verify the following equations involving cylindrical coordinates ρ, ϕ and z:

(a) $\nabla\phi + \nabla\wedge(\hat{\mathbf{k}}\ln\rho) = 0$ (b) $\nabla(\ln\rho) - \nabla\wedge(\hat{\mathbf{k}}\phi) = 0$

6.6 Verify the following equations for general orthogonal curvilinear coordinates:

(a) $\hat{\mathbf{e}}_1 = h_2h_3\,\nabla u_2\wedge\nabla u_3$ (b) $\hat{\mathbf{e}}_2 = h_3h_1\,\nabla u_3\wedge\nabla u_1$

(c) $\hat{\mathbf{e}}_3 = h_1h_2\,\nabla u_1\wedge\nabla u_2$

6.3 Line, surface and volume elements

In the previous section we obtained the result that, if $\hat{\mathbf{t}}$ is a unit vector to a curve C, then

$$\hat{\mathbf{t}}\,ds = h_1\hat{\mathbf{e}}_1\,du_1 + h_2\hat{\mathbf{e}}_2\,du_2 + h_3\hat{\mathbf{e}}_3\,du_3$$

(see equation 6.24). The quantity $\mathrm{d}s$ is called the **differential element** of arc length along the curve (or sometimes the **line element**). Clearly, if the curve C is along the u_1-coordinate curve, then

$$\mathrm{d}s \equiv \mathrm{d}s_1 = h_1\,\mathrm{d}u_1 \tag{6.30a}$$

We can similarly define differential elements of length along the u_2- and u_3-coordinate curves as

$$\mathrm{d}s_2 = h_2\,\mathrm{d}u_2 \quad \text{and} \quad \mathrm{d}s_3 = h_3\,\mathrm{d}u_3 \tag{6.30b}$$

For example, in Cartesians the differential elements of arc length are (since $h_1 = h_2 = h_3 = 1$)

$$\mathrm{d}s_1 = \mathrm{d}x, \qquad \mathrm{d}s_2 = \mathrm{d}y, \qquad \mathrm{d}s_3 = \mathrm{d}z$$

In spherical polars the differential elements of length are (using $h_r = 1$, $h_\theta = r$ and $h_\phi = r\sin\theta$)

$$\mathrm{d}s_1 = \mathrm{d}r, \qquad \mathrm{d}s_2 = r\,\mathrm{d}\theta, \qquad \mathrm{d}s_3 = r\sin\theta\,\mathrm{d}\phi \tag{6.31a}$$

and in cylindrical polars the elements are

$$\mathrm{d}s_1 = \mathrm{d}\rho, \qquad \mathrm{d}s_2 = \rho\,\mathrm{d}\phi, \qquad \mathrm{d}s_3 = \mathrm{d}z \tag{6.31b}$$

The determination of a differential element of *area* on a surface is a much more complicated exercise, although the use of vectors and vector notation is invaluable. Consider a surface S, which we assume has been parametrized by (u, v) (see Section 2.2 and Figure 2.7). As explained in that section, the u-constant and v-constant parameter lines are transformed into curves (lying in the surface) through the equations which define the surface:

$$u_1 = u_1(u, v), \qquad u_2 = u_2(u, v), \qquad u_3 = u_3(u, v)$$

If $\delta\mathbf{r}$ denotes the vector joining two 'close' parameter curves, $u = \text{constant}$ and $u + \delta u = \text{constant}$, then

$$\delta\mathbf{r} = \mathbf{r}(u + \delta u, v) - \mathbf{r}(u, v) \approx \frac{\partial \mathbf{r}}{\partial u}\,\delta u$$

and the length of this vector is

$$|\delta\mathbf{r}| = \left|\frac{\partial \mathbf{r}}{\partial u}\right| \delta u$$

Similarly, the distance between two closely related points on the curves $v = \text{constant}$ and $v + \delta v = \text{constant}$ on S is

$$\left|\frac{\partial \mathbf{r}}{\partial v}\right| \delta v$$

(see Figure 6.10a).

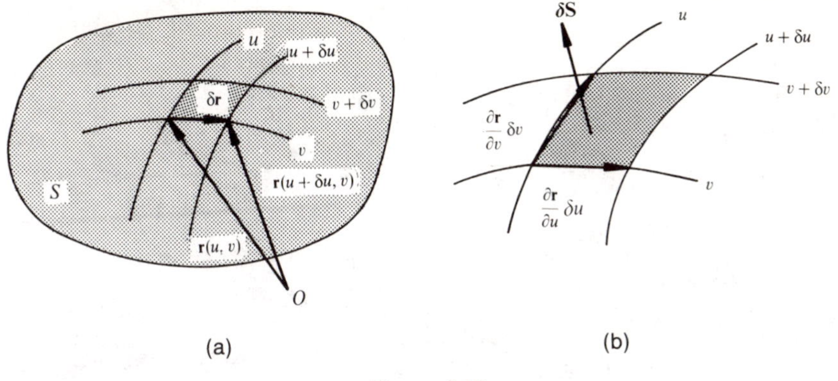

Figure 6.10

We showed in Section 2.9 that tangent vectors to the $v = \text{constant}$ and $u = \text{constant}$ curves are $\partial\mathbf{r}/\partial u$ and $\partial\mathbf{r}/\partial v$ respectively, so that an expression for an element of area $\delta\mathbf{S}$ on the surface S is

$$\delta\mathbf{S} = \left(\frac{\partial\mathbf{r}}{\partial u}\,\delta u\right) \wedge \left(\frac{\partial\mathbf{r}}{\partial v}\,\delta v\right) = \left(\frac{\partial\mathbf{r}}{\partial u} \wedge \frac{\partial\mathbf{r}}{\partial v}\right)\delta u\,\delta v$$

(see Figure 6.10b). This quantity is easily expressed in terms of curvilinear coordinates. Firstly,

$$\frac{\partial\mathbf{r}}{\partial u} = \frac{\partial\mathbf{r}}{\partial u_1}\frac{\partial u_1}{\partial u} + \frac{\partial\mathbf{r}}{\partial u_2}\frac{\partial u_2}{\partial u} + \frac{\partial\mathbf{r}}{\partial u_3}\frac{\partial u_3}{\partial u} = h_1\hat{\mathbf{e}}_1\frac{\partial u_1}{\partial u} + h_2\hat{\mathbf{e}}_2\frac{\partial u_2}{\partial u} + h_3\hat{\mathbf{e}}_3\frac{\partial u_3}{\partial u}$$

with a similar expression for $\partial\mathbf{r}/\partial v$. Since $\delta\mathbf{S}$ is given by a cross product, we can conveniently write it in determinant form as

$$\mathrm{d}\mathbf{S} = \begin{vmatrix} \hat{\mathbf{e}}_1 & \hat{\mathbf{e}}_2 & \hat{\mathbf{e}}_3 \\ h_1\dfrac{\partial u_1}{\partial u} & h_2\dfrac{\partial u_2}{\partial u} & h_3\dfrac{\partial u_3}{\partial u} \\ h_1\dfrac{\partial u_1}{\partial v} & h_2\dfrac{\partial u_2}{\partial v} & h_3\dfrac{\partial u_3}{\partial v} \end{vmatrix}\,\mathrm{d}u\,\mathrm{d}v \tag{6.32}$$

which can be expanded if necessary. (Here we have used the notation $\mathrm{d}\mathbf{S}$, $\mathrm{d}u$ and $\mathrm{d}v$ for $\delta\mathbf{S}$, δu and δv.)

The (rather complicated) expression (6.32) is greatly simplified if the surface of interest coincides with one of the three coordinate surfaces. For example, an expression for the element of area on the u_3-constant surface, which we shall call $\mathrm{d}\mathbf{S}_3$, is obtained from (6.32) by putting $u \equiv u_1$ and $v \equiv u_2$. Then,

$$\mathrm{d}\mathbf{S}_3 = \begin{vmatrix} \hat{\mathbf{e}}_1 & \hat{\mathbf{e}}_2 & \hat{\mathbf{e}}_3 \\ h_1 & 0 & 0 \\ 0 & h_2 & 0 \end{vmatrix}\,\mathrm{d}u_1\,\mathrm{d}u_2 = \hat{\mathbf{e}}_3 h_1 h_2\,\mathrm{d}u_1\,\mathrm{d}u_2$$

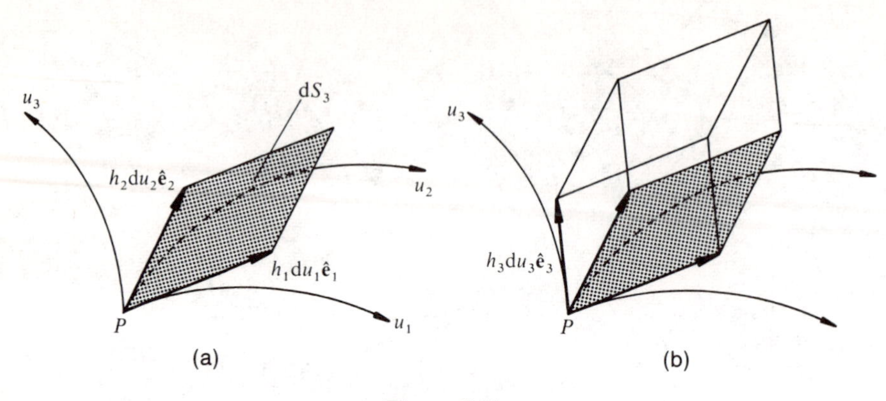

Figure 6.11

with magnitude

$$dS_3 = h_1 h_2 \, du_1 \, du_2 \tag{6.33a}$$

(see Figure 6.11a). Expressions for the surface elements of area on the u_2-constant and u_1-constant surfaces are, similarly,

$$dS_2 = h_1 h_3 \, du_1 \, du_3 \quad \text{and} \quad dS_1 = h_2 h_3 \, du_2 \, du_3 \tag{6.33b}$$

For the three major coordinate systems, the surface elements on the coordinate surfaces are readily obtained from (6.33):

- Cartesians

$$dS_x = dy\,dz, \qquad dS_y = dx\,dz, \qquad dS_z = dx\,dy$$

- Cylindrical polars (see Figure 6.12a)

$$dS_\rho = \rho\, d\phi\, dz, \qquad dS_\phi = d\rho\, dz, \qquad dS_z = \rho\, d\rho\, d\phi$$

- Spherical polars (see Figure 6.12b)

$$dS_r = r^2 \sin\theta\, d\theta\, d\phi, \qquad dS_\theta = r\sin\theta\, dr\, d\phi, \qquad dS_\phi = r\, dr\, d\theta$$

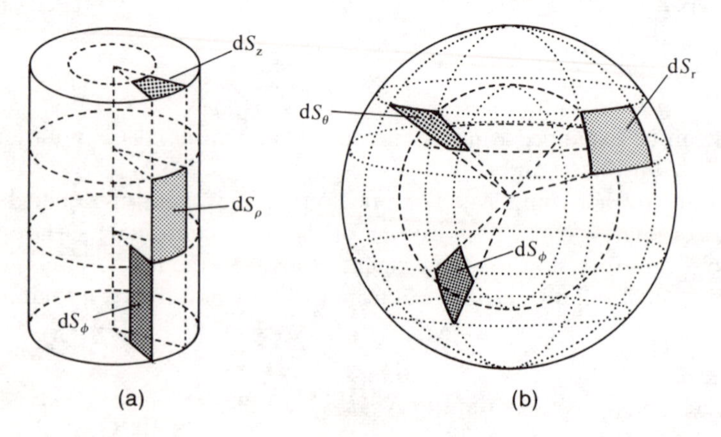

Figure 6.12

We have already used some of these surface elements in evaluating surface integrals in Chapter 4.

Finally, an expression for the differential volume element $\mathrm{d}V$ at a point P is

$$\mathrm{d}V = h_1 h_2 h_3 \,\mathrm{d}u_1 \,\mathrm{d}u_2 \,\mathrm{d}u_3 \tag{6.34}$$

(see Figure 6.11b). This simple form follows directly from the property that the coordinate surfaces are mutually perpendicular. So the Cartesian, cylindrical polar and spherical polar coordinate systems have volume elements

$$\mathrm{d}V = \mathrm{d}x\,\mathrm{d}y\,\mathrm{d}z, \qquad \mathrm{d}V = \rho\,\mathrm{d}\rho\,\mathrm{d}\phi\,\mathrm{d}z, \qquad \mathrm{d}V = r^2 \sin\theta\,\mathrm{d}r\,\mathrm{d}\theta\,\mathrm{d}\phi$$

respectively (see Figure 4.28). We have, of course, already used these volume elements in evaluating volume integrals in Chapter 4.

Example 6.4

Determine the surface area S of a cone of slant height a with apex at the origin and whose curved surface is defined by $\theta = \pi/4$.

Solution

We first construct a differential surface element on the cone (see Figure 6.13). This surface element is part of the θ-constant coordinate surface in spherical polars and so

$$\mathrm{d}S_\theta = r\sin\theta\,\mathrm{d}r\,\mathrm{d}\phi \quad \text{with } \theta = \pi/4$$

that is, $\mathrm{d}S_\theta = \dfrac{r}{\sqrt{2}}\,\mathrm{d}r\,\mathrm{d}\phi$

The total surface area is simply the 'sum' of this expression as the variables r and ϕ vary over the range $0 \leqslant r \leqslant a$, $0 \leqslant \phi < 2\pi$. Hence

$$S = \iint_{\text{cone}} \mathrm{d}S_\theta = \int_0^{2\pi}\int_0^a \frac{r}{\sqrt{2}}\,\mathrm{d}r\,\mathrm{d}\phi = \int_0^{2\pi}\left[\frac{r^2}{2\sqrt{2}}\right]_0^a \mathrm{d}\phi = \frac{\sqrt{2}}{2}a^2\pi$$

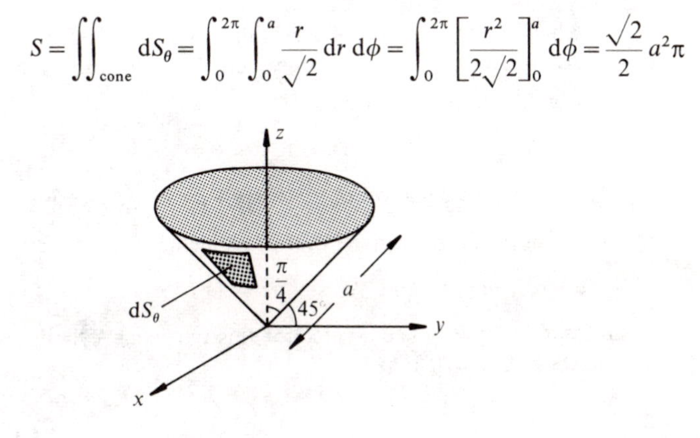

Figure 6.13

Example 6.5

Determine the volume of the elliptical cylinder

$$\frac{x^2}{4}+\frac{y^2}{1}=1$$

between $z=0$ and $z=3$.

Solution

Although there are a number of simple ways in which this volume may be determined, we shall take the opportunity to introduce a less well-known orthogonal coordinate system: **elliptical cylindrical coordinates** (u, v, z). These are defined by the transformation of coordinates

$$x=a\cosh u\cos v, \qquad y=a\sinh u\sin v, \qquad z=z$$

in which a is a constant (yet to be specified) and the ranges of the new variables are

$$u\geqslant 0, \qquad 0\leqslant v\leqslant 2\pi, \qquad -\infty<z<\infty$$

We investigate the properties of these coordinates:

- A v-constant coordinate surface has representation

  $$\frac{x^2}{a^2\cos^2 v}-\frac{y^2}{a^2\sin^2 v}=\cosh^2 u-\sinh^2 u=1, \qquad -\infty<z<\infty$$

  which represents a hyperbola drawn out along the z-direction, there being one such hyperbola for each value of v.
- A u-constant coordinate surface has representation

  $$\frac{x^2}{a^2\cosh^2 u}+\frac{y^2}{a^2\sinh^2 u}=1, \qquad -\infty<z<\infty$$

  which is the equation of a cylinder with an elliptical cross-section, there being one such ellipse for each value of u. For the particular case $u=0$,

  $$x=a\cos v, \qquad y=0, \qquad -\infty<z<\infty$$

  which is the equation of the strip $-a<x<a$ (see Figure 6.14).

For the elliptical cylinder given in this problem, we must have

$$a^2\cosh^2 u=4 \quad\text{and}\quad a^2\sinh^2 u=1$$

or, subtracting, $a^2=3$. The u-value for this particular cylinder is clearly such that $3\sinh^2 u=1$, that is

$$u=\sinh^{-1}\left(\frac{1}{\sqrt{3}}\right)$$

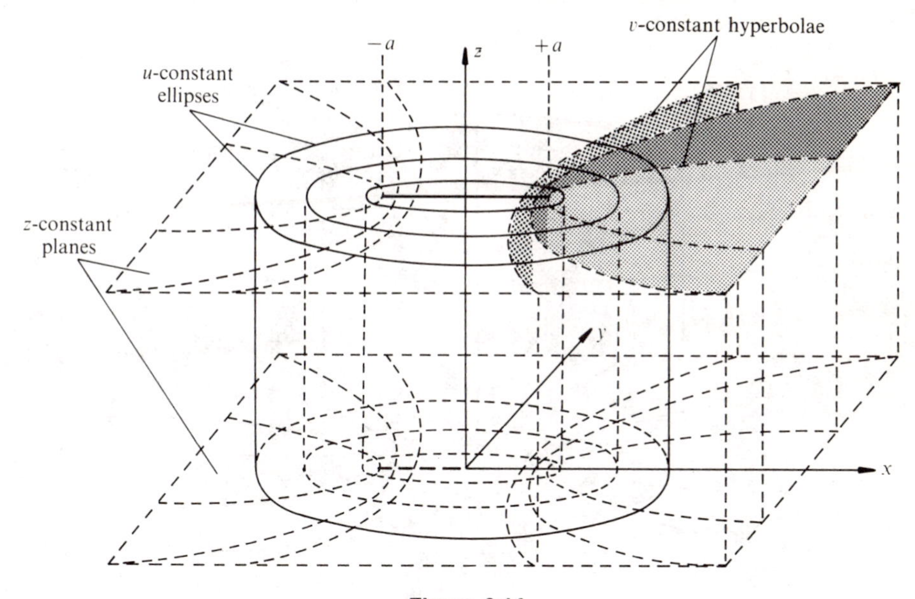

Figure 6.14

In order to obtain the required volume, we must clearly determine the volume element for this coordinate system and then integrate. This implies that we must first determine the scale factors. It is easily shown that this coordinate system is orthogonal (this is left as an exercise for the reader), and the scale factors are obtained using

$$\mathbf{r} = a \cosh u \cos v\, \hat{\mathbf{i}} + a \sinh u \sin v\, \hat{\mathbf{j}} + z\hat{\mathbf{k}}$$

We find

$$h_u = \left|\frac{\partial \mathbf{r}}{\partial u}\right| = |(a \sinh u \cos v)\hat{\mathbf{i}} + (a \cosh u \sin v)\hat{\mathbf{j}}|$$

$$= a\sqrt{(\sinh^2 u \cos^2 v + \cosh^2 u \sin^2 v)}$$

$$= a\sqrt{(\sinh^2 u + \sin^2 v)} \quad (\text{or } a\sqrt{(\cosh^2 u + \cos^2 v)})$$

Similarly,

$$h_v = \left|\frac{\partial \mathbf{r}}{\partial v}\right| = |(-a \cosh u \sin v)\hat{\mathbf{i}} + (a \sinh u \cos v)\hat{\mathbf{j}}|$$

$$= a\sqrt{(\sinh^2 u + \sin^2 v)}$$

and $\quad h_z = \left|\dfrac{\partial \mathbf{r}}{\partial z}\right| = |\hat{\mathbf{k}}| = 1$

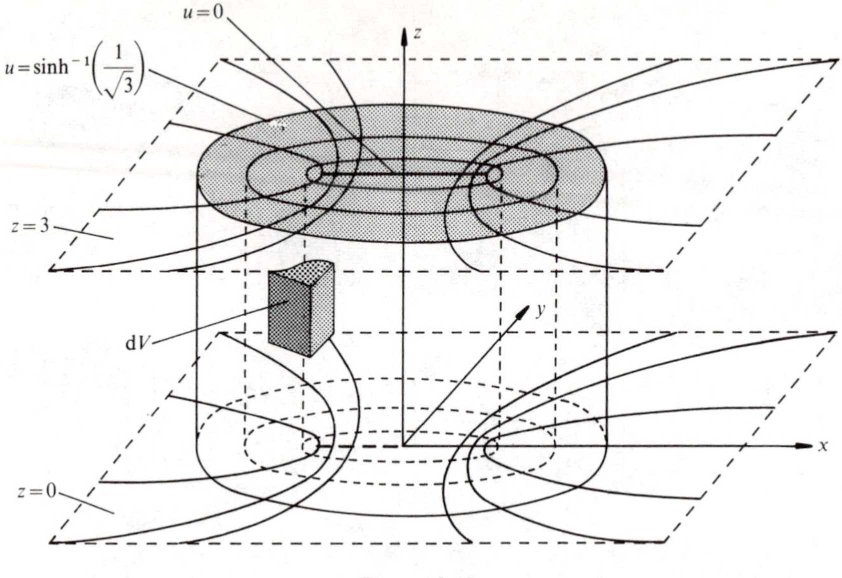

Figure 6.15

Thus the volume element in elliptical cylindrical coordinates is, using (6.34),

$$\begin{aligned} \mathrm{d}V &= h_u h_v h_z \,\mathrm{d}u\,\mathrm{d}v\,\mathrm{d}z = a^2(\sinh^2 u + \sin^2 v)\,\mathrm{d}u\,\mathrm{d}v\,\mathrm{d}z \\ &= 3(\sinh^2 u + \sin^2 v)\,\mathrm{d}u\,\mathrm{d}v\,\mathrm{d}z \end{aligned}$$

for our problem. The total volume V of the given elliptical cylinder is then obtained by integrating this element over the cylinder as z ranges from 0 to 3, u ranges from 0 to $\sinh^{-1}(1/\sqrt{3})$ and v ranges from 0 to 2π (see Figure 6.15). Therefore

$$V = 3\int_0^3 \int_0^{2\pi} \int_0^{\sinh^{-1}(1/\sqrt{3})} (\sinh^2 u + \sin^2 v)\,\mathrm{d}u\,\mathrm{d}v\,\mathrm{d}z$$

Now $\displaystyle\int_0^{\sinh^{-1}(1/\sqrt{3})} (\sinh^2 u + \sin^2 v)\,\mathrm{d}u$

$$\begin{aligned} &= \int_0^{\sinh^{-1}(1/\sqrt{3})} \left(\frac{-1+\cosh 2u}{2} + \sin^2 v\right)\mathrm{d}u \\ &= \left[u(\sin^2 v - \tfrac{1}{2}) + \frac{\sinh 2u}{4}\right]_0^{\sinh^{-1}(1/\sqrt{3})} \\ &= \sinh^{-1}\left(\frac{1}{\sqrt{3}}\right)(\sin^2 v - \tfrac{1}{2}) + \tfrac{1}{2}\left(\frac{1}{\sqrt{3}}\right)\sqrt{(1+\tfrac{1}{3})} \\ &\qquad\qquad \text{(using } \sinh 2u = 2\sinh u \cosh u\text{)} \\ &= \sinh^{-1}\left(\frac{1}{\sqrt{3}}\right)(\sin^2 v - \tfrac{1}{2}) + \tfrac{1}{3} \end{aligned}$$

Also $\displaystyle\int_0^{2\pi} (\sin^2 v - \tfrac{1}{2})\,\mathrm{d}v = \int_0^{2\pi} (-\tfrac{1}{2}\cos 2v)\,\mathrm{d}v = 0$

Therefore the required volume is

$$V = 3\int_0^3 \int_0^{2\pi} \tfrac{1}{3}\,\mathrm{d}v\,\mathrm{d}z = 6\pi$$

6.4 Divergence and curl in orthogonal curvilinear coordinates

In this section we derive expressions for the divergence and curl of a vector field in orthogonal curvilinear coordinates. There are two ways in which this can be done. Firstly, we could choose to use the coordinate-free integral definitions of these quantities obtained in Sections 5.4 and 5.9 and work consistently in curvilinear coordinates to deduce the required results; we shall, in fact, use this approach in our treatment of the divergence. Secondly, we could regard ∇ as a vector differential operator (as we did for Cartesians) and simply determine the scalar or vector product of ∇ with a vector field **A**, expressing all quantities in curvilinear coordinates; we shall use this approach in the case of the curl.

Divergence

The coordinate-free definition of the divergence of a vector field **A** is

$$\nabla\cdot\mathbf{A} = \lim_{V\to 0}\left(\frac{1}{V}\oint\!\!\oint_S \mathbf{A}\cdot\hat{\mathbf{n}}\,\mathrm{d}S\right) \tag{6.35}$$

where V is an elementary volume bounded by a closed surface S.

Let P be a point with curvilinear coordinates (u_1, u_2, u_3) and V a small parallelepiped with edges parallel to the unit vectors $\hat{\mathbf{e}}_1, \hat{\mathbf{e}}_2, \hat{\mathbf{e}}_3$ at P (see Figure 6.16). We calculate the flux integral $\oint\!\!\oint_S \mathbf{A}\cdot\hat{\mathbf{n}}\,\mathrm{d}S$ over the six faces of the box. We shall assume that the dimensions of the box are so small that the vector field **A** may be regarded as a constant over each bounding face.

On the face $PEHD$ we assume that **A** is a constant vector equal to the actual vector at P, $\mathbf{A}^{(P)}$ say. On this face, $\hat{\mathbf{n}} = -\hat{\mathbf{e}}_1$ (outward-pointing unit normal), hence

$$\begin{aligned}\iint_{PEHD} \mathbf{A}\cdot\hat{\mathbf{n}}\,\mathrm{d}S &\approx -\mathbf{A}^{(P)}\cdot\hat{\mathbf{e}}_1 \iint_{PEHD} \mathrm{d}S_1 \\ &= -\mathbf{A}^{(P)}\cdot\hat{\mathbf{e}}_1 (h_2h_3)^{(P)}\,\mathrm{d}u_2\,\mathrm{d}u_3 \quad \text{(using (6.33b))} \\ &= -A_1^{(P)}(h_2h_3)^{(P)}\,\mathrm{d}u_2\,\mathrm{d}u_3\end{aligned}$$

where we have evaluated h_2h_3 at the point P.

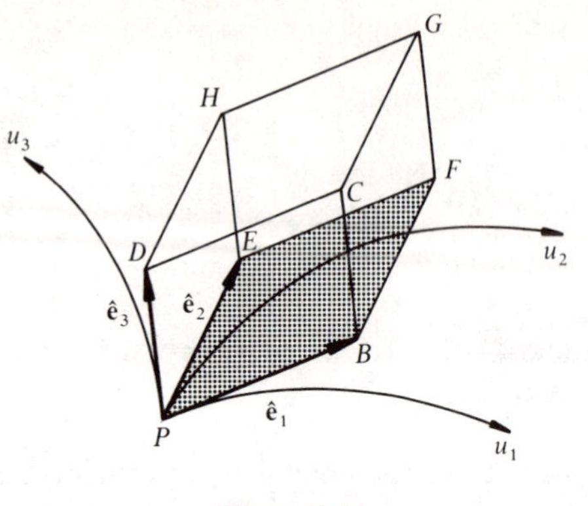

Figure 6.16

On the face $BFGC$ we assume that $\mathbf{A}$ is a constant vector equal to the actual vector at B, $\mathbf{A}^{(B)}$ say. On this face, $\hat{\mathbf{n}} = +\hat{\mathbf{e}}_1$, hence

$$\iint_{BFGC} \mathbf{A}\cdot\hat{\mathbf{n}}\,\mathrm{d}S \approx A_1^{(B)}(h_2h_3)^{(B)}\,\mathrm{d}u_2\,\mathrm{d}u_3$$

where h_2h_3 is evaluated at the point B.

But, to a good approximation,

$$(A_1h_2h_3)^{(B)} \approx (A_1h_2h_3)^{(P)} + \frac{\partial}{\partial u_1}(A_1h_2h_3)^{(P)}\,\mathrm{d}u_1$$

Adding these two expressions, we obtain the approximate value

$$\frac{\partial}{\partial u_1}(A_1h_2h_3)^{(P)}\,\mathrm{d}u_1\,\mathrm{d}u_2\,\mathrm{d}u_3$$

for the flux across these two faces.

A similar calculation of the surface integral over the other two pairs of faces ($PBCD$ and EFGH, and $PEFB$ and $DHGC$) gives the further contributions

$$\frac{\partial}{\partial u_2}(A_2h_1h_3)^{(P)}\,\mathrm{d}u_1\,\mathrm{d}u_2\,\mathrm{d}u_3 \quad\text{and}\quad \frac{\partial}{\partial u_3}(A_3h_1h_2)^{(P)}\,\mathrm{d}u_1\,\mathrm{d}u_2\,\mathrm{d}u_3$$

respectively. Adding all the contributions gives

$$\oiint_S \mathbf{A}\cdot\hat{\mathbf{n}}\,\mathrm{d}S \approx \left(\frac{\partial}{\partial u_1}(A_1h_2h_3)^{(P)} + \frac{\partial}{\partial u_2}(A_2h_1h_3)^{(P)} + \frac{\partial}{\partial u_3}(A_3h_1h_2)^{(P)}\right)\mathrm{d}u_1\,\mathrm{d}u_2\,\mathrm{d}u_3$$

Since P is an arbitrary point, we may drop the superscript. Then, from the definition (6.35) and the volume element (6.34),

$$\nabla\cdot\mathbf{A} = \lim_{du_1\,du_2\,du_3\to\infty}\left(\frac{\dfrac{\partial}{\partial u_1}(A_1h_2h_3)+\dfrac{\partial}{\partial u_2}(A_2h_1h_3)+\dfrac{\partial}{\partial u_3}(A_3h_1h_2)}{h_1h_2h_3\,du_1\,du_2\,du_3}\right)du_1\,du_2\,du_3$$

or
$$\nabla\cdot\mathbf{A} = \frac{1}{h_1h_2h_3}\left[\frac{\partial}{\partial u_1}(A_1h_2h_3)+\frac{\partial}{\partial u_2}(A_2h_1h_3)+\frac{\partial}{\partial u_3}(A_3h_1h_2)\right] \quad \textbf{(6.36)}$$

Equation (6.36) is the expression for the divergence in orthogonal curvilinear coordinates.

We can use (6.36) and the scale factors already calculated to write down immediately the form of the divergence in the main coordinate systems:

- Cartesians

$$\nabla\cdot\mathbf{A} = \frac{\partial A_x}{\partial x}+\frac{\partial A_y}{\partial y}+\frac{\partial A_z}{\partial z} \quad \text{(as originally defined in Chapter 3)}$$

- Cylindrical polars

$$\nabla\cdot\mathbf{A} = \frac{1}{\rho}\left[\frac{\partial}{\partial\rho}(\rho A_\rho)+\frac{\partial}{\partial\phi}(A_\phi)+\frac{\partial}{\partial z}(\rho A_z)\right]$$
$$= \frac{1}{\rho}\frac{\partial}{\partial\rho}(\rho A_\rho)+\frac{1}{\rho}\frac{\partial A_\phi}{\partial\phi}+\frac{\partial A_z}{\partial z} \quad \textbf{(6.37)}$$

- Spherical polars

$$\nabla\cdot\mathbf{A} = \frac{1}{r^2\sin\theta}\left[\frac{\partial}{\partial r}(r^2\sin\theta\,A_r)+\frac{\partial}{\partial\theta}(r\sin\theta\,A_\theta)+\frac{\partial}{\partial\phi}(rA_\phi)\right]$$
$$= \frac{1}{r^2}\frac{\partial}{\partial r}(r^2A_r)+\frac{1}{r\sin\theta}\frac{\partial}{\partial\theta}(\sin\theta\,A_\theta)+\frac{1}{r\sin\theta}\frac{\partial}{\partial\phi}(A_\phi) \quad \textbf{(6.38)}$$

The expression for the Laplacian ∇^2 in curvilinear coordinates is now easily obtained if we remember that $\nabla^2G=\nabla\cdot(\nabla G)$ and simply use the expressions for **grad** and div in (6.25) and (6.36). We find, after a straightforward calculation,

$$\nabla^2G = \frac{1}{h_1h_2h_3}\left[\frac{\partial}{\partial u_1}\left(\frac{h_2h_3}{h_1}\frac{\partial G}{\partial u_1}\right)+\frac{\partial}{\partial u_2}\left(\frac{h_1h_3}{h_2}\frac{\partial G}{\partial u_2}\right)+\frac{\partial}{\partial u_3}\left(\frac{h_1h_2}{h_3}\frac{\partial G}{\partial u_3}\right)\right] \quad \textbf{(6.39)}$$

from which we can obtain the Laplacian in the three most important coordinate systems:

- Cartesian

$$\nabla^2G = \frac{\partial^2G}{\partial x^2}+\frac{\partial^2G}{\partial y^2}+\frac{\partial^2G}{\partial z^2} \quad \textbf{(6.40a)}$$

- Cylindrical polars

$$\nabla^2 G = \frac{1}{\rho}\frac{\partial}{\partial \rho}\left(\rho \frac{\partial G}{\partial \rho}\right) + \frac{1}{\rho^2}\frac{\partial^2 G}{\partial \phi^2} + \frac{\partial^2 G}{\partial z^2} \qquad \textbf{(6.40b)}$$

- Spherical polars

$$\nabla^2 G = \frac{1}{r^2}\frac{\partial}{\partial r}\left(r^2 \frac{\partial G}{\partial r}\right) + \frac{1}{r^2 \sin\theta}\frac{\partial}{\partial \theta}\left(\sin\theta \frac{\partial G}{\partial \theta}\right) + \frac{1}{r^2 \sin^2\theta}\frac{\partial^2 G}{\partial \phi^2} \qquad \textbf{(6.40c)}$$

Curl

We first note that by choosing the scalar field G in (6.25) to be each of the generalized coordinates u_1, u_2 and u_3 in turn, we obtain

$$\nabla u_1 = \frac{\hat{\mathbf{e}}_1}{h_1}, \qquad \nabla u_2 = \frac{\hat{\mathbf{e}}_2}{h_2}, \qquad \nabla u_3 = \frac{\hat{\mathbf{e}}_3}{h_3} \qquad \textbf{(6.41)}$$

Hence, if $\mathbf{A} = A_1\hat{\mathbf{e}}_1 + A_2\hat{\mathbf{e}}_2 + A_3\hat{\mathbf{e}}_3$ is a vector field expressed in an orthogonal coordinate system,

$$\mathbf{A} = A_1 h_1 \nabla u_1 + A_2 h_2 \nabla u_2 + A_3 h_3 \nabla u_3 \qquad \textbf{(6.42)}$$

Therefore

$$\nabla \wedge \mathbf{A} = \sum_{i=1}^{3} \nabla \wedge (A_i h_i \nabla u_i) = \sum_{i=1}^{3} [\nabla(A_i h_i) \wedge \nabla u_i + A_i h_i \nabla \wedge \nabla u_i]$$

where we have used a standard result for the curl (see Section 3.6). Using the fact that **curl grad** is always zero, we have

$$\nabla \wedge \mathbf{A} = \sum_{i=1}^{3} \nabla(A_i h_i) \wedge \nabla u_i = \sum_{i=1}^{3} \nabla(A_i h_i) \wedge \frac{\hat{\mathbf{e}}_i}{h_i}$$

again using (6.41). Expanding the three gradient terms on the right-hand side using (6.25),

$$\nabla \wedge \mathbf{A} = \sum_{i=1}^{3} \left[\frac{\hat{\mathbf{e}}_1}{h_1}\frac{\partial}{\partial u_1}(A_i h_i) + \frac{\hat{\mathbf{e}}_2}{h_2}\frac{\partial}{\partial u_2}(A_i h_i) + \frac{\hat{\mathbf{e}}_3}{h_3}\frac{\partial}{\partial u_3}(A_i h_i)\right] \wedge \frac{\hat{\mathbf{e}}_i}{h_i}$$

which is a total of nine terms. Three terms are immediately zero, however, because $\hat{\mathbf{e}}_i \wedge \hat{\mathbf{e}}_i = 0$. Also, $\hat{\mathbf{e}}_1 \wedge \hat{\mathbf{e}}_2 = \hat{\mathbf{e}}_3$, etc. for orthogonal coordinate systems (see Figure 6.5). Therefore

$$\begin{aligned}\nabla \wedge \mathbf{A} = &-\frac{\hat{\mathbf{e}}_3}{h_1 h_2}\frac{\partial}{\partial u_2}(A_1 h_1) + \frac{\hat{\mathbf{e}}_2}{h_1 h_3}\frac{\partial}{\partial u_3}(A_1 h_1) + \frac{\hat{\mathbf{e}}_3}{h_1 h_2}\frac{\partial}{\partial u_1}(A_2 h_2) \\ &-\frac{\hat{\mathbf{e}}_1}{h_2 h_3}\frac{\partial}{\partial u_3}(A_2 h_2) - \frac{\hat{\mathbf{e}}_2}{h_1 h_3}\frac{\partial}{\partial u_1}(A_3 h_3) + \frac{\hat{\mathbf{e}}_1}{h_2 h_3}\frac{\partial}{\partial u_2}(A_3 h_3)\end{aligned}$$

Collecting like components,

$$\nabla \wedge \mathbf{A} = \frac{\hat{\mathbf{e}}_1}{h_2 h_3}\left[\frac{\partial}{\partial u_2}(A_3 h_3) - \frac{\partial}{\partial u_3}(A_2 h_2)\right] + \frac{\hat{\mathbf{e}}_2}{h_1 h_3}\left[\frac{\partial}{\partial u_3}(A_1 h_1) - \frac{\partial}{\partial u_1}(A_3 h_3)\right] + \frac{\hat{\mathbf{e}}_3}{h_1 h_2}\left[\frac{\partial}{\partial u_1}(A_2 h_2) - \frac{\partial}{\partial u_2}(A_1 h_1)\right] \tag{6.43}$$

Equation (6.43) is the general form for the curl of a vector field **A** in orthogonal curvilinear coordinates. As already mentioned, the same expression could also be derived using the integral definition given in (5.14).

We note that (6.43) may be expressed in the more concise determinant form

$$\nabla \wedge \mathbf{A} = \frac{1}{h_1 h_2 h_3}\begin{vmatrix} h_1\hat{\mathbf{e}}_1 & h_2\hat{\mathbf{e}}_2 & h_3\hat{\mathbf{e}}_3 \\ \dfrac{\partial}{\partial u_1} & \dfrac{\partial}{\partial u_2} & \dfrac{\partial}{\partial u_3} \\ A_1 h_1 & A_2 h_2 & A_3 h_3 \end{vmatrix} \tag{6.44}$$

For the three main coordinate systems, we find immediately:

- Cartesians

$$\nabla \wedge \mathbf{A} = \hat{\mathbf{i}}\left[\frac{\partial}{\partial y}(A_z) - \frac{\partial}{\partial z}(A_y)\right] + \hat{\mathbf{j}}\left[\frac{\partial}{\partial z}(A_x) - \frac{\partial}{\partial x}(A_z)\right] + \hat{\mathbf{k}}\left[\frac{\partial}{\partial x}(A_y) - \frac{\partial}{\partial y}(A_x)\right]$$

$$\equiv \begin{vmatrix} \hat{\mathbf{i}} & \hat{\mathbf{j}} & \hat{\mathbf{k}} \\ \dfrac{\partial}{\partial x} & \dfrac{\partial}{\partial y} & \dfrac{\partial}{\partial z} \\ A_x & A_y & A_z \end{vmatrix} \tag{6.45}$$

- Cylindrical polars

$$\nabla \wedge \mathbf{A} = \hat{\boldsymbol{\rho}}\left[\frac{1}{\rho}\frac{\partial}{\partial \phi}(A_z) - \frac{\partial}{\partial z}(A_\phi)\right] + \hat{\boldsymbol{\phi}}\left[\frac{\partial}{\partial z}(A_\rho) - \frac{\partial}{\partial \rho}(A_z)\right] + \frac{1}{\rho}\hat{\mathbf{k}}\left[\frac{\partial}{\partial \rho}(\rho A_\phi) - \frac{\partial}{\partial \phi}(A_\rho)\right]$$

$$\equiv \frac{1}{\rho}\begin{vmatrix} \hat{\boldsymbol{\rho}} & \rho\hat{\boldsymbol{\phi}} & \hat{\mathbf{k}} \\ \dfrac{\partial}{\partial \rho} & \dfrac{\partial}{\partial \phi} & \dfrac{\partial}{\partial z} \\ A_\rho & \rho A_\phi & A_z \end{vmatrix} \tag{6.46}$$

- Spherical polars

$$\nabla \wedge \mathbf{A} = \frac{\hat{\mathbf{r}}}{r \sin\theta}\left[\frac{\partial}{\partial\theta}(A_\phi \sin\theta) - \frac{\partial}{\partial\phi}(A_\theta)\right]$$
$$+ \hat{\boldsymbol{\theta}}\left[\frac{1}{r\sin\theta}\frac{\partial}{\partial\phi}(A_r) - \frac{1}{r}\frac{\partial}{\partial r}(rA_\phi)\right]$$
$$+ \hat{\boldsymbol{\phi}}\left[\frac{1}{r}\frac{\partial}{\partial r}(rA_\theta) - \frac{1}{r}\frac{\partial}{\partial\theta}(A_r)\right]$$

$$\equiv \frac{1}{r^2\sin\theta}\begin{vmatrix} \hat{\mathbf{r}} & r\hat{\boldsymbol{\theta}} & r\sin\theta\,\hat{\boldsymbol{\phi}} \\ \dfrac{\partial}{\partial r} & \dfrac{\partial}{\partial\theta} & \dfrac{\partial}{\partial\phi} \\ A_r & rA_\theta & r\sin\theta\, A_\phi \end{vmatrix} \tag{6.47}$$

These rather formidable expressions for divergence, Laplacian and curl are considerably simplified for fields where symmetry is present, as we shall see in the following example.

Example 6.6

The magnetic field produced by an infinite current-carrying conductor of radius a whose axis is along the z-axis (see Figure 6.17) is given by

$$\mathbf{H} = \begin{cases} \dfrac{I\rho}{2\pi a^2}\,\hat{\boldsymbol{\phi}} & \text{if } \rho < a \\ \dfrac{I}{2\pi\rho}\,\hat{\boldsymbol{\phi}} & \text{if } \rho > a \end{cases}$$

(a) Deduce $\nabla \wedge \mathbf{H}$ for (i) $\rho < a$, and (ii) $\rho > a$.

(b) Show that $\nabla \cdot \mathbf{H} = 0$.

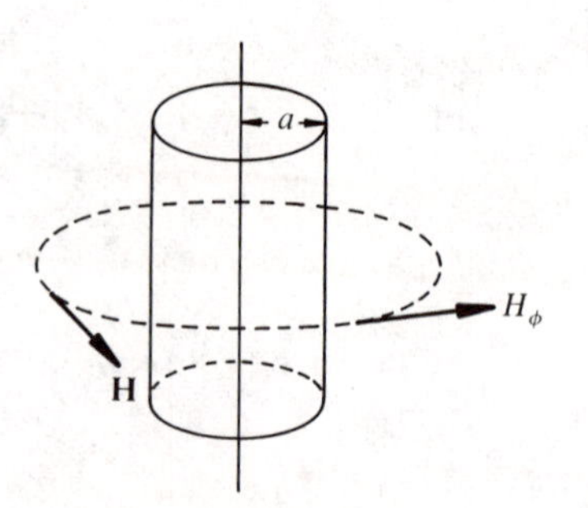

Figure 6.17

Solution

(a) As the only component of $\mathbf{H}$ is parallel to $\hat{\boldsymbol{\phi}}$ and the magnitude depends only on ρ, (6.46) reduces in both cases to

$$\nabla \wedge \mathbf{H} = \frac{1}{\rho}\begin{vmatrix} \hat{\boldsymbol{\rho}} & \rho\hat{\boldsymbol{\phi}} & \hat{\mathbf{k}} \\ \dfrac{\partial}{\partial \rho} & \dfrac{\partial}{\partial \phi} & \dfrac{\partial}{\partial z} \\ 0 & \rho H_\phi(\rho) & 0 \end{vmatrix} = \frac{1}{\rho}\frac{\mathrm{d}}{\mathrm{d}\rho}(\rho H_\phi(\rho))\hat{\mathbf{k}}$$

(i) $$\nabla \wedge \mathbf{H} = \frac{1}{\rho}\frac{\mathrm{d}}{\mathrm{d}\rho}\left(\frac{I\rho^2}{2\pi a^2}\right)\hat{\mathbf{k}} = \frac{I}{\pi a^2}\hat{\mathbf{k}}$$

(ii) $$\nabla \wedge \mathbf{H} = \frac{1}{\rho}\frac{\mathrm{d}}{\mathrm{d}\rho}\left(\frac{I\rho}{2\pi\rho}\right)\hat{\mathbf{k}} = 0$$

(b) This follows immediately from (6.37) in both cases.

EXERCISES

6.7 Find the divergence of each of the following vector fields:

(a) $\mathbf{A} = \dfrac{K}{r^n}\hat{\mathbf{r}}$ (b) $\mathbf{B} = \dfrac{K}{\rho^n}\hat{\boldsymbol{\rho}}$

In each case, find the value of n that makes the field divergenceless.

6.8 Use appropriate coordinates to deduce the Laplacian $\nabla^2\Phi$ if

(a) $\Phi = \ln\sqrt{(x^2+y^2)}$ (b) $\Phi = \ln\sqrt{(x^2+y^2+z^2)}$

6.9 Find $|\nabla \wedge \mathbf{A}|$ at the point $x=0$, $y=1$, $z=0$ if

(a) $\mathbf{A} = \rho^2 z\hat{\boldsymbol{\phi}}$ (b) $\mathbf{A} = r^2\sin\theta\,\hat{\boldsymbol{\phi}}$

6.10 Calculate $\nabla\cdot\mathbf{F}$ if

(a) $\mathbf{F} = \rho\sin\phi\,\hat{\boldsymbol{\rho}} + 2\rho\cos\phi\,\hat{\boldsymbol{\phi}} + 2z^2\hat{\mathbf{k}}$

(b) $\mathbf{F} = \dfrac{5}{r^2}\sin\theta\,\hat{\mathbf{r}} + r\cot\theta\,\hat{\boldsymbol{\theta}} + r\sin\theta\cos\phi\,\hat{\boldsymbol{\phi}}$

6.11 Calculate

(a) $\nabla \wedge \mathbf{H}$ at the origin if $\mathbf{H} = \left[\dfrac{1}{\rho} - \left(40 + \dfrac{1}{\rho}\right)\mathrm{e}^{-40\rho}\right]\hat{\boldsymbol{\phi}}$,

(b) $\nabla \wedge \mathbf{F}$ at the point $(2.0, \pi/6, 0)$ if $\mathbf{F} = 2.5\hat{\boldsymbol{\theta}} + 5.0\hat{\boldsymbol{\phi}}$.

6.12 Verify the following equations involving spherical coordinates:

(a) $\nabla\left(\frac{1}{r}\right)=\nabla\wedge(\cos\theta\,\nabla\phi)$ (b) $\nabla\phi=\nabla\wedge\left(\frac{r\,\nabla\theta}{\sin\theta}\right)$

6.13 Show that

$$\mathbf{A}\equiv\frac{x\hat{\mathbf{i}}+y\hat{\mathbf{j}}+z\hat{\mathbf{k}}}{(x^2+y^2+z^2)^{3/2}}=\frac{\hat{\mathbf{r}}}{r^2}=\frac{\rho\hat{\boldsymbol{\rho}}+z\hat{\mathbf{k}}}{(\rho^2+z^2)^{3/2}}$$

Using (a) Cartesians, (b) spherical polars, and (c) cylindrical polars, verify that $\nabla\cdot\mathbf{A}=0$ and $\nabla\wedge\mathbf{A}=0$.

6.14 A sphere of radius R and centre the origin contains charge of constant density σ. Using symmetry arguments and Gauss' theorem (see Section 4.7), show that the electric field $\mathbf{E}$ is

$$\mathbf{E}=\begin{cases}\dfrac{\sigma R^3}{3\varepsilon_0 r^2}\hat{\mathbf{r}} & \text{outside the sphere}\\[2ex] \dfrac{\sigma r}{3\varepsilon_0}\hat{\mathbf{r}} & \text{within the sphere}\end{cases}$$

ADDITIONAL EXERCISES

1 In cylindrical polar coordinates (ρ,ϕ,z) find $\nabla\wedge\mathbf{A}$ if $\mathbf{A}=f(\rho)\hat{\mathbf{k}}$, f being a differentiable function of ρ only. What is the direction of $\nabla\wedge\mathbf{A}$?

2 In spherical polar coordinates a scalar field with spherical symmetry is denoted by $f(r)$.

(a) Show that $\nabla f(r)=\dfrac{\mathrm{d}f}{\mathrm{d}r}\hat{\mathbf{r}}$.

(b) Show that $\nabla^2 f(r)=\dfrac{\mathrm{d}^2 f}{\mathrm{d}r^2}+\dfrac{2}{r}\dfrac{\mathrm{d}f}{\mathrm{d}r}$.

(c) Find explicitly the form of $f(r)$ if it is to satisfy Laplace's equation $\nabla^2 f(r)=0$.

3 Verify that the curvilinear coordinates (u,v,w) defined by the transformation equations

$$x=u^2-v^2,\qquad y=2uv,\qquad z=w$$

form an orthogonal system of coordinates. Determine the scale factors in this system.

4 In Example 6.5 we introduced elliptical cylindrical coordinates (u, v, z). Determine the form taken by gradient, divergence and curl in this coordinate system.

5 Verify the divergence theorem for the vector field $\mathbf{A} = \frac{10}{3}\rho\hat{\boldsymbol{\rho}}$ using the annulus enclosed by $\rho = 1$, $\rho = 2$, $z = 0$ and $z = 10$.

6 Verify Stokes' theorem for the vector field $\mathbf{F} = 2\rho^2(z+1)\sin^2\phi\,\hat{\boldsymbol{\phi}}$ using the cylindrical surface $\rho = 2$, $\pi/4 < \phi < \pi/2$, $1 < z < 1.5$. Draw a careful diagram of the surface of integration S and its perimeter C.

SUMMARY

- The curvilinear coordinates (u_1, u_2, u_3) are defined by the **transformation equations**

$$x = x(u_1, u_2, u_3), \quad y = y(u_1, u_2, u_3), \quad z = z(u_1, u_2, u_3)$$

together with their inverses

$$u_1 = u_1(x, y, z), \quad u_2 = u_2(x, y, z), \quad u_3 = u_3(x, y, z)$$

- The **basis vectors** of a curvilinear system are unit vectors $\hat{\mathbf{e}}_1$, $\hat{\mathbf{e}}_2$ and $\hat{\mathbf{e}}_3$ defined by

$$\hat{\mathbf{e}}_i = \frac{\partial r}{\partial u_i} \bigg/ \left|\frac{\partial \mathbf{r}}{\partial u_i}\right|, \quad i = 1, 2 \text{ or } 3$$

- The system (u_1, u_2, u_3) is **orthogonal** if $\hat{\mathbf{e}}_1$, $\hat{\mathbf{e}}_2$ and $\hat{\mathbf{e}}_3$ are mutually perpendicular.

- The **scale factors** are defined by

$$h_i = \left|\frac{\partial \mathbf{r}}{\partial u_i}\right|, \quad i = 1, 2 \text{ or } 3$$

- The **differential element of arc length** along a u_i-coordinate curve is

$$h_i\,du_i, \quad i = 1, 2 \text{ or } 3$$

- The **differential elements of area** on the coordinate surfaces $u_1 = \text{constant}$, $u_2 = \text{constant}$ and $u_3 = \text{constant}$ are

$$h_2h_3\,du_2\,du_3, \quad h_1h_3\,du_1\,du_3, \quad h_1h_2\,du_1\,du_2$$

respectively.

- The **differential element of volume** is

$$h_1 h_2 h_3 \; du_1 \, du_2 \, du_3$$

- If Φ is a scalar function and $\mathbf{A} = A_1\hat{\mathbf{e}}_1 + A_2\hat{\mathbf{e}}_2 + A_3\hat{\mathbf{e}}_3$ is a vector function of orthogonal curvilinear coordinates, then

$$\nabla\Phi = \sum_{i=1}^{3} \frac{1}{h_i}\frac{\partial \Phi}{\partial u_i}\hat{\mathbf{e}}_i$$

$$\nabla\cdot\mathbf{A} = \frac{1}{h_1h_2h_3}\left[\frac{\partial}{\partial u_1}(h_2h_3A_1) + \frac{\partial}{\partial u_2}(h_1h_3A_2) + \frac{\partial}{\partial u_3}(h_1h_2A_3)\right]$$

$$\nabla\wedge\mathbf{A} = \frac{1}{h_1h_2h_3}\begin{vmatrix} h_1\hat{\mathbf{e}}_1 & h_2\hat{\mathbf{e}}_2 & h_3\hat{\mathbf{e}}_3 \\ \dfrac{\partial}{\partial u_1} & \dfrac{\partial}{\partial u_2} & \dfrac{\partial}{\partial u_3} \\ h_1A_1 & h_2A_2 & h_3A_3 \end{vmatrix}$$

$$\nabla^2\Phi = \frac{1}{h_1h_2h_3}\left[\frac{\partial}{\partial u_1}\left(\frac{h_2h_3}{h_1}\frac{\partial\Phi}{\partial u_1}\right) + \frac{\partial}{\partial u_2}\left(\frac{h_1h_3}{h_2}\frac{\partial\Phi}{\partial u_2}\right) + \frac{\partial}{\partial u_3}\left(\frac{h_1h_2}{h_3}\frac{\partial\Phi}{\partial u_3}\right)\right]$$

- The three major orthogonal coordinate systems are:
  - Cartesians $(u_1, u_2, u_3) \to (x, y, z)$

    $h_1 = 1, \quad h_2 = 1, \quad h_3 = 1; \qquad \hat{\mathbf{e}}_1 = \hat{\mathbf{i}}, \quad \hat{\mathbf{e}}_2 = \hat{\mathbf{j}}, \quad \hat{\mathbf{e}}_3 = \hat{\mathbf{k}}$

  - Cylindrical polars $(u_1, u_2, u_3) \to (\rho, \phi, z)$

    $h_1 = 1, \quad h_2 = \rho, \quad h_3 = 1; \qquad \hat{\mathbf{e}}_1 = \hat{\boldsymbol{\rho}}, \quad \hat{\mathbf{e}}_2 = \hat{\boldsymbol{\phi}}, \quad \hat{\mathbf{e}}_3 = \hat{\mathbf{k}}$

  - Spherical polars $(u_1, u_2, u_3) \to (r, \theta, \phi)$

    $h_1 = 1, \quad h_2 = r, \quad h_3 = r\sin\theta;$

    $\hat{\mathbf{e}}_1 = \hat{\mathbf{r}}, \quad \hat{\mathbf{e}}_2 = \hat{\boldsymbol{\theta}}, \quad \hat{\mathbf{e}}_3 = \hat{\boldsymbol{\phi}}$

Applications of Vector Field Theory in Electromagnetism

PREVIEW

Here we study one particular area where the basic describing equations can be succinctly represented using the vector field concepts already developed. Although the material is of most relevance to the electrical engineer or physicist, other readers are invited to work through the chapter as it brings together many of the threads from preceding chapters, both manipulative and conceptual. Maxwell's description of electromagnetic theory displays the strength and the beauty of vector fields and was perhaps the main impetus behind the acceptance of vectors as being of widespread applicability in science and engineering.

7.1 Introduction

In some respects, our chapter heading is a misnomer. The theory of electromagnetic fields and waves is not just an area of knowledge where vector field concepts are a useful tool; rather, it is a subject where the basic experimental laws concerning electric and magnetic fields can be fully described by five equations whose most concise form utilizes the divergence and curl concepts of vector field theory. These equations have equivalent integral forms involving the line, surface and volume integrals that we have discussed in earlier chapters. It is also possible, using vector identities, to manipulate the basic equations so as mathematically to predict phenomena, such as electromagnetic waves, which are of major importance in, for example, communications engineering. Since electromagnetic field theory is also useful at the 'heavy' end of electrical engineering in connection with electrical machines, is of interest to the physicist dealing with applications of high-intensity magnetic fields and to the biologist concerned with modelling the electrical behaviour of the heart and the brain, it is clear that it is a very wide-ranging area of application for us to discuss.

It is possible to study electromagnetic phenomena without using vector analysis, but the mathematical equations quickly become long and involved and the underlying physics is easily lost. With the aid of vector field theory, however, we can readily represent the basic equations in an elegant and concise manner and, hopefully, understand their significance.

After further discussion of the continuity equation in the context of electromagnetism, our approach will be to state the basic electromagnetic equations in vector form and to use vector identities and integral theorems to interpret the equations and to extend and apply them. The reader who has little knowledge of electric and magnetic fields should still be able to read this chapter with profit, as it illustrates much of the theory from earlier chapters in a specific context.

7.2 The continuity equation again

Electric current is the flow of electric charge. The electric current across a surface S is defined as the rate at which electric charge flows across S. If ρ_V denotes the electric charge density, that is, the amount of electric charge per unit volume, and $\mathbf{v}$ is the velocity of the electric charge, then the vector field

$$\mathbf{J} = \rho_V \mathbf{v}$$

is called the **current density**.

The current I crossing a surface element ΔS in an electric conductor is given by

$$I = \rho_V |\mathbf{v}| \cos\theta \, \Delta S = \rho_V \mathbf{v} \cdot \hat{\mathbf{n}} \, \Delta S = \mathbf{J} \cdot \hat{\mathbf{n}} \, \Delta S$$

because this is the amount of charge that will cross ΔS every second (see Figure 7.1). For a non-elementary surface S, the current will then be given by

$$I = \iint_S \mathbf{J} \cdot \hat{\mathbf{n}} \, \mathrm{d}S \equiv \iint_S \mathbf{J} \cdot \mathrm{d}\mathbf{S} \tag{7.1}$$

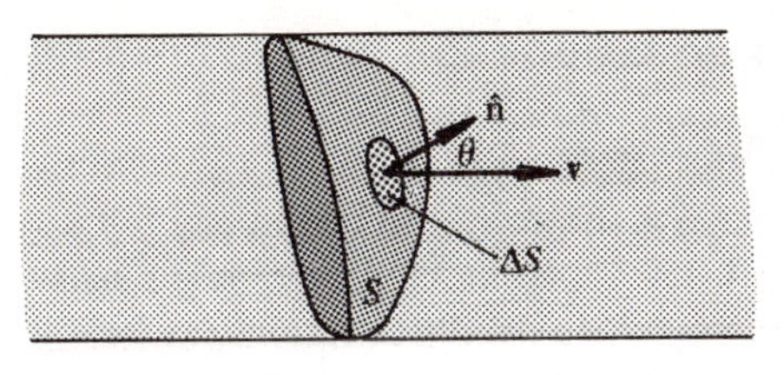

Figure 7.1

which is easily recognized as a flux integral. Clearly, the sign of I depends on the sense of the unit normal $\hat{\mathbf{n}}$. If S is a closed surface, $\hat{\mathbf{n}}$ is chosen, as previously, to be the outwardly-directed unit normal.

As in the corresponding calculation for fluid flow (see Section 5.5), the continuity equation for current flow is derived using a simple physical principle, in this case the principle of conservation of electric charge. The total electric charge Q in a volume V is given by the volume integral of the charge density,

$$Q = \iiint_V \rho_V \, \mathrm{d}V$$

so that the rate at which charge decreases in this volume is

$$-\frac{\mathrm{d}Q}{\mathrm{d}t} = -\frac{\mathrm{d}}{\mathrm{d}t} \iiint_V \rho_V \, \mathrm{d}V = -\iiint_V \frac{\partial \rho_V}{\partial t} \, \mathrm{d}V$$

But the total electric charge in V can only change if charge is flowing across the bounding surface S. Hence, we must equate the rate of flow of charge across S as given by (7.1) with the rate of decrease of charge in V. In other words,

$$\oiint_S \mathbf{J} \cdot \hat{\mathbf{n}} \, \mathrm{d}S = -\iiint_V \frac{\partial \rho_V}{\partial t} \, \mathrm{d}V$$

Transforming the left-hand side with the aid of the divergence theorem, we obtain

$$\iiint_V \nabla \cdot \mathbf{J} \, \mathrm{d}V = -\iiint_V \frac{\partial \rho_V}{\partial t} \, \mathrm{d}V$$

or

$$\iiint_V \left(\nabla \cdot \mathbf{J} + \frac{\partial \rho_V}{\partial t} \right) \mathrm{d}V = 0 \tag{7.2}$$

However, since we demand that (7.2) must hold for an arbitrary region V, the integrand must be zero at each point, that is,

$$\nabla \cdot \mathbf{J} = -\frac{\partial \rho_V}{\partial t} \tag{7.3}$$

This is the continuity equation for current flow. As we would expect, it is identical in form with the continuity equation for fluid flow.

7.3 Maxwell's equations for the electromagnetic field

As we stated in Section 7.1, we can summarize electromagnetism in five equations, one of which is the continuity equation discussed above. We shall reverse the historical development of electromagnetism and treat the other four equations, known as **Maxwell's equations**, as postulates.

Maxwell's equations involve four basic field vectors:

- the electric field **E**,
- the electric flux density **D** (called the displacement vector in older textbooks),
- the magnetic field **H**,
- the magnetic flux density **B**.

Of course, the observable physical phenomena are forces rather than fields. However, the observed fact that static electric charges exert forces on each other is ascribed to a hypothetical electric field **E** which is defined, at any point P, as the force that would be exerted on a unit charge (one coulomb) placed at P (see Figure 7.2a). Similarly, the experimental observation that current-carrying conductors exert forces, additional to electrostatic forces, on one another is ascribed to the existence of magnetic fields. Reduced to the case of the force on a single charge of unit magnitude moving with a velocity **v**, the magnetic vector **B** is defined in such a way that the observed force is $\mathbf{v} \wedge \mathbf{B}$ (see Figure 7.2b).

The field vectors **D** and **H** are related to **E** and **B** respectively by functional relationships which are characteristic of the medium. The simplest situation is where the relationships are

$$\mathbf{D} = \varepsilon \mathbf{E} \tag{7.4a}$$

and

$$\mathbf{H} = \frac{1}{\mu} \mathbf{B} \tag{7.4b}$$

ε and μ being constants, called the **permittivity** and **permeability** of the medium respectively. Formally, these relations hold for linear isotropic homogeneous media.

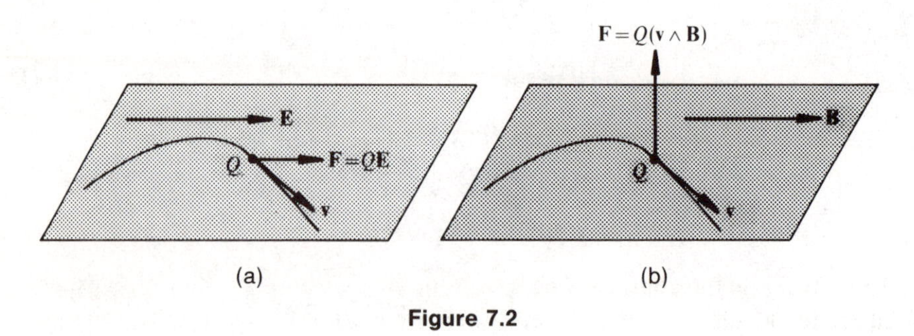

Figure 7.2

After this preamble, we can state Maxwell's equations in general form:

(1) Divergence equations:

$$\nabla \cdot \mathbf{D} = \rho_V \tag{7.5a}$$

$$\nabla \cdot \mathbf{B} = 0 \tag{7.5b}$$

(2) Curl equations:

$$\nabla \wedge \mathbf{E} = -\frac{\partial \mathbf{B}}{\partial t} \tag{7.6a}$$

$$\nabla \wedge \mathbf{H} = \mathbf{J} + \frac{\partial \mathbf{D}}{\partial t} \tag{7.6b}$$

where $\mathbf{J}$ is the electric current density, ρ_V the electric charge density and t the time.

Both the curl equations (7.6) relate an electric vector ($\mathbf{E}$ or $\mathbf{D}$) with a magnetic vector ($\mathbf{B}$ or $\mathbf{H}$), justifying the use of the term 'electromagnetic'. However, for fields which are classed as static or time-independent, the time-derivative terms are zero and the curl equations become

$$\nabla \wedge \mathbf{E} = 0 \tag{7.7a}$$

$$\nabla \wedge \mathbf{H} = \mathbf{J} \tag{7.7b}$$

That is, the electric and magnetic vectors are separate or **uncoupled** in the static situation.

The theory of static fields involves the study of equations (7.7) and also of equations (7.5), which are the same for static and time-varying fields.

The interpretation of (7.5a) for electric fields, if we recall the discussion in Chapter 5 on the divergence concept, is that positive charge is a source of the vector $\mathbf{D}$ and negative charge is a sink ($\nabla \cdot \mathbf{D} > 0$ when $\rho_V > 0$ and $\nabla \cdot \mathbf{D} < 0$ when $\rho_V < 0$). At points not occupied by electric charge, $\nabla \cdot \mathbf{D}$ is zero.

Equation (7.7a), for static fields, tells us that an electrostatic field $\mathbf{E}$ is an example of a conservative vector field, so that

$$\oint_C \mathbf{E} \cdot d\mathbf{l} = 0$$

for any closed path C. Physically, since $\mathbf{E}$ is the force on unit charge, the zero value for this line integral tells us that the work done in moving a unit charge around a closed path is always zero. It follows that we can neither gain nor lose energy by such a process. Also, since $\mathbf{E}$ is conservative, we know from Section 4.4 that we can represent $\mathbf{E}$ as the gradient of a scalar potential Φ. Thus we write

$$\mathbf{E} = -\nabla \Phi \quad \text{(static fields)} \tag{7.8}$$

The advantage of using Φ in electrostatic field problems is that it is usually easier to calculate a scalar quantity than a vector. To actually find Φ (and then **E**, using (7.8)), we must determine the equation satisfied by Φ. This is readily done using (7.5a) and assuming that the medium is such that (7.4a) holds:

$$\nabla \cdot \mathbf{D} = \nabla \cdot (\varepsilon \mathbf{E}) = \varepsilon \nabla \cdot \mathbf{E} = -\varepsilon \nabla \cdot \nabla \Phi = \rho_V$$

or

$$\nabla^2 \Phi = -\frac{\rho_V}{\varepsilon} \tag{7.9}$$

Equation (7.9), which is called **Poisson's equation**, is the relation between the potential at a point and the electric charge density at that point. In particular, at points where no charge is present ($\rho_V = 0$), the potential satisfies

$$\nabla^2 \Phi = 0 \tag{7.10}$$

which is Laplace's equation. Practical electrostatic problems often involve solving Laplace's equation in a region with known boundary conditions, for example calculating the potential and field between the conductors in a capacitor. The equation satisfied by the electric field **E** at points where $\rho_V = 0$ is, of course,

$$\nabla \cdot \mathbf{E} = 0 \tag{7.11}$$

Example 7.1

Given the vector field

$$\mathbf{F} = \frac{K}{r^n}\hat{\mathbf{r}} \qquad (r \neq 0)$$

(in spherical coordinates), deduce the value of n that makes the field solenoidal. Show that **F** is conservative for any value of n. Give an interpretation for **F** in an electrostatic context.

Solution

Using equation (6.38) for divergence in spherical coordinates

$$\nabla \cdot \mathbf{F} = \frac{1}{r^2}\frac{\mathrm{d}}{\mathrm{d}r}\left(\frac{Kr^2}{r^n}\right) = \frac{K}{r^2}(2-n)r^{1-n} = K\left(\frac{2-n}{r^{n+1}}\right)$$

which is solenoidal ($\nabla \cdot \mathbf{F} = 0$) if $n = 2$.

Using equation (6.47) for curl in spherical coordinates

$$\nabla \wedge \mathbf{F} = \frac{1}{r^2 \sin\theta} \begin{vmatrix} \hat{\mathbf{r}} & r\hat{\boldsymbol{\theta}} & r\sin\theta\,\hat{\boldsymbol{\phi}} \\ \dfrac{\partial}{\partial r} & \dfrac{\partial}{\partial \theta} & \dfrac{\partial}{\partial \phi} \\ \dfrac{K}{r^n} & 0 & 0 \end{vmatrix}$$

$$= 0 \quad \text{(all components zero)}$$

Since $\mathbf{F}$ has zero divergence when $n = 2$ and also zero curl, it follows from (7.7a) and (7.11) that

$$\mathbf{F} = \frac{K}{r^2}\hat{\mathbf{r}}$$

is a possible electrostatic field at any point $r \neq 0$. In fact, $\mathbf{F}$ represents the field $\mathbf{E}$ due to a point charge at $r = 0$, the value of the constant K depending on the system of units being employed.

Turning now to equations (7.5b) and (7.7b) for the magnetostatic field, the first of these,

$$\nabla \cdot \mathbf{B} = 0$$

tells us, as already mentioned in Section 5.4, that a magnetic field is always solenoidal, so that magnetic field lines neither start nor end. This implies that isolated magnetic poles do not exist (that is, there is no magnetic equivalent of an electric charge). The equation

$$\nabla \wedge \mathbf{H} = \mathbf{J}$$

links the magnetic field $\mathbf{H}$ and the electric current density $\mathbf{J}$ at a point (see Example 6.6 for a specific problem). Since $\nabla \wedge \mathbf{H}$ is only zero at points where $\mathbf{J} = 0$, that is outside current-carrying conductors, it is clear that $\mathbf{H}$, unlike $\mathbf{E}$, is in general a non-conservative vector field. Therefore, we cannot normally represent $\mathbf{H}$ as the gradient of a scalar potential, although a magnetic scalar potential function is occasionally used to solve magnetic field problems in regions where $\mathbf{J}$ is zero everywhere. We shall not, however, pursue the concept here.

Example 7.2

Deduce the corresponding integral forms of the Maxwell equations (7.5) and (7.7) for static fields, and interpret the results.

Solution

For the divergence equations (7.5), we integrate both sides of each equation over an arbitrary volume V bounded by a surface S and use the divergence theorem. In the electrostatic case we obtain

$$\iiint_V \nabla \cdot \mathbf{D}\, \mathrm{d}V = \iiint_V \rho_V\, \mathrm{d}V$$

giving $$\oiint_S \mathbf{D} \cdot \hat{\mathbf{n}}\, \mathrm{d}S = Q \qquad \textbf{(7.12)}$$

where $Q = \iiint_V \rho_V\, \mathrm{d}V$ is the total charge in the volume V. Equation (7.12) tells us that the electric flux over a closed surface equals the charge inside that surface. This is known as **Gauss' law**, and it can be used for finding **D** (or **E**) in field problems where a high degree of symmetry is present. We have already discussed Gauss' law in Section 4.7.

Similarly, equation (7.5b) gives

$$\iiint_V \nabla \cdot \mathbf{B}\, \mathrm{d}V = 0 \quad \text{for any volume } V$$

which transforms to

$$\oiint_S \mathbf{B} \cdot \hat{\mathbf{n}}\, \mathrm{d}S = 0 \qquad \textbf{(7.13)}$$

This is Gauss' law in magnetostatics. It tells us that the magnetic flux over *any* closed surface is zero, confirming the non-existence of sources or sinks for magnetic fields. In fact, (7.13) holds for any magnetic field, static or time-varying, since the result $\nabla \cdot \mathbf{B} = 0$ is valid in both cases.

For the curl equations (7.7), we find the flux of both sides over an arbitrary surface S bounded by a closed curve C. The magnetic field equation (7.7b) gives

$$\iint_S \nabla \wedge \mathbf{H} \cdot \hat{\mathbf{n}}\, \mathrm{d}S = \iint_S \mathbf{J} \cdot \hat{\mathbf{n}}\, \mathrm{d}S$$

or, using Stokes' theorem to transform the left-hand side,

$$\oint_C \mathbf{H} \cdot \mathrm{d}\mathbf{l} = I \qquad \textbf{(7.14)}$$

where we have also used (7.1) to identify the right-hand side as the electric current enclosed by the path C. Equation (7.14) is known as **Ampere's law in magnetostatics**; it can be used to find the magnitude of a magnetic field in a problem where a high degree of symmetry is present.

Similarly, for the electrostatic field, equation (7.7a) transforms by Stokes' theorem into

$$\oint_C \mathbf{E} \cdot \mathrm{d}\mathbf{l} = 0$$

which merely confirms that the electrostatic field is conservative.

EXERCISES

7.1 In cylindrical coordinates, the electric flux density vector **D** in a region is given by

$$\mathbf{D} = 4\rho^3\hat{\boldsymbol{\rho}} \quad \mathrm{C\,m^{-2}}$$

Verify that the total flux of **D** over the surface of the cylinder $\rho = 2$, $|z| < 5$ is equal to the total charge insider the cylinder.

7.2 If $\mathbf{D} = \frac{10}{4}\rho^3\hat{\boldsymbol{\rho}}\ \mathrm{C\,m^{-2}}$ in the region $0 \leqslant \rho \leqslant 3$ m and $\mathbf{D} = (700/\rho)\hat{\boldsymbol{\rho}}\ \mathrm{C\,m^{-2}}$ elsewhere, obtain expressions for the charge density.

7.3 In spherical coordinates, a certain charge distribution produces a flux density

$$\mathbf{D} = \begin{cases} \dfrac{Qr}{4\pi a^3}\hat{\mathbf{r}} & \text{if } r < a \\ \dfrac{Q}{4\pi r^2}\hat{\mathbf{r}} & \text{if } r > a \end{cases}$$

Determine expressions for the charge density and interpret the situation physically.

7.4 In a cylindrical conductor of radius 2 mm, the current density **J** is given by

$$\mathbf{J} = 10^3\,\mathrm{e}^{-400\rho}\,\hat{\mathbf{k}} \quad \mathrm{A\,m^{-2}}$$

Calculate the total current I in the conductor.

7.5 In spherical coordinates, a magnetic field **H** is given by

$$\mathbf{H} = 10^6 r \sin\theta\,\hat{\boldsymbol{\phi}} \quad \mathrm{A\,m^{-1}}$$

Deduce the r-component of $\nabla \wedge \mathbf{H}$, and hence the current flowing across the spherical 'cap'

$$r = 1\,\mathrm{mm}, \qquad 0 < \theta < \pi/6, \qquad 0 < \phi < 2\pi$$

7.6 Calculate the total magnetic flux crossing the $z = 0$ plane for $\rho \leqslant 5 \times 10^{-2}$ m when

$$\mathbf{B} = \frac{0.2}{\rho}\sin^2\phi\,\hat{\mathbf{k}} \quad \mathrm{Wb\,m^{-2}}$$

7.7 Show that the fields

$$\mathbf{E} = E_0 \sin x \sin t\,\hat{\mathbf{j}} \quad \text{and} \quad \mathbf{H} = \frac{E_0}{\mu_0}\cos x \cos t\,\hat{\mathbf{k}}$$

satisfy only three of the four Maxwell equations and thus are not valid solutions.

7.4 Time-varying fields and potentials

We now investigate the Maxwell equations (7.5) and (7.6) for the case where all the field vectors are functions not only of the space coordinates, say (x, y, z), but also of the time t. The divergence equations are the same as in the static situation, so there is little new to say about them at this stage.

The curl equations, however, now possess additional terms on the right-hand side, so the interpretations made for static fields need modifying. For example, the equation

$$\nabla \wedge \mathbf{E} = -\frac{\partial \mathbf{B}}{\partial t} \tag{7.15}$$

tells us immediately that the vector field **E** ceases to be conservative ($\nabla \wedge \mathbf{E} \neq 0$) when a time-varying magnetic field is present. Thus, we can no longer simply use the potential Φ as in (7.8). Before considering the question of the use of potentials for the time-varying case, let us obtain an insight into the significance of (7.15). To do this, we obtain the flux of both sides over a surface S bounded by a closed curve C, that is,

$$\iint_S \nabla \wedge \mathbf{E} \cdot \hat{\mathbf{n}}\, dS = -\iint_S \frac{\partial \mathbf{B}}{\partial t} \cdot \hat{\mathbf{n}}\, dS$$

Applying Stokes' theorem to the left-hand side,

$$\oint_C \mathbf{E} \cdot d\mathbf{l} = -\frac{d}{dt} \iint_S \mathbf{B} \cdot \hat{\mathbf{n}}\, dS \tag{7.16}$$

where we have also, as is permissible for a *fixed* path C, interchanged the order of integration over S and differentiation with respect to t. The integral on the right-hand side is the magnetic flux across S, while the left-hand side is the circulation of the non-conservative vector **E** around C. This latter quantity is usually called the **electromotive force** (e.m.f.) around C. Hence (7.16) tells us that the e.m.f. around a closed curve C equals the (negative) time rate of change of the magnetic flux across any surface S bounded by C. This is the mathematical statement of **Faraday's law** of electromagnetic induction, it being Faraday who, in 1831, first showed that a time-changing magnetic field could induce an e.m.f. (and hence an electric current if a closed conducting circuit is present).

The term $\partial \mathbf{D}/\partial t$ in the other curl equation, (7.6b), was Maxwell's main contribution to the equations that bear his name. He pointed out that this term was needed to produce consistency between equations (7.5) and (7.6) and the continuity equation. To see this, we take the divergence of both sides of (7.6b):

$$\nabla \cdot \nabla \wedge \mathbf{H} = \nabla \cdot \left(\mathbf{J} + \frac{\partial \mathbf{D}}{\partial t} \right) = \nabla \cdot \mathbf{J} + \nabla \cdot \frac{\partial \mathbf{D}}{\partial t}$$

or, since $\nabla\cdot\nabla\wedge\mathbf{F}\equiv 0$ for any vector field,

$$\nabla\cdot\mathbf{J} = -\nabla\cdot\frac{\partial\mathbf{D}}{\partial t} = -\frac{\partial}{\partial t}(\nabla\cdot\mathbf{D})$$

$$= -\frac{\partial\rho_V}{\partial t} \quad \text{using (7.5a)} \tag{7.17}$$

This is indeed the continuity equation (7.3). Clearly, the absence of the $\partial\mathbf{D}/\partial t$ term in (7.6b) would imply $\nabla\cdot\mathbf{J}\equiv 0$, which is only true in the static case.

The integral form of (7.6b) is found by the same process as used for the other Maxwell curl equation, namely integrating both sides over a surface S and converting the left-hand side to a line integral using Stokes' theorem. We obtain

$$\oint_C \mathbf{H}\cdot\mathrm{d}\mathbf{l} = \iint_S \mathbf{J}\cdot\hat{\mathbf{n}}\,\mathrm{d}S + \iint_S \frac{\partial\mathbf{D}}{\partial t}\cdot\hat{\mathbf{n}}\,\mathrm{d}S$$

$$= I + \iint_S \frac{\partial\mathbf{D}}{\partial t}\cdot\hat{\mathbf{n}}\,\mathrm{d}S \tag{7.18}$$

The second term on the right-hand side was called the displacement current by Maxwell. In the absence of any conduction current I, equation (7.18) has a pleasing symmetry with (7.16). Indeed, engineers often refer to the circulation $\oint_C \mathbf{H}\cdot\mathrm{d}\mathbf{l}$ as the **magnetomotive force**.

Example 7.3

Show that for time-varying electromagnetic fields we must use two potentials – a scalar Φ and a vector **A**. Determine equations satisfied by these potentials.

Solution

We have already pointed out that, since $\nabla\wedge\mathbf{E}$ is not identically zero in the time-varying case, we cannot put $\mathbf{E} = -\nabla\Phi$. However, the use of a magnetic vector potential **A**, which we first mentioned in Section 5.11, is still permissible because this depends only on the vanishing divergence of the magnetic vector **B**. Thus we put

$$\mathbf{B} = \nabla\wedge\mathbf{A} \tag{7.19}$$

so that (7.15) becomes

$$\nabla\wedge\mathbf{E} = -\frac{\partial}{\partial t}(\nabla\wedge\mathbf{A}) = -\nabla\wedge\left(\frac{\partial\mathbf{A}}{\partial t}\right)$$

or $$\nabla\wedge\left(\mathbf{E}+\frac{\partial\mathbf{A}}{\partial t}\right) = 0$$

In other words, the vector field $\mathbf{E} + \partial\mathbf{A}/\partial t$ *is* conservative, and we can put

$$\mathbf{E} + \frac{\partial \mathbf{A}}{\partial t} = -\nabla\Phi$$

or $$\mathbf{E} = -\frac{\partial \mathbf{A}}{\partial t} - \nabla\Phi \tag{7.20}$$

Thus, if we can determine $\mathbf{A}$ and Φ we can find $\mathbf{B}$ and $\mathbf{E}$ from (7.19) and (7.20). However, as we pointed out in Section 5.11, defining only $\nabla \wedge \mathbf{A}$ does not uniquely define $\mathbf{A}$ – by Helmholtz's theorem, we must specify $\nabla \cdot \mathbf{A}$ as well. To do this, and hence find the equations satisfied by $\mathbf{A}$ and Φ, we use the remaining Maxwell equations, (7.5a) and (7.6b), in the form

$$\nabla \cdot \mathbf{E} = \frac{\rho_V}{\varepsilon} \quad \text{and} \quad \frac{1}{\mu}\nabla \wedge \mathbf{B} = \mathbf{J} + \varepsilon\frac{\partial \mathbf{E}}{\partial t}$$

respectively. Using the potentials defined by (7.19) and (7.20), these equations become

$$\nabla \cdot \left(-\frac{\partial \mathbf{A}}{\partial t} - \nabla\Phi\right) = \frac{\rho_V}{\varepsilon} \tag{7.21}$$

and $$\frac{1}{\mu}\nabla \wedge (\nabla \wedge \mathbf{A}) = \mathbf{J} + \varepsilon\frac{\partial}{\partial t}\left(-\frac{\partial \mathbf{A}}{\partial t} - \nabla\Phi\right) \tag{7.22}$$

Interchanging the order of the differentiation with respect to time and the divergence operation in (7.21), we obtain the scalar equation

$$\frac{\partial}{\partial t}(\nabla \cdot \mathbf{A}) + \nabla \cdot \nabla\Phi = -\frac{\rho_V}{\varepsilon}$$

or $$\frac{\partial}{\partial t}(\nabla \cdot \mathbf{A}) + \nabla^2\Phi = -\frac{\rho_V}{\varepsilon} \tag{7.23}$$

(for the time-independent case, $\partial/\partial t \equiv 0$, this reduces to Poisson's equation, (7.9)).

In (7.22) we use the identity for curl curl $\mathbf{A}$ (proved in Supplement 3S.1) on the left-hand side and interchange the order of the time differentiation and gradient operations on the right-hand side to give the vector equation

$$\nabla(\nabla \cdot \mathbf{A}) - \nabla^2\mathbf{A} = \mu\mathbf{J} - \mu\varepsilon\frac{\partial^2 \mathbf{A}}{\partial t^2} - \mu\varepsilon\,\nabla\left(\frac{\partial \Phi}{\partial t}\right) \tag{7.24}$$

This is a formidable equation so we now make our delayed choice of $\nabla \cdot \mathbf{A}$ in such a way as to simplify it. We choose

$$\nabla \cdot \mathbf{A} = -\mu\varepsilon\frac{\partial \Phi}{\partial t}$$

(which is sometimes called the **Lorentz condition**) so that the gradients of these two scalar quantities are the same, and hence (7.24) reduces to

$$\nabla^2 \mathbf{A} = -\mu \mathbf{J} + \mu\varepsilon \frac{\partial^2 \mathbf{A}}{\partial t^2} \tag{7.25}$$

Similarly, for this choice of $\nabla \cdot \mathbf{A}$, (7.23) becomes

$$\nabla^2 \Phi = -\frac{\rho_V}{\varepsilon} + \mu\varepsilon \frac{\partial^2 \Phi}{\partial t^2} \tag{7.26}$$

These are the governing equations for the potentials $\mathbf{A}$ and Φ. Moreover, (7.25) is a vector equation, each of whose three components has the same form as (7.26). This can be of considerable use when the solution of these equations is sought; however, this is beyond the scope of the present text.

7.5 The Poynting vector

The Poynting vector $\mathbf{P}$ at any point in an electromagnetic field is defined as the cross product

$$\mathbf{P} = \mathbf{E} \wedge \mathbf{H} \tag{7.27}$$

The divergence of this vector turns out to give us results of considerable significance in electrical engineering. We have

$$\begin{aligned}\nabla \cdot \mathbf{P} &= \nabla \cdot (\mathbf{E} \wedge \mathbf{H}) \\ &= \mathbf{H} \cdot (\nabla \wedge \mathbf{E}) - \mathbf{E} \cdot (\nabla \wedge \mathbf{H})\end{aligned}$$

by a standard identity (see Section 3.6). Using the Maxwell curl equations (7.6),

$$\nabla \cdot \mathbf{P} = \mathbf{H} \cdot \left(-\frac{\partial \mathbf{B}}{\partial t} \right) - \mathbf{E} \cdot \left(\mathbf{J} + \frac{\partial \mathbf{D}}{\partial t} \right)$$

$$\therefore \quad \nabla \cdot \mathbf{P} + \mathbf{H} \cdot \frac{\partial \mathbf{B}}{\partial t} + \mathbf{E} \cdot \frac{\partial \mathbf{D}}{\partial t} = -\mathbf{E} \cdot \mathbf{J}$$

or, assuming that equations (7.4) hold,

$$\nabla \cdot \mathbf{P} + \mu \mathbf{H} \cdot \frac{\partial \mathbf{H}}{\partial t} + \varepsilon \mathbf{E} \cdot \frac{\partial \mathbf{E}}{\partial t} = -\mathbf{E} \cdot \mathbf{J} \tag{7.28}$$

But we saw in Section 2.5 that for any vectors $\mathbf{F}$ and $\mathbf{G}$,

$$\frac{\partial}{\partial t}(\mathbf{F} \cdot \mathbf{G}) = \mathbf{F} \cdot \frac{\partial \mathbf{G}}{\partial t} + \mathbf{G} \cdot \frac{\partial \mathbf{F}}{\partial t}$$

or $$\frac{\partial}{\partial t}(\mathbf{F} \cdot \mathbf{F}) = \frac{\partial}{\partial t}(|\mathbf{F}|^2) = 2\mathbf{F} \cdot \frac{\partial \mathbf{F}}{\partial t}$$

Using this result twice in (7.28),

$$\nabla \cdot \mathbf{P} + \tfrac{1}{2}\mu \frac{\partial}{\partial t}(|\mathbf{H}|^2) + \tfrac{1}{2}\varepsilon \frac{\partial}{\partial t}(|\mathbf{E}|^2) = -\mathbf{E} \cdot \mathbf{J}$$

or
$$\nabla \cdot \mathbf{P} + \frac{\partial \mathscr{E}}{\partial t} = -\mathbf{E} \cdot \mathbf{J} \qquad \textbf{(7.29)}$$

where $\mathscr{E} = \frac{1}{2}\mu|\mathbf{H}|^2 + \frac{1}{2}\varepsilon|\mathbf{E}|^2$ is known, for reasons that we do not have space to discuss, as the **energy density** (energy per unit volume) in the electromagnetic field.

In the special case of a perfect insulator, the current density $\mathbf{J}$ would be zero, in which case (7.29) reduces to

$$\nabla \cdot \mathbf{P} = -\frac{\partial \mathscr{E}}{\partial t} \qquad \textbf{(7.30)}$$

which has exactly the same form as the continuity equation

$$\nabla \cdot \mathbf{J} = -\frac{\partial \rho_V}{\partial t}$$

This similarity suggests that just as the current density vector $\mathbf{J}$ is associated with flow of electric charge of density ρ_V, so the Poynting vector $\mathbf{P}$ can be associated with flow of energy of density $\mathscr{E}$. Indeed, we can obtain an energy balance equation from (7.30) by integrating both sides over a volume V:

$$\iiint_V \nabla \cdot \mathbf{P} \, \mathrm{d}V = -\iiint_V \frac{\partial \mathscr{E}}{\partial t} \, \mathrm{d}V$$

or
$$\oiint_S \mathbf{P} \cdot \hat{\mathbf{n}} \, \mathrm{d}S = -\frac{\partial}{\partial t} \iiint_V \mathscr{E} \, \mathrm{d}V \qquad \textbf{(7.31)}$$

Using the interpretation of $\mathscr{E}$ as an energy density, the right-hand side of (7.31) is the rate of change of electromagnetic energy within the volume V. Hence the left-hand side must be the rate at which energy crosses the surface S. Thus the Poynting vector $\mathbf{P}$ has the significance of being a **power density**, provided that the interpretation implied in (7.31) is used, namely that the flux integral of $\mathbf{P}$ gives us the power crossing S.

If we integrate (7.29) over the volume V, by a similar process, we obtain, using the divergence theorem immediately,

$$\oiint_S \mathbf{P} \cdot \hat{\mathbf{n}} \, \mathrm{d}S = -\frac{\partial}{\partial t} \iiint_V \mathscr{E} \, \mathrm{d}V - \iiint_V \mathbf{E} \cdot \mathbf{J} \, \mathrm{d}V$$

or
$$-\frac{\partial}{\partial t} \iiint_V \mathscr{E} \, \mathrm{d}V = \oiint_S \mathbf{P} \cdot \hat{\mathbf{n}} \, \mathrm{d}S + \iiint_V \mathbf{E} \cdot \mathbf{J} \, \mathrm{d}V \qquad \textbf{(7.32)}$$

Since the left-hand side here is the rate of loss of energy within V and the first term on the right-hand side is the power (or rate of energy flow) outwards across S, it follows that the second term must also contribute to a loss of electromagnetic energy. It is in fact the rate at which energy is *dissipated* by the passage of current – the Joule heating loss.

For many materials, the current density vector $\mathbf{J}$ is related to the electric field $\mathbf{E}$ by

$$\mathbf{J} = \sigma \mathbf{E} \tag{7.33}$$

where σ is called the electrical conductivity of the medium. Equation (7.33) is, in fact, a field form of the well-known Ohm's law. It follows from (7.32) and (7.33) that the Joule heating loss occurs at a rate $\sigma|\mathbf{E}|^2$ per unit volume.

7.6 Wave solutions of the Maxwell equations

One consequence of the Maxwell equations for time-varying electromagnetic fields is the prediction that variations in the electric and magnetic fields $\mathbf{E}$ and $\mathbf{H}$ propagate as waves. The formal analysis leading to this prediction is readily carried out using the tools of vector field theory. The experimental verification of the predictions by Hertz was powerful evidence for the validity of the Maxwell equations.

The simplest model to consider is where the Maxwell equations (7.5) and (7.6) take on the forms

$$\nabla \cdot \mathbf{D} = 0$$

$$\nabla \cdot \mathbf{B} = 0$$

$$\nabla \wedge \mathbf{E} = -\frac{\partial \mathbf{B}}{\partial t}$$

$$\nabla \wedge \mathbf{H} = \frac{\partial \mathbf{D}}{\partial t}$$

or, using the linear relationships (7.4) to eliminate $\mathbf{D}$ and $\mathbf{B}$,

$$\nabla \cdot \mathbf{E} = 0 \tag{7.34a}$$

$$\nabla \cdot \mathbf{H} = 0 \tag{7.34b}$$

$$\nabla \wedge \mathbf{E} = -\mu \frac{\partial \mathbf{H}}{\partial t} \tag{7.35a}$$

$$\nabla \wedge \mathbf{H} = \varepsilon \frac{\partial \mathbf{E}}{\partial t} \tag{7.35b}$$

In other words, we are considering a linear homogeneous medium with no electric charges or electric currents at any point. The symmetry of the two divergence equations (7.34) and of the two curl equations (7.35) is readily seen.

To obtain predictions of wave behaviour, we take the **curl** of both sides of equation (7.35a) and use the identity for **curl curl A**. We obtain

$$\nabla(\nabla \cdot \mathbf{E}) - \nabla^2 \mathbf{E} = -\mu \frac{\partial}{\partial t} (\nabla \wedge \mathbf{H})$$

or
$$\nabla^2 \mathbf{E} = \mu\varepsilon \frac{\partial^2 \mathbf{E}}{\partial t^2} \qquad \textbf{(7.36)}$$

using (7.34a) and (7.35b). A similar procedure using the curl of equation (7.35b) leads to

$$\nabla^2 \mathbf{H} = \mu\varepsilon \frac{\partial^2 \mathbf{H}}{\partial t^2} \qquad \textbf{(7.37)}$$

Equations (7.36) and (7.37) are both vector equations and are thus each equivalent to three scalar equations. In Cartesian coordinates, these equations are

$$\nabla^2 E_x = \mu\varepsilon \frac{\partial^2 E_x}{\partial t^2} \quad \text{and} \quad \nabla^2 H_x = \mu\varepsilon \frac{\partial^2 H_x}{\partial t^2} \qquad \textbf{(7.38)}$$

with similar relationships for the y- and z-components. Equations (7.38) are known as **wave equations**.

Many different solutions of the wave equations are possible. We consider here only the simplest solutions, which are referred to as **plane waves**. In this case, the magnitudes of the vectors **E** and **H** are assumed to be constant at all points on a plane perpendicular to a given direction, which we shall take to be the z-axis of a Cartesian coordinate system. Mathematically, this implies that all partial derivatives with respect to x and y may be set equal to zero.

The Maxwell equation $\nabla \cdot \mathbf{E} = 0$ reduces, for this case, to $\partial E_z/\partial z = 0$, which tells us that E_z does not vary with z. Hence E_z is constant in space and we can, without loss of generality, take it to be zero. Hence we need only consider an electric field vector of the form

$$\mathbf{E}(z, t) = E_x(z, t)\hat{\mathbf{i}} + E_y(z, t)\hat{\mathbf{j}}$$

and, by similar reasoning since $\nabla \cdot \mathbf{H} = 0$, we assume a magnetic field vector

$$\mathbf{H}(z, t) = H_x(z, t)\hat{\mathbf{i}} + H_y(z, t)\hat{\mathbf{j}}$$

The equations satisfied by the components of **E** and **H** are, using (7.38),

$$\frac{\partial^2 E_x}{\partial z^2} = \mu\varepsilon \frac{\partial^2 E_x}{\partial t^2} \qquad \textbf{(7.39a)}$$

$$\frac{\partial^2 H_x}{\partial z^2} = \mu\varepsilon \frac{\partial^2 H_x}{\partial t^2} \qquad \textbf{(7.39b)}$$

and similarly for the y-components. These equations are all examples of the *one-dimensional* wave equation.

It is easy to show, by substitution, that

$$E_x = E_{x0}\, f(z - vt) \tag{7.40}$$

(where $v = 1/\sqrt{(\mu\varepsilon)}$) is one possible solution of equation (7.39a), where f is any well-behaved (differentiable) function of the variable $(z - vt)$ and E_{x0} is a constant. Physically, such a solution represents a wave travelling without change of shape in the positive z-direction with a speed v (see Figure 7.3). Similarly, one solution for the y-component of $\mathbf{E}$ is

$$E_y = E_{y0}\, f(z - vt)$$

We see that the **E** vector in a plane electromagnetic wave is normal or *transverse* to the direction in which the wave is propagating.

We now calculate the corresponding components of **H** using the Maxwell equation (7.35a) and the Cartesian form of the **curl**:

$$\boldsymbol{\nabla} \wedge \mathbf{E} = -\frac{\partial E_y}{\partial z}\hat{\mathbf{i}} + \frac{\partial E_x}{\partial z}\hat{\mathbf{j}}$$

Hence, since

$$\frac{\partial \mathbf{H}}{\partial t} = \frac{\partial H_x}{\partial t}\hat{\mathbf{i}} + \frac{\partial H_y}{\partial t}\hat{\mathbf{j}}$$

we obtain, on comparing components,

$$\frac{\partial E_y}{\partial z} = \mu \frac{\partial H_x}{\partial t} \tag{7.41a}$$

$$\frac{\partial E_x}{\partial z} = -\mu \frac{\partial H_y}{\partial t} \tag{7.41b}$$

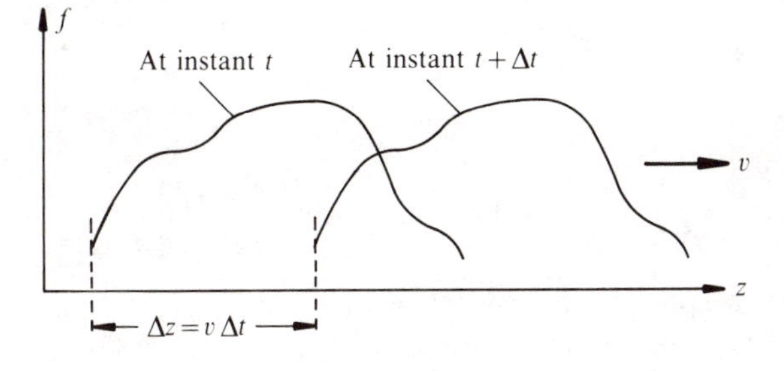

Figure 7.3

But, from (7.40),

$$\frac{\partial E_x}{\partial z} = \frac{\partial E_x}{\partial (z - vt)} \frac{\partial (z - vt)}{\partial z} = \frac{\partial E_x}{\partial (z - vt)}$$

and
$$\frac{\partial E_x}{\partial t} = \frac{\partial E_x}{\partial (z - vt)} \frac{\partial (z - vt)}{\partial t} = -v \frac{\partial E_x}{\partial (z - vt)}$$

from which

$$\frac{\partial E_x}{\partial z} = -\frac{1}{v} \frac{\partial E_x}{\partial t}$$

Hence (7.41b) becomes

$$\frac{\partial H_y}{\partial t} = \frac{1}{\mu v} \frac{\partial E_x}{\partial t} = \sqrt{\left(\frac{\varepsilon}{\mu}\right)} \frac{\partial E_x}{\partial t}$$

and integrating,

$$H_y = \sqrt{\left(\frac{\varepsilon}{\mu}\right)} E_x = \sqrt{\left(\frac{\varepsilon}{\mu}\right)} E_{x0} f(z - vt)$$

Similar reasoning applied to (7.41a) shows that

$$H_x = -\sqrt{\left(\frac{\varepsilon}{\mu}\right)} E_y = -\sqrt{\left(\frac{\varepsilon}{\mu}\right)} E_{y0} f(z - vt)$$

We see that variations of **H**, like those of **E**, propagate as transverse travelling waves with speed $v = 1/\sqrt{(\mu\varepsilon)}$.

The ratio

$$\frac{|\mathbf{E}|}{|\mathbf{H}|} = \frac{\sqrt{(E_x^2 + E_y^2)}}{\sqrt{(H_x^2 + H_y^2)}} = \frac{\sqrt{(E_x^2 + E_y^2)}}{\sqrt{\left(\frac{\varepsilon}{\mu} E_x^2 + \frac{\varepsilon}{\mu} E_y^2\right)}} = \sqrt{\left(\frac{\mu}{\varepsilon}\right)}$$

depends, like the speed, on the electrical constants μ and ε. This ratio is referred to as the **characteristic impedance** Z of the medium of propagation. For vacuum, or free space, the numerical values are

$$\varepsilon \equiv \varepsilon_0 = 8.85 \times 10^{-12}\ \mathrm{F\ m^{-1}} \quad \text{and} \quad \mu \equiv \mu_0 = 4\pi \times 10^{-7}\ \mathrm{H\ m^{-1}}$$

from which

$$Z \equiv Z_0 = \sqrt{\left(\frac{\mu_0}{\varepsilon_0}\right)} \approx 120\pi \text{ ohms} \quad \text{and} \quad v = \frac{1}{\sqrt{(\mu\varepsilon)}} \approx 3 \times 10^8\ \mathrm{m\ s^{-1}}$$

We see that the predicted speed of propagation v of electromagnetic waves in free space is equal to the experimental speed of light. This result was early evidence that light waves are electromagnetic in nature.

Finally, we can readily show that **E** and **H** are mutually perpendicular vectors and that **E** ∧ **H** is a vector in the direction of propagation of the electromagnetic wave. The former is shown by noting that

$$\mathbf{E}\cdot\mathbf{H} = E_x H_x + E_y H_y = -E_x\sqrt{\left(\frac{\varepsilon}{\mu}\right)}E_y + E_y\sqrt{\left(\frac{\varepsilon}{\mu}\right)}E_x = 0$$

while
$$\mathbf{E}\wedge\mathbf{H} = \begin{vmatrix} \hat{\mathbf{i}} & \hat{\mathbf{j}} & \hat{\mathbf{k}} \\ E_x & E_y & 0 \\ H_x & H_y & 0 \end{vmatrix}$$

$$= (E_x H_y - E_y H_x)\hat{\mathbf{k}}$$

$$= (E_x^2 + E_y^2)\sqrt{\left(\frac{\varepsilon}{\mu}\right)}\hat{\mathbf{k}}$$

$$= \frac{|\mathbf{E}|^2}{Z}\hat{\mathbf{k}}$$

so that **E** ∧ **H** is indeed directed along the propagation direction (positive z).

We recall, from Section 7.5, that **E** ∧ **H** is the Poynting vector **P**, whose direction gives us the direction of flow of electromagnetic energy. We have proved therefore, as we would expect, that a travelling plane wave transports energy in the direction in which it is propagating. A diagram of the vectors in a plane electromagnetic wave (for the case where E_y, and hence H_x, are zero) is shown in Figure 7.4.

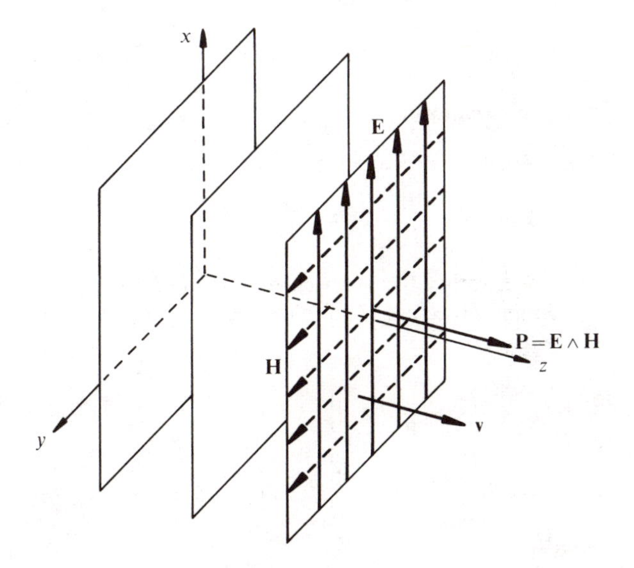

Figure 7.4

Example 7.4

(a) A plane sinusoidal electromagnetic wave in free space has an electric field vector

$$\mathbf{E} = E_0 \cos(\beta z - \omega t)\,\hat{\mathbf{i}}$$

Deduce the corresponding magnetic field $\mathbf{H}$ and show that the phase velocity of the wave is given by $v = 1/\sqrt{(\mu_0 \varepsilon_0)}$.

(b) If $E_0 = 200\pi$ V m^{-1}, find the amplitude of the magnetic field vector and the value of the angular frequency ω if the wavelength is 1.89 m.

Solution

In this problem we are considering perhaps the most important type of plane wave, namely one where the general function f of the earlier theory is sinusoidal, so that instead of $f(z - vt)$ we have $\cos 2\pi(z - vt)/\lambda$, where λ is called the **wavelength**. The name arises because, for a given t,

$$\cos\frac{2\pi}{\lambda}[(z+\lambda) - vt] = \cos\frac{2\pi}{\lambda}(z - vt)$$

that is, the wave repeats after a distance λ. Also if, at a given value of z, t increases by λ/v,

$$\cos\frac{2\pi}{\lambda}\left[z - v\left(t + \frac{\lambda}{v}\right)\right] = \cos\left[\frac{2\pi}{\lambda}(z - vt) - 2\pi\right] = \cos\frac{2\pi}{\lambda}(z - vt)$$

That is, the period T of the oscillation at a point is λ/v and the frequency f is $1/T = v/\lambda$. The **angular frequency** ω is defined as $2\pi f$. If we also put $2\pi/\lambda = \beta$, we obtain

$$\cos\frac{2\pi}{\lambda}(z - vt) = \cos(\beta z - \omega t) \quad \text{(or, equivalently, } \cos(\omega t - \beta z))$$

which is the given form.

(a) We see that, for the given field $\mathbf{E}$,

$$\nabla \cdot \mathbf{E} = \frac{\partial E_x}{\partial x} + \frac{\partial E_y}{\partial y} + \frac{\partial E_z}{\partial z} = \frac{\partial E_x}{\partial x} = 0$$

so that $\mathbf{E}$ does satisfy the Maxwell equation (7.34a).

To find $\mathbf{H}$ we use (7.35a), that is,

$$\frac{\partial \mathbf{H}}{\partial t} = -\frac{1}{\mu}\nabla \wedge \mathbf{E} = -\frac{1}{\mu}\begin{vmatrix} \hat{\mathbf{i}} & \hat{\mathbf{j}} & \hat{\mathbf{k}} \\ \dfrac{\partial}{\partial x} & \dfrac{\partial}{\partial y} & \dfrac{\partial}{\partial z} \\ E_0\cos(\omega t - \beta z) & 0 & 0 \end{vmatrix}$$

$$= -\frac{E_0}{\mu}\beta \sin(\omega t - \beta z)\,\hat{\mathbf{j}}$$

Integrating with respect to t,

$$\mathbf{H} = \frac{\beta E_0}{\omega\mu}\cos(\omega t - \beta z)\,\hat{\mathbf{j}}$$

$$= H_0 \cos(\omega t - \beta z)\,\hat{\mathbf{j}}, \quad \text{say}$$

Thus, as in the general theory, we have an **H** vector with a y-component associated with an **E** vector with an x-component, both transverse to the direction of propagation.

Clearly,

$$\nabla \cdot \mathbf{H} = \frac{\partial H_y}{\partial y} = 0$$

that is, **H** satisfies (7.34b). Also

$$\nabla \wedge \mathbf{H} = \begin{vmatrix} \hat{\mathbf{i}} & \hat{\mathbf{j}} & \hat{\mathbf{k}} \\ \dfrac{\partial}{\partial x} & \dfrac{\partial}{\partial y} & \dfrac{\partial}{\partial z} \\ 0 & H_0 \cos(\omega t - \beta z) & 0 \end{vmatrix}$$

$$= \beta H_0 \sin(\omega t - \beta z)\,\hat{\mathbf{i}}$$

But $\partial \mathbf{E}/\partial t = \omega E_0 \sin(\omega t - \beta z)\,\hat{\mathbf{i}}$, so that for the final Maxwell equation (7.35b) to be satisfied we must have

$$\beta H_0 = \frac{\beta^2 E_0}{\omega\mu} = \varepsilon_0 \omega E_0$$

or
$$\frac{\omega^2}{\beta^2} = \frac{1}{\varepsilon_0 \mu_0} \tag{7.42}$$

But for a sinusoidal wave of the form

$$\mathbf{E} = E_0 \cos(\omega t - \beta z)\,\hat{\mathbf{i}}$$

the **phase velocity** is determined by putting

$$\omega t - \beta z = \text{constant} \quad \text{or} \quad \frac{\mathrm{d}z}{\mathrm{d}t} = \frac{\omega}{\beta}$$

But we have shown in the general theory that $1/\sqrt{(\mu_0 \varepsilon_0)}$ is the speed of a plane electromagnetic wave in free space, so that (7.42) is valid and (7.35b) is indeed satisfied by our calculated field **H**.

(b) We know from the general theory that the amplitudes of the **E** and **H** vectors are linked in free space by

$$\frac{E_0}{H_0} = Z_0 = 120\pi \text{ ohms}$$

Here, $H_0 = 200\pi/120\pi = 1.667\ \mathrm{A\ m^{-1}}$.

The frequency of the sinusoidal oscillations is

$$f = \frac{v}{\lambda} = \frac{3 \times 10^8}{1.89} = 1.587 \times 10^8 \text{ Hz}$$

so the required angular frequency is

$$\omega = 2\pi f = 9.97 \times 10^8 \text{ rad s}^{-1}$$

In this section we have demonstrated the properties of simple electromagnetic wave solutions of Maxwell's equations. We have assumed that the propagation medium is of infinite extent and is a perfect insulator. The interested reader should refer to standard texts on electromagnetic waves for information on such topics as waves in non-ideal insulators, reflection and refraction of waves at boundaries, guided waves and sources of electromagnetic radiation (antenna theory). A sound knowledge of vector fields should aid the understanding of all these specialist areas.

EXERCISES

7.8 In free space, a time-dependent magnetic vector potential $\mathbf{A} = \exp(-2\rho)\,\hat{\mathbf{k}}$ is present. Calculate expressions for the vector fields **B**, **H** and **J**. Calculate also the total current crossing the $z = 0$ plane between $\rho = 0$ and $\rho = 10$ mm.

7.9 An electric charge of 3 C is moved from the point P_1 (1, 2, 0) to the point P_2 (2, 8, 0) (distances are in metres). Find the work done if the electric field is

$$\mathbf{E} = 4xy\hat{\mathbf{i}} + 2x^2\hat{\mathbf{j}} \quad \text{V m}^{-1}$$

(a) by direct integration, and

(b) by finding the associated scalar potential Φ.

7.10 Electric charge is distributed within a spherical region of radius a such that the electric field is

$$\mathbf{E} = Ar^3\hat{\mathbf{r}} \quad \text{V m}^{-1}$$

where A is a constant. Calculate

(a) the total charge within the sphere,

(b) the total electrostatic energy stored within this region. (Recall, from Section 7.5, that $\frac{1}{2}\varepsilon_0|\mathbf{E}|^2$ is the electrostatic energy density.)

7.11 The fields in a spherical wave at a large distance r from a dipole antenna in free space are, using spherical coordinates,

$$\mathbf{E} = \frac{150}{r} \sin\theta \cos(\omega t - \beta r)\,\hat{\boldsymbol{\theta}} \quad \text{V m}^{-1}$$

$$\mathbf{H} = \frac{0.2}{r} \sin\theta \cos(\omega t - \beta r)\,\hat{\boldsymbol{\phi}} \quad \text{A m}^{-1}$$

Calculate the instantaneous and average powers crossing a hemisphere $0 < \theta < \pi/2$ at a distance of 1 km from the antenna.

7.12 Show that the average value of the Poynting vector $\mathbf{E} \wedge \mathbf{H}$ for sinusoidal plane waves in a perfect dielectric is $E_0^2/2Z$, where E_0 is the amplitude of the $\mathbf{E}$ vector and $Z = \sqrt{(\mu/\varepsilon)}$. (*Hint:* see Example 7.4.)

7.13 Show, by substitution, that the fields

$$\mathbf{E} = E_0 \cos(\omega t + \beta z)\,\hat{\mathbf{i}} \quad \text{and} \quad \mathbf{H} = -\frac{E_0}{Z} \cos(\omega t + \beta z)\,\hat{\mathbf{j}}$$

(where $Z = \sqrt{(\mu/\varepsilon)}$) satisfy Maxwell's four equations for a perfect charge-free dielectric characterized by μ and ε.

7.14 In free space, an electromagnetic wave has an $\mathbf{E}$ vector

$$\mathbf{E}(z, t) = 150 \sin(\omega t - \beta z)\,\hat{\mathbf{i}} \quad \text{V m}^{-1}$$

Calculate the total average power passing through a rectangular area of sides 30 mm and 15 mm in the $z = 0$ plane.

ADDITIONAL EXERCISES

1 Using spherical coordinates, the flux density $\mathbf{D}$ in a region is given by

$$\mathbf{D} = \frac{0.1}{r} \cos\theta\,\hat{\boldsymbol{\theta}} \quad \text{C m}^{-2}$$

Calculate the total electric charge in the truncated cone defined by $2 \leqslant r \leqslant 5$ m, $0 < \theta < \pi/4$, $0 \leqslant \phi \leqslant 2\pi$.

2 A cylindrical conductor of radius 1 cm has internal magnetic flux density

$$\mathbf{B} = 0.06\left(\frac{\rho}{2} - \frac{\rho^2}{3 \times 10^{-2}}\right)\hat{\boldsymbol{\phi}} \quad \text{Wb m}^{-2}$$

Calculate the total current in the conductor using (a) the differential form, and (b) the integral form of Ampere's law. (Assume that $\mathbf{B} = \mu_0 \mathbf{H}$, where $\mu_0 = 4\pi \times 10^{-7}$ units.)

3 A cylindrical conductor of radius a, with its axis along the z-axis, has a current density $\mathbf{J} = J_0(\rho/a)^n \hat{\mathbf{k}}$, where J_0 and n ($\geqslant 1$) are constants. Deduce the magnetic field $\mathbf{B}(\rho)$ inside the conductor, using the integral form of Ampere's law, and find its curl. Verify that $\nabla \wedge \mathbf{B} = \mu_0 \mathbf{J}$. Given that the magnetic vector potential $\mathbf{A}$ is parallel to the current density vector, find an expression for $\mathbf{A}(\rho)$ inside the conductor. Verify that $\nabla^2 \mathbf{A} = -\mu_0 \mathbf{J}$.

4 The displacement current density in a perfect dielectric with $\varepsilon = 4\varepsilon_0$ and $\mu = 5\mu_0$ is given by

$$\frac{\partial \mathbf{D}}{\partial t} = 2\cos(\omega t - 5z)\,\hat{\mathbf{i}} \quad \mu\text{A m}^{-2}$$

Find expressions for the fields $\mathbf{D}$, $\mathbf{E}$, $\mathbf{B}$ and $\mathbf{H}$, ignoring all constant terms. Recalculate $\partial \mathbf{D}/\partial t$, and hence find the value of ω.

SUMMARY

- Electromagnetic fields are governed by a set of four equations, called **Maxwell's equations**, and a **continuity equation**. The equations are

$$\nabla \cdot \mathbf{D} = \rho_V, \quad \nabla \cdot \mathbf{B} = 0, \quad \nabla \wedge \mathbf{E} = -\frac{\partial \mathbf{B}}{\partial t},$$

$$\nabla \wedge \mathbf{H} = \mathbf{J} + \frac{\partial \mathbf{D}}{\partial t}, \quad \text{and} \quad \nabla \cdot \mathbf{J} = -\frac{\partial \rho_V}{\partial t}$$

- The relationship between the vectors depends on the medium under consideration. The simplest relationships are

$$\mathbf{D} = \varepsilon \mathbf{E}, \quad \mathbf{B} = \mu \mathbf{H} \quad \text{and} \quad \mathbf{J} = \sigma \mathbf{E}$$

where ε, μ and σ are constants.

- For static fields ($\partial/\partial t \equiv 0$), the Maxwell equations 'decouple' into two equations for electric vectors and two for magnetic vectors. A static electric field $\mathbf{E}$ is always conservative ($\nabla \wedge \mathbf{E} = 0$) and the scalar potential Φ is a useful concept. A static magnetic field $\mathbf{H}$ is only conservative in a region where $\mathbf{J} \equiv 0$.

- For time-varying fields, we can use a scalar potential Φ and a (magnetic) vector potential **A**, where

$$\mathbf{B} = \nabla \wedge \mathbf{A} \quad \text{and} \quad \mathbf{E} = -\frac{\partial \mathbf{A}}{\partial t} - \nabla \Phi$$

- The rate at which energy crosses a surface S in an electromagnetic field is given by the flux integral $\oiint_S \mathbf{P} \cdot \hat{\mathbf{n}} \, dS$, where $\mathbf{P} = \mathbf{E} \wedge \mathbf{H}$ is called the **Poynting vector**.

- Maxwell's equations predict that variations in **E** and **H** can propagate as waves where the fields are transverse to the direction of propagation. In an ideal dielectric, the speed of propagation of the waves is given by $1/\sqrt{(\varepsilon\mu)}$. In free space, this quantity has the same numerical value as the speed of light.

Vectors and Rotation

PREVIEW

In this chapter, we explore the applications of vectors to 'rotation' in its many guises. We consider infinitesimal and finite rotations and also study the 'effects' of rotating coordinate systems in mechanics. The material is independent of preceding chapters apart from Chapters 1 and 2.

8.1 Finite rotations

A single finite rotation from P to P' (see Figure 8.1) is completely specified by

- the direction and location in space of the axis of rotation, and
- the angle θ turned through.

We use the word 'finite' when θ is not small (usually $\theta > 5°$). A natural question to ask is whether a finite rotation can be represented by a vector. It is tempting to think that such a representation is possible – after all, rotation 'seems' vectorial in nature and it is uniquely characterized by three pieces of information.

Whether or not a finite rotation can be represented by a vector is easily checked. We know that vector addition is commutative, that is, if **a** and **b** are vectors, then

$$\mathbf{a} + \mathbf{b} = \mathbf{b} + \mathbf{a}$$

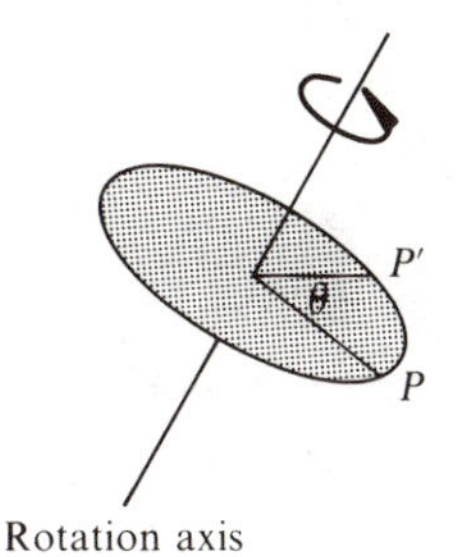

Figure 8.1

We investigate, therefore, whether a similar result holds for rotations: if we perform two successive rotations R_1 and R_2 in that order, do we get the same result as if we apply R_2 first and then R_1?

To answer this question we set up a Cartesian coordinate system and consider an initial point P with coordinates $(0, 0, 1)$ (see Figure 8.2a). We apply two simple right-handed rotations:

R_1: about the x-axis through an angle of 90°

R_2: about the y-axis through an angle of 90°

(The term 'right-handed' here simply means that the sense of the rotation is the same as that of a right-handed screw advancing in the positive direction along a coordinate axis.)

Applying R_1 first takes P into the point P' $(0, -1, 0)$ on the y-axis, as shown in Figure 8.2b. Application of R_2 will then leave P' unchanged.

Conversely, applying R_2 first takes P into P'' $(1, 0, 0)$ on the x-axis, as shown in Figure 8.2c. Application of R_1 will then leave P'' unchanged.

We can see clearly that the order in which the two rotations, about different axes, are performed *is* significant; that is, R_1 followed by R_2 has a different resultant from R_2 followed by R_1. In other words, the commutative law *fails* for finite rotations and, perhaps contrary to our intuition, a finite rotation *cannot* be fully characterized by a single vector. This does not imply

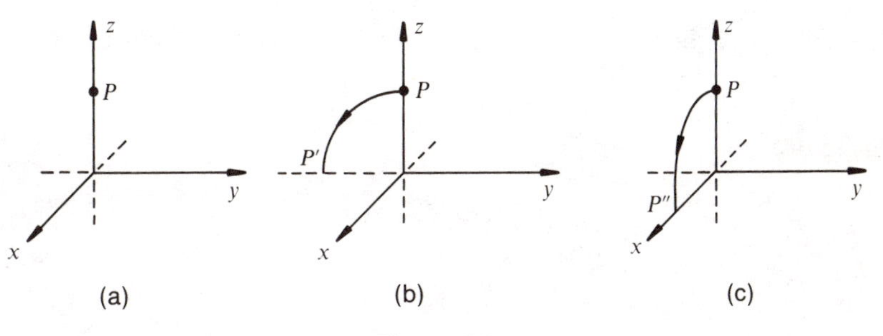

Figure 8.2

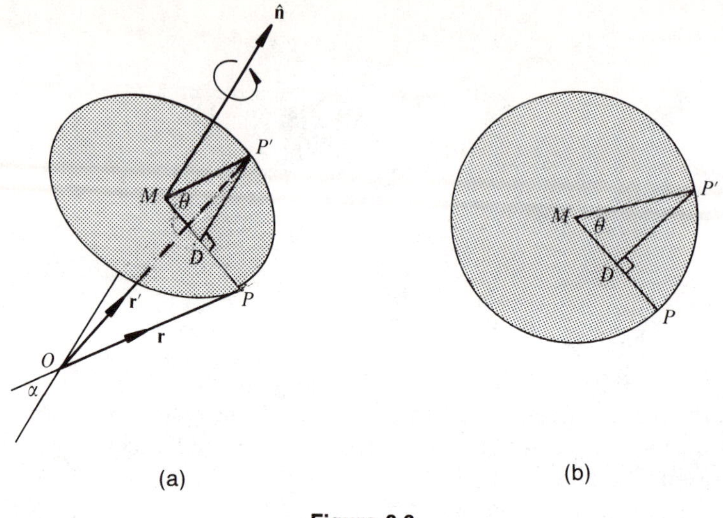

Figure 8.3

that a vector treatment cannot be applied to the analysis of rotation: it simply means that we shall need more than one vector to describe a finite rotation.

Consider a rigid body undergoing a finite rotation through an angle θ about an axis in the direction of the unit vector $\hat{\mathbf{n}}$. If $\mathbf{r}$ denotes the position vector of a particle originally at P (choosing an arbitrary point O on the rotation axis as origin), we shall determine an expression for the position vector $\mathbf{r}'$ of this particle after the rotation (see Figure 8.3a). Clearly,

$$\mathbf{r}' = \mathbf{r} + \overrightarrow{PD} + \overrightarrow{DP'} \tag{8.1}$$

where D is the point on MP such that $P'D$ and MP are perpendicular (see Figure 8.3b). Also,

$$\overrightarrow{OM} = |\overrightarrow{OM}|\hat{\mathbf{n}} = (\mathbf{r}\cdot\hat{\mathbf{n}})\hat{\mathbf{n}}$$

therefore

$$\overrightarrow{MP} = \mathbf{r} - \overrightarrow{OM} = \mathbf{r} - (\mathbf{r}\cdot\hat{\mathbf{n}})\hat{\mathbf{n}}$$

Since P and P' lie on a circle with centre on the rotation axis at M,

$$|\overrightarrow{P'M}| \equiv |\overrightarrow{PM}|$$

Therefore

$$\begin{aligned}\overrightarrow{PD} = |\overrightarrow{PD}|\frac{\overrightarrow{PM}}{|\overrightarrow{PM}|} &= |\overrightarrow{PM}|(1-\cos\theta)\frac{\overrightarrow{PM}}{|\overrightarrow{PM}|}\\ &= \overrightarrow{PM}\,(1-\cos\theta)\\ &= [(\mathbf{r}\cdot\hat{\mathbf{n}})\hat{\mathbf{n}} - \mathbf{r}](1-\cos\theta) \quad \text{(note the sign change)}\end{aligned}$$

Also $|\overrightarrow{DP'}| = |\overrightarrow{PM}| \sin \theta$, and $\overrightarrow{DP'}$ is in the direction of $\hat{\mathbf{n}} \wedge \mathbf{r}$, so

$$\overrightarrow{DP'} = |\overrightarrow{PM}| \sin \theta \frac{\hat{\mathbf{n}} \wedge \mathbf{r}}{|\hat{\mathbf{n}} \wedge \mathbf{r}|} = |\overrightarrow{PM}| \sin \theta \frac{\hat{\mathbf{n}} \wedge \mathbf{r}}{|\mathbf{r}| \sin \alpha}$$

But, from triangle OPM,

$$|\overrightarrow{PM}| = |\mathbf{r}| \sin \alpha$$

$$\therefore \quad \overrightarrow{DP'} = (\hat{\mathbf{n}} \wedge \mathbf{r}) \sin \theta$$

Finally, substituting for $\overrightarrow{PD}$ and $\overrightarrow{DP'}$ in (8.1),

$$\mathbf{r}' = \mathbf{r} + [(\mathbf{r} \cdot \hat{\mathbf{n}})\hat{\mathbf{n}} - \mathbf{r}](1 - \cos \theta) + (\hat{\mathbf{n}} \wedge \mathbf{r}) \sin \theta$$

or

$$\mathbf{r}' = \mathbf{r} \cos \theta + (\mathbf{r} \cdot \hat{\mathbf{n}})\hat{\mathbf{n}}(1 - \cos \theta) + (\hat{\mathbf{n}} \wedge \mathbf{r}) \sin \theta \qquad \mathbf{(8.2)}$$

Equation (8.2) is known as **Rodrigue's formula**. To illustrate the formula, we use it to confirm the visually obvious results of Figure 8.2.

Example 8.1

Determine the final coordinates of a point P at (0, 0, 1) after (a) R_1 followed by R_2, and (b) R_2 followed by R_1, in which R_1 and R_2 are the rotations discussed earlier.

Solution

(a) R_1 is a rotation about the x-axis through 90°, hence $\hat{\mathbf{n}} = \hat{\mathbf{i}}$ and $\theta = \pi/2$. For the initial point P, $\mathbf{r} = \hat{\mathbf{k}}$. Therefore $\mathbf{r} \cdot \hat{\mathbf{n}} = 0$ and

$$\hat{\mathbf{n}} \wedge \mathbf{r} = \hat{\mathbf{i}} \wedge \hat{\mathbf{k}} = -\hat{\mathbf{j}}$$

Hence, using (8.2), P moves to P', whose position vector is given by $\mathbf{r}' = -\hat{\mathbf{j}}$.

R_2 is a rotation about the y-axis through 90°, hence $\hat{\mathbf{n}}' = \hat{\mathbf{j}}$ and $\theta' = \pi/2$. Applied to P' with $\mathbf{r}' = -\hat{\mathbf{j}}$, we have $\mathbf{r}' \cdot \hat{\mathbf{n}}' = -1$ and $\hat{\mathbf{n}}' \wedge \mathbf{r}' = 0$. If the final position vector is $\mathbf{r}''$, then using (8.2) in the form

$$\mathbf{r}'' = \mathbf{r}' \cos \theta' + (\mathbf{r}' \cdot \hat{\mathbf{n}}')\hat{\mathbf{n}}'(1 - \cos \theta') + (\hat{\mathbf{n}}' \wedge \mathbf{r}') \sin \theta'$$

gives $\mathbf{r}'' = -\hat{\mathbf{j}} = \mathbf{r}'$

That is, the point does not change its position during the second rotation. Thus, after the rotations in this order, the coordinates of P become $(0, -1, 0)$.

(b) Performing R_2 first,

$$\hat{\mathbf{n}} = \hat{\mathbf{j}}, \qquad \mathbf{r} = \hat{\mathbf{k}}, \qquad \theta = \frac{\pi}{2}$$

therefore

$$\mathbf{r} \cdot \hat{\mathbf{n}} = 0 \quad \text{and} \quad \hat{\mathbf{n}} \wedge \mathbf{r} = \hat{\mathbf{i}}$$

and so, from (8.2),

$$\mathbf{r}' = \hat{\mathbf{i}}$$

Now applying R_1,

$$\hat{\mathbf{n}}' = \hat{\mathbf{i}}, \qquad \mathbf{r}' = \hat{\mathbf{i}}, \qquad \theta' = \frac{\pi}{2}$$

therefore

$$\hat{\mathbf{n}}' \cdot \mathbf{r}' = 1 \quad \text{and} \quad \hat{\mathbf{n}}' \wedge \mathbf{r}' = 0$$

Hence (8.2) gives the final position vector $\mathbf{r}''$ as

$$\mathbf{r}'' = \hat{\mathbf{i}}$$

That is, the coordinates of P become $(1, 0, 0)$ after performing R_2 first and then R_1.

These results obtained using (8.2) are, of course, entirely consistent with the earlier 'visual' discussion based on Figure 8.2.

Example 8.2

Show that two finite rotations through angles θ and θ' *about the same axis* $\hat{\mathbf{n}}$ are equivalent to a single rotation through an angle $\theta + \theta'$ about $\hat{\mathbf{n}}$.

Solution

After the first rotation, the position vector of a point originally at $\mathbf{r}$ is, by Rodrigue's formula,

$$\mathbf{r}' = \mathbf{r} \cos\theta + (\mathbf{r} \cdot \hat{\mathbf{n}})\hat{\mathbf{n}}(1 - \cos\theta) + (\hat{\mathbf{n}} \wedge \mathbf{r}) \sin\theta \tag{8.3}$$

After the second rotation, let the position vector of the point under consideration become $\mathbf{r}''$. Then

$$\mathbf{r}'' = \mathbf{r}' \cos\theta' + (\mathbf{r}' \cdot \hat{\mathbf{n}})\hat{\mathbf{n}}(1 - \cos\theta') + (\hat{\mathbf{n}} \wedge \mathbf{r}') \sin\theta' \tag{8.4}$$

using the same value of $\hat{\mathbf{n}}$ because the rotation is about the same axis.

Clearly, we need to substitute (8.3) into (8.4). From (8.3), on taking the scalar product of each side with $\hat{\mathbf{n}}$,

$$\mathbf{r}' \cdot \hat{\mathbf{n}} = \mathbf{r} \cdot \hat{\mathbf{n}} \cos\theta + \mathbf{r} \cdot \hat{\mathbf{n}}(1 - \cos\theta) = \mathbf{r} \cdot \hat{\mathbf{n}} \tag{8.5}$$

It also follows from (8.3) that

$$\hat{\mathbf{n}} \wedge \mathbf{r}' = (\hat{\mathbf{n}} \wedge \mathbf{r}) \cos\theta + \hat{\mathbf{n}} \wedge (\hat{\mathbf{n}} \wedge \mathbf{r}) \sin\theta$$

But for any three vectors the vector triple product is given by (1.61):

$$\mathbf{a} \wedge (\mathbf{b} \wedge \mathbf{c}) = \mathbf{b}(\mathbf{a} \cdot \mathbf{c}) - \mathbf{c}(\mathbf{a} \cdot \mathbf{b})$$

Hence $\hat{\mathbf{n}} \wedge (\hat{\mathbf{n}} \wedge \mathbf{r}) = \hat{\mathbf{n}}(\mathbf{r} \cdot \hat{\mathbf{n}}) - \mathbf{r}(\hat{\mathbf{n}} \cdot \hat{\mathbf{n}}) = \hat{\mathbf{n}}(\mathbf{r} \cdot \hat{\mathbf{n}}) - \mathbf{r}$

$$\therefore \qquad \hat{\mathbf{n}} \wedge \mathbf{r}' = (\hat{\mathbf{n}} \wedge \mathbf{r}) \cos\theta + [\hat{\mathbf{n}}(\mathbf{r} \cdot \hat{\mathbf{n}}) - \mathbf{r}] \sin\theta \tag{8.6}$$

Substituting from (8.3), (8.5) and (8.6) into (8.4), we find

$$\mathbf{r}'' = [\mathbf{r}\cos\theta + (\mathbf{r}\cdot\hat{\mathbf{n}})\hat{\mathbf{n}}(1-\cos\theta) + (\hat{\mathbf{n}}\wedge\mathbf{r})\sin\theta]\cos\theta' + \hat{\mathbf{n}}(\mathbf{r}\cdot\hat{\mathbf{n}})(1-\cos\theta') + \sin\theta'[(\hat{\mathbf{n}}\wedge\mathbf{r})\cos\theta + (\hat{\mathbf{n}}(\mathbf{r}\cdot\hat{\mathbf{n}})-\mathbf{r})\sin\theta] \quad \textbf{(8.7)}$$

Grouping similar terms together (that is, terms involving $\mathbf{r}$, $(\mathbf{r}\cdot\hat{\mathbf{n}})\hat{\mathbf{n}}$ and $\hat{\mathbf{n}}\wedge\mathbf{r}$), we find

$$\mathbf{r}'' = \mathbf{r}(\cos\theta\cos\theta' - \sin\theta'\sin\theta) + (\mathbf{r}\cdot\hat{\mathbf{n}})\hat{\mathbf{n}}(\cos\theta' - \cos\theta\cos\theta' + 1 - \cos\theta' + \sin\theta'\sin\theta) + (\hat{\mathbf{n}}\wedge\mathbf{r})(\sin\theta\cos\theta' + \cos\theta\sin\theta')$$

Using various trigonometric addition formulae, we finally obtain

$$\mathbf{r}'' = \mathbf{r}\cos(\theta+\theta') + (\mathbf{r}\cdot\hat{\mathbf{n}})\hat{\mathbf{n}}[1-\cos(\theta+\theta')] + (\hat{\mathbf{n}}\wedge\mathbf{r})\sin(\theta+\theta') \quad \textbf{(8.8)}$$

which, by (8.2), clearly corresponds to a *single* rotation about the axis $\hat{\mathbf{n}}$ through an angle $\theta+\theta'$. We note that order does *not* matter for rotations about the same axis (θ and θ' occur symmetrically in (8.8)). This, of course, is visually obvious.

Following the result of the above example, it is natural to ask whether we can replace *any* two finite rotations by an equivalent single rotation. To answer this question we shall find it convenient to study a finite rotation from a slightly different point of view.

It follows immediately from (8.2) that

$$\mathbf{r} + \mathbf{r}' = \mathbf{r}(1+\cos\theta) + (\mathbf{r}\cdot\hat{\mathbf{n}})\hat{\mathbf{n}}(1-\cos\theta) + (\hat{\mathbf{n}}\wedge\mathbf{r})\sin\theta \quad \textbf{(8.9)}$$

and

$$\mathbf{r}' - \mathbf{r} = [(\mathbf{r}\cdot\hat{\mathbf{n}})\hat{\mathbf{n}} - \mathbf{r}](1-\cos\theta) + (\hat{\mathbf{n}}\wedge\mathbf{r})\sin\theta \quad \textbf{(8.10)}$$

Hence, from (8.9),

$$\hat{\mathbf{n}}\wedge(\mathbf{r}+\mathbf{r}') = (\hat{\mathbf{n}}\wedge\mathbf{r})(1+\cos\theta) + \hat{\mathbf{n}}\wedge(\hat{\mathbf{n}}\wedge\mathbf{r})\sin\theta = (\hat{\mathbf{n}}\wedge\mathbf{r})(1+\cos\theta) + [\hat{\mathbf{n}}(\mathbf{r}\cdot\hat{\mathbf{n}}) - \mathbf{r}]\sin\theta \quad \textbf{(8.11)}$$

using again the general result for the vector triple product.

The right-hand side of (8.11) is very similar to the right-hand side of (8.10), except that the sine and cosine terms are interchanged. We can obtain the precise relationship using the half-angle formulae

$$\sin\theta = 2\sin\left(\frac{\theta}{2}\right)\cos\left(\frac{\theta}{2}\right), \qquad 1+\cos\theta = 2\cos^2\left(\frac{\theta}{2}\right),$$

$$1-\cos\theta = 2\sin^2\left(\frac{\theta}{2}\right) \quad \textbf{(8.12)}$$

Hence, (8.11) becomes

$$\hat{\mathbf{n}} \wedge (\mathbf{r} + \mathbf{r}') = 2\cos^2\left(\frac{\theta}{2}\right)(\hat{\mathbf{n}} \wedge \mathbf{r}) + 2\sin\left(\frac{\theta}{2}\right)\cos\left(\frac{\theta}{2}\right)[\hat{\mathbf{n}}(\mathbf{r}\cdot\hat{\mathbf{n}}) - \mathbf{r}]$$

$$= \cot\left(\frac{\theta}{2}\right)[\sin\theta\,(\hat{\mathbf{n}} \wedge \mathbf{r}) + (1 - \cos\theta)(\hat{\mathbf{n}}(\mathbf{r}\cdot\hat{\mathbf{n}}) - \mathbf{r})]$$

Comparing this with (8.10), we see that

$$\hat{\mathbf{n}} \wedge (\mathbf{r} + \mathbf{r}') = \cot\left(\frac{\theta}{2}\right)(\mathbf{r}' - \mathbf{r}) \qquad \textbf{(8.13)}$$

If we define a new vector $\boldsymbol{\Omega}$ by

$$\boldsymbol{\Omega} = \tan\left(\frac{\theta}{2}\right)\hat{\mathbf{n}} \qquad \textbf{(8.14)}$$

we can write (8.13) in the form

$$\mathbf{r}' - \mathbf{r} = \boldsymbol{\Omega} \wedge (\mathbf{r} + \mathbf{r}') \qquad \textbf{(8.15)}$$

We emphasize that (8.15) contains all the information present in Rodrique's formula (8.2) but in a more concise form.

In a sense, the vector $\boldsymbol{\Omega}$ can be used to 'label' a rotation through an angle θ about an axis $\hat{\mathbf{n}}$. To define the rotation fully, however, equation (8.15) (or its equivalent (8.2)) is required.

If we now perform a second rotation through an angle θ' about an axis $\hat{\mathbf{n}}'$, this can be labelled by the vector $\boldsymbol{\Omega}'$, where

$$\boldsymbol{\Omega}' = \tan\left(\frac{\theta'}{2}\right)\hat{\mathbf{n}}'$$

and the change in position vector produced by this rotation would be

$$\mathbf{r}'' - \mathbf{r}' = \boldsymbol{\Omega}' \wedge (\mathbf{r}' + \mathbf{r}'') \qquad \textbf{(8.16)}$$

Adding (8.15) and (8.16) gives

$$\mathbf{r}'' - \mathbf{r} = \boldsymbol{\Omega} \wedge \mathbf{r} + \boldsymbol{\Omega} \wedge \mathbf{r}' + \boldsymbol{\Omega}' \wedge \mathbf{r}' + \boldsymbol{\Omega}' \wedge \mathbf{r}''$$

$$= (\boldsymbol{\Omega} + \boldsymbol{\Omega}') \wedge (\mathbf{r} + \mathbf{r}'') + \boldsymbol{\Omega} \wedge (\mathbf{r}' - \mathbf{r}'') + \boldsymbol{\Omega}' \wedge (\mathbf{r}' - \mathbf{r}) \qquad \textbf{(8.17)}$$

By using (8.15) and (8.16) on the right-hand side of (8.17), we obtain

$$\mathbf{r}'' - \mathbf{r} = (\boldsymbol{\Omega} + \boldsymbol{\Omega}') \wedge (\mathbf{r} + \mathbf{r}'') - \boldsymbol{\Omega} \wedge [\boldsymbol{\Omega}' \wedge (\mathbf{r}' + \mathbf{r}'')] + \boldsymbol{\Omega}' \wedge [\boldsymbol{\Omega} \wedge (\mathbf{r} + \mathbf{r}')]$$

$$= (\boldsymbol{\Omega} + \boldsymbol{\Omega}') \wedge (\mathbf{r} + \mathbf{r}'') - \boldsymbol{\Omega}'[\boldsymbol{\Omega}\cdot(\mathbf{r}' + \mathbf{r}'')] + (\mathbf{r}' + \mathbf{r}'')(\boldsymbol{\Omega}'\cdot\boldsymbol{\Omega})$$

$$+ \boldsymbol{\Omega}[\boldsymbol{\Omega}'\cdot(\mathbf{r}' + \mathbf{r})] - (\mathbf{r} + \mathbf{r}')(\boldsymbol{\Omega}\cdot\boldsymbol{\Omega}') \qquad \textbf{(8.18)}$$

where we have used the result for the vector triple product twice.

But, from (8.15),

$$\boldsymbol{\Omega}\cdot(\mathbf{r}' - \mathbf{r}) = \boldsymbol{\Omega}\cdot\boldsymbol{\Omega} \wedge (\mathbf{r} + \mathbf{r}') = 0$$

using the properties of the scalar triple product. Hence

$$\mathbf{\Omega}\cdot\mathbf{r} = \mathbf{\Omega}\cdot\mathbf{r}'$$

and similarly, from (8.16),

$$\mathbf{\Omega}'\cdot\mathbf{r}' = \mathbf{\Omega}'\cdot\mathbf{r}'' \qquad \textbf{(8.19)}$$

Hence (8.18) becomes

$$\mathbf{r}'' - \mathbf{r} = (\mathbf{\Omega} + \mathbf{\Omega}') \wedge (\mathbf{r} + \mathbf{r}'') - \mathbf{\Omega}'[\mathbf{\Omega}\cdot(\mathbf{r}'' + \mathbf{r})] + \mathbf{\Omega}[\mathbf{\Omega}'\cdot(\mathbf{r}'' + \mathbf{r})]$$
$$+ (\mathbf{r}'' - \mathbf{r})\mathbf{\Omega}\cdot\mathbf{\Omega}'$$

$$\therefore \quad (\mathbf{r}'' - \mathbf{r})(1 - \mathbf{\Omega}'\cdot\mathbf{\Omega}) = (\mathbf{\Omega} + \mathbf{\Omega}') \wedge (\mathbf{r} + \mathbf{r}'') + (\mathbf{r}'' + \mathbf{r}) \wedge (\mathbf{\Omega} \wedge \mathbf{\Omega}')$$
$$= [(\mathbf{\Omega} + \mathbf{\Omega}') - \mathbf{\Omega} \wedge \mathbf{\Omega}'] \wedge (\mathbf{r}'' + \mathbf{r})$$

and finally

$$\mathbf{r}'' - \mathbf{r} = \left(\frac{\mathbf{\Omega} + \mathbf{\Omega}' - \mathbf{\Omega} \wedge \mathbf{\Omega}'}{1 - \mathbf{\Omega}'\cdot\mathbf{\Omega}}\right) \wedge (\mathbf{r}'' + \mathbf{r}) \qquad \textbf{(8.20)}$$

which implies, on comparison with (8.15), that the two rotations are indeed equivalent to a single rotation, the latter being labelled by the vector

$$\frac{\mathbf{\Omega} + \mathbf{\Omega}' - \mathbf{\Omega} \wedge \mathbf{\Omega}'}{1 - \mathbf{\Omega}'\cdot\mathbf{\Omega}} \qquad \textbf{(8.21)}$$

From this vector we can extract a unit vector $\hat{\mathbf{n}}''$ as well as an angle θ'' since, by definition,

$$\tan\left(\frac{\theta''}{2}\right)\hat{\mathbf{n}}'' = \frac{\mathbf{\Omega} + \mathbf{\Omega}' - \mathbf{\Omega} \wedge \mathbf{\Omega}'}{1 - \mathbf{\Omega}'\cdot\mathbf{\Omega}} \qquad \textbf{(8.22)}$$

Clearly, two finite rotations may be replaced by a single equivalent rotation through an angle θ'' about an axis $\hat{\mathbf{n}}''$. The result (8.22) emphasizes once again that, in general, the order in which the rotations are performed is important: order can only be ignored if the right-hand side of (8.22) is symmetric in $\mathbf{\Omega}$ and $\mathbf{\Omega}'$, that is, when

$$\mathbf{\Omega} \wedge \mathbf{\Omega}' = 0$$

This will be the case if $\hat{\mathbf{n}}$ and $\hat{\mathbf{n}}'$, and hence $\mathbf{\Omega}$ and $\mathbf{\Omega}'$, are parallel (see Example 8.2). The other occasion when order is immaterial is when the term $\mathbf{\Omega} \wedge \mathbf{\Omega}'$ is so small that it may be ignored in comparison with $\mathbf{\Omega} + \mathbf{\Omega}'$. We shall discuss this situation later, when we study infinitesimal rotations.

Example 8.3

Determine a single rotation equivalent to the two successive rotations:

R_1: about the x-axis through an angle of 90°

R_2: about the z-axis through an angle of 45°

Solution

Using (8.14), the labelling vectors are,

$$\mathbf{\Omega} = \tan\left(\frac{\pi}{4}\right)\hat{\mathbf{i}} = \hat{\mathbf{i}} \quad \text{and} \quad \mathbf{\Omega}' = \tan\left(\frac{\pi}{8}\right)\hat{\mathbf{k}} = (-1+\sqrt{2})\hat{\mathbf{k}}$$

Hence, by (8.22), the single resultant rotation is by an angle θ'' about an axis $\hat{\mathbf{n}}''$, where

$$\tan\left(\frac{\theta''}{2}\right)\hat{\mathbf{n}}'' = \hat{\mathbf{i}} + (-1+\sqrt{2})\hat{\mathbf{k}} + (-1+\sqrt{2})\hat{\mathbf{j}} \quad (\text{as } \mathbf{\Omega}'\cdot\mathbf{\Omega} = 0 \text{ here})$$

$$\therefore \qquad \hat{\mathbf{n}}'' = \frac{\hat{\mathbf{i}} + (-1+\sqrt{2})\hat{\mathbf{j}} + (-1+\sqrt{2})\hat{\mathbf{k}}}{\tan\left(\frac{\theta''}{2}\right)}$$

But, since $\hat{\mathbf{n}}''$ is a unit vector,

$$\tan\left(\frac{\theta''}{2}\right) = |\hat{\mathbf{i}} + (-1+\sqrt{2})\hat{\mathbf{j}} + (-1+\sqrt{2})\hat{\mathbf{k}}| = (7-4\sqrt{2})^{1/2}$$

from which

$$\theta'' \approx 98.4°$$

The rotations R_1 and R_2 and the equivalent single rotation, applied to an initial point P on the z-axis, are shown in Figure 8.4.

The reader might appreciate the usefulness of this approach to rotation more fully if we rework Example 8.2 for rotations about the same axis $\hat{\mathbf{n}}$. If the two rotations are labelled by the vectors

$$\mathbf{\Omega} = \tan\left(\frac{\theta}{2}\right)\hat{\mathbf{n}} \quad \text{and} \quad \mathbf{\Omega}' = \tan\left(\frac{\theta'}{2}\right)\hat{\mathbf{n}}$$

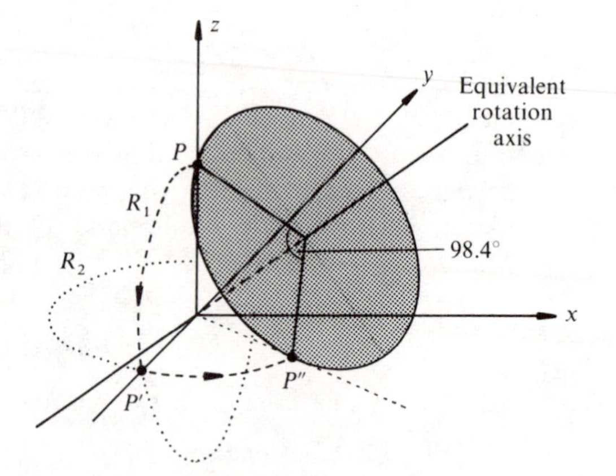

Figure 8.4

then the combined rotation is labelled by the vector $\mathbf{\Omega}''$ where, by (8.21),

$$\mathbf{\Omega}'' = \frac{\mathbf{\Omega} + \mathbf{\Omega}' - \mathbf{\Omega} \wedge \mathbf{\Omega}'}{1 - \mathbf{\Omega}' \cdot \mathbf{\Omega}} = \frac{\tan\left(\dfrac{\theta}{2}\right)\hat{\mathbf{n}} + \tan\left(\dfrac{\theta'}{2}\right)\hat{\mathbf{n}}}{1 - \tan\left(\dfrac{\theta}{2}\right)\tan\left(\dfrac{\theta'}{2}\right)}$$

since $\mathbf{\Omega} \wedge \mathbf{\Omega}' = 0$ and $\hat{\mathbf{n}} \cdot \hat{\mathbf{n}} = 1$ for this case. Hence

$$\mathbf{\Omega}'' = \tan\left(\frac{\theta + \theta'}{2}\right)\hat{\mathbf{n}}$$

which proves, much more readily than previously, that the two rotations are equivalent to a single rotation through an angle $\theta + \theta'$.

8.2 Infinitesimal rotations and angular velocity

In Section 8.1, we found that a finite rotation *cannot* in general be described by a single vector. However, there is an important special case where a single vector description *is* possible. This is when the angle of rotation θ is small, say less than about 5°.

For small values of θ, the labelling vector $\mathbf{\Omega}$ defined earlier by

$$\mathbf{\Omega} = \tan\left(\frac{\theta}{2}\right)\hat{\mathbf{n}}$$

may be approximated as

$$\mathbf{\Omega} \approx \frac{\theta}{2}\hat{\mathbf{n}} \tag{8.23}$$

Similarly, the general expression (8.21)

$$\mathbf{\Omega}'' = \frac{\mathbf{\Omega} + \mathbf{\Omega}' - \mathbf{\Omega} \wedge \mathbf{\Omega}'}{1 - \mathbf{\Omega}' \cdot \mathbf{\Omega}}$$

for the labelling vector of two successive rotations by angles θ and θ' may be approximated by

$$\mathbf{\Omega}'' \approx \mathbf{\Omega} + \mathbf{\Omega}' \tag{8.24}$$

when these angles are small, because both the scalar and vector product terms will involve products of the form $\theta\theta'$ which can be neglected. We can see immediately from (8.24) that for small rotations the order is irrelevant, because

$$\mathbf{\Omega} + \mathbf{\Omega}' = \mathbf{\Omega}' + \mathbf{\Omega}$$

To summarize, we have obtained the important result that an infinitesimal rotation may be fully characterized by the vector

$$\mathbf{\Omega} \approx \frac{\theta}{2}\hat{\mathbf{n}} \tag{8.25}$$

where θ is the rotation angle and $\hat{\mathbf{n}}$ is a unit vector along the axis of rotation.

We can apply these results to the motion of a rigid body. There are two distinct cases:

- motion in which one point of the body remains fixed,
- motion in which no point of the body remains fixed.

In the first case, if O is the fixed point then it is easily shown that an arbitrary displacement of the rigid body is equivalent to a single rotation about an axis through O. Clearly, in this type of motion an arbitrary displacement can be considered to be made up of a number of finite rotations. Since O is a fixed point of the body, the axis of rotation of each of these rotations must pass through O (otherwise O would not be fixed).

From the discussion in the previous section, an arbitrary number of successive rotations is equivalent to a single rotation (because any two are), and since O is fixed, the axis of this single rotation must pass through O.

If the motion is continuous, then an angular velocity may be defined. We assume that the continuous motion is made up of a large number of infinitesimal rotations possibly involving different rotation angles and different rotation axes. In a small interval of time δt, let the angle turned through, about an axis $\hat{\mathbf{n}}$, be $\delta\theta$. The average **angular speed** during this interval is $\delta\theta/\delta t$ rad s^{-1}. As $\delta t \to 0$ we may define an instantaneous angular speed $d\theta/dt$ by

$$\frac{d\theta}{dt} = \lim_{\delta t \to 0} \frac{\delta\theta}{\delta t} \tag{8.26}$$

The **instantaneous angular velocity** is defined by the vector

$$\boldsymbol{\omega} = \frac{d\theta}{dt}\hat{\mathbf{n}}$$

in which $\hat{\mathbf{n}}$ is the instantaneous axis of rotation (Clearly, using (8.25), $\boldsymbol{\omega} = 2\, d\mathbf{\Omega}/dt$.)

The instantaneous velocity of any particle in the rigid body may now be obtained. If $\mathbf{r}$ is the position vector of the particle at time t (see Figure 8.5), then the motion of the particle at this instant is in the direction of $\boldsymbol{\omega} \wedge \mathbf{r}$ (see Section 1.6). Since the distance moved through in time δt is $|\mathbf{r}| \sin\alpha\, \delta\theta$, the instantaneous speed is

$$\lim_{\delta t \to 0} |\mathbf{r}| \sin\alpha \frac{\delta\theta}{\delta t} = |\mathbf{r}| \sin\alpha \frac{d\theta}{dt}$$

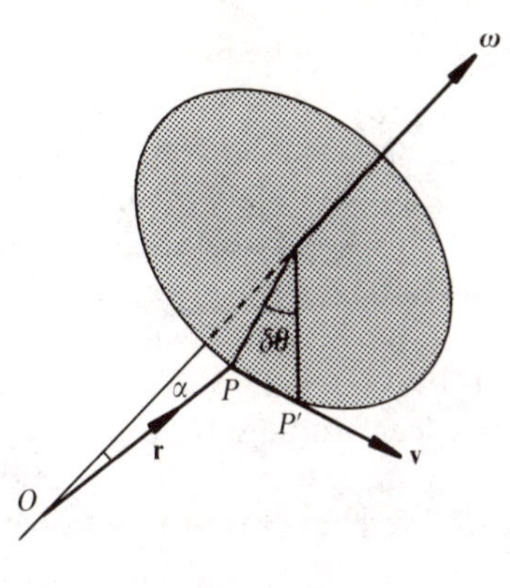

Figure 8.5

Hence, the instantaneous velocity $\mathbf{v}$ is given by

$$\begin{aligned}\mathbf{v} &= |\mathbf{r}| \sin \alpha \frac{\mathrm{d}\theta}{\mathrm{d}t} \frac{\boldsymbol{\omega} \wedge \mathbf{r}}{|\boldsymbol{\omega} \wedge \mathbf{r}|} \\ &= |\mathbf{r}| \sin \alpha \frac{\mathrm{d}\theta}{\mathrm{d}t} \frac{\boldsymbol{\omega} \wedge \mathbf{r}}{|\boldsymbol{\omega}|\,|\mathbf{r}| \sin \alpha} \\ &= \boldsymbol{\omega} \wedge \mathbf{r} \qquad \textbf{(8.27)}\end{aligned}$$

The treatment of a general motion of a rigid body in which *no* point is fixed is deferred until Section 8.3, after a result first obtained by Coriolis has been developed.

EXERCISES

8.1 Determine the final coordinates of a point P initially at $(1, 0, 0)$ after successive rotations of 45° about the x-axis followed by 90° about the y-axis. What would be the position if the order of the rotations were reversed?

8.2 Determine a single rotation equivalent to the two successive rotations:

R_1: about the x-axis through an angle of 90°

R_2: about the z-axis through an angle of 90°

8.3 Explain carefully the differences between finite rotations and infinitesimal rotations. When can finite rotations be treated as if they were infinitesimal rotations?

8.4 A rigid body with a fixed point at O has an angular velocity of magnitude $3\ \mathrm{rad\ s^{-1}}$ parallel to the vector $\hat{\mathbf{i}} + \hat{\mathbf{j}} + 3\hat{\mathbf{k}}$. Determine the instantaneous velocity of the body at the point $(1, 0, 0)$.

8.3 Rotating frames of reference

Imagine a frame of reference (a Cartesian coordinate system) that is fixed. In fact, a fixed frame of reference is a purely theoretical concept which cannot be fully realized in practice. This is because apparently stationary objects on the Earth are in reality moving, since the Earth is rotating about its axis and also about the Sun. Even the Sun is not stationary, because it is in orbit on the outer rim of the Milky Way galaxy. To a very good approximation, however, the distant stars appear to be stationary with respect to observers in the Solar System, and for most applications these distant stars are used to define a 'fixed' frame of reference. From this frame we can observe the Universe – from the motions of planets and their satellites to the flight of a tennis ball and the trajectory of a rocket. It is with respect to such a fixed system that **Newton's second law** has been postulated. For a particle with constant mass, this law relates the acceleration $\mathbf{a}$ of the particle to the force $\mathbf{F}$ applied to it:

$$\mathbf{F} = m\mathbf{a} \tag{8.28}$$

An **inertial system** is defined to be any system in which (8.28) is valid.

We can readily show that any frame of reference moving with *constant velocity* with respect to the fixed system is also inertial (see Figure 8.6). To do this we examine the position of a moving point P with position vectors $\mathbf{r}$ from O in the fixed system and $\mathbf{r}'$ from O' in the moving system. If the two systems are coincident at some time $t = t_0$, then

$$\mathbf{r}' = \mathbf{r} - \mathbf{v}(t - t_0) \tag{8.29}$$

since after a time $(t - t_0)$ the position vector of O' with respect to O is $\mathbf{v}(t - t_0)$.

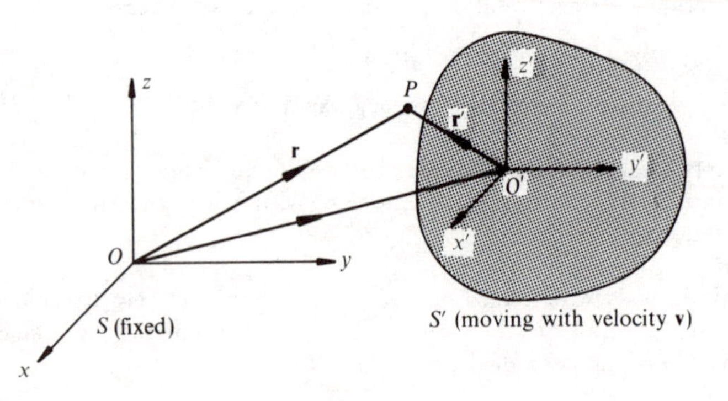

Figure 8.6

Differentiating (8.29),

$$\frac{d\mathbf{r}'}{dt} = \frac{d\mathbf{r}}{dt} - \mathbf{v} \quad \text{since } \mathbf{v} \text{ is constant}$$

and
$$\frac{d^2\mathbf{r}'}{dt^2} = \frac{d^2\mathbf{r}}{dt^2} \tag{8.30}$$

that is, the *acceleration* of the point P is the same in both systems. Newton's law will thus be valid in the moving system S' if the force $\mathbf{F}$ has the same form in both systems. This can be shown to be the case for many applications arising in mechanics. (It is, in fact, a matter of common experience: for example, playing tennis on the deck of a ship moving with a constant velocity is no different from playing tennis on dry land – the laws of motion appear to be exactly the same.)

Frames of reference that are rotating or accelerating with respect to an inertial frame are not inertial and the effect (if any) on Newton's law must therefore be considered.

For some purposes a frame of reference fixed on the Earth may be regarded as inertial. However, there are circumstances in which the rotation of the Earth (with respect to a truly inertial frame) must be taken into account, and it is this aspect of rotation that we now consider. We shall examine the motion of a particle viewed from an inertial frame and from a frame of reference rotating with respect to the inertial frame.

Consider two Cartesian coordinate systems, which we shall label C_1 and C_2 (see Figure 8.7). We imagine C_1 to be fixed and C_2 to be rotating with angular velocity $\boldsymbol{\omega}$ about the origin of C_1. Let A be the instantaneous position of a particle which is moving in space and whose position vector $\mathbf{a}$ with respect to the fixed system is thus a function of the parameter t, the time. We shall examine the rate of change of the position of this particle as viewed from both systems.

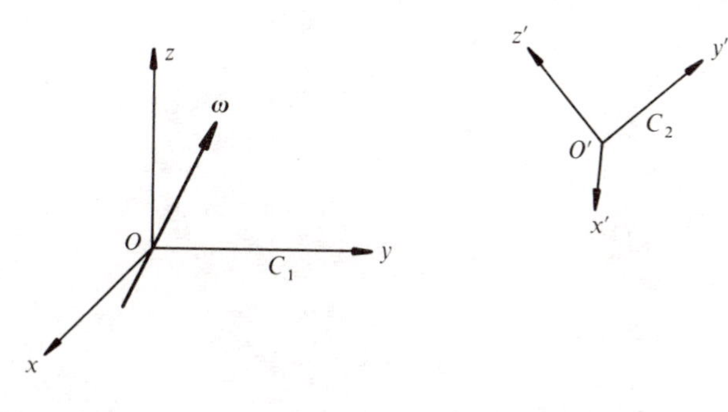

Figure 8.7

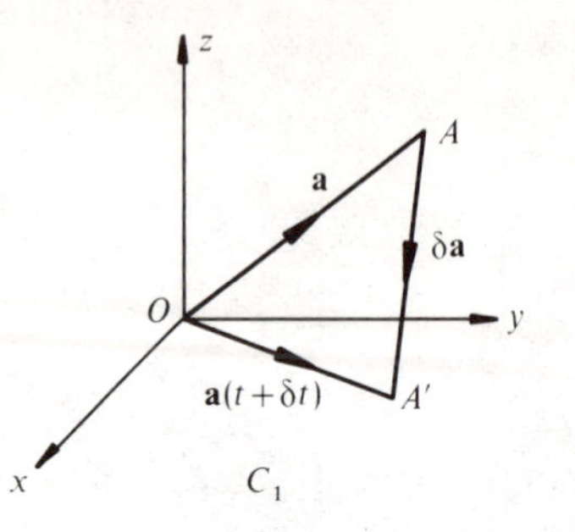

Figure 8.8

With respect to the fixed system, the particle originally at a point A has moved to a point A' after a time δt has elapsed. The rate of change of the position vector of the particle is

$$\frac{\mathbf{a}(t+\delta t)-\mathbf{a}(t)}{\delta t} \quad \text{or} \quad \frac{\delta \mathbf{a}}{\delta t} \tag{8.31}$$

(see Figure 8.8). An observer in the C_2 system, however, will regard the point A as a fixed point, always having (from O') the same orientation and separation. However, since C_2 is rotating with respect to C_1, after an interval of time δt, O' has moved to O'' (with respect to the fixed system). The same observer, now at O'', will regard the point B as the original position of the particle (because B has the same orientation and separation from O'' as A did from O'). Clearly, with respect to the fixed system,

$$\overrightarrow{AB} = (\boldsymbol{\omega} \wedge \mathbf{a})\,\delta t \tag{8.32}$$

(see Figure 8.9). The change in the position vector of the particle as seen by the observer in the C_2 system would be $\overrightarrow{O''A'} - \overrightarrow{O''B}$. The rate of change of the position of the particle viewed from C_2 would therefore be

$$\frac{\overrightarrow{O''A'} - \overrightarrow{O''B}}{\delta t} \tag{8.33}$$

But, from Figure 8.9,

$$\overrightarrow{O''A'} - \overrightarrow{O''B} = \overrightarrow{BA'} = \overrightarrow{BA} + \delta\mathbf{a}$$

$$\therefore \quad \delta\mathbf{a} = \overrightarrow{O''A'} - \overrightarrow{O''B} + \overrightarrow{AB}$$

which gives

$$\frac{\delta\mathbf{a}}{\delta t} = \frac{\overrightarrow{O''A'} - \overrightarrow{O''B}}{\delta t} + \boldsymbol{\omega} \wedge \mathbf{a} \tag{8.34}$$

In the limit, we can write

$$\left(\frac{\mathrm{d}\mathbf{a}}{\mathrm{d}t}\right)_{C_1} = \left(\frac{\mathrm{d}\mathbf{a}}{\mathrm{d}t}\right)_{C_2} + \boldsymbol{\omega} \wedge \mathbf{a} \tag{8.35}$$

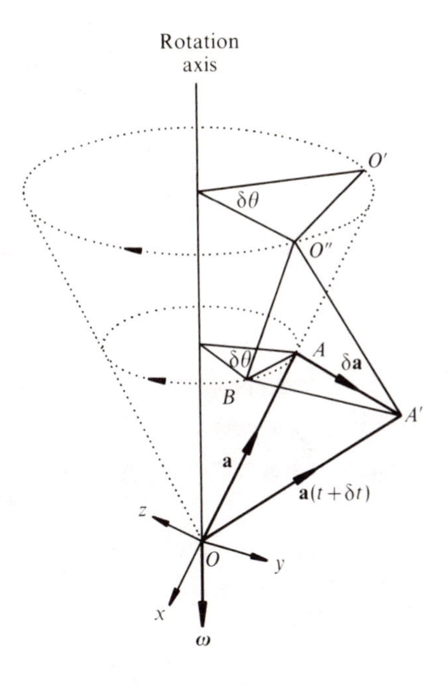

Figure 8.9

where $\left(\frac{\mathrm{d}\mathbf{a}}{\mathrm{d}t}\right)_{C_1}$ and $\left(\frac{\mathrm{d}\mathbf{a}}{\mathrm{d}t}\right)_{C_2}$ stand for the rate of change of the position vector as measured in the fixed system C_1 and the rotating system C_2 respectively. These quantities are thus the velocities, say $\mathbf{v}_1$ and $\mathbf{v}_2$, of the same particle viewed from the two systems. Hence

$$\mathbf{v}_1 = \mathbf{v}_2 + \boldsymbol{\omega} \wedge \mathbf{a} \tag{8.36}$$

The result (8.36) is known as **Coriolis' theorem**. This relation (or (8.35)) can be regarded as an expression for the relation between the time rates of change $\left(\frac{\mathrm{d}}{\mathrm{d}t}\right)_{C_1}$ and $\left(\frac{\mathrm{d}}{\mathrm{d}t}\right)_{C_2}$ in the two systems, and we shall find it useful to write (8.35) in the 'operator' form

$$\left(\frac{\mathrm{d}}{\mathrm{d}t}\right)_{C_1} = \left(\frac{\mathrm{d}}{\mathrm{d}t}\right)_{C_2} + \boldsymbol{\omega} \wedge \tag{8.37}$$

We now return to complete the analysis of the motion of a rigid body begun in the previous section for the case where no point of the body remains fixed during the motion.

Consider a fixed frame of reference C_1 and a frame of reference C_2 fixed in the body (and hence moving with it, see Figure 8.10). If P is a point in the body, then its position with respect to O' is fixed (because we are

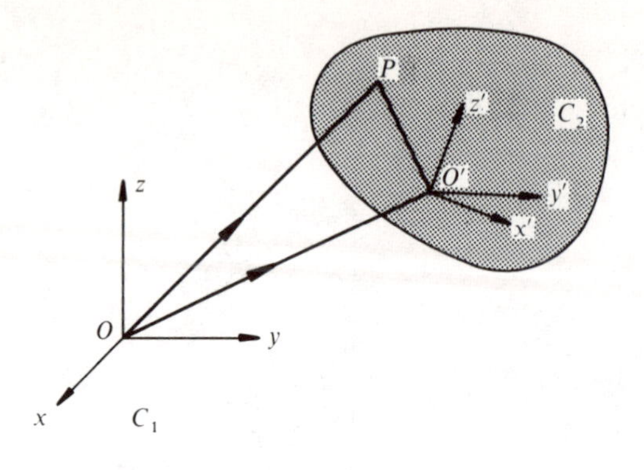

Figure 8.10

dealing with a rigid body). Clearly,

$$\overrightarrow{OO'} + \overrightarrow{O'P} = \overrightarrow{OP}$$

From the viewpoint of the fixed frame, the velocity of the particle at P is $\left(\frac{d}{dt}(\overrightarrow{OP})\right)_{C_1}$, and

$$\left(\frac{d}{dt}(\overrightarrow{OP})\right)_{C_1} = \left(\frac{d}{dt}(\overrightarrow{OO'})\right)_{C_1} + \left(\frac{d}{dt}(\overrightarrow{O'P})\right)_{C_1} \tag{8.38}$$

Using the Coriolis theorem, the second term on the right-hand side is given by

$$\left(\frac{d}{dt}(\overrightarrow{O'P})\right)_{C_1} = \left(\frac{d}{dt}(\overrightarrow{O'P})\right)_{C_2} + \boldsymbol{\omega} \wedge \overrightarrow{O'P} \tag{8.39}$$

where $\boldsymbol{\omega}$ is the (instantaneous) angular velocity of C_2 about C_1. But $\left(\frac{d}{dt}(\overrightarrow{O'P})\right)_{C_2} = 0$, since P is fixed relative to O', so (8.38) becomes

$$\left(\frac{d}{dt}(\overrightarrow{OP})\right)_{C_1} = \left(\frac{d}{dt}(\overrightarrow{OO'})\right)_{C_1} + \boldsymbol{\omega} \wedge \overrightarrow{O'P}$$

or $$\mathbf{v}_P = \mathbf{v}_{O'} + \boldsymbol{\omega} \wedge \overrightarrow{O'P} \tag{8.40a}$$

That is, the velocity of a particle at P is made up of two components – a velocity $\mathbf{v}_{O'}$ of O' relative to O (a translational velocity of the body as a whole), and a velocity $\boldsymbol{\omega} \wedge \overrightarrow{O'P}$ which is the velocity of P relative to O', due to the rotation of the body.

If we consider another point Q in the rigid body, then its velocity is, by similar reasoning,

$$\mathbf{v}_Q = \mathbf{v}_{O'} + \boldsymbol{\omega} \wedge \overrightarrow{O'Q}$$

Writing (8.40a) in the form

$$\mathbf{v}_P = \mathbf{v}_{O'} + \boldsymbol{\omega} \wedge \overrightarrow{O'Q} + \boldsymbol{\omega} \wedge (\overrightarrow{O'P} - \overrightarrow{O'Q})$$

we obtain

$$\begin{aligned}\mathbf{v}_P &= \mathbf{v}_{O'} + \boldsymbol{\omega} \wedge \overrightarrow{O'Q} + \boldsymbol{\omega} \wedge \overrightarrow{QP} \\ &= \mathbf{v}_Q + \boldsymbol{\omega} \wedge \overrightarrow{QP} \qquad \textbf{(8.40b)}\end{aligned}$$

We see that the velocity of P relative to Q, $(\mathbf{v}_P - \mathbf{v}_Q)$, is due to the angular velocity $\boldsymbol{\omega}$ of the body about Q. But since P and Q are arbitrary points of the body, this shows that $\boldsymbol{\omega}$ is a characteristic of the body as a whole. In addition, since the formula (8.40b) does not make any reference to a (possibly instantaneous) rotation axis in any way, it illustrates that $\boldsymbol{\omega}$ may be regarded as a **free vector**.

Example 8.4

A rod of length l rests with one end A on the floor and the other end B in contact with the wall (see Figure 8.11). End B begins to slide downwards with speed v whilst maintaining contact with the wall. End A slides and maintains contact with the floor. Calculate the velocity of A and the angular velocity of the rod.

Solution

We can, without loss of generality, assume that the angular velocity vector $\boldsymbol{\omega}$ is parallel to the unit vector $\hat{\mathbf{k}}$, that is, perpendicular to the plane of the rod. Then $\mathbf{v}_B = -v\hat{\mathbf{j}}$ and $\mathbf{v}_A = \alpha\hat{\mathbf{i}}$, where α is to be determined. But, by (8.40b),

$$\mathbf{v}_A = \mathbf{v}_B + \boldsymbol{\omega} \wedge \overrightarrow{BA}$$

$$\therefore \qquad \alpha\hat{\mathbf{i}} = -v\hat{\mathbf{j}} + \boldsymbol{\omega} \wedge \overrightarrow{BA}$$

We can eliminate $\boldsymbol{\omega}$ by taking the dot product of both sides of this equation with the vector

$$\overrightarrow{BA} = l\cos\theta\,\hat{\mathbf{i}} - l\sin\theta\,\hat{\mathbf{j}}$$

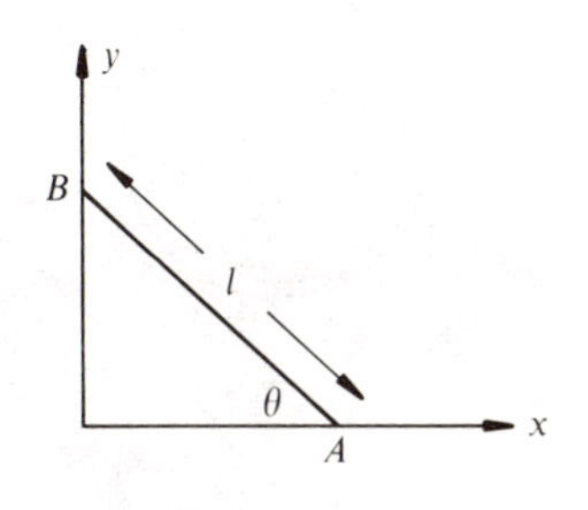

Figure 8.11

We obtain

$$\alpha l \cos \theta = +vl \sin \theta \quad \text{or} \quad \alpha = v \tan \theta$$

That is, the velocity of A is $\mathbf{v}_A = v \tan \theta \, \hat{\mathbf{i}}$. Hence

$$\begin{aligned} v \tan \theta \, \hat{\mathbf{i}} &= -v\hat{\mathbf{j}} + \boldsymbol{\omega} \wedge \overrightarrow{BA} \\ &= -v\hat{\mathbf{j}} + \omega\hat{\mathbf{k}} \wedge (l \cos \theta \, \hat{\mathbf{i}} - l \sin \theta \, \hat{\mathbf{j}}) \\ &= -v\hat{\mathbf{j}} + \omega(l \cos \theta \, \hat{\mathbf{j}} + l \sin \theta \, \hat{\mathbf{i}}) \end{aligned}$$

Comparing x- or y-components,

$$\omega l \cos \theta = v$$

so that the required angular velocity vector is

$$\boldsymbol{\omega} = \frac{v}{l \cos \theta} \hat{\mathbf{k}}$$

Another simple application of the basic result (8.40b) is in the determination of the constraints to be applied when one rigid body rolls over another. If two rigid bodies are in contact, then **rolling** is said to occur between the two if the velocities at the point of contact are equal.

Example 8.5

Two rollers of radii 0.5 m and 1 m, with fixed axes, are rotating with angular speeds ω_1 and ω_2 rad s^{-1} respectively. Determine the relation between ω_1 and ω_2 if the motion is to be a rolling motion.

Solution

We choose a fixed inertial system, and let the centres of the rollers be at positions $(1, 0)$ and $(1, 1.5)$ as shown in Figure 8.12.

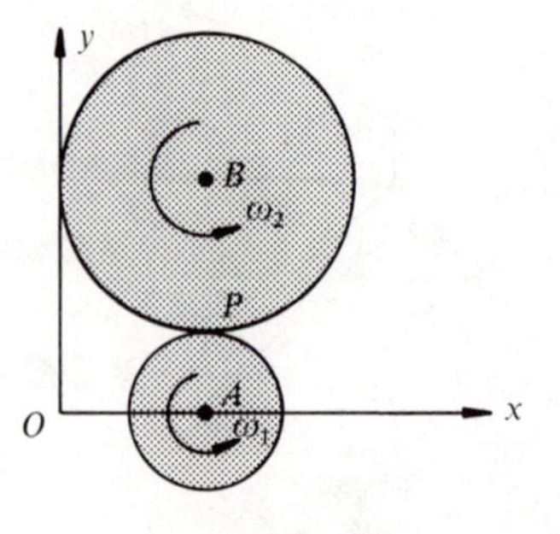

Figure 8.12

Viewed from the inertial system, the velocity of roller 1 at the point of contact P is, by (8.40b),

$$\mathbf{v}_P^{(1)} = \mathbf{v}_A + \boldsymbol{\omega}_1 \wedge \overrightarrow{AP}$$

But $\mathbf{v}_A = 0$, since the axis of roller 1 is fixed. Also, $\boldsymbol{\omega}_1 = \omega_1 \hat{\mathbf{k}}$ and $\overrightarrow{AP} = 0.5\hat{\mathbf{j}}$, so

$$\mathbf{v}_P^{(1)} = \omega_1 \hat{\mathbf{k}} \wedge (0.5\hat{\mathbf{j}}) = -0.5\omega_1 \hat{\mathbf{i}}$$

For roller 2 at the same point P,

$$\mathbf{v}_P^{(2)} = \mathbf{v}_B + \boldsymbol{\omega}_2 \wedge \overrightarrow{BP}$$

The axis of roller 2 is also fixed, so $\mathbf{v}_B = 0$. Also, $\boldsymbol{\omega}_2 = \omega_2 \hat{\mathbf{k}}$ and $\overrightarrow{BP} = -\hat{\mathbf{j}}$, so

$$\mathbf{v}_P^{(2)} = \omega_2 \hat{\mathbf{k}} \wedge (-\hat{\mathbf{j}}) = \omega_2 \hat{\mathbf{i}}$$

For the motion to be rolling we must have $\mathbf{v}_P^{(1)} = \mathbf{v}_P^{(2)}$, or

$$-0.5\omega_1 = \omega_2$$

that is, roller 1 must rotate twice as fast as roller 2 and in the opposite sense.

8.4 The Coriolis force

A second application of Coriolis' theorem is to the analysis of the motion of a particle with respect to a (non-inertial) frame of reference rotating with the Earth. It is only after this analysis has been carried out that we can decide whether or not the rotation of the Earth affects the validity of Newton's second law. To examine the effect, we simply take the time derivative of both sides of the equation

$$\mathbf{v}_1 = \mathbf{v}_2 + \boldsymbol{\omega} \wedge \mathbf{r} \tag{8.41}$$

where $\mathbf{r}$ is the position vector of a particle viewed from the inertial system. Using(8.37), we find that

$$\left(\frac{d\mathbf{v}_1}{dt}\right)_{C_1} = \left(\frac{d}{dt}(\mathbf{v}_2 + \boldsymbol{\omega} \wedge \mathbf{r})\right)_{C_2} + \boldsymbol{\omega} \wedge (\mathbf{v}_2 + \boldsymbol{\omega} \wedge \mathbf{r})$$

$$= \left(\frac{d\mathbf{v}_2}{dt}\right)_{C_2} + \left(\frac{d\boldsymbol{\omega}}{dt}\right)_{C_2} \wedge \mathbf{r} + \boldsymbol{\omega} \wedge \left(\frac{d\mathbf{r}}{dt}\right)_{C_2} + \boldsymbol{\omega} \wedge \mathbf{v}_2 + \boldsymbol{\omega} \wedge (\boldsymbol{\omega} \wedge \mathbf{r})$$

But, again using (8.37),

$$\left(\frac{d\boldsymbol{\omega}}{dt}\right)_{C_1} = \left(\frac{d\boldsymbol{\omega}}{dt}\right)_{C_2} + \boldsymbol{\omega} \wedge \boldsymbol{\omega} = \left(\frac{d\boldsymbol{\omega}}{dt}\right)_{C_2}$$

That is, the rate of change of $\boldsymbol{\omega}$ is the same in both systems and therefore we can drop the subscript.

Finally, if $\mathbf{a}_1 = \left(\frac{d\mathbf{v}_1}{dt}\right)_{C_1}$ and $\mathbf{a}_2 = \left(\frac{d\mathbf{v}_2}{dt}\right)_{C_2}$, then

$$\mathbf{a}_1 = \mathbf{a}_2 + \frac{d\boldsymbol{\omega}}{dt} \wedge \mathbf{r} + 2\boldsymbol{\omega} \wedge \mathbf{v}_2 + \boldsymbol{\omega} \wedge (\boldsymbol{\omega} \wedge \mathbf{r}) \tag{8.42}$$

But $\mathbf{a}_1$ and $\mathbf{a}_2$ are the accelerations measured in the two systems, and their different values clearly emphasize the non-inertial nature of the rotating frame of reference. We now examine the consequences of this result for an observer in the rotating system of the Earth.

Newton's second law,

$$\mathbf{F} = m\frac{\mathrm{d}^2\mathbf{r}}{\mathrm{d}t^2} = m\mathbf{a}_1 \quad \text{(in an inertial system)}$$

becomes

$$\mathbf{F} = m\mathbf{a}_2 + m\frac{\mathrm{d}\boldsymbol{\omega}}{\mathrm{d}t} \wedge \mathbf{r} + 2m\boldsymbol{\omega} \wedge \mathbf{v}_2 + m\boldsymbol{\omega} \wedge (\boldsymbol{\omega} \wedge \mathbf{r}) \tag{8.43}$$

Therefore, to an observer moving in the rotating system, the particle appears to have an acceleration $\mathbf{a}_2$ resulting from a force

$$\mathbf{F} - m\frac{\mathrm{d}\boldsymbol{\omega}}{\mathrm{d}t} \wedge \mathbf{r} - 2m\boldsymbol{\omega} \wedge \mathbf{v}_2 - m\boldsymbol{\omega} \wedge (\boldsymbol{\omega} \wedge \mathbf{r}) \tag{8.44}$$

We shall choose the origin of an inertial system of coordinates with respect to the rotating Earth at the centre of the Earth, with the Oz-axis pointing north (see Figure 8.13). P is the position of a particle near the Earth's surface.

We can fix our non-inertial system at any point on the Earth's surface. We choose $O'x'$ to point due east, $O'y'$ to point due north and $O'z'$ to point vertically out of the ground. With this choice, it is known that the Earth rotates in a positive direction about Oz, and so the angular velocity is

$$\boldsymbol{\omega} = \omega\hat{\mathbf{k}} \quad \text{where } \omega \text{ is a constant}$$

$$\therefore \quad \frac{\mathrm{d}\boldsymbol{\omega}}{\mathrm{d}t} = 0$$

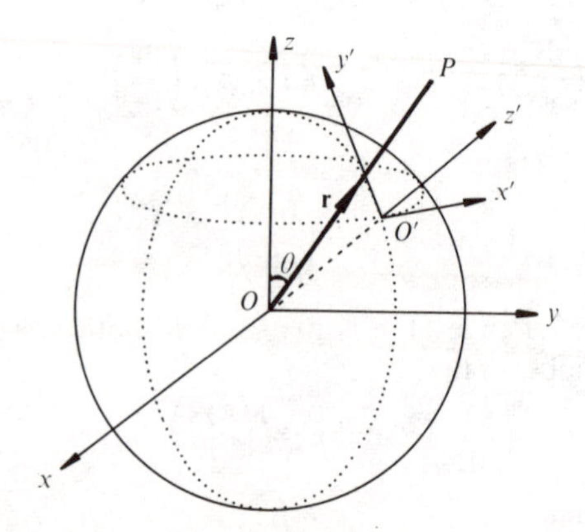

Figure 8.13

and the first of the 'extra' force terms in (8.44) is thus accounted for. The last of these extra terms, namely $m\boldsymbol{\omega} \wedge (\boldsymbol{\omega} \wedge \mathbf{r})$, is a vector normal to $\boldsymbol{\omega}$ and of magnitude $|\boldsymbol{\omega}|\,|\boldsymbol{\omega} \wedge \mathbf{r}| \sin 90° = |\boldsymbol{\omega}|\,|\boldsymbol{\omega} \wedge \mathbf{r}|$. But since $\boldsymbol{\omega}$ and $\boldsymbol{\omega} \wedge \mathbf{r}$ are perpendicular,

$$|\boldsymbol{\omega} \wedge (\boldsymbol{\omega} \wedge \mathbf{r})| = \omega^2 r \sin\theta$$

where θ is the angle between the vectors $\mathbf{r}$ (measured from the origin at the centre of the Earth) and $\boldsymbol{\omega}$. This term is clearly the familiar centrifugal force, $r \sin\theta$ being the perpendicular distance from the particle to the rotation axis.

Now the magnitude of the Earth's angular velocity $\boldsymbol{\omega}$ is approximately 7.3×10^{-5} rad s^{-1} relative to the stars. The value of the centrifugal acceleration due to the Earth's rotation for a particle near the surface ($r = 6.37 \times 10^3$ km) and at the Equator ($\theta = 90°$) would be approximately

$$(7.3 \times 10^{-5})^2 (6.37 \times 10^8) \text{ cm s}^{-2} \approx 3.4 \text{ cm s}^{-2}$$

This value is about 0.3% of the acceleration due to gravity (≈ 981 cm s^{-2}) and, though small, would need to be taken into consideration if accuracy were at a premium. Having said this, it should be realized that the 'local' acceleration due to gravity is generally accepted as being due to the *combined* effects resulting from gravitational attraction and from the Earth's rotation. Hence, for most applications, we can ignore the acceleration arising from centrifugal effects.

The remaining 'extra' term in (8.44), namely $-2m\boldsymbol{\omega} \wedge \mathbf{v}_2$, is known as the **Coriolis force**. It is clearly zero if the particle is stationary in the rotating system ($\mathbf{v}_2 = 0$).

Any velocity $\mathbf{v}_2$ has three components in the $O'x'y'z'$ system: a component along $O'x'$ (easterly), a component along $O'y'$ (northerly) and a component along $O'z'$ (vertically). We may consider the effect of the Coriolis force on the three components separately.

(1) Particle projected in an easterly direction. Here

$$\mathbf{v}_2 = v_2 \hat{\mathbf{i}}' \quad \text{and} \quad -\boldsymbol{\omega} \wedge \mathbf{v}_2 = -\omega v_2 \hat{\mathbf{k}} \wedge \hat{\mathbf{i}}'$$

But the angle between $\hat{\mathbf{k}}$ and $\hat{\mathbf{i}}'$ is fixed all over the Earth's surface at 90°, hence

$$|\boldsymbol{\omega} \wedge \mathbf{v}_2| = \omega v_2$$

The direction of this acceleration is always to the right of the local vertical axis in the northern hemisphere and to the left of it in the southern hemisphere (see Figure 8.14).

(2) Particle projected in a northerly direction. Here $\mathbf{v}_2 = v_2 \hat{\mathbf{j}}'$ and $-|\boldsymbol{\omega} \wedge \mathbf{v}_2| = -\omega v_2 \cos\theta$ (see Figure 8.14). The direction of $\mathbf{v}_2 \wedge \boldsymbol{\omega}$ is clearly in an easterly direction in the northern hemisphere and in a westerly direction in the southern hemisphere.

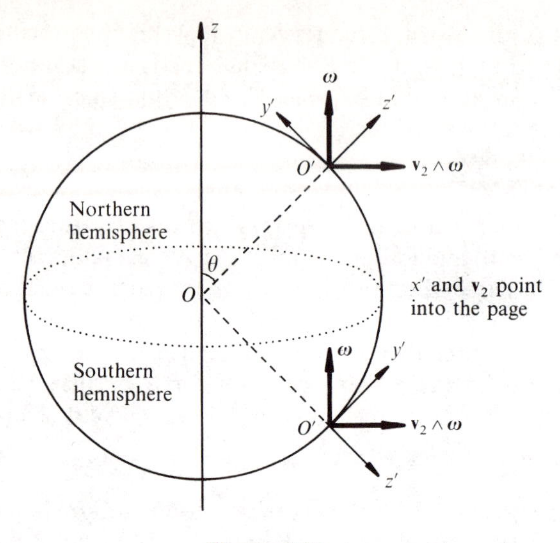

Figure 8.14

Combining the two motions in (1) and (2), we conclude that the effect of the Coriolis force on a particle moving in a horizontal plane is to deflect that particle to the right in the northern hemisphere and to the left in the southern hemisphere.

(3) Particle projected vertically (upwards). Here $\mathbf{v}_2 = v_2\hat{\mathbf{k}}'$ and the direction of $\mathbf{v}_2 \wedge \boldsymbol{\omega}$ is clearly in a westerly direction in both the northern and the southern hemispheres. It has its maximum effect at the Equator (where $\theta = 90°$). For a falling particle, the direction of $\mathbf{v}_2 \wedge \boldsymbol{\omega}$ is reversed, and so the deflection will be in an easterly direction.

Example 8.6

(a) A particle at a height h is allowed to fall from rest. Show that the effect of the Coriolis force is to deflect the particle from its vertical path by a factor

$$\frac{\omega}{3}\sqrt{\left(\frac{(2h)^3}{g}\right)}\sin\theta$$

where ω is the angular speed of the Earth and θ is the colatitude angle (see Figure 8.14).

(b) Show further that if the particle were projected vertically upwards from ground level with sufficient initial speed to attain a maximum height h and then allowed to return to Earth, the effect of the Coriolis force would be to produce a deflection

$$\frac{-4\omega}{3}\sqrt{\left(\frac{(2h)^3}{g}\right)}\sin\theta$$

Solution

(a) We again choose local non-inertial axes $O'x'y'z'$ as in Figure 8.14, with the $O'z'$-axis pointing vertically upwards. We will perform the analysis appropriate to the northern hemisphere, and will assume that the motion of the particle comprises two independent accelerations (which is not strictly true): an acceleration controlling the vertical motion due entirely to gravity, and an acceleration controlling movement in an east–west direction due entirely to the Coriolis force. If (x', z') are the coordinates of a particle of mass m at time t, then this assumption implies that

$$m\frac{d^2z'}{dt^2} = -mg \qquad \textbf{(8.45)}$$

and $$m\frac{d^2x'}{dt^2} = -2m\omega v_2 \sin\theta \qquad \textbf{(8.46)}$$

where v_2 is the speed of the particle (which may be positive or negative, according to whether the motion is up or down). Here v_2 is the vertical speed, which may be obtained from (8.45) as dz'/dt. Integrating (8.45),

$$m\frac{dz'}{dt} = -mgt \quad \text{and} \quad z' = -\tfrac{1}{2}gt^2 \qquad \textbf{(8.47)}$$

(the speed of the particle is zero at $t = 0$ since it is released from rest). Therefore

$$v_2 = -gt$$

and $$\frac{d^2x'}{dt^2} = 2g\omega t \sin\theta \qquad \textbf{(8.48)}$$

Integrating twice,

$$x' = 2g\omega\frac{t^3}{6}\sin\theta \qquad \textbf{(8.49)}$$

From (8.47), when $z' = -h$, $t = \sqrt{(2h/g)}$, which is thus the time taken to fall a distance h. Substituting this value into (8.49) shows that the deflection experienced by the particle in falling through a height h is

$$\frac{\omega}{3}\sqrt{\left(\frac{(2h)^3}{g}\right)}\sin\theta$$

as required. (The deflection produced by the Coriolis force in a 100 m fall at the Equator is then approximately 2.2 cm.)

The deflection in the east–west direction is essentially dependent on the time of flight, which can be obtained from (8.45).

(b) In this case, (8.45) integrates to

$$\frac{dz'}{dt} = -gt + v_0$$

where v_0 is the speed with which the particle is projected. Equation (8.46) now becomes

$$\frac{d^2x'}{dt^2} = -2\omega \sin\theta\,(-gt + v_0)$$

Integrating twice,

$$x' = -2\omega \sin\theta\left(-g\frac{t^3}{6} + v_0\frac{t^2}{2}\right) \qquad \textbf{(8.50)}$$

The total time of flight is $2\sqrt{(2h/g)}$, while v_0 and h are related through $v_0 = \sqrt{(2hg)}$; substituting into (8.50) gives the time of flight as

$$\frac{-4\omega}{3}\sqrt{\left(\frac{(2h)^3}{g}\right)}\sin\theta$$

The Coriolis force helps to explain the generally observed fact that winds in the northern hemisphere are deflected to the right of their course and those in the southern hemisphere are deflected to the left.

In a simplified view, 'wind' is observed when air moves from regions of high pressure to regions of low pressure. Owing to the effects of heating by the Sun there are, relatively speaking, a number of well-defined bands of pressure on the Earth's surface (see Figure 8.15).

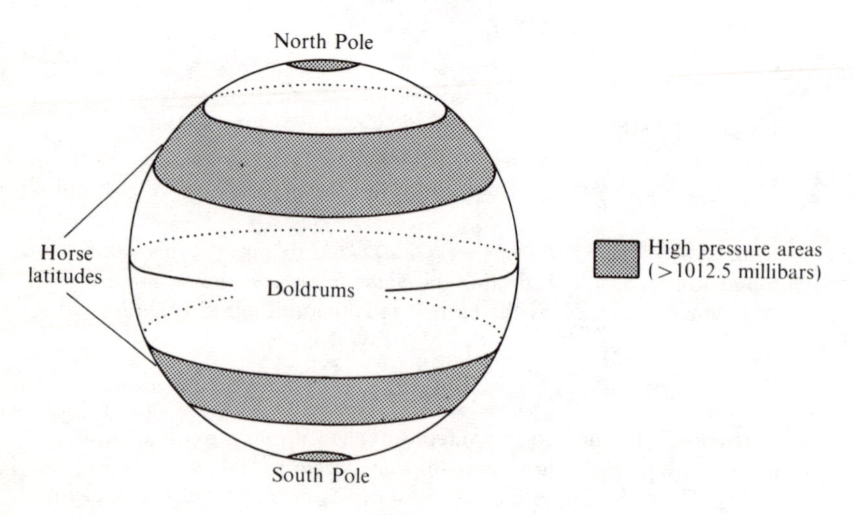

Figure 8.15

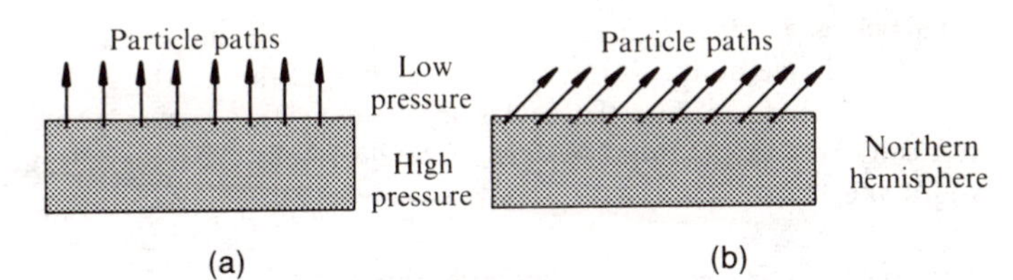

Figure 8.16

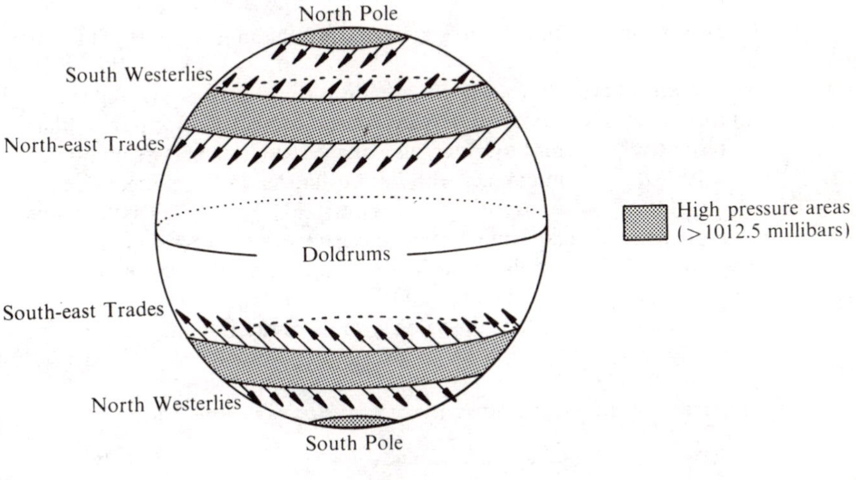

Figure 8.17

If the Earth were perfectly still, then the motion of the air from high pressure to low pressure areas would be as indicated in Figure 8.16a. But owing to the Coriolis effect these 'perfect' paths are deflected to the right in the northern hemisphere, as shown in Figure 8.16b, with a reverse effect in the southern hemisphere. Figure 8.17 gives a schematic view of winds observed at the Earth's surface. Obviously, there will be numerous local variations to be superimposed on this global picture.

EXERCISES

8.5 Making the usual assumptions about the magnitude of the Earth's angular velocity, show that the equation of motion of a particle of mass m, with position vector $\mathbf{r}$ relative to axes fixed in the Earth and rotating with it, is

$$\text{'sum of all forces acting on the particle'} = m\left(\frac{\mathrm{d}^2\mathbf{r}}{\mathrm{d}t^2} + 2\boldsymbol{\omega} \wedge \frac{\mathrm{d}\mathbf{r}}{\mathrm{d}t}\right)$$

8.6 A particle is dropped from a height of 100 m above the ground at a colatitude angle of 45°. If the angular velocity of the Earth is 7.30×10^{-5} rad s^{-1}, find the deviation of the particle from a point vertically beneath its initial position. State carefully the assumptions that you make.

8.7 A particle of mass m is projected with a velocity $\mathbf{v}$ over a smooth horizontal plane on the Earth's surface at colatitude angle θ. Show that the reaction of the plane on the particle is

$$R = mg - 2m\omega v_y \sin\theta$$

where g is the acceleration due to gravity and ω is the angular speed of the Earth.

8.8 Two concentric cylinders A and B, of radii a and b ($b > a$) and aligned with axes along the z-axis, rotate with constant angular velocities $\omega_A\hat{\mathbf{k}}$ and $\omega_B\hat{\mathbf{k}}$ respectively. A third cylinder, also aligned along the z-axis and of radius $\frac{1}{2}(b-a)$, rolls in contact with both cylinders. Is this possible?

8.9 A 2 m diameter circular lamina rolls between two parallel rods 2 m apart, which are aligned along the x-axis and moving with velocities of $7\hat{\mathbf{i}}$ and $9\hat{\mathbf{i}}$ m s^{-1} respectively (see Figure 8.18). Show that the angular speed of the lamina is 1 rad s^{-1} and that the velocity of its centre is $8\hat{\mathbf{i}}$ m s^{-1}.

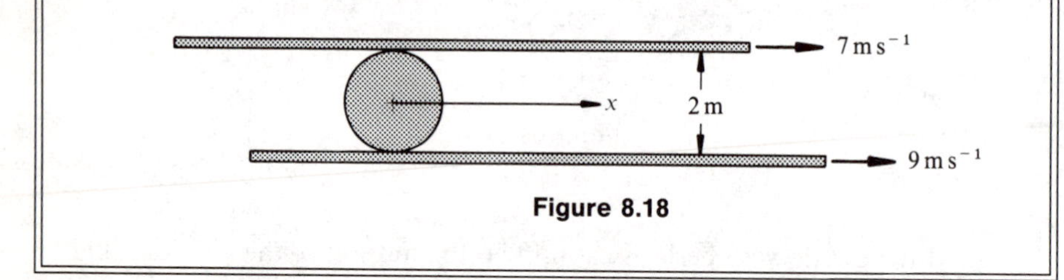

Figure 8.18

ADDITIONAL EXERCISES

1 A particle is projected with speed V on a smooth horizontal table at colatitude angle θ. Show that the rotation of the Earth induces the particle to move on an arc of a circle of radius $V/(2\omega\cos\theta)$.

2 Two concentric spheres A and B, of radii a and b ($b > a$), rotate with constant angular velocities $\boldsymbol{\omega}_A$ and $\boldsymbol{\omega}_B$ respectively about fixed diameters. A third sphere C, of radius $\frac{1}{2}(b-a)$, rolls in contact with both spheres. Show that this is only possible if $\boldsymbol{\omega}_A = \boldsymbol{\omega}_B$ and that in this case the centre of C describes a circle with constant angular velocity equal to this common value.

3 A frame of three rods is in the form of an equilateral triangle of side l. Two of the corners are attached to bearings A and B at $(0, l/\sqrt{2}, 0)$ and $(0, 0, l/\sqrt{2})$ respectively. The bearings slide along the grooves as shown in Figure 8.19. If bearing B is given a velocity $-V\hat{\mathbf{k}}$, determine the velocity of C and show that the angular velocity of the frame is $\boldsymbol{\omega} = (\sqrt{2}/l)V(\hat{\mathbf{i}} - \hat{\mathbf{j}} + \hat{\mathbf{k}})$. You may assume that node C remains in the xy-plane.

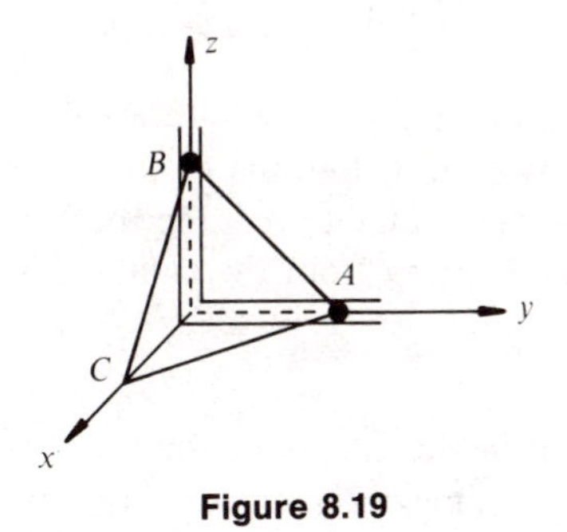

Figure 8.19

4 Rewrite Rodrigue's formula (8.2) in terms of indices. Determine the form taken in the special case of 'small' rotations, and verify that infinitesimal rotations can be characterized by a single vector.

SUMMARY

- A **finite rotation** cannot be fully characterized by a single vector.

- After rotation through an angle θ about an axis $\hat{\mathbf{n}}$, the position vector of a point originally at $\mathbf{r}$ is

$$\mathbf{r}' = \mathbf{r}\cos\theta + (\mathbf{r}\cdot\hat{\mathbf{n}})\hat{\mathbf{n}}(1 - \cos\theta) + (\hat{\mathbf{n}} \wedge \mathbf{r})\sin\theta$$

  Equivalently, the effect of a rotation is given implicitly through the equation

$$\mathbf{r}' - \mathbf{r} = \boldsymbol{\Omega} \wedge (\mathbf{r} + \mathbf{r}') \quad \text{where} \quad \boldsymbol{\Omega} = \tan\left(\frac{\theta}{2}\right)\hat{\mathbf{n}}$$

- An **infinitesimal rotation** can be fully represented by a single vector proportional to $\theta\hat{\mathbf{n}}$.

- The **instantaneous angular velocity** of a rotating body is

$$\boldsymbol{\omega} = \frac{\mathrm{d}\theta}{\mathrm{d}t}\hat{\mathbf{n}}$$

- An **inertial frame of reference** is one in which Newton's second law is valid.

- The rates of change in an inertial system C_1 and in a frame of reference C_2 rotating with angular velocity $\boldsymbol{\omega}$ with respect to C_1 are related by

$$\left(\frac{\mathrm{d}}{\mathrm{d}t}\right)_{C_1} = \left(\frac{\mathrm{d}}{\mathrm{d}t}\right)_{C_2} + \boldsymbol{\omega} \wedge$$

- There are two contributions to the velocity of a **rigid body** – a translational velocity of the body as a whole, and a contribution arising from the rotation of the body:

$$\mathbf{v}_P = \mathbf{v}_{O'} + \boldsymbol{\omega} \wedge \overrightarrow{O'P}$$

- The **Coriolis force** is an 'extra' force acting on a particle moving in a rotating frame of reference. The Coriolis force on a particle of mass m is $-2m\boldsymbol{\omega} \wedge \mathbf{v}_2$, where $\mathbf{v}_2$ is the velocity of the mass with respect to the rotating frame.

- For a particle moving in a horizontal plane at the Earth's surface, the effect of the Coriolis force is to deflect the particle to the right of its path in the northern hemisphere and to the left of its path in the southern hemisphere. For a particle projected vertically upwards (downwards) at the Earth's surface, the effect of the Coriolis force is to deflect the particle to the west (east) whether it be in the northern or the southern hemisphere.

APPENDIX 1

Determinants

A1.1 Second-order determinants

A second-order determinant D is a square array of four numbers, written as

$$D = \begin{vmatrix} a_{11} & a_{12} \\ a_{21} & a_{22} \end{vmatrix}$$

and defined as $a_{11}a_{22} - a_{21}a_{12}$. In other words, the value of a second-order determinant is the product of the diagonal terms minus the product of the off-diagonal terms. For example,

$$\begin{vmatrix} 3 & 4 \\ -2 & 1 \end{vmatrix} = 3 - (-8) = 11$$

Note that interchanging rows or interchanging columns introduces a minus sign:

$$\begin{vmatrix} a_{21} & a_{22} \\ a_{11} & a_{12} \end{vmatrix} = a_{21}a_{12} - a_{11}a_{22} = -D$$

$$\begin{vmatrix} a_{12} & a_{11} \\ a_{22} & a_{21} \end{vmatrix} = a_{12}a_{21} - a_{11}a_{22} = -D$$

It is an immediate consequence of the definition that if the two rows or the two columns of the determinant are identical (or proportional), then that determinant must have zero value. For example,

$$\begin{vmatrix} 3 & -4 \\ 9 & -12 \end{vmatrix} = -36 - (-36) = 0$$

A1.2 Third-order determinants

A third-order determinant is written as a square array of nine numbers,

$$D = \begin{vmatrix} a_{11} & a_{12} & a_{13} \\ a_{21} & a_{22} & a_{23} \\ a_{31} & a_{32} & a_{33} \end{vmatrix}$$

and is defined as

$$a_{11}\begin{vmatrix} a_{22} & a_{23} \\ a_{32} & a_{33} \end{vmatrix} - a_{12}\begin{vmatrix} a_{21} & a_{23} \\ a_{31} & a_{33} \end{vmatrix} + a_{13}\begin{vmatrix} a_{21} & a_{22} \\ a_{31} & a_{32} \end{vmatrix}$$

That is, a third-order determinant is defined in terms of three second-order determinants. Obviously this could be expanded further, to give

$$D = a_{11}(a_{22}a_{33} - a_{32}a_{23}) - a_{12}(a_{21}a_{33} - a_{23}a_{31}) + a_{13}(a_{21}a_{32} - a_{31}a_{22}) \quad \textbf{(A1.1)}$$

As with second-order determinants, interchanging any two rows or any two columns changes the sign of the value of the determinant. Also, if a third-order determinant has any two rows or any two columns proportional to each other, then the value of the determinant is zero.

Determinants of higher orders may be similarly defined and enjoy similar properties to second- and third-order determinants. However, such general determinants will not be discussed here, since they have no immediate application to two- or three-dimensional vector field theory.

The following section will only be of interest to those readers familiar within the index notation covered as supplementary material in earlier chapters.

A1.3 Determinants and index notation

The astute reader might have surmised from the previous section that there is a close correspondence between the permutation symbol introduced in Supplement 1S.1 and a determinant. In fact, it can be shown by direct expansion that a third-order determinant may be written more concisely as follows:

$$\begin{vmatrix} a_{11} & a_{12} & a_{13} \\ a_{21} & a_{22} & a_{23} \\ a_{31} & a_{32} & a_{33} \end{vmatrix} = \tfrac{1}{6}\varepsilon_{ijk}\varepsilon_{lmn}a_{il}a_{jm}a_{kn} \quad \textbf{(A1.2)}$$

To prove (A1.2), one needs to expand the right-hand side (which is easier than it looks) and check that the expression (A1.1) is obtained.

We can use (A1.2) to obtain some useful results. For example, since a_{ij} is an arbitrary array, we can choose a_{ij} as the Kronecker delta δ_{ij}, so that

$$\begin{vmatrix} a_{11} & a_{12} & a_{13} \\ a_{21} & a_{22} & a_{23} \\ a_{31} & a_{32} & a_{33} \end{vmatrix} = \begin{vmatrix} 1 & 0 & 0 \\ 0 & 1 & 0 \\ 0 & 0 & 1 \end{vmatrix} = 1$$

The right-hand side of (A1.2) for this case is

$$\tfrac{1}{6}\varepsilon_{ijk}\varepsilon_{lmn}\delta_{il}\delta_{jm}\delta_{kn} = \tfrac{1}{6}\varepsilon_{lmn}\varepsilon_{lmn}$$

hence $\varepsilon_{lmn}\varepsilon_{lmn} = 6$

(see exercise 1S.4).

The ε–δ identity referred to in Supplement 1S.1 may be derived by considering the following special determinant

$$\begin{vmatrix} \delta_{il} & \delta_{im} & \delta_{in} \\ \delta_{jl} & \delta_{jm} & \delta_{jn} \\ \delta_{kl} & \delta_{km} & \delta_{kn} \end{vmatrix}$$

On expansion of this determinant we obtain a collection of terms, each of which has the six 'free' indices i, j, k, l, m and n. Now if we put $i = j$ or $i = k$ or $j = k$, then two of the rows of the determinant are identical and so for these special cases the value of the determinant will be zero. Also, if we interchange the first and second rows (or equivalently interchange i and j), we introduce a minus sign because we are dealing with a determinant. Similarly, interchanging the indices i and k or j and k (which are equivalent to interchanging the first and third or the second and third rows respectively) also introduces a minus sign. But these are precisely the properties enjoyed by the permutation symbol ε_{ijk}. This implies the relationship

$$\begin{vmatrix} \delta_{il} & \delta_{im} & \delta_{in} \\ \delta_{jl} & \delta_{jm} & \delta_{jn} \\ \delta_{kl} & \delta_{km} & \delta_{kn} \end{vmatrix} = D_{lmn}\varepsilon_{ijk}$$

in which D_{lmn} are yet to be found. By a similar argument dealing with the columns of the determinant and the indices l, m and n, we find that

$$\begin{vmatrix} \delta_{il} & \delta_{im} & \delta_{in} \\ \delta_{jl} & \delta_{jm} & \delta_{jn} \\ \delta_{kl} & \delta_{km} & \delta_{kn} \end{vmatrix} = C\varepsilon_{lmn}\varepsilon_{ijk}$$

where C is a constant whose value is easily determined by giving the indices specific values. For example, if $i = 1, j = 2, k = 3, l = 1, m = 2$ and $n = 3$, we find $C = 1$. We finally arrive at the general identity connecting the Kronecker delta with the permutation symbol:

$$\begin{vmatrix} \delta_{il} & \delta_{im} & \delta_{in} \\ \delta_{jl} & \delta_{jm} & \delta_{jn} \\ \delta_{kl} & \delta_{km} & \delta_{kn} \end{vmatrix} = \varepsilon_{lmn}\varepsilon_{ijk}$$

To obtain the ε–δ identity we simply contract the indices i and l to give

$$\varepsilon_{ijk}\varepsilon_{imn} = \begin{vmatrix} 3 & \delta_{im} & \delta_{in} \\ \delta_{ji} & \delta_{jm} & \delta_{jn} \\ \delta_{ki} & \delta_{km} & \delta_{kn} \end{vmatrix}$$

$$= 3(\delta_{jm}\delta_{kn} - \delta_{km}\delta_{jn}) - \delta_{im}(\delta_{ji}\delta_{kn} - \delta_{ki}\delta_{jn}) + \delta_{in}(\delta_{ji}\delta_{km} - \delta_{ki}\delta_{jm})$$

$$= \delta_{jm}\delta_{kn} - \delta_{km}\delta_{jn}$$

which is the required identity.

APPENDIX 2

Double Integration

A2.1 Introduction

Before we study the subject of double integration we shall find it useful to review briefly our knowledge of ordinary integration.

An ordinary integral of the form $\int_a^b f(x)\,\mathrm{d}x$ may be considered in two ways. Firstly, and perhaps most importantly as far as this text is concerned, we have a 'physical' interpretation of the integral as the *area* under the curve $y = f(x)$ between the points $x = a$ and $x = b$ (see Figure A2.1). The shaded area represents $\int_a^b f(x)\,\mathrm{d}x$, and we remember that areas above the x-axis make a positive contribution and those below the x-axis a negative contribution to the value of the integral.

The second approach to evaluation of a definite integral is through the use of the **fundamental theorem of calculus**, which basically tells us that integration is the reverse process to differentiation: if $F(x)$ is such that $\mathrm{d}F/\mathrm{d}x = f(x)$, then

$$\int_a^b f(x)\,\mathrm{d}x = F(b) - F(a) = \Big[F(x) \Big]_a^b \qquad \textbf{(A2.1)}$$

and so the determination of the area under a curve is directly related to the search for a function $F(x)$ – the **indefinite integral** of $f(x)$.

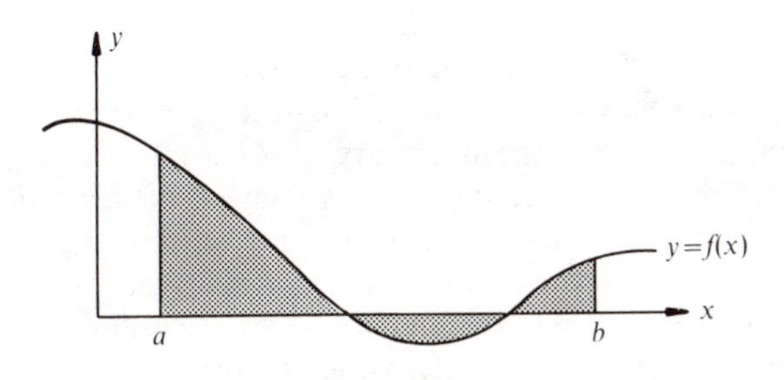

Figure A2.1

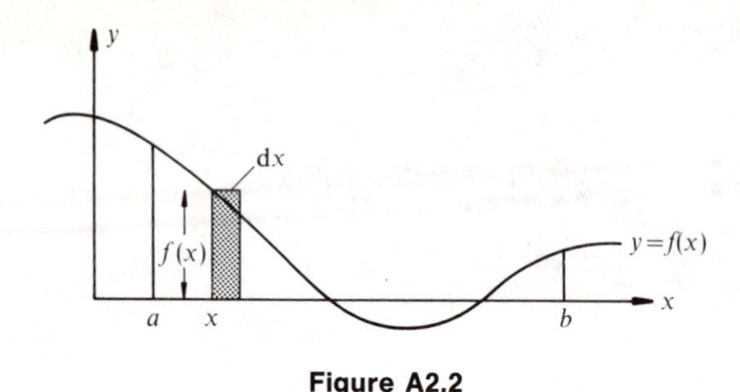

Figure A2.2

In physical applications involving the formation or use of integrals, the reader will find it useful to *read* an integral. Remembering that the integral sign stands for a sum (of an infinite number of terms), we can say

'$\int_a^b f(x)\,dx$ is the sum of all terms of the form $f(x)\,dx$ as x changes from a to b.'

This statement allows us to calculate an approximate value for the integral. We simply divide the range $a \leqslant x \leqslant b$ into a large number of segments, each of width dx (see Figure A2.2). As long as dx is sufficiently small, then the shaded area shown (the area of a rectangle of sides $f(x)$ and dx) will be a good approximation to the area under the curve above the element dx. Clearly, if we sum all terms of this form from $x = a$ to $x = b$ we shall end up with a good approximation to the area under the curve $y = f(x)$ from $x = a$ to $x = b$.

It should be noted that most definite integrals that arise in practice have to be approximated in this way, either with the rather crude 'rectangular division' approach outlined above or with some more sophisticated numerical technique, the details of which need not concern us here.

A2.2 Evaluation of double integrals

We shall now extend the subject of integration to two dimensions. Consider a plane area A (with an arbitrarily shaped boundary) in the xy-plane and a surface S defined by $z = f(x, y)$ (see Figure A2.3a).

We can form a cylinder (whose cross-section is A) by projecting lines parallel to the z-axis from the boundary of A to intersect S as shown in Figure A2.3b. We wish to determine the volume V of this cylinder. This is a similar problem to calculating the area under a curve above an axis, but in one dimension higher. The curve $y = f(x)$ has been generalized to a surface $z = f(x, y)$ and the one-dimensional region on the x-axis has been generalized to a two-dimensional region in the xy-plane, A. Our approach to determining the volume V will therefore parallel that for finding the area under the curve.

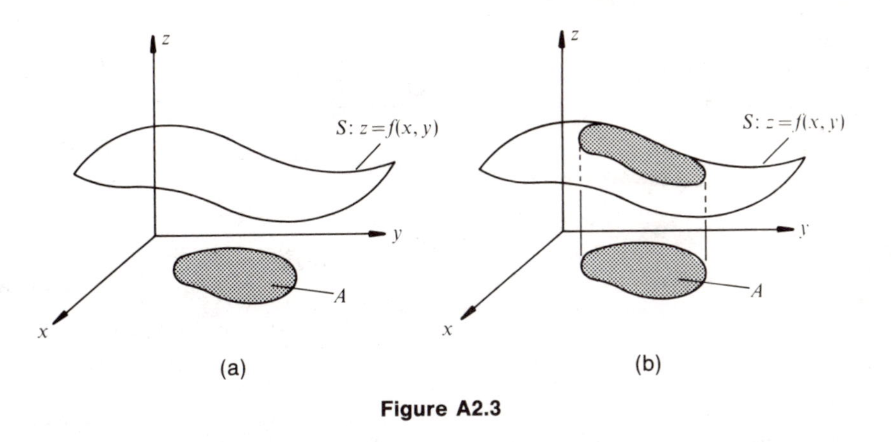

Figure A2.3

We denote the **double integral** of $f(x, y)$ over A by $\iint_A f(x, y)\,\mathrm{d}x\,\mathrm{d}y$ and define it to be the volume V required. The notation is not arbitrary: the two integral signs imply the summation over the two-dimensional region A of all terms of the form $f(x, y)\,\mathrm{d}x\,\mathrm{d}y$.

To show how to evaluate $\iint_A f(x, y)\,\mathrm{d}x\,\mathrm{d}y$, we simply have to determine V. To do this we slice the cylinder into a large number of 'slabs', each of thickness $\mathrm{d}x$, by planes parallel to the zy-plane. In Figure A2.4 we have highlighted one such slab. To a good approximation, the volume $\mathrm{d}V$ of this slab is

$$\mathrm{d}V = \text{area}(PQRT) \times \mathrm{d}x \qquad \textbf{(A2.2)}$$

Now consider $PQRT$ in more detail by projecting it onto the zy-plane as shown in Figure A2.5. The equation of the curve joining R and T can be obtained from the equation of the surface $z = f(x, y)$, since the curve lies on this surface, provided we note that everywhere along this slab the value of

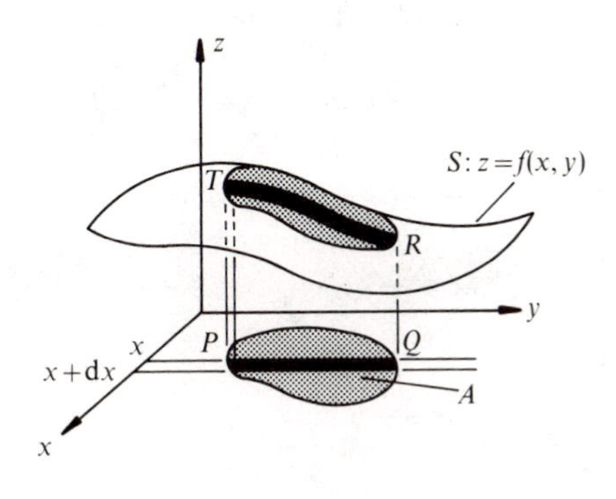

Figure A2.4

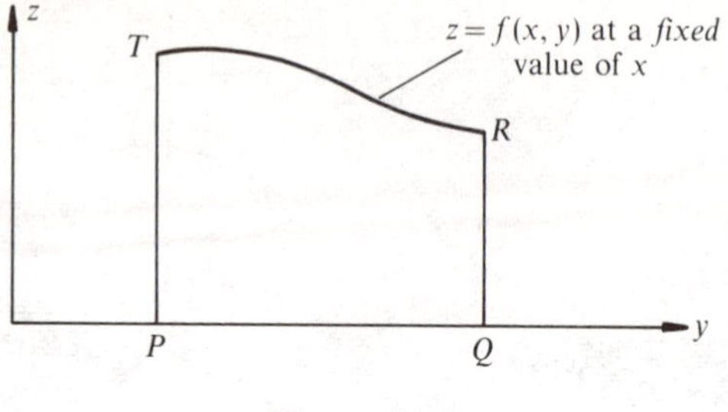

Figure A2.5

x is kept constant. But we already know how to evaluate the area between this curve and the region between P and Q on the y-axis. It is

$$\text{area}(PQRT) = \int_{y=P}^{y=Q} f(x, y)\,\mathrm{d}y \tag{A2.3}$$

with the constraint that in evaluating this integral the variable x is held fixed.

To perform this integration we need more precise information on the values of P and Q. We obtain this by analysing the area A in more detail. The region A is *defined* by its boundary curve, the major characteristic of which is that it is closed and, as such, will be (at least) double valued, that is, to each value of x there are *two* values of y (see Figure A2.6). We can eliminate this difficulty by splitting the boundary curve into two single-valued parts (called branches) by the lines $x = a$ and $x = b$. These lines segment the boundary curve into two parts, each of which may be described by a single-valued function. The upper part may be denoted by $y = \phi_2(x)$ and the lower part by $y = \phi_1(x)$ (see Figure A2.7). For example, the circle $x^2 + y^2 = a^2$ which is a double-valued curve may be split into two single-valued parts – $y = +\sqrt{(a^2 - x^2)}$, which describes the upper half, and $y = -\sqrt{(a^2 - x^2)}$, which defines the lower half. Figure A2.7 immediately implies that P has coordinates $(x, \phi_1(x))$ since it lies on the lower curve and that Q, lying on

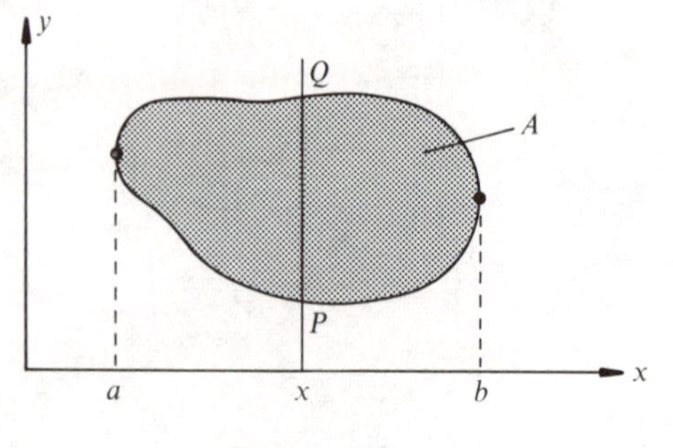

Figure A2.6

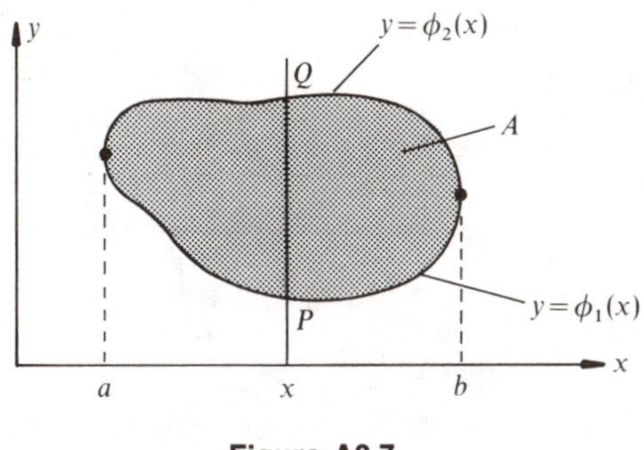

Figure A2.7

the upper curve, has coordinates $(x, \phi_2(x))$. Hence, using (A2.3),

$$\text{area}(PQRT) = \int_{\phi_1(x)}^{\phi_2(x)} f(x, y)\,\mathrm{d}y \quad \text{(keeping } x \text{ fixed)}$$

Now to find the total volume of the cylinder in Figure A2.3b, we simply sum all terms of the form $\mathrm{d}V$ as given by (A2.2) as x changes from its extreme left-hand value a to its extreme right-hand value b:

volume of the cylinder
= sum from $x=a$ to $x=b$ of all terms of the form $\mathrm{d}V$

that is, total volume $V = \int_a^b \mathrm{d}V$

or
$$\iint_A f(x, y)\,\mathrm{d}x\,\mathrm{d}y = \int_a^b \mathrm{d}V = \int_a^b (\text{area}(PQRT))\,\mathrm{d}x$$

$$= \int_a^b \left\{ \int_{y=\phi_1(x)}^{y=\phi_2(x)} f(x, y)\,\mathrm{d}y \right\} \mathrm{d}x \qquad \textbf{(A2.4)}$$

where we must emphasize that in the inner integral (which is performed first), the variable x is held fixed. The integration with respect to y is carried through in the usual manner. When substituting in the limits (again as usual) the 'y-terms' are replaced by $\phi_2(x)$ at the upper limit and by $\phi_1(x)$ at the lower limit. This inner integral thus results not in a constant, but in a function of x. This function of x is now integrated in the normal way from $x=a$ to $x=b$. The integral on the right-hand side of (A2.4) is sometimes called an **iterated integral**.

In conclusion, to evaluate a double integral $\iint_A f(x, y)\,\mathrm{d}x\,\mathrm{d}y$ in terms of an iterated integral, we need to describe the region A by the use of four pieces of information:

- $x=a$ the equation of the line bounding A to the left,

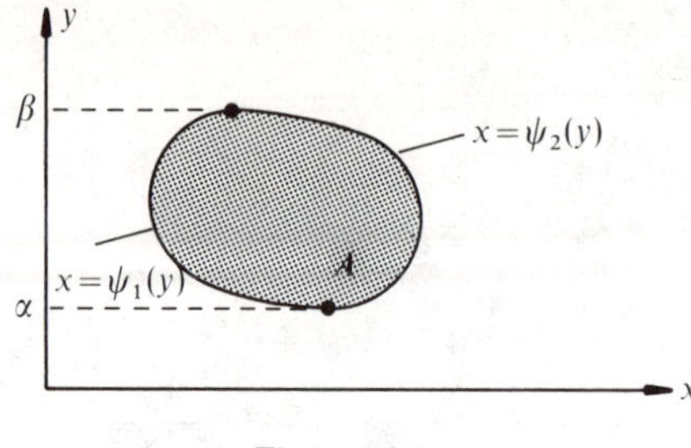

Figure A2.8

- $x = b$ the equation of the line bounding A to the right,
- the curve $y = \phi_1(x)$ bounding A below,
- the curve $y = \phi_2(x)$ bounding A above.

In obtaining (A2.4) we have chosen to segment the cylinder by planes parallel to the zy-plane. The alternative approach is to segment the cylinder by planes parallel to the zx-plane. Proceeding precisely as before, we are led to an alternative (but equivalent) prescription for a double integral in terms of an iterated integral:

$$\iint_A f(x, y)\,\mathrm{d}x\,\mathrm{d}y = \int_\alpha^\beta \left\{ \int_{x=\psi_1(y)}^{x=\psi_2(y)} f(x, y)\,\mathrm{d}x \right\} \mathrm{d}y \tag{A2.5}$$

where α, β, $\psi_1(y)$ and $\psi_2(y)$ are as shown in Figure A2.8.

Although both forms (A2.4) and (A2.5) are equivalent (that is, they lead to identical numerical values), one form may be easier to evaluate than the other, depending on the problem under consideration.

It is clear from the meaning of a double integral outlined above that if A comprises two disjoint parts $A = A_1 \cup A_2$ (see Figure A2.9), then

$$\iint_A f(x, y)\,\mathrm{d}x\,\mathrm{d}y = \iint_{A_1} f(x, y)\,\mathrm{d}x\,\mathrm{d}y + \iint_{A_2} f(x, y)\,\mathrm{d}x\,\mathrm{d}y \tag{A2.6}$$

Also, if $f(x, y)$ is a surface, part of which lies above the xy-plane and part

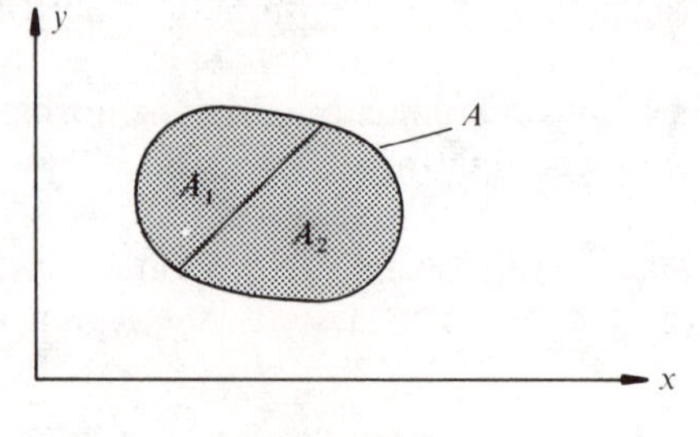

Figure A2.9

of which lies below the xy-plane, then those surface elements above the plane make a positive contribution to $\iint_A f(x, y)\,dx\,dy$ whilst those below make a negative contribution.

Example A2.1

Evaluate $\displaystyle\iint_A (x^2 - y^2)\,dx\,dy$, where A is the triangle with vertices $(-1, 1)$, $(0, 0)$ and $(1, 1)$.

Solution

From Figure A2.10 we see that we must proceed with caution because of the complicated nature of the 'lower' curve.

Using (A2.4),

$$\iint_A (x^2 - y^2)\,dx\,dy = \int_{-1}^{1} \left\{ \int_{|x|}^{1} (x^2 - y^2)\,dy \right\} dx$$

Now $\displaystyle\int_{|x|}^{1} (x^2 - y^2)\,dy = \left[x^2 y - \tfrac{1}{3}y^3 \right]_{|x|}^{1}$, treating x as fixed

$$= (x^2 - \tfrac{1}{3}) - (x^2|x| - \tfrac{1}{3}|x|^3)$$

$$= x^2 - \tfrac{1}{3} - \tfrac{2}{3}x^2|x|$$

since $|x|^3 = x^2|x|$. (We note that $|x|^2$ is always positive and has the same numerical value as x^2.) Therefore

$$\iint_A (x^2 - y^2)\,dx\,dy = \int_{-1}^{1} (x^2 - \tfrac{1}{3} - \tfrac{2}{3}x^2|x|)\,dx$$

For the integration with respect to x, we split the range of integration from -1 to 0 and then from 0 to 1 in order to accommodate the

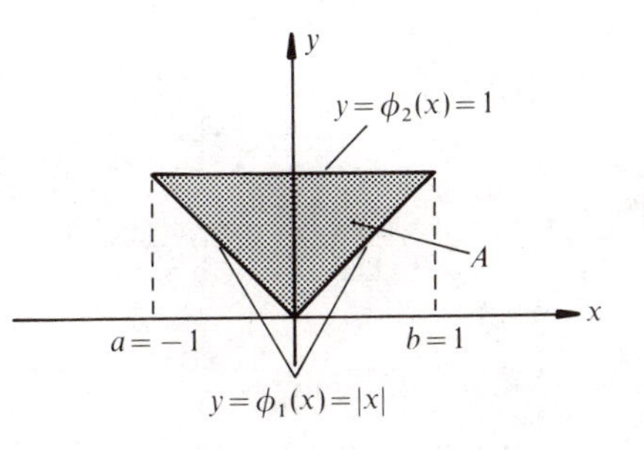

Figure A2.10

appearance of $|x|$ in the integrand. Since

$$|x| = \begin{cases} -x & \text{if } x < 0 \\ x & \text{if } x > 0 \end{cases}$$

we can write

$$\iint_A (x^2 - y^2)\,\mathrm{d}x\,\mathrm{d}y = \int_{-1}^{0} (x^2 - \tfrac{1}{3} + \tfrac{2}{3}x^3)\,\mathrm{d}x + \int_0^1 (x^2 - \tfrac{1}{3} - \tfrac{2}{3}x^3)\,\mathrm{d}x$$

$$= \left[\tfrac{1}{3}x^3 - \tfrac{1}{3}x + \tfrac{2}{12}x^4\right]_{-1}^{0} + \left[\tfrac{1}{3}x^3 - \tfrac{1}{3}x - \tfrac{2}{12}x^4\right]_0^1$$

$$= -\tfrac{1}{3}$$

For this particular problem, the integration is less taxing if we use the alternative formulation (A2.5). For this, area A needs to be re-described as in Figure A2.11. Then

$$\iint_A (x^2 - y^2)\,\mathrm{d}x\,\mathrm{d}y = \int_0^1 \left\{\int_{-y}^{y} (x^2 - y^2)\,\mathrm{d}x\right\} \mathrm{d}y$$

$$= \int_0^1 \left[\tfrac{1}{3}x^3 - y^2 x\right]_{-y}^{y} \mathrm{d}y$$

$$= \int_0^1 (\tfrac{1}{3}y^3 - y^3 + \tfrac{1}{3}y^3 - y^3)\,\mathrm{d}y$$

$$= \int_0^1 (-\tfrac{4}{3}y^3)\,\mathrm{d}y = \left[-\tfrac{1}{3}y^4\right]_0^1 = -\tfrac{1}{3}$$

The negative value of the final answer is a consequence of the fact that the expression $(x^2 - y^2)$ is negative within the region A, so that the surface S defined by $z = x^2 - y^2$ within A lies below the xy-plane.

We should note that the first approach to the evaluation of this double integral could have been made considerably easier with a little thought, by utilizing the inherent symmetry in the problem. The integrand $(x^2 - y^2)$ is symmetric in x; also, the region over which the integration is performed is symmetric with respect to the y-axis. Hence the required double integral is given by twice the value of the double

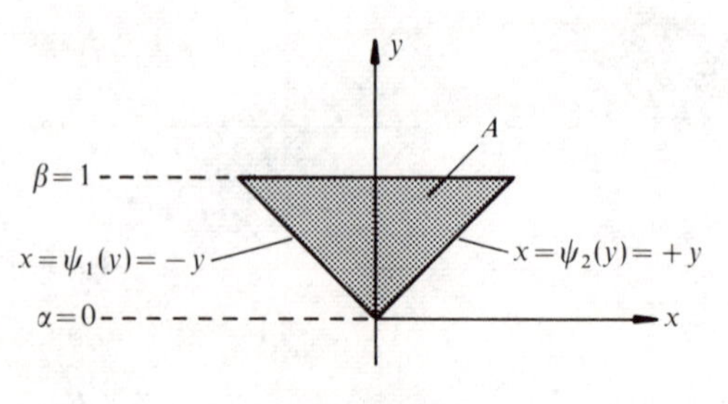

Figure A2.11

integral taken over that half of the region A which is in the positive quadrant:

$$\iint_A (x^2 - y^2)\,dx\,dy = 2\int_0^1 \left\{\int_x^1 (x^2 - y^2)\,dy\right\} dx$$

$$= 2\int_0^1 \left[x^2 y - \tfrac{1}{3}y^3\right]_x^1 dx$$

$$= 2\int_0^1 (x^2 - \tfrac{1}{3} - \tfrac{2}{3}x^3)\,dx$$

$$= -\tfrac{1}{3}$$

Example A2.2

A lamina A of total mass M is bounded by the parabola $y^2 = 4ax$ and the line $x = a$. Its mass density is $\rho(x, y) = kx^2$. Use double integrals to find the centre of mass of the lamina.

Solution

The centre of mass is defined to be that point at which a point mass, equal to the total mass M of the lamina, would have to be placed so as to have the same moments about the x- and y-axes as does the actual lamina.

We first segment the lamina into a large number of elements, each of area $dx\,dy$ (see Figure A2.12). The mass of such an element is $\rho(x, y)\,dx\,dy$, where ρ is the density of the lamina (mass per unit area). In this case,

$$dm = kx^2\,dx\,dy$$

The moment of this mass about the y-axis is approximately $x\,dm$ and its moment about the x-axis approximately $y\,dm$. Therefore if $(\bar{x}, \bar{y})$ denotes the position of the centre of mass, then $M\bar{x}$ is the sum of all moments of the form $x\,dm$ as (x, y) ranges over A, that is

$$M\bar{x} = \iint_A x(kx^2\,dx\,dy) = \iint_A kx^3\,dx\,dy$$

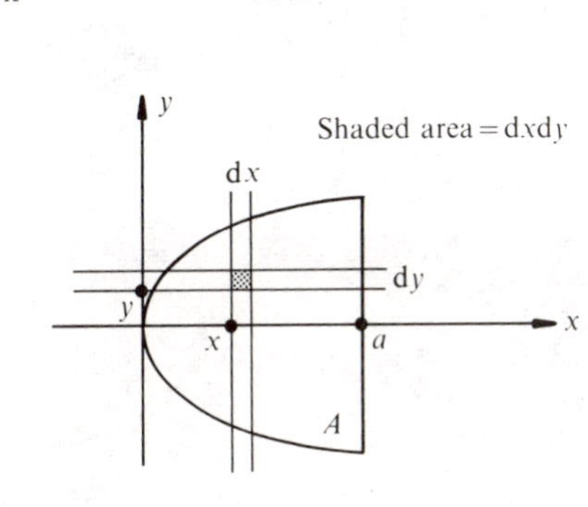

Figure A2.12

Now M is the sum of all terms of the form dm as (x, y) ranges over the lamina, that is,

$$M = \iint_A kx^2 \, \mathrm{d}x \, \mathrm{d}y$$

Therefore

$$\bar{x} = \frac{\displaystyle\iint_A kx^3 \, \mathrm{d}x \, \mathrm{d}y}{\displaystyle\iint_A kx^2 \, \mathrm{d}x \, \mathrm{d}y}$$

Clearly, $\bar{y} = 0$ (by symmetry).

To evaluate these integrals, we analyse the lamina A as in (A2.5) (see Figure A2.13), although we could equally well use (A2.4). We have

$$\begin{aligned} M = \iint_A kx^2 \, \mathrm{d}x \, \mathrm{d}y &= \int_{-2a}^{2a} \left(\int_{y^2/4a}^{a} kx^2 \, \mathrm{d}x \right) \mathrm{d}y = \int_{-2a}^{2a} \left[\tfrac{1}{3}kx^3 \right]_{y^2/4a}^{a} \mathrm{d}y \\ &= \int_{-2a}^{2a} \left(\tfrac{1}{3}ka^3 - \tfrac{1}{3}k \frac{y^6}{64a^3} \right) \mathrm{d}y \\ &= \tfrac{2}{3}k \left[a^3 y - \frac{y^7}{7(64)a^3} \right]_0^{2a} \end{aligned}$$

where we have utilized the symmetry of the region and the integrand about the x-axis. Hence

$$M = \tfrac{2}{3}k[2a^4 - \tfrac{2}{7}a^4] = \tfrac{8}{7}ka^4$$

Also,

$$\begin{aligned} \iint_A kx^3 \, \mathrm{d}x \, \mathrm{d}y &= \int_{-2a}^{2a} \left(\int_{y^2/4a}^{a} kx^3 \, \mathrm{d}x \right) \mathrm{d}y = \int_{-2a}^{2a} \left[\tfrac{1}{4}kx^4 \right]_{y^2/4a}^{a} \mathrm{d}y \\ &= \int_{-2a}^{2a} \left(\tfrac{1}{4}ka^4 - \tfrac{1}{4}k \frac{y^8}{256a^4} \right) \mathrm{d}y \\ &= \tfrac{1}{2}k \left[a^4 y - \frac{y^9}{9(256)a^4} \right]_0^{2a} \\ &= \tfrac{1}{2}k[2a^5 - \tfrac{2}{9}a^5] = \tfrac{8}{9}ka^5 \end{aligned}$$

Therefore

$$\bar{x} = \frac{\tfrac{8}{9}ka^5}{\tfrac{8}{7}ka^4} = \tfrac{7}{9}a$$

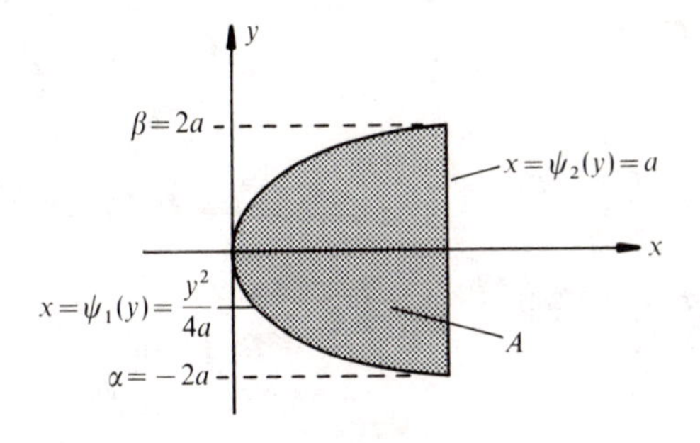

Figure A2.13

In this particular problem we should note that, with a little ingenuity, the position of the centre of mass could have been obtained using the techniques of single integration. This is because the surface density $\rho(x, y)$ depended only on x. If ρ also depended on y, then the more basic double integration technique that we have used would be essential.

APPENDIX 3

Some Historical Notes on Vector Analysis

The word 'vector' is derived from the Latin for 'to carry', and the idea that a force is a vector quantity – in that the resultant of two forces can be found using the parallelogram law – was known to Aristotle. Galileo Galilei (1564–1642) stated the parallelogram law explicitly. Thus the roots of our subject lie far back in history, but vector analysis as we know it today is largely a nineteenth-century creation.

By the year 1830, the use of complex numbers to represent vectors in a plane was well known as a result of the work of Wessel in the late eighteenth century and Argand and Gauss in the early nineteenth. Essentially, complex numbers provided an algebra to represent vectors in a plane, as well as operations with such vectors. Clearly, however, if several forces act on a body these forces do not necessarily all lie in one plane, and this led to a search for a 'three-dimensional complex number' and an associated algebra. The two most significant developments in this regard were the invention of **quaternions** by W. R. Hamilton (1805–65) in Ireland in 1843 and the publication of the book *Die lineale Ausdehnungslehre* (*The Calculus of Extension*) by H. G. Grassmann (1809–77) in Germany in 1844. Though these events were almost contemporaneous, they were independent and were motivated by very different considerations. Hamilton wished to develop mathematical tools for use in astronomy and physics, whereas Grassmann's aims were more philosophic and abstract, and were concerned with developing a theoretical algebraic structure for multidimensional geometry.

A3.1 Hamilton's quaternions

A quaternion is an entity of the form

$$a + b\mathbf{i} + c\mathbf{j} + d\mathbf{k}$$

where a, b, c and d are scalars and $\mathbf{i}$, $\mathbf{j}$ and $\mathbf{k}$ are somewhat analogous to the quantity i in complex number theory. The quantity a is referred to as the scalar part of the quaternion and the remainder as the vector part, the three coefficients b, c and d being the Cartesian coordinates of a point P.

The operation of adding two quaternions is defined in an obvious way, and the criterion for equality of two quaternions is as one would expect. The usual algebraic rules of multiplication are assumed to be valid, except that the rules for products of i, j and k are

(1) $\mathbf{i}^2 = \mathbf{j}^2 = \mathbf{k}^2 = -1$ (and hence the similarity to complex numbers)

(2) $\mathbf{ij} = \mathbf{k}, \quad \mathbf{ji} = -\mathbf{k}$

$\mathbf{jk} = \mathbf{i}, \quad \mathbf{kj} = -\mathbf{i}$

$\mathbf{ki} = \mathbf{j}, \quad \mathbf{ik} = -\mathbf{j}$

where the commutative law is abandoned (and where we can see the germination of the familiar cross product results).

Hence, if

$$\alpha = \alpha_0 + \alpha_1\mathbf{i} + \alpha_2\mathbf{j} + \alpha_3\mathbf{k} \quad \text{and} \quad \beta = \beta_0 + \beta_1\mathbf{i} + \beta_2\mathbf{j} + \beta_3\mathbf{k}$$

are two quaternions, then their product is

$$\begin{aligned}\alpha\beta = \alpha_0\beta_0 &- (\alpha_1\beta_1 + \alpha_2\beta_2 + \alpha_3\beta_3) \\ &+ (\alpha_2\beta_3 - \beta_2\alpha_3)\mathbf{i} + (\alpha_3\beta_1 - \beta_3\alpha_1)\mathbf{j} + (\alpha_1\beta_2 - \beta_1\alpha_2)\mathbf{k} \\ &+ (\alpha_0\beta_1 + \beta_0\alpha_1)\mathbf{i} + (\alpha_0\beta_2 + \beta_0\alpha_2)\mathbf{j} + (\alpha_0\beta_3 + \beta_0\alpha_3)\mathbf{k} \quad \textbf{(A3.1)}\end{aligned}$$

Hamilton also introduced the symbol ∇, standing for the differential operator

$$\mathbf{i}\frac{\partial}{\partial x} + \mathbf{j}\frac{\partial}{\partial y} + \mathbf{k}\frac{\partial}{\partial z}$$

which when applied to a scalar point function $u(x, y, z)$ produces a vector

$$\nabla u = \mathbf{i}\frac{\partial u}{\partial x} + \mathbf{j}\frac{\partial u}{\partial y} + \mathbf{k}\frac{\partial u}{\partial z}$$

The application of this operator to a vector point function

$$\mathbf{v} = v_1\mathbf{i} + v_2\mathbf{j} + v_3\mathbf{k}$$

was defined as producing a quaternion:

$$\begin{aligned}\nabla\mathbf{v} &= \left(\mathbf{i}\frac{\partial}{\partial x} + \mathbf{j}\frac{\partial}{\partial y} + \mathbf{k}\frac{\partial}{\partial z}\right)(v_1\mathbf{i} + v_2\mathbf{j} + v_3\mathbf{k}) \\ &= -\left(\frac{\partial v_1}{\partial x} + \frac{\partial v_2}{\partial y} + \frac{\partial v_3}{\partial z}\right) + \left(\frac{\partial v_3}{\partial y} - \frac{\partial v_2}{\partial z}\right)\mathbf{i} + \left(\frac{\partial v_1}{\partial z} - \frac{\partial v_3}{\partial x}\right)\mathbf{j} \\ &\quad + \left(\frac{\partial v_2}{\partial x} - \frac{\partial v_1}{\partial y}\right)\mathbf{k}\end{aligned}$$

in accordance with the multiplication rules. Today we would recognize the scalar part of $\nabla\mathbf{v}$ as the negative of the divergence of $\mathbf{v}$ and the vector part as **curl v**.

Hamilton used quaternions in solving physical problems and strongly advocated their use to other physicists for the remainder of his life. Later, the quaternion concept was championed by P. G. Tait (1831–1901) in Scotland, who pressed the case for them as a fundamental tool in both physics and geometry. However, they were not extensively taken up by physicists or applied mathematicians, conventional Cartesian coordinates continuing to be preferred.

A3.2 Grassmann's calculus of extension

Grassmann's basic concept, which he called an extensive quantity, has n components,

$$\alpha_1 \mathrm{e}_1 + \alpha_2 \mathrm{e}_2 + \cdots + \alpha_n \mathrm{e}_n$$

where the α_i are scalars and the e_i are primary units with, in the case $n = 3$, geometric representations as directed line segments of unit length drawn from a common origin to determine a right-handed orthogonal system of axes. As well as an obvious definition of the sum of two such quantities, Grassmann introduced two kinds of multiplication:

(1) An inner product, such that

$$\mathrm{e}_i/\mathrm{e}_i = 1 \quad \text{and} \quad \mathrm{e}_i/\mathrm{e}_j = 0, \quad i \neq j$$

so that if

$$\alpha = \alpha_1 \mathrm{e}_1 + \alpha_2 \mathrm{e}_2 + \cdots + \alpha_n \mathrm{e}_n$$

and $$\beta = \beta_1 \mathrm{e}_1 + \beta_2 \mathrm{e}_2 + \cdots + \beta_n \mathrm{e}_n$$

the inner product α/β is

$$\alpha/\beta = \alpha_1 \beta_1 + \alpha_2 \beta_2 + \cdots + \alpha_n \beta_n \tag{A3.2}$$

and $\beta/\alpha = \alpha/\beta$. The numerical value of one of Grassmann's extensive quantities was defined as

$$\sqrt{(\alpha/\alpha)} = \sqrt{(\alpha_1^2 + \alpha_2^2 + \cdots + \alpha_n^2)}$$

(2) An outer product $[\alpha\beta]$, defined such that

$$[\mathrm{e}_i \mathrm{e}_i] = 0 \quad \text{and} \quad [\mathrm{e}_i \mathrm{e}_j] = -[\mathrm{e}_j \mathrm{e}_i], \quad i \neq j$$

so that, in the $n = 3$ case,

$$\begin{aligned}[\alpha\beta] = {} & (\alpha_2\beta_3 - \alpha_3\beta_2)[\mathrm{e}_2\mathrm{e}_3] + (\alpha_3\beta_1 - \alpha_1\beta_3)[\mathrm{e}_3\mathrm{e}_1] \\ & + (\alpha_1\beta_2 - \alpha_2\beta_1)[\mathrm{e}_1\mathrm{e}_2]\end{aligned} \tag{A3.3}$$

If we compare Grassmann's inner product (A3.2) in the case where $n=3$ with Hamilton's quaternion product (A3.1), we see clearly that the former is equivalent to the negative of the scalar part of Hamilton's quaternion product when the two quaternions have no scalar part (that is, when they are vectors). Similarly, and again in the three-dimensional case, if we replace $[e_2e_3]$ by e_1, and so on, Grassmann's outer product (A3.3) is equivalent to the vector part of the quaternion product of two vectors. However, whereas in the theory of quaternions the vector is just one part of the quaternion, in Grassmann's algebra the vector is the basic quantity.

For the general n-dimensional case, Grassmann defined products other than the two discussed here, and his thinking helped to lead on to more advanced concepts, such as tensors. However, his work, although highly original, remained little known for many years after its publication.

A3.3 The work of Maxwell

With Grassmann's work little known and the quaternion concept (despite the advocacy of Hamilton and Tait) considered unsuitable for applications, the next significant development in vector analysis was made in England by J. C. Maxwell (1831–79) in his theoretical analysis of electricity and magnetism. Maxwell was familiar with Hamilton's ideas, and essentially he separated out the scalar and vector parts of a quaternion. In Hamilton's quaternion,

$$\nabla\mathbf{v} = -\left(\frac{\partial v_1}{\partial x}+\frac{\partial v_2}{\partial y}+\frac{\partial v_3}{\partial z}\right)+\left(\frac{\partial v_3}{\partial y}-\frac{\partial v_2}{\partial z}\right)\mathbf{i}+\left(\frac{\partial v_1}{\partial z}-\frac{\partial v_3}{\partial x}\right)\mathbf{j}$$
$$+\left(\frac{\partial v_2}{\partial x}-\frac{\partial v_1}{\partial y}\right)\mathbf{k}$$

Maxwell denoted the scalar part by $S\,\nabla\mathbf{v}$ and called it the **convergence** of **v** because the expression was already known in fluid dynamics, where **v** was a velocity. The modern term **divergence** was coined shortly afterwards by W. L. Clifford (1845–79), who defined

$$\text{div}\,\mathbf{v} = -S\,\nabla\mathbf{v} = \frac{\partial v_1}{\partial x}+\frac{\partial v_2}{\partial y}+\frac{\partial v_3}{\partial z}$$

The vector part of $\nabla\mathbf{v}$ was denoted by $V\,\nabla\mathbf{v}$ by Maxwell, who called this quantity the **curl** or **rotation** of **v** because this expression too had already appeared in fluid dynamics as twice the rate of rotation of the fluid at a point. The term **rot v** still appears occasionally as a synonym for both **curl v** and, as we have mainly used in this text, $\nabla \wedge \mathbf{v}$.

Other contributions made by Maxwell were to point out that the operator ∇ repeated gives

$$\nabla^2 = -\left(\frac{\partial^2}{\partial x^2} + \frac{\partial^2}{\partial y^2} + \frac{\partial^2}{\partial z^2}\right)$$

which he called the Laplacian operator. He also noted the results

$$\mathbf{curl}\,\mathbf{grad}\,u = 0 \quad \text{for any scalar function } u$$

$$\text{div}\,\mathbf{curl}\,\mathbf{v} = 0 \quad \text{for any vector field } \mathbf{v}$$

A3.4 Modern vector analysis

The final break from quaternion theory and the beginning of modern vector analysis was due to J. W. Gibbs (1839–1903) in the USA and O. Heaviside (1850–1925), an electrical engineer, in Britain. Their work was independent but the end results were effectively the same, apart from differences in notation and terminology. Gibbs, a professor of mathematical physics, was motivated by the need for a simpler mathematical framework than quaternions, both for teaching and research purposes. In 1881 he produced a pamphlet *Elements of Vector Analysis* for private circulation among his students and other interested parties, and only later (in 1901) consented to the formal presentation of his work in book form, the actual writing of this book being carried out by one of his pupils, E. B. Wilson. Most texts on vector analysis adhere to the form set down by Gibbs, although he himself was somewhat reluctant to publish his ideas:

> 'The reluctance of Professor Gibbs to publish his system of vector analysis certainly did not arise from any doubt in his own mind as to its utility, or the desirability of its being more widely employed; it seemed rather to be due to the feeling that it was not an original contribution to mathematics, but was an adaptation, for special purposes, of the work of others. Of many portions of the work this is of course necessarily true; and it is rather by the selection of methods and by systemisation of the presentation that the author has served the cause of vector analysis. But in the treatment of the linear vector function and the theory of dyadics to which this leads, a distinct advance was made which was of consequence not only in the more restricted field of vector analysis, but also in the broader theory of multiple algebra in general.'

Heaviside, like Maxwell, was primarily interested in electricity and magnetism. He rejected quaternions as being an unsuitable mathematical tool for this area and developed his own vector analysis, which he regarded as merely a shorthand. The correspondence between Gibbs and Heaviside on the one hand and Tait, as a 'quaternionist', on the other makes for interesting and even amusing reading today, Heaviside in particular being acerbic in his remarks about quaternions. Some of the correspondence is quoted in the introduction to the classic text of C. E. Weatherburn (first published in 1921), from which the following comment of Heaviside is reproduced, together with a postscript by Weatherburn.

> 'Suppose a sufficiently competent mathematician desired to find out from the Cartesian mathematics what vector algebra was like, and its laws. He could do so by careful inspection and comparison of the Cartesian formulae. He would find certain combinations of symbols and quantities occurring again and again, usually in systems of threes. He might introduce tentatively an abbreviated notation for these combinations. After a little practice he would perceive the laws according to which these combinations arose and how they operated. Finally, he would come to a very compact system in which vectors themselves and certain simple functions of vectors appeared, and would be delighted to find that the rules for the multiplication and general manipulation of these vectors were, considering the complexity of the Cartesian mathematics out of which he had discovered them, of an almost incredible simplicity. But there would be no sign of a quaternion in his result, for one thing; and for another, there would be no metaphysics or abstruse reasoning required to establish the rules of manipulation of his vectors.' This is the manner in which one would expect Vector Analysis to have originated. But it did not; and its parentage has in many quarters counted against it.'

The notational differences between Gibbs and Heaviside are worth noting:

	Scalar product	Vector product	Gradient	Divergence	Curl
Gibbs:	$\mathbf{a}\cdot\mathbf{b}$	$\mathbf{a}\times\mathbf{b}$	∇	$\nabla\cdot$	$\nabla\times$
Heaviside:	$\mathbf{ab}$	$V\mathbf{ab}$	∇	div	**curl**

It is interesting to note that Maxwell's equations for the electromagnetic field were usually written out by Maxwell himself in component form: it was Heaviside who wrote them in the vector form that we have used in Chapter 7.

The notation $\mathbf{a}\wedge\mathbf{b}$ for the vector product and $\nabla\wedge\mathbf{a}$ for the curl, as used in this text, was used by Italian vector analysts, particularly R. Marcolongo (1862–1943) and C. Burali-Forti (1861–1931). In point of fact, the problem of notation in vector analysis was a subject for heated debate in the early years of the twentieth century.

A3.5 Other contributions

The two basic integral theorems – that of Stokes, and the divergence theorem – were not originally developed in the concise language of vector analysis, indeed they predated the work of Gibbs and Heaviside by many years. In a paper published in 1831 on solving the heat conduction equation, the Russian M. Ostogradsky (1801–61) converted a volume integral of the form

$$\iiint_V \left(\frac{\partial P}{\partial x} + \frac{\partial Q}{\partial y} + \frac{\partial R}{\partial z}\right) \mathrm{d}x\, \mathrm{d}y\, \mathrm{d}z$$

into a surface integral

$$\iint_S (P \cos \lambda + Q \cos \mu + R \cos v)\, \mathrm{d}S$$

where P, Q and R are scalar functions of x, y and z and are components of a vector, and λ, μ and v are the direction cosines of the normal to the surface S enclosing the volume V. However, the theorem, which in modern notation is of course

$$\iiint_V \nabla \cdot \mathbf{v}\, \mathrm{d}V = \oiint_S \mathbf{v} \cdot \hat{\mathbf{n}}\, \mathrm{d}S$$

is also attributed to C. F. Gauss (1777–1855) in Germany, while George Green (1793–1841) in England also made systematic use of integral identities equivalent to the divergence theorem in 1828. Stokes' theorem was probably first stated in a letter to Stokes by Lord Kelvin (Sir William Thomson) in England in 1850. Stokes (1819–1903) used the theorem as a question in a prize examination at Cambridge in 1854. Interestingly enough, one of the candidates for the prize was J. C. Maxwell!

References

Crowe M. J. (1967). *History of Vector Analysis*. Indiana: University of Notre Dame Press

Gibbs J. W. and Wilson E. B. (1960). *Vector Analysis* (reprint). New York: Dover

Heaviside O. (1925). *Electromagnetic Theory* (reprint). New York: Dover

Kline M. (1972). *Mathematical Thought from Ancient to Modern Times*. Oxford University Press

Weatherburn C. E. (1955). *Elementary Vector Analysis* (2nd edn). Bell

Answers and Hints to Exercises

Chapter 1

1.1 $(-4, -5.333, -4)$, $(-1.333, 0, 1.333)$

1.3 $6\hat{\mathbf{i}} + \hat{\mathbf{j}} + 16\hat{\mathbf{k}}$; $\sqrt{74}, \sqrt{145}, \sqrt{98}$; $\sqrt{293}$

1.4 $-2.3\hat{\mathbf{i}} + 3\hat{\mathbf{k}}$, $3.7\hat{\mathbf{i}} + \hat{\mathbf{j}}$, $1.4\hat{\mathbf{i}} + \hat{\mathbf{j}} + 3\hat{\mathbf{k}}$; 3.78, 3.83, 3.30

1.10 $\frac{14}{3}, \frac{5}{3}$

1.14 $2.439\hat{\mathbf{i}} + 2.439\hat{\mathbf{j}} + 4.33\hat{\mathbf{k}}$; 5.536 N

1.15 $(2, 1, -1.05)$

1.16 $\mathbf{r} = \mathbf{a} + t(\mathbf{b} - \mathbf{a})$, where $\mathbf{a} = \hat{\mathbf{i}} + 2\hat{\mathbf{j}} - \hat{\mathbf{k}}$ and $\mathbf{b} = 2\hat{\mathbf{i}} + \hat{\mathbf{k}}$; $\dfrac{x-1}{1} = \dfrac{y-2}{-2} = \dfrac{z+1}{2} = t$

1.17 The points are not collinear.

1.18 13 min 36 s

1.20 $\mathbf{r} = (3 + 4t)\hat{\mathbf{i}} + (-2 + 6t)\hat{\mathbf{j}} + (5 - 7t)\hat{\mathbf{k}}$

1.21 $b + d = a + e$

1.27 5.1, 3.61, 10, 10

1.28 $P_1: \mathbf{r} \cdot \mathbf{b} = \mathbf{a} \cdot \mathbf{b}$, where $\mathbf{a} = 2\hat{\mathbf{i}} + 3\hat{\mathbf{j}} - \hat{\mathbf{k}}$ and $\mathbf{b} = 3\hat{\mathbf{i}} - 4\hat{\mathbf{j}} + \hat{\mathbf{k}}$; $P_2: \mathbf{r} \cdot \mathbf{b} = \mathbf{c} \cdot \mathbf{b}$, where $\mathbf{c} = \hat{\mathbf{i}} - \hat{\mathbf{j}} - \hat{\mathbf{k}}$
Distance between planes = 2.55 (planes on opposite sides of the origin)

1.29 120.56°, 150.56°, 30° (one of the angles is outside the triangle)

1.30 39 joules

1.31 Choose $\mathbf{c} = \hat{\mathbf{i}}$, then choose $\mathbf{c} = \hat{\mathbf{j}}$, then choose $\mathbf{c} = \hat{\mathbf{k}}$.

1.33 86 joules

1.35 70.89°

1.40 1.732, 5, $4\hat{\mathbf{i}} - 7\hat{\mathbf{j}} + 3\hat{\mathbf{k}}$, 8.6

1.41 $0.309\hat{\mathbf{i}} + 0.722\hat{\mathbf{j}} - 0.62\hat{\mathbf{k}}$

1.42 $57.5\ \mathrm{m\ s^{-1}}$

1.43 1.225

1.44 2.86

1.48 $-5\hat{\mathbf{i}}+8\hat{\mathbf{j}}+4.5\hat{\mathbf{k}}$, 4.5, 2.97

1.50 Resultant of magnitude 5.099 N and making angles of 56.3°, 46.1° and 65.4° with the edges of the cube

1.51 $(60/\sqrt{50})(-\hat{\mathbf{i}}+2\hat{\mathbf{j}}+\hat{\mathbf{k}})$ N m

1.57 **(a)** -4 **(b)** 20 **(c)** and **(d)** $-5\hat{\mathbf{i}}+5\hat{\mathbf{j}}-2\hat{\mathbf{k}}$

Solutions to the additional exercises may be found on p. 364.

1S.1 $a_i + a_k b_k b_i = 0$

1S.1 **(a)** Meaningful **(b)** Not meaningful (i, j free in first term and i free in second) **(c)** Not meaningful (i free in first term and k, i free in second) **(d)** Meaningful **(e)** Meaningful

1S.3 **(a)** For each value of i, k, evaluate the left-hand side and check that this is the same as the right-hand side.
(b) First sum over i (or j), then sum over j (or i); use the defined values of δ_{ij} to obtain the result.

1S.4 **(a)** $\varepsilon_{iik} = \varepsilon_{11k} + \varepsilon_{22k} + \varepsilon_{33k} = 0$ since the permutation symbol is zero if any two indices are equal.
(b) First summing over j,

$$\varepsilon_{ijk} a_j a_k = \varepsilon_{i1k} a_1 a_k + \varepsilon_{i2k} a_2 a_k + \varepsilon_{i3k} a_3 a_k$$

Now summing over k and using the properties of the permutation symbol,

$$\begin{aligned}\varepsilon_{ijk} a_j a_k &= \varepsilon_{i12} a_1 a_2 + \varepsilon_{i13} a_1 a_3 + \varepsilon_{i21} a_2 a_1 + \varepsilon_{i23} a_2 a_3 + \varepsilon_{i31} a_3 a_1 \\ &\quad + \varepsilon_{i32} a_3 a_2 \\ &= a_1 a_2(\varepsilon_{i12} + \varepsilon_{i21}) + a_1 a_3(\varepsilon_{i13} + \varepsilon_{i31}) + a_2 a_3(\varepsilon_{i23} + \varepsilon_{i32})\end{aligned}$$

But since $\varepsilon_{i12} = -\varepsilon_{i21}$, $\varepsilon_{i13} = -\varepsilon_{i31}$ and $\varepsilon_{i23} = -\varepsilon_{i32}$, we obtain the required result.

(c)
$$\begin{aligned}\varepsilon_{ijk}\varepsilon_{ijk} &= \varepsilon_{1jk}\varepsilon_{1jk} + \varepsilon_{2jk}\varepsilon_{2jk} + \varepsilon_{3jk}\varepsilon_{3jk} \\ &= 0 + \varepsilon_{12k}\varepsilon_{12k} + \varepsilon_{13k}\varepsilon_{13k} + \varepsilon_{21k}\varepsilon_{21k} + 0 + \varepsilon_{23k}\varepsilon_{23k} + \varepsilon_{31k}\varepsilon_{31k} \\ &\quad + \varepsilon_{32k}\varepsilon_{32k} + 0\end{aligned}$$

Now choosing k (in each term) to be such as to give a non-zero value for the permutation symbol, we obtain

$$\varepsilon_{ijk}\varepsilon_{ijk} = (\varepsilon_{123})^2 + (\varepsilon_{132})^2 + (\varepsilon_{213})^2 + (\varepsilon_{231})^2 + (\varepsilon_{312})^2 + (\varepsilon_{321})^2$$

Each of the terms on the right-hand side is $+1$, so finally $\varepsilon_{ijk}\varepsilon_{ijk} = 6$.

1S.5 Let a_i, b_i and c_i correspond to **a**, **b** and **c** respectively. Then

$$(\mathbf{b} \wedge \mathbf{c})_i = \varepsilon_{ijk} b_j c_k$$

$$\therefore \quad \mathbf{a} \cdot (\mathbf{b} \wedge \mathbf{c}) = a_i (\mathbf{b} \wedge \mathbf{c})_i = a_i \varepsilon_{ijk} b_j c_k = \varepsilon_{ijk} a_i b_j c_k$$

Similarly,

$$(\mathbf{a} \wedge \mathbf{b})\cdot\mathbf{c} = (\mathbf{a} \wedge \mathbf{b})_i c_i = \varepsilon_{ijk} a_j b_k c_i$$

But $\varepsilon_{ijk} = \varepsilon_{jki} = \varepsilon_{kij}$, so

$$\mathbf{a}\cdot(\mathbf{b} \wedge \mathbf{c}) = \varepsilon_{kij} a_i b_j c_k$$

which by relabelling the dummy indices as $i \to p$, $j \to q$ and $k \to r$ is clearly equal to $\varepsilon_{rpq} a_p b_q c_r$, and with a second relabelling $r \to i$, $p \to j$ and $q \to k$ is equal to $\varepsilon_{ijk} a_j b_k c_i$. But this is now identical to the indexed version of $(\mathbf{a} \wedge \mathbf{b})\cdot\mathbf{c}$, which proves the result. (After some practice, the index relabelling outlined here could be accomplished in one step.)

Chapter 2

2.1 **(a)** $\dfrac{x^2}{9} + \dfrac{y^2}{16} = 1$, $z = 3$; ellipse on the plane $z = 3$

(b) $8x = y^2$, $z = 0$; parabola on the plane $z = 0$

(c) $\dfrac{x^2}{9} - \dfrac{z^2}{16} = 1$, $y = 3$; hyperbola on the plane $y = 3$

2.2 $\mathbf{r} = a\{\sqrt{2}\cos(\theta - \frac{1}{4}\pi) + \sqrt{[\cos(2\theta - \frac{1}{2}\pi)]}\}\{\cos\theta\,\hat{\mathbf{i}} + \sin\theta\,\hat{\mathbf{j}}\}$, $0 \leqslant \theta \leqslant \pi/2$

2.3 **(b)** $-x - y + \sqrt{2}\,z - 1 = 0$

2.4 Any point on the curve has position vector $a\hat{\mathbf{i}} + b\hat{\mathbf{j}} + c\hat{\mathbf{k}}$, where $a = t\cos t$, $b = t\sin t$ and $c = t$. The angle this vector makes with the z-axis is $\theta = \cos^{-1}[c/\sqrt{(a^2 + b^2 + c^2)}]$. But for each point on the curve, $a^2 + b^2 = c^2$ and so $\theta = \cos^{-1}(1/\sqrt{2}) = 45°$, implying that the surface on which the curve lies can be a cone with apex at $t = 0$, the origin.

2.5

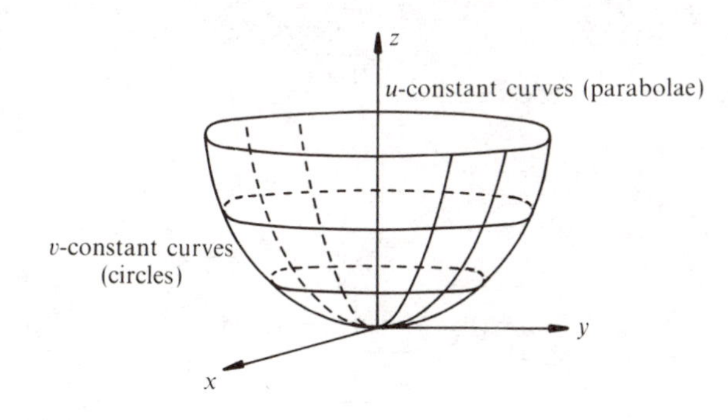

Figure E1

2.9 **(a)** $-3\cos\theta\,\hat{\mathbf{i}} - \cos 2\theta\,\hat{\mathbf{j}} + \frac{1}{4}\theta^4\hat{\mathbf{k}} + \mathbf{c}$ **(b)** $\frac{4}{15}t^{5/2}\hat{\mathbf{i}} + \frac{1}{2}t^2\hat{\mathbf{j}} + \frac{1}{2}t^2\hat{\mathbf{k}} + \mathbf{c}$

2.10 $\mathbf{v} = -0.141\hat{\mathbf{i}} - 0.989\hat{\mathbf{j}} + \hat{\mathbf{k}}$, $\mathbf{a} = +0.989\hat{\mathbf{i}} - 0.141\hat{\mathbf{j}}$

Speed at separation $= 1.4142$; position after 1 s is $(-1.13, -0.849, 4)$

2.16 **(a)** $\dfrac{1}{\sqrt{362}}(-19\hat{\mathbf{i}} + \hat{\mathbf{j}})$, $\kappa = 8.7 \times 10^{-4}$ **(b)** $\dfrac{1}{\sqrt{13}}(2\hat{\mathbf{i}} + 3\hat{\mathbf{j}})$, $\kappa = 0.362$

2.17 $\mathbf{n} = 16.63\hat{\mathbf{i}} - 6.89\hat{\mathbf{j}} - 3\hat{\mathbf{k}}$; $(\mathbf{r} - \mathbf{r}_0)\cdot\mathbf{n} = 0$, where $\mathbf{r} = x\hat{\mathbf{i}} + y\hat{\mathbf{j}} + z\hat{\mathbf{k}}$;
$\mathbf{r}_0 = 2.77\hat{\mathbf{i}} + 1.15\hat{\mathbf{j}} + 9\hat{\mathbf{k}}$

2.18 $0.707(\hat{\mathbf{j}} + \hat{\mathbf{k}})$

2.19 The angle between the surfaces is the angle between the normals at this point, namely 164.2°.

Solutions to the additional exercises may be found on p. 369.

Chapter 3

3.1 **(a)** Ellipsoids;

$$\hat{\mathbf{n}} = \frac{2x\hat{\mathbf{i}} + y\hat{\mathbf{j}} + 0.222z\hat{\mathbf{k}}}{\sqrt{(4x^2 + y^2 + 0.049z^2)}}$$

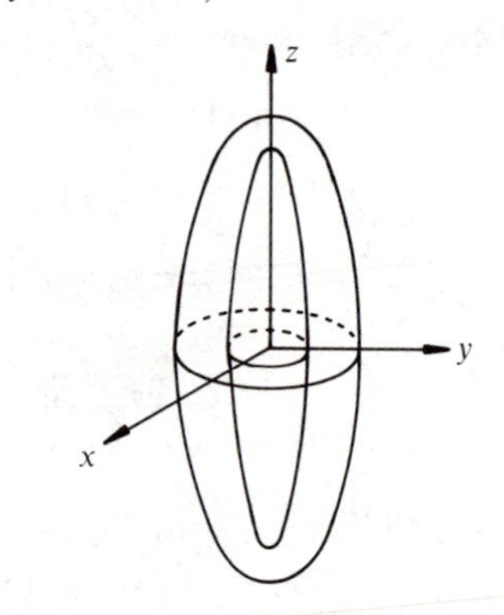

Figure E2

(b) A cone (single sheet) with apex on the z-axis at $z = k$;

$$\hat{\mathbf{n}} = \frac{1}{\sqrt{2}}\left(-\frac{x}{\sqrt{(x^2 + y^2)}}\hat{\mathbf{i}} - \frac{y}{\sqrt{(x^2 + y^2)}}\hat{\mathbf{j}} + \hat{\mathbf{k}}\right)$$

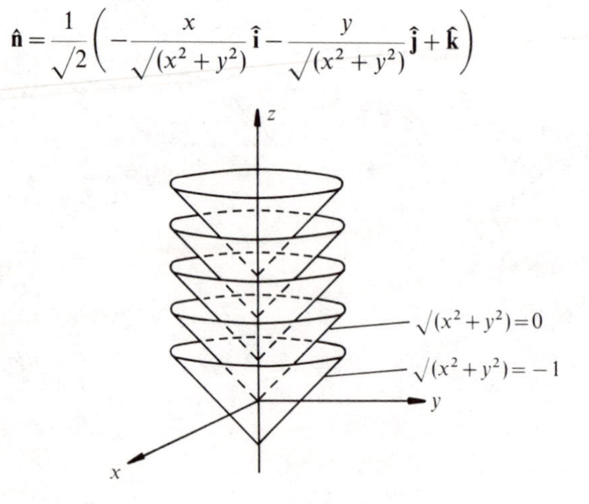

Figure E3

(c) Parabolae of revolution (upper sheets only);

$$\hat{\mathbf{n}} = \frac{-2x\hat{\mathbf{i}} - 2y\hat{\mathbf{j}} + \hat{\mathbf{k}}}{\sqrt{(4x^2 + 4y^2 + 1)}}$$

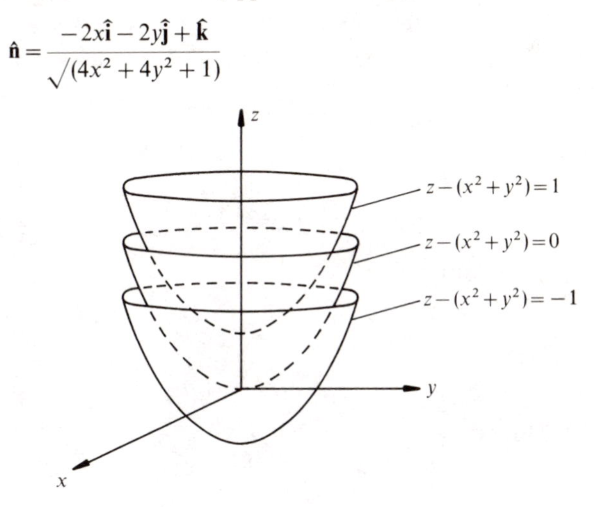

Figure E4

(d) Hyperbolae of revolution;

$$\hat{\mathbf{n}} = \frac{-x\hat{\mathbf{i}} - y\hat{\mathbf{j}} + z\hat{\mathbf{k}}}{\sqrt{(x^2 + y^2 + z^2)}}$$

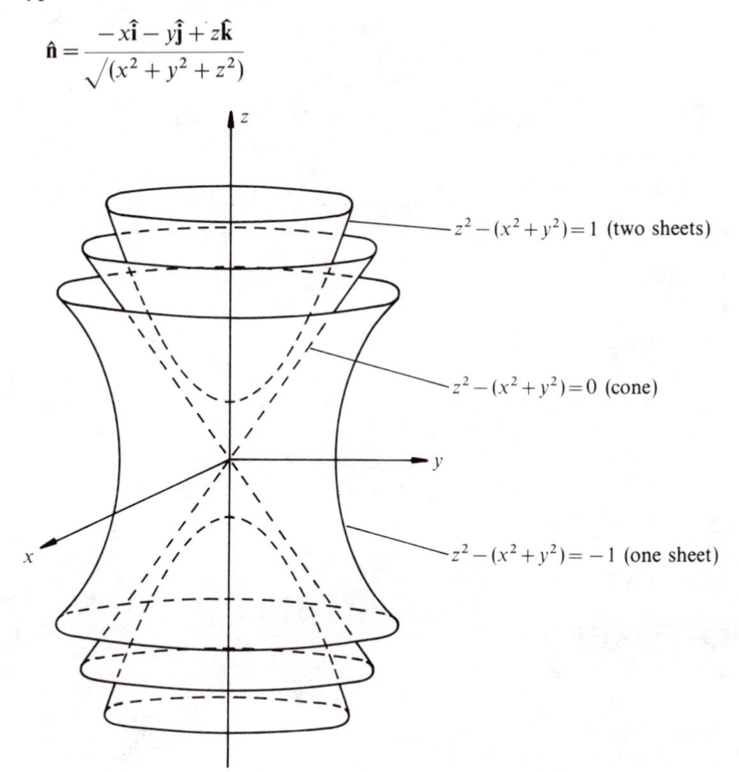

Figure E5

3.2 (a) $2\hat{\mathbf{i}}+2\hat{\mathbf{j}}+2\hat{\mathbf{k}}$ (b) $-\hat{\mathbf{i}}-\hat{\mathbf{k}}$

3.3 $(17-12\sqrt{24})/7$; along the direction of $\nabla\Phi$, that is $3\hat{\mathbf{i}}+2\hat{\mathbf{j}}-12\hat{\mathbf{k}}$

3.4 Along the negative x-direction; top of mountain is that point at which $\nabla\Phi=0$, in this case (0, 0).

3.7 (a) $4x+4y-4z=24$, $\mathbf{r}=t(2\hat{\mathbf{i}}+2\hat{\mathbf{j}}-2\hat{\mathbf{k}})$
(b) $4x+6y+4z=18$, $\mathbf{r}=\hat{\mathbf{i}}+\hat{\mathbf{j}}+2\hat{\mathbf{k}}+t(4\hat{\mathbf{i}}+6\hat{\mathbf{j}}+4\hat{\mathbf{k}})$

3.10 (b)
$$a_x\left(\frac{\partial b_x}{\partial x}\hat{\mathbf{i}}+\frac{\partial b_y}{\partial x}\hat{\mathbf{j}}+\frac{\partial b_z}{\partial x}\hat{\mathbf{k}}\right)+a_y\left(\frac{\partial b_x}{\partial y}\hat{\mathbf{i}}+\frac{\partial b_y}{\partial y}\hat{\mathbf{j}}+\frac{\partial b_z}{\partial y}\hat{\mathbf{k}}\right)+a_z\left(\frac{\partial b_x}{\partial z}\hat{\mathbf{i}}+\frac{\partial b_y}{\partial z}\hat{\mathbf{j}}+\frac{\partial b_z}{\partial z}\hat{\mathbf{k}}\right)$$

3.13 (a) $-\dfrac{(\mathbf{r}-\mathbf{a})}{|\mathbf{r}-\mathbf{a}|}$ (b) $\dfrac{\mathbf{a}}{r^3}-\dfrac{3\mathbf{r}(\mathbf{a}\cdot\mathbf{r})}{r^5}$

3.14 (b) (i) $\hat{\mathbf{i}}\left(-2-\dfrac{\pi}{2}\right)-\hat{\mathbf{j}}+\hat{\mathbf{k}}\left(1+\dfrac{\pi}{2}\right)$ (ii) $-\dfrac{\pi}{2}(-2\hat{\mathbf{i}}+\hat{\mathbf{k}})$ (iii) $-\dfrac{2}{\pi}(-2\hat{\mathbf{i}}+\hat{\mathbf{k}})$

3.15 See Section 3.6, where each of these vector fields is discussed in some detail.

3.18 No

3.19 (a) $yz\hat{\mathbf{i}}+xz\hat{\mathbf{j}}+xy\hat{\mathbf{k}}$ (b) $3xz\hat{\mathbf{i}}-3zy\hat{\mathbf{j}}$ (c) 0 (d) $2x+2y+2z$
(e) $3z(x^3-y^3)$ (f) $3(y\hat{\mathbf{i}}+x\hat{\mathbf{j}})$

Solutions to the additional exercises may be found on p. 372.

Chapter 4

4.1 (a) $\frac{3}{2}\sin 1$ (b) $\sin 1$ (c) $\frac{1}{2}+\sin 1-\frac{1}{2}\cos 1$

4.2 6π

4.3 27π

4.4 (a) $\frac{32}{3}$ (b) $\frac{28}{3}$

4.5 The fields which are conservative are (a), (c), (d) and (e); associated potentials are
(a) $\Phi=x+2x^2-yx+e^y-y^2+c$ (c) $\Phi=e^{xy}+c$
(d) $\Phi=xz+y+c$ (e) $\Phi=x^3-xy^2-z+c$

4.6 (a) 9 joules (b) 36 joules (c) $(45-45)=0$ joules. Yes

4.7 (a) $\Phi_1+\Phi_2$ (b) $\Phi_1=\dfrac{x^2+y^2}{2}+$ a constant; $\Phi_2=\dfrac{x^3y}{3}+$ a constant

4.8 (a) 0 (b) $\frac{48}{5}$ joules

4.9 (a) -2 joules (b) -2 joules Potential $\Phi=(xyz+2xz+\text{a constant})$ N m

4.10 Surface integral is $\iint_S f\,dS$, where f is a scalar function, often the normal component $\mathbf{F}\cdot\hat{\mathbf{n}}$ of a vector field; area of a surface $=\iint_S 1\,dS$.

4.11 16

4.12 Flux $= \oiint_S \mathbf{D}\cdot\hat{\mathbf{n}}\,\mathrm{d}S = 600\pi\,\mathrm{e}^{-1}$ (curved surface) $- 40\pi$ (ends)

4.13 $\pi a^2 b(b+2)$

4.14 **(a)** 4π **(b)** 4π

4.15 $9(1+\frac{3}{4}\pi)$

4.16 81

4.17 -1

4.18 Volume integral is $\iiint_V f\,\mathrm{d}V$, where f is a scalar function; volume $V \equiv \iiint_V 1\,\mathrm{d}V$.

4.19 $\frac{8}{3}$

4.20 13.5 C

4.21 $\nabla\cdot\mathbf{F} = x$; $\iiint_V x\,\mathrm{d}V = \frac{81}{4}\pi$.
Units are those of **F** multiplied by (units of distance)^2.

4.22 $\pi q_0^2 l R^2/4EI$

4.23 **(a)** 0 **(b)** 0

Solutions to the additional exercises may be found on p. 378.

Chapter 5

5.1 **(a)** $4\pi a^3$ **(b)** 3

5.2 -8π

5.3 $5832\pi/5$

5.4 Common values of the integrals are: **(a)** $\frac{3}{2}$ **(b)** $\frac{5}{4}\pi a^4$ **(c)** 0

5.7 $-2ab^2$

5.8 Common value of the integrals is $\frac{45}{8}$.

5.9 **(a)** $2\pi a^2$ **(b)** $2\pi a^2$

5.11 0

5.12 Common values of the integrals are: **(a)** π **(b)** -12π

5.13 -24

Solutions to the additional exercises may be found on p. 388.

Chapter 6

6.2 **(a)** $\hat{\mathbf{i}} = \hat{\mathbf{r}}\sin\theta\cos\phi + \hat{\boldsymbol{\theta}}\cos\theta\cos\phi - \hat{\boldsymbol{\phi}}\sin\phi$,
$\hat{\mathbf{j}} = \hat{\mathbf{r}}\sin\theta\sin\phi + \hat{\boldsymbol{\theta}}\cos\theta\sin\phi + \hat{\boldsymbol{\phi}}\cos\phi$,
$\hat{\mathbf{k}} = \hat{\mathbf{r}}\cos\theta - \hat{\boldsymbol{\theta}}\sin\theta$

(i) $\mathbf{A} = \hat{\mathbf{r}}(3\sin\theta\cos\phi + 5\sin\theta\sin\phi - \cos\theta)$
$+ \hat{\boldsymbol{\theta}}(3\cos\theta\cos\phi + 5\cos\theta\sin\phi + \sin\theta)$
$+ \hat{\boldsymbol{\phi}}(-3\sin\phi + 5\cos\phi)$

(ii) $\mathbf{r} = r\hat{\mathbf{r}}$

(b) $\hat{\mathbf{i}} = \hat{\boldsymbol{\rho}}\cos\phi - \hat{\boldsymbol{\phi}}\sin\phi, \quad \hat{\mathbf{j}} = \hat{\boldsymbol{\rho}}\sin\phi + \hat{\boldsymbol{\phi}}\cos\phi, \quad \hat{\mathbf{k}} = \hat{\mathbf{k}}$

(i) $\mathbf{A} = \hat{\boldsymbol{\rho}}(3\cos\phi + 5\sin\phi) + \hat{\boldsymbol{\phi}}(-3\sin\phi + 5\cos\phi) - \hat{\mathbf{k}}$

(ii) $\mathbf{r} = \rho\hat{\boldsymbol{\rho}} + z\hat{\mathbf{k}}$

6.3 No

6.4 (a) $2\mathbf{r}$ (b) $-\dfrac{\mathbf{r}}{r^3}$ (c) $\mathbf{a}$ (d) $\dfrac{\mathbf{a}}{r^3} - \dfrac{3\mathbf{r}(\mathbf{a}\cdot\mathbf{r})}{r^5}$

6.7 (a) $n = 2$ (b) $n = 1$

6.8 (a) 0 (b) $1/r^2$

6.9 (a) 1 (b) 3

6.10 (a) $4z$ (b) $-1 - \sin\phi$

6.11 (a) $1600\hat{\mathbf{k}}$ (b) $4.33\hat{\mathbf{r}} - 2.50\hat{\boldsymbol{\theta}} + 1.25\hat{\boldsymbol{\phi}}$

Solutions to the additional exercises may be found on p. 393.

Chapter 7

7.1 Flux $= \oiint_S \mathbf{D}\cdot\hat{\mathbf{n}}\,\mathrm{d}S = 1280\pi$ C; charge density $\rho_V = \nabla\cdot\mathbf{D} = 16\rho^2$;

total charge $= \iiint_V \rho_V\,\mathrm{d}V = 1280\pi$ C

7.2 $\rho_V = \nabla\cdot\mathbf{D} = \begin{cases} 10\rho^2\,\mathrm{C\,m^{-3}}, & 0 \leqslant \rho \leqslant 3\,\mathrm{m} \\ 0, & \text{elsewhere} \end{cases}$

7.3 $\rho_V = \nabla\cdot\mathbf{D} = \begin{cases} \dfrac{3Q}{4\pi a^3}, & r < a \\ 0, & r > a \end{cases}$ (uniform charge distribution inside a sphere of radius a)

7.4 $I = \iint_S \mathbf{J}\cdot\hat{\mathbf{n}}\,\mathrm{d}S = 7.51 \times 10^{-3}$ A

7.5 $\nabla \wedge \mathbf{H} = 2 \times 10^6 \cos\theta\,\hat{\mathbf{r}}$ + non-radial terms; $I = \iint_S \mathbf{J}\cdot\hat{\mathbf{n}}\,\mathrm{d}S = \frac{1}{2}\pi$ A

(putting $\mathrm{d}S = 10^{-6}\sin\theta\,\mathrm{d}\theta\,\mathrm{d}\phi\ \mathrm{m}^2$)

7.6 $\iint_S \mathbf{B}\cdot\hat{\mathbf{n}}\,\mathrm{d}S = \frac{1}{100}\pi$ Wb

7.8 $\mathbf{B} = 2\,e^{-2\rho}\,\hat{\boldsymbol{\phi}}$; $\mathbf{H} = \dfrac{2}{\mu_0}\,e^{-2\rho}\,\hat{\boldsymbol{\phi}}$; $\mathbf{J}(=\nabla \wedge \mathbf{H}) = \dfrac{2e^{-2\rho}}{\mu_0}\left(\dfrac{1}{\rho} - 2\right)\hat{\mathbf{k}}$;

$$I = \iint_S \mathbf{J}\cdot\hat{\mathbf{n}}\,dS = 9.80 \times 10^4\ \text{A}$$

7.9 -180 J

7.10 **(a)** $4\pi A\varepsilon_0 a^5$ **(b)** $Q^2/72\pi\varepsilon_0 a$

7.11 Instantaneous power $= 120\dfrac{\pi}{3}\cos^2(\omega t - \beta r)$ W;

average power $= 120\dfrac{\pi}{3} \times \dfrac{1}{2} = 62.8$ W

7.14 13.4 mW

Solutions to the additional exercises may be found on p. 396.

Chapter 8

8.1 No change after the first rotation; final position is $(0, 0, -1)$. If the order of rotations is reversed, the final position is $(0, -0.707, 0.707)$.

8.2 Rotation of 120° about $\hat{\mathbf{i}} + \hat{\mathbf{j}} + \hat{\mathbf{k}}$

8.3 If the angle of rotation is less than about 5° or if the rotations are about the same axis.

8.4 Velocity $= 2.71(-\hat{\mathbf{i}} + \hat{\mathbf{j}})$ m s^{-1}

8.6 1.55 cm to the east

8.8 Only possible if $\omega_A = \omega_B$, and then all three cylinders rotate as a rigid body, that is, the enclosed cylinder does not move relative to inner and outer cylinders (see additional exercise 2).

Solutions to the additional exercises may be found on p. 399.

Worked Solutions to Additional Exercises

Chapter 1

1 (a) The vector **a** may be considered to emanate from the origin (see Figure 1).

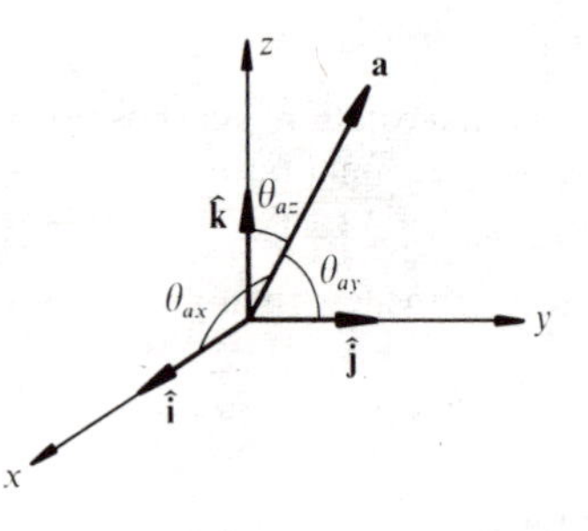

Figure 1

Now

$$\cos\theta_{ax} = \frac{\mathbf{a}\cdot\hat{\mathbf{i}}}{|\mathbf{a}|\,|\hat{\mathbf{i}}|} = a_x, \quad \text{since } \mathbf{a} \text{ is a unit vector}$$

Similarly,

$$\cos\theta_{ay} = a_y, \qquad \cos\theta_{az} = a_z$$

A knowledge of the direction cosines of a vector is sufficient to specify its direction uniquely.

(b) (i) $|\hat{\mathbf{i}}+\hat{\mathbf{j}}+\hat{\mathbf{k}}| = \sqrt{3}$, hence $\hat{\mathbf{i}}+\hat{\mathbf{j}}+\hat{\mathbf{k}} = \sqrt{3}\left(\dfrac{\hat{\mathbf{i}}+\hat{\mathbf{j}}+\hat{\mathbf{k}}}{\sqrt{3}}\right)$. The term in brackets is a unit vector. The direction cosines of this vector are $1/\sqrt{3}$, $1/\sqrt{3}$ and $1/\sqrt{3}$ respectively.
(ii) $|3\hat{\mathbf{i}}+4\hat{\mathbf{j}}| = \sqrt{(9+16)} = 5$, hence $3\hat{\mathbf{i}}+4\hat{\mathbf{j}} = 5(\frac{3}{5}\hat{\mathbf{i}}+\frac{4}{5}\hat{\mathbf{j}})$. The term in brackets is a unit vector. The direction cosines are 0.6, 0.8 and 0 respectively.

2 The vector equation of a straight line has the form $\mathbf{r} = \mathbf{a} + \mathbf{b}t$, where **a** is the position vector of a point on the line, **b** denotes the direction of the line and t is a parameter. In this problem we shall take t to be the time in hours after 12 noon, allowing the paths taken by the three ships to be written in the form

$$\mathbf{r} \equiv \mathbf{r}_1 = 2\hat{\mathbf{i}} + 6\hat{\mathbf{j}} + (5\hat{\mathbf{i}} + 4\hat{\mathbf{j}})t$$

$$\mathbf{r} \equiv \mathbf{r}_2 = 6\hat{\mathbf{i}} + 9\hat{\mathbf{j}} + (4\hat{\mathbf{i}} + 3\hat{\mathbf{j}})(t - \tfrac{1}{2})$$

$$\mathbf{r} \equiv \mathbf{r}_3 = 11\hat{\mathbf{i}} + 6\hat{\mathbf{j}} + (2\hat{\mathbf{i}} + 7\hat{\mathbf{j}})(t - 1)$$

Two of the ships will collide if there is a value of the time parameter t for which

$$\mathbf{r}_1 = \mathbf{r}_2 \quad \text{or} \quad \mathbf{r}_1 = \mathbf{r}_3 \quad \text{or} \quad \mathbf{r}_2 = \mathbf{r}_3$$

For the first possibility, we find

$$2\hat{\mathbf{i}} + 6\hat{\mathbf{j}} + (5\hat{\mathbf{i}} + 4\hat{\mathbf{j}})t = 6\hat{\mathbf{i}} + 9\hat{\mathbf{j}} + (4\hat{\mathbf{i}} + 3\hat{\mathbf{j}})(t - \tfrac{1}{2})$$

Equating coefficients of $\hat{\mathbf{i}}$ and $\hat{\mathbf{j}}$, we find $t = 2$ from the $\hat{\mathbf{i}}$ coefficient and $t = \frac{3}{2}$ from the $\hat{\mathbf{j}}$ coefficient, implying that these two ships do not collide. For the second possibility, we have

$$2\hat{\mathbf{i}} + 6\hat{\mathbf{j}} + (5\hat{\mathbf{i}} + 4\hat{\mathbf{j}})t = 11\hat{\mathbf{i}} + 6\hat{\mathbf{j}} + (2\hat{\mathbf{i}} + 7\hat{\mathbf{j}})(t - 1)$$

This time, the value of $t = \frac{7}{3}$ implies equality of both the $\hat{\mathbf{i}}$ and $\hat{\mathbf{j}}$ coefficients. Hence ship 1 and ship 3 collide after two hours and 20 minutes.

At the time of collision, ship 2 is at position

$$\mathbf{r} \equiv \mathbf{r}_2 = 6\hat{\mathbf{i}} + 9\hat{\mathbf{j}} + (4\hat{\mathbf{i}} + 3\hat{\mathbf{j}})(\tfrac{7}{3} - \tfrac{1}{2}) = \tfrac{40}{3}\hat{\mathbf{i}} + \tfrac{29}{2}\hat{\mathbf{j}}$$

whilst both ships 1 and 3 are at position

$$\mathbf{r}_3 = 11\hat{\mathbf{i}} + 6\hat{\mathbf{j}} + (2\hat{\mathbf{i}} + 7\hat{\mathbf{j}})(\tfrac{7}{3} - 1) = \tfrac{41}{3}\hat{\mathbf{i}} + \tfrac{46}{3}\hat{\mathbf{j}}$$

The distance between ship 2 and the collision is therefore

$$\sqrt{[(\tfrac{41}{3} - \tfrac{40}{3})^2 + (\tfrac{46}{3} - \tfrac{29}{2})^2]} = \tfrac{1}{18}\sqrt{261} \text{ nautical miles}$$

Since the speed of ship 2 is $\sqrt{(4^2 + 3^2)}$ knots per hour, it will arrive 10.77 minutes after the collision.

3 $\mathbf{m} = (5/\sqrt{2})(\hat{\mathbf{i}} + \hat{\mathbf{j}})$. This is a vector of magnitude 5 in the direction of $(\hat{\mathbf{i}} + \hat{\mathbf{j}})$. This magnet is placed at $(-1, 2, 3)$. Thus the position vector of $(1, 0, 0)$ with respect to this point is

$$\mathbf{r} = -\hat{\mathbf{i}} + 2\hat{\mathbf{j}} + 3\hat{\mathbf{k}} - (\hat{\mathbf{i}} + 0\hat{\mathbf{j}} + 0\hat{\mathbf{k}}) = -2\hat{\mathbf{i}} + 2\hat{\mathbf{j}} + 3\hat{\mathbf{k}}$$

$$\therefore \quad |\mathbf{r}| = \sqrt{(4 + 4 + 9)} = \sqrt{17}$$

and $$\mathbf{m} \cdot \mathbf{r} = (5/\sqrt{2})(\hat{\mathbf{i}} + \hat{\mathbf{j}}) \cdot (-2\hat{\mathbf{i}} + 2\hat{\mathbf{j}} + 3\hat{\mathbf{k}}) = (5/\sqrt{2})(-2 + 2) = 0$$

$$\therefore \quad \mathbf{H} = -\frac{(5/\sqrt{2})(\hat{\mathbf{i}} + \hat{\mathbf{j}})}{17\sqrt{17}} = -\frac{5}{17\sqrt{34}}(\hat{\mathbf{i}} + \hat{\mathbf{j}})$$

is the vector at the point $(1, 0, 0)$

4 The point P $(3, 4, 5)$ lies on the sphere since $3^2 + 4^2 + 5^2 = 50$. A normal to the tangent plane is clearly a normal to the sphere at P. If (l, m, n) is any

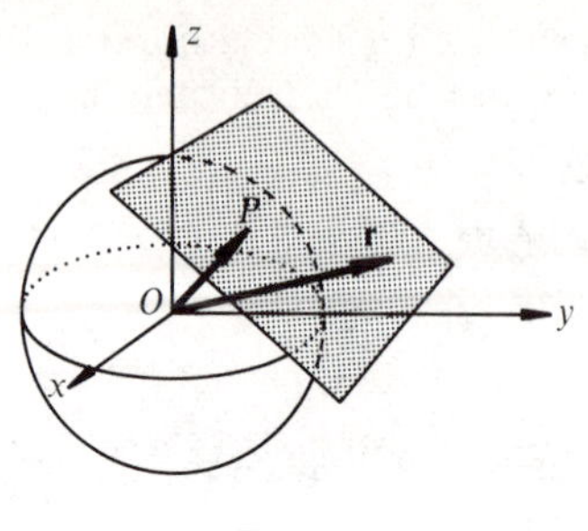

Figure 2

point on the sphere, then clearly the position vector of this point, $l\hat{\mathbf{i}}+m\hat{\mathbf{j}}+n\hat{\mathbf{k}}$, is a normal to the sphere at this point (see Figure 2). Thus a normal at P is $\mathbf{n}=3\hat{\mathbf{i}}+4\hat{\mathbf{j}}+5\hat{\mathbf{k}}$. If $\mathbf{r}$ is the position vector of any point on the tangent plane, then since P lies on the plane its equation is $\mathbf{r}\cdot\mathbf{n}=\overrightarrow{OP}\cdot\mathbf{n}$. In Cartesian form, we find

$$(x\hat{\mathbf{i}}+y\hat{\mathbf{j}}+z\hat{\mathbf{k}})\cdot(3\hat{\mathbf{i}}+4\hat{\mathbf{j}}+5\hat{\mathbf{k}})=(3\hat{\mathbf{i}}+4\hat{\mathbf{j}}+5\hat{\mathbf{k}})\cdot(3\hat{\mathbf{i}}+4\hat{\mathbf{j}}+5\hat{\mathbf{k}})$$

so that $3x+4y+5z=50$ is the equation of the tangent plane.

5 Let the velocity of A be represented by $\mathbf{v}_A=v_x\hat{\mathbf{i}}+v_y\hat{\mathbf{j}}+v_z\hat{\mathbf{k}}$. Since A travels in a north-easterly direction (see Figure 3), then $v_x=v_y$. The vertical speed of this plane is 4 km h^{-1} and so we may write $v_z=4$. Now, since the speed of the plane is 1000 km h^{-1}, we may write

$$1000=\sqrt{(v_x^2+v_y^2+v_z^2)}=\sqrt{(2v_x^2+16)}$$

leading to $v_x=707.10$ km h^{-1}. Therefore $\mathbf{v}_A=707.10(\hat{\mathbf{i}}+\hat{\mathbf{j}})+4\hat{\mathbf{k}}$.

The relative velocity of B with respect to A is

$$\mathbf{v}_B-\mathbf{v}_A=500\hat{\mathbf{i}}+8\hat{\mathbf{k}}-v_x(\hat{\mathbf{i}}+\hat{\mathbf{j}})-4\hat{\mathbf{k}}=-207.10\hat{\mathbf{i}}-707.10\hat{\mathbf{j}}+4\hat{\mathbf{k}}$$

B appears to be moving away from A with a speed

$$\sqrt{(207.10^2+707.10^2+16)}=736.82 \text{ km h}^{-1}$$

and so the distance apart after two hours is 1473.6 km.

6 This problem is independent of the orientation of the triangle, so without loss of generality we can consider the triangle to be placed in the position shown in Figure 4. $\mathbf{F}_1$ is in the direction of AB and so is parallel to the

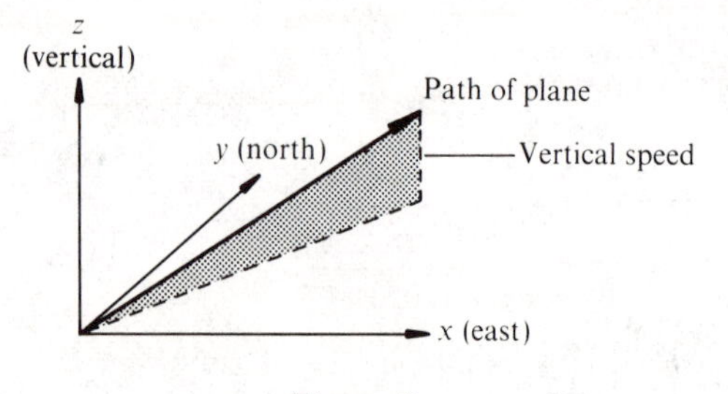

Figure 3

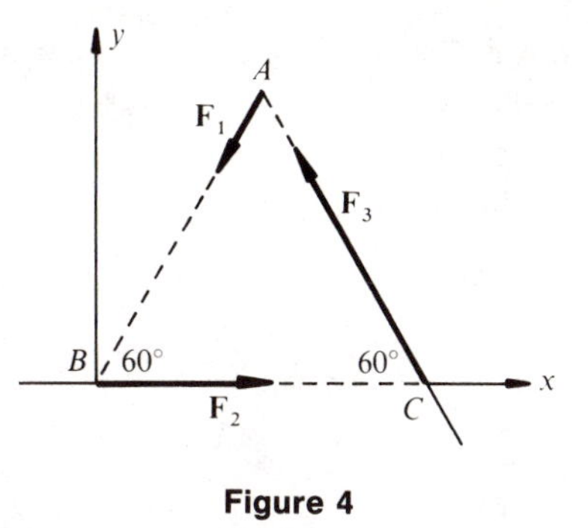

Figure 4

vector $-\cos 60°\,\hat{\mathbf{i}} - \sin 60°\,\hat{\mathbf{j}}$ (both signs are negative because the components of $\mathbf{F}_1$ are in the negative x- and y-directions). Therefore

$$\mathbf{F}_1 = F(-\cos 60°\,\hat{\mathbf{i}} - \sin 60°\,\hat{\mathbf{j}}) = F(-\tfrac{1}{2}\hat{\mathbf{i}} - \tfrac{1}{2}\sqrt{3}\hat{\mathbf{j}})$$

Similarly,

$$\mathbf{F}_2 = 2F\hat{\mathbf{i}} \quad \text{and} \quad \mathbf{F}_3 = 3F(-\tfrac{1}{2}\hat{\mathbf{i}} + \tfrac{1}{2}\sqrt{3}\hat{\mathbf{j}})$$

The resultant force $\mathbf{F}$ is

$$\mathbf{F} = \mathbf{F}_1 + \mathbf{F}_2 + \mathbf{F}_3 = F[\hat{\mathbf{i}}(-\tfrac{1}{2} + 2 - \tfrac{3}{2}) + \hat{\mathbf{j}}(-\tfrac{1}{2}\sqrt{3} + \tfrac{3}{2}\sqrt{3})] = \sqrt{3}\,F\hat{\mathbf{j}}$$

The torque of the forces about B is $\mathbf{r} \wedge \mathbf{F}_3$ (since the lines of action of $\mathbf{F}_1$ and $\mathbf{F}_2$ pass through B), where $\mathbf{r}$ is the position vector of a point on the line of action of $\mathbf{F}_3$. We choose the point A.

$$\therefore \qquad \text{torque} = AB(\cos 60°\,\hat{\mathbf{i}} + \sin 60°\,\hat{\mathbf{j}}) \wedge \mathbf{F}_3$$

$$= AB\begin{vmatrix} \hat{\mathbf{i}} & \hat{\mathbf{j}} & \hat{\mathbf{k}} \\ \tfrac{1}{2} & \tfrac{1}{2}\sqrt{3} & 0 \\ -\tfrac{3}{2}F & \tfrac{3}{2}\sqrt{3}\,F & 0 \end{vmatrix} = AB\,\tfrac{3}{2}\sqrt{3}\,F\hat{\mathbf{k}}$$

The resultant force should have the same torque as this. Let its line of action pass through P on the line BC (possibly extended). Then

$$\overrightarrow{BP} = |\overrightarrow{BP}|\hat{\mathbf{i}}$$

and the torque is

$$|\overrightarrow{BP}|\hat{\mathbf{i}} \wedge \mathbf{F} = \begin{vmatrix} \hat{\mathbf{i}} & \hat{\mathbf{j}} & \hat{\mathbf{k}} \\ |\overrightarrow{BP}| & 0 & 0 \\ 0 & \sqrt{3}\,F & 0 \end{vmatrix} = \sqrt{3}\,|\overrightarrow{BP}|F\hat{\mathbf{k}}$$

For equality, we require that

$$|\overrightarrow{BP}|\sqrt{3}\,F = \tfrac{3}{2}\sqrt{3}(AB)F = \tfrac{3}{2}\sqrt{3}(BC)F$$

since $AB = BC$ for an equilateral triangle. Thus

$$|\overrightarrow{BP}| = \tfrac{3}{2}BC$$

7 If the vectors **a**, **b**, **c**, **d**, **e** and **f** are vectors representing the six edges of the tetrahedron (see Figure 5), then the four vector areas

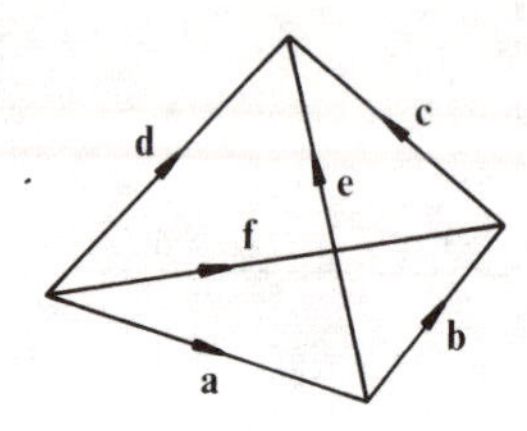

Figure 5

(with outward-pointing normals) representing the four faces are

$$\mathbf{A}_1 = \tfrac{1}{2}\mathbf{b} \wedge \mathbf{a}, \qquad \mathbf{A}_2 = \tfrac{1}{2}\mathbf{a} \wedge \mathbf{d}, \qquad \mathbf{A}_3 = \tfrac{1}{2}\mathbf{b} \wedge \mathbf{e}, \qquad \mathbf{A}_4 = \tfrac{1}{2}\mathbf{d} \wedge \mathbf{f}$$

so that $\mathbf{A}_1 + \mathbf{A}_2 + \mathbf{A}_3 + \mathbf{A}_4 = \tfrac{1}{2}(\mathbf{b} \wedge \mathbf{a} + \mathbf{a} \wedge \mathbf{d} + \mathbf{d} \wedge \mathbf{f} + \mathbf{b} \wedge \mathbf{e})$

But $\mathbf{d} = \mathbf{a} + \mathbf{e}$ and $\mathbf{f} = \mathbf{a} + \mathbf{b}$; also

$$\mathbf{a} \wedge \mathbf{d} = \mathbf{a} \wedge (\mathbf{a} + \mathbf{e}) = \mathbf{a} \wedge \mathbf{e}$$

and $\quad \mathbf{d} \wedge \mathbf{f} = (\mathbf{a} + \mathbf{e}) \wedge (\mathbf{a} + \mathbf{b}) = \mathbf{a} \wedge \mathbf{b} + \mathbf{e} \wedge \mathbf{a} + \mathbf{e} \wedge \mathbf{b}$

so that $\mathbf{A}_1 + \mathbf{A}_2 + \mathbf{A}_3 + \mathbf{A}_4 = \tfrac{1}{2}(\mathbf{b} \wedge \mathbf{a} + \mathbf{a} \wedge \mathbf{e} + \mathbf{a} \wedge \mathbf{b} + \mathbf{e} \wedge \mathbf{a} + \mathbf{e} \wedge \mathbf{b} + \mathbf{b} \wedge \mathbf{e})$

$$= 0$$

8 (a) If we again consider the tetrahedron of Figure 5, the area of the base is $\tfrac{1}{2}|(\mathbf{a} \wedge \mathbf{f})|$, whilst the perpendicular height h is the projection of **d** onto the unit normal to the base formed by **a** and **f**; that is

$$h = \mathbf{d} \cdot \frac{\mathbf{a} \wedge \mathbf{f}}{|\mathbf{a} \wedge \mathbf{f}|}$$

Hence the volume of the tetrahedron is

$$V = \tfrac{1}{6}|\mathbf{d} \cdot (\mathbf{a} \wedge \mathbf{f})|$$

(b) The lines connecting $ABCD$ form a tetrahedron, with the sides meeting at A represented by the vectors $(\mathbf{b} - \mathbf{a})$, $(\mathbf{c} - \mathbf{a})$ and $(\mathbf{d} - \mathbf{a})$. If the four points A, B, C and D are coplanar, then the volume of this tetrahedron will be zero; that is, the required condition is

$$(\mathbf{b} - \mathbf{a}) \cdot (\mathbf{c} - \mathbf{a}) \wedge (\mathbf{d} - \mathbf{a}) = 0$$

Expanding the vector product and then the scalar product, we obtain

$$(\mathbf{b} - \mathbf{a}) \cdot (\mathbf{c} \wedge \mathbf{d} - \mathbf{c} \wedge \mathbf{a} - \mathbf{a} \wedge \mathbf{d})$$

$$= \mathbf{b} \cdot (\mathbf{c} \wedge \mathbf{d}) - \mathbf{b} \cdot (\mathbf{c} \wedge \mathbf{a}) - \mathbf{b} \cdot (\mathbf{a} \wedge \mathbf{d}) - \mathbf{a} \cdot (\mathbf{c} \wedge \mathbf{d}) = 0$$

which, after repeatedly using (1.57), becomes

$$\mathbf{a} \cdot (\mathbf{b} \wedge \mathbf{c}) + \mathbf{a} \cdot (\mathbf{c} \wedge \mathbf{d}) + \mathbf{a} \cdot (\mathbf{d} \wedge \mathbf{b}) = \mathbf{b} \cdot (\mathbf{c} \wedge \mathbf{d})$$

Chapter 2

1 The gravitational force keeps the satellite in a circular orbit. The centripetal force required is $m\kappa v^2 = mv^2/r$, where m is the mass of the satellite, v is its speed and r is the radius of the circle on which it moves. The magnitude of the gravitational force is

$$\frac{GmM}{|\mathbf{r}|^2} = \frac{GmM}{r^2}$$

Hence, for the motion of the satellite to be circular

$$\frac{GmM}{r^2} = \frac{mv^2}{r}$$

$$\therefore \qquad v^2 = \frac{GM}{r}$$

The distance travelled by the satellite in one period is $2\pi r$. Thus the period of the satellite is

$$T = \frac{2\pi r}{v} = 2\pi r\sqrt{\left(\frac{r}{GM}\right)} = \frac{2\pi r^{3/2}}{\sqrt{(GM)}}$$

Using the given values of G and M, and with $r = (6.37 \times 10^6 + 500\,000) = 6\,870\,000$ m, we find $T = 5665$ s.

2 (a) Since $\hat{\boldsymbol{\rho}}$ and $\hat{\boldsymbol{\phi}}$ are unit vectors, we have (directly from Figure 2.25)

$$\hat{\boldsymbol{\rho}} = \cos\phi\,\hat{\mathbf{i}} + \sin\phi\,\hat{\mathbf{j}} \quad \text{and} \quad \hat{\boldsymbol{\phi}} = -\sin\phi\,\hat{\mathbf{i}} + \cos\phi\,\hat{\mathbf{j}}$$

(This could also be shown algebraically, by writing $\hat{\boldsymbol{\rho}} = a\hat{\mathbf{i}} + b\hat{\mathbf{j}}$ and $\hat{\boldsymbol{\phi}} = c\hat{\mathbf{i}} + d\hat{\mathbf{j}}$, and then using the properties $|\hat{\boldsymbol{\rho}}| = 1$, $|\hat{\boldsymbol{\phi}}| = 1$, $\hat{\boldsymbol{\rho}}\cdot\hat{\boldsymbol{\phi}} = 0$ and (less obviously) $\hat{\boldsymbol{\rho}} \wedge \hat{\boldsymbol{\phi}} = \hat{\mathbf{k}}$ to determine a, b, c and d.)

(b) We note that $\hat{\mathbf{i}}$ and $\hat{\mathbf{j}}$ are fixed vectors, unchanging in both magnitude and direction, and so $\mathrm{d}\hat{\mathbf{i}}/\mathrm{d}t = 0$ and $\mathrm{d}\hat{\mathbf{j}}/\mathrm{d}t = 0$, giving

$$\frac{\mathrm{d}\hat{\boldsymbol{\rho}}}{\mathrm{d}t} = \frac{\mathrm{d}}{\mathrm{d}t}(\cos\phi\,\hat{\mathbf{i}} + \sin\phi\,\hat{\mathbf{j}}) = -\sin\phi\,\frac{\mathrm{d}\phi}{\mathrm{d}t}\,\hat{\mathbf{i}} + \cos\phi\,\frac{\mathrm{d}\phi}{\mathrm{d}t}\,\hat{\mathbf{j}} = \hat{\boldsymbol{\phi}}\,\frac{\mathrm{d}\phi}{\mathrm{d}t}$$

Similarly,

$$\frac{\mathrm{d}\hat{\boldsymbol{\phi}}}{\mathrm{d}t} = \frac{\mathrm{d}}{\mathrm{d}t}(-\sin\phi\,\hat{\mathbf{i}} + \cos\phi\,\hat{\mathbf{j}}) = -\cos\phi\,\frac{\mathrm{d}\phi}{\mathrm{d}t}\,\hat{\mathbf{i}} - \sin\phi\,\frac{\mathrm{d}\phi}{\mathrm{d}t}\,\hat{\mathbf{j}} = -\hat{\boldsymbol{\rho}}\,\frac{\mathrm{d}\phi}{\mathrm{d}t}$$

(c) The position vector of the particle is $\rho\hat{\boldsymbol{\rho}}$ (this quantity being a vector of magnitude ρ – the distance from the origin – and in the direction of $\hat{\boldsymbol{\rho}}$, that is, pointing directly away from the origin).

$$\therefore \qquad \text{velocity} = \frac{\mathrm{d}}{\mathrm{d}t}(\rho\hat{\boldsymbol{\rho}}) = \frac{\mathrm{d}\rho}{\mathrm{d}t}\,\hat{\boldsymbol{\rho}} + \rho\,\frac{\mathrm{d}\hat{\boldsymbol{\rho}}}{\mathrm{d}t} = \frac{\mathrm{d}\rho}{\mathrm{d}t}\,\hat{\boldsymbol{\rho}} + \rho\hat{\boldsymbol{\phi}}\,\frac{\mathrm{d}\phi}{\mathrm{d}t}$$

Also, $\quad \text{acceleration} = \dfrac{\mathrm{d}}{\mathrm{d}t}(\text{velocity})$

$$= \frac{\mathrm{d}}{\mathrm{d}t}\left(\frac{\mathrm{d}\rho}{\mathrm{d}t}\hat{\boldsymbol{\rho}} + \rho\hat{\boldsymbol{\phi}}\frac{\mathrm{d}\phi}{\mathrm{d}t}\right)$$

$$= \frac{\mathrm{d}^2\rho}{\mathrm{d}t^2}\hat{\boldsymbol{\rho}} + \frac{\mathrm{d}\rho}{\mathrm{d}t}\hat{\boldsymbol{\phi}}\frac{\mathrm{d}\phi}{\mathrm{d}t} + \frac{\mathrm{d}\rho}{\mathrm{d}t}\hat{\boldsymbol{\phi}}\frac{\mathrm{d}\phi}{\mathrm{d}t} + \rho\frac{\mathrm{d}\phi}{\mathrm{d}t}\left(-\hat{\boldsymbol{\rho}}\frac{\mathrm{d}\phi}{\mathrm{d}t}\right) + \rho\hat{\boldsymbol{\phi}}\frac{\mathrm{d}^2\phi}{\mathrm{d}t^2}$$

$$= \left(\frac{\mathrm{d}^2\rho}{\mathrm{d}t^2} - \rho\left(\frac{\mathrm{d}\phi}{\mathrm{d}t}\right)^2\right)\hat{\boldsymbol{\rho}} + \left(2\frac{\mathrm{d}\rho}{\mathrm{d}t}\frac{\mathrm{d}\phi}{\mathrm{d}t} + \rho\frac{\mathrm{d}^2\phi}{\mathrm{d}t^2}\right)\hat{\boldsymbol{\phi}}$$

Note that, for circular motion, $\rho = \text{constant}$ and so

$$\frac{\mathrm{d}\rho}{\mathrm{d}t} = 0 \quad \text{and} \quad \frac{\mathrm{d}^2\rho}{\mathrm{d}t^2} = 0$$

so in this case

$$\text{acceleration} = -\rho\left(\frac{\mathrm{d}\phi}{\mathrm{d}t}\right)^2\hat{\boldsymbol{\rho}} + \rho\frac{\mathrm{d}^2\phi}{\mathrm{d}t^2}\hat{\boldsymbol{\phi}}$$

and so the force directed towards the centre is

$$\rho\left(\frac{\mathrm{d}\phi}{\mathrm{d}t}\right)^2 = \rho\frac{v^2}{\rho^2} = \frac{v^2}{\rho}, \quad \text{in agreement with the well-known result}$$

3 We choose an origin at O (see Figure 6). Now $\mathrm{d}\phi/\mathrm{d}t = 3$. Using a well-known result for the geometry of the circle, $\theta = 2\phi$. Also $\rho = |\overrightarrow{OA}|$ may be determined from the cosine rule:

$$\rho^2 = 1^2 + 1^2 - 2 \times 1 \times 1 \times \cos(180° - 2\phi) = 2 + 2\cos 2\phi = 4\cos^2\phi$$

$$\therefore \qquad \rho = 2\cos\phi \quad \text{and} \quad \frac{\mathrm{d}\rho}{\mathrm{d}t} = -2\sin\phi\frac{\mathrm{d}\phi}{\mathrm{d}t} = -6\sin\phi$$

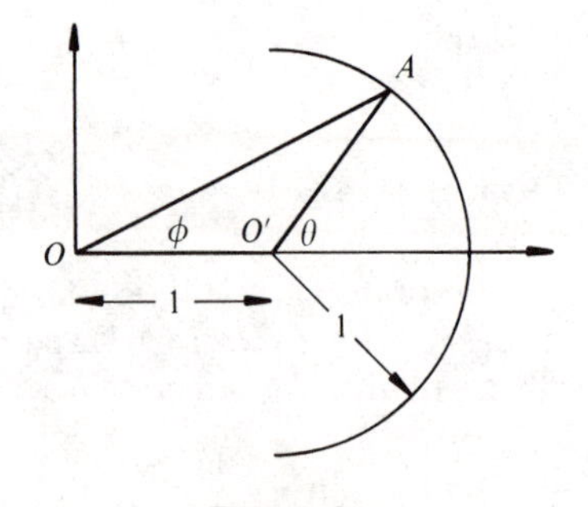

Figure 6

Now the velocity of A is (from the results of additional exercise 2)

$$\mathbf{v}=\frac{\mathrm{d}\rho}{\mathrm{d}t}\hat{\boldsymbol{\rho}}+\rho\hat{\boldsymbol{\phi}}\frac{\mathrm{d}\phi}{\mathrm{d}t}=-6\sin\phi\,\hat{\boldsymbol{\rho}}+(2\cos\phi)(3)\hat{\boldsymbol{\phi}}$$

$$\therefore \qquad \text{speed}=|\mathbf{v}|=\sqrt{(36\sin^2\phi+36\cos^2\phi)}=6$$

(this follows since $|\mathbf{v}|=\sqrt{(\mathbf{v}\cdot\mathbf{v})}$ and $\hat{\boldsymbol{\rho}}$ and $\hat{\boldsymbol{\phi}}$ are mutually orthogonal.

The acceleration is

$$\mathbf{a}=\left(\frac{\mathrm{d}^2\rho}{\mathrm{d}t^2}-\rho\left(\frac{\mathrm{d}\phi}{\mathrm{d}t}\right)^2\right)\hat{\boldsymbol{\rho}}+\left(2\frac{\mathrm{d}\rho}{\mathrm{d}t}\frac{\mathrm{d}\phi}{\mathrm{d}t}+\rho\frac{\mathrm{d}^2\phi}{\mathrm{d}t^2}\right)\hat{\boldsymbol{\phi}}$$

Now $$\frac{\mathrm{d}^2\phi}{\mathrm{d}t^2}=\frac{\mathrm{d}}{\mathrm{d}t}\left(\frac{\mathrm{d}\phi}{\mathrm{d}t}\right)=0$$

and $$\frac{\mathrm{d}^2\rho}{\mathrm{d}t^2}=\frac{\mathrm{d}}{\mathrm{d}t}\left(\frac{\mathrm{d}\rho}{\mathrm{d}t}\right)=\frac{\mathrm{d}}{\mathrm{d}t}(-6\sin\phi)=-6\cos\phi\frac{\mathrm{d}\phi}{\mathrm{d}t}=-18\cos\phi$$

$$\therefore \qquad \mathbf{a}=[-18\cos\phi-(2\cos\phi)(9)]\hat{\boldsymbol{\rho}}+2(-6\sin\phi)(3)\hat{\boldsymbol{\phi}}$$

and so $|\mathbf{a}|=36\text{ m s}^{-2}$

4 Let the altitude be h. Then the distance from the centre of the Earth is $(6.37\times10^6+h)$m. In order for it to appear stationary, the period of the satellite must be exactly the same as that of the Earth, that is 24 hours or 86 400 s. Thus, from the solution to additional exercise 1,

$$\frac{GM}{r^2}=\frac{v^2}{r}=\frac{1}{r}\left(\frac{4\pi^2r^2}{T^2}\right)$$

$$\therefore \qquad \frac{GMT}{4\pi^2}=r^3=(6.37\times10^6+h)^3$$

leading to $h=35\,880$ km.

5 The mass m of the particle is a function of time and is such that

$$\frac{\mathrm{d}m}{\mathrm{d}t}=m_0k, \quad \text{with } m=m_0 \text{ at } t=0$$

It follows that $m=m_0(tk+1)$. The only force on the particle is that due to gravity, which acts in a downward direction and has magnitude mg (see Figure 7). The momentum of the particle at any time is $m\,\mathrm{d}\mathbf{r}/\mathrm{d}t$, where $\mathbf{r}$ is the position vector of the particle at time t. By Newton's second law,

rate of change of momentum = applied force

that is, $$\frac{\mathrm{d}}{\mathrm{d}t}\left(m\frac{\mathrm{d}\mathbf{r}}{\mathrm{d}t}\right)=-mg\hat{\mathbf{j}}=-gm_0(tk+1)\hat{\mathbf{j}}$$

Integrating,

$$m\frac{\mathrm{d}\mathbf{r}}{\mathrm{d}t}=-gm_0(\tfrac{1}{2}t^2k+t)\hat{\mathbf{j}}+\mathbf{d}$$

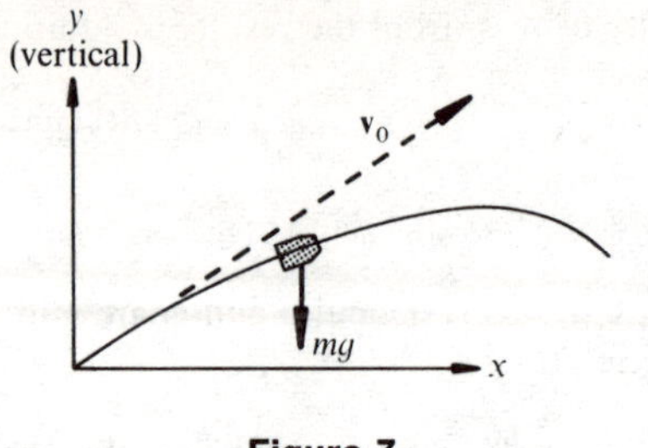

Figure 7

Now when $t = 0$, $\mathrm{d}\mathbf{r}/\mathrm{d}t = \mathbf{v}_0$ and $m = m_0$, so

$$\mathbf{d} = m_0\mathbf{v}_0$$

$$\therefore \qquad m\frac{\mathrm{d}\mathbf{r}}{\mathrm{d}t} = -gm_0(\tfrac{1}{2}t^2k + t)\hat{\mathbf{j}} + m_0\mathbf{v}_0$$

or

$$\frac{\mathrm{d}\mathbf{r}}{\mathrm{d}t} = -\frac{g}{2}\frac{(t^2k + 2t)}{(tk+1)}\hat{\mathbf{j}} + \frac{\mathbf{v}_0}{(tk+1)}$$

$$= -\frac{g}{2}\left(t + \frac{1}{k} - \frac{1}{k(tk+1)}\right)\hat{\mathbf{j}} + \frac{\mathbf{v}_0}{(tk+1)}$$

Integrating again,

$$\mathbf{r} = -\frac{g}{2}\left[\frac{t^2}{2} + \frac{t}{k} - \frac{1}{k^2}\ln\left(t + \frac{1}{k}\right)\right]\hat{\mathbf{j}} + \frac{\mathbf{v}_0}{k}\ln\left(t + \frac{1}{k}\right) + \mathbf{e}$$

When $t = 0$, $\mathbf{r} = 0$ and so

$$\mathbf{e} = \frac{g}{2k^2}\ln k\,\hat{\mathbf{j}} + \frac{\mathbf{v}_0}{k}\ln k$$

Substituting for $\mathbf{e}$ and $\mathbf{v}_0$ gives the required answer.

Chapter 3

1 By property (2a) of Section 3.6,

$$\nabla\cdot(\mathbf{r}(\mathbf{a}\cdot\mathbf{r})) = (\nabla\cdot\mathbf{r})(\mathbf{a}\cdot\mathbf{r}) + \mathbf{r}\cdot\nabla(\mathbf{a}\cdot\mathbf{r})$$

But $\quad \nabla\cdot\mathbf{r} = 3$

and

$$\nabla(\mathbf{a}\cdot\mathbf{r}) = \hat{\mathbf{i}}\frac{\partial}{\partial x}(a_x x) + \hat{\mathbf{j}}\frac{\partial}{\partial y}(a_y y) + \hat{\mathbf{k}}\frac{\partial}{\partial z}(a_z z)$$

$$= \hat{\mathbf{i}}a_x + \hat{\mathbf{j}}a_y + \hat{\mathbf{k}}a_z \quad \text{(because } a_x,\ a_y \text{ and } a_z \text{ are constants)}$$

$$= \mathbf{a}$$

so

$$\nabla\cdot(\mathbf{r}(\mathbf{a}\cdot\mathbf{r})) = 3(\mathbf{a}\cdot\mathbf{r}) + \mathbf{r}\cdot\mathbf{a} = 4(\mathbf{a}\cdot\mathbf{r})$$

From property (2b) of Section 3.6,

$$\nabla \wedge \mathbf{v} = \nabla \wedge (\mathbf{r}(\mathbf{a}\cdot\mathbf{r})) = \nabla(\mathbf{a}\cdot\mathbf{r}) \wedge \mathbf{r} + (\mathbf{a}\cdot\mathbf{r})\nabla \wedge \mathbf{r}$$

But, from above, $\nabla(\mathbf{a}\cdot\mathbf{r}) = \mathbf{a}$

also $$\nabla \wedge \mathbf{r} = \begin{vmatrix} \hat{\mathbf{i}} & \hat{\mathbf{j}} & \hat{\mathbf{k}} \\ \dfrac{\partial}{\partial x} & \dfrac{\partial}{\partial y} & \dfrac{\partial}{\partial z} \\ x & y & z \end{vmatrix} \doteq 0$$

so $$\nabla \wedge \mathbf{r}(\mathbf{a}\cdot\mathbf{r}) = \mathbf{a} \wedge \mathbf{r}$$

2 $$\nabla \wedge \mathbf{G} = \begin{vmatrix} \hat{\mathbf{i}} & \hat{\mathbf{j}} & \hat{\mathbf{k}} \\ \dfrac{\partial}{\partial x} & \dfrac{\partial}{\partial y} & \dfrac{\partial}{\partial z} \\ 2xye^z & x^2e^z & x^2ye^z + z^2 \end{vmatrix}$$

$$= \hat{\mathbf{i}}(x^2e^z - e^zx^2) - \hat{\mathbf{j}}(2xye^z - 2xye^z) + \hat{\mathbf{k}}(2xe^z - 2xe^z)$$

$$= 0$$

If $$G = \nabla\Phi = \hat{\mathbf{i}}\frac{\partial\Phi}{\partial x} + \hat{\mathbf{j}}\frac{\partial\Phi}{\partial y} + \hat{\mathbf{k}}\frac{\partial\Phi}{\partial z}$$

then $$\frac{\partial\Phi}{\partial x} = 2xye^z \qquad \textbf{(1)}$$

$$\frac{\partial\Phi}{\partial y} = x^2e^z \qquad \textbf{(2)}$$

$$\frac{\partial\Phi}{\partial z} = x^2ye^z + z^2 \qquad \textbf{(3)}$$

Integrating (1),

$$\Phi = x^2ye^z + h(y, z), \quad \text{where } h \text{ is some function of } y \text{ and } z$$

Substituting in (2),

$$x^2e^z + \frac{\partial h}{\partial y} = x^2e^z$$

$\therefore$ $$\frac{\partial h}{\partial y} = 0, \quad \text{implying that } h = h(z), \text{ a function of } z \text{ only}$$

Substituting in (3),

$$x^2ye^z + \frac{dh}{dz} = x^2ye^z + z^2$$

$$\therefore \qquad \frac{\mathrm{d}h}{\mathrm{d}z} = z^2, \quad \text{leading to} \quad h = \tfrac{1}{3}z^3 + c$$

where c is a constant of integration

Finally,

$$\Phi(x, y, z) = x^2 y e^z + \tfrac{1}{3}z^3 + c$$

3 Since $\nabla\Phi/|\nabla\Phi|$ is a unit normal to a level surface of Φ, then clearly the component magnitude of $\mathbf{A}$ in the normal direction is $\mathbf{A}\cdot\nabla\Phi/|\nabla\Phi|$, leading to an expression for the vector component of $\mathbf{A}$ in the normal direction as

$$\mathbf{A}\cdot\frac{\nabla\Phi}{|\nabla\Phi|}\left(\frac{\nabla\Phi}{|\nabla\Phi|}\right) = \frac{(\mathbf{A}\cdot\nabla\Phi)\nabla\Phi}{(\nabla\Phi)^2} \quad \text{where } (\nabla\Phi)^2 \equiv |\nabla\Phi|^2$$

The vector sum of the tangential component and the normal component is the vector $\mathbf{A}$. Hence the tangential component of $\mathbf{A}$ is

$$\mathbf{A} - \frac{(\mathbf{A}\cdot\nabla\Phi)\nabla\Phi}{(\nabla\Phi)^2} = \frac{(\nabla\Phi)^2\mathbf{A} - (\mathbf{A}\cdot\nabla\Phi)\nabla\Phi}{(\nabla\Phi)^2} = \frac{(\nabla\Phi)\wedge(\mathbf{A}\wedge\nabla\Phi)}{(\nabla\Phi)^2}$$

using the vector triple product identity (1.61).

4 (a)

$$\mathbf{F}\wedge\mathbf{G} = \begin{vmatrix} \hat{\mathbf{i}} & \hat{\mathbf{j}} & \hat{\mathbf{k}} \\ 2x & x^2y & z^2 \\ x & y & 0 \end{vmatrix} = \hat{\mathbf{i}}(-z^2y) - \hat{\mathbf{j}}(-xz^2) + \hat{\mathbf{k}}(2xy - x^3y)$$

$$\therefore \qquad \nabla\cdot(\mathbf{F}\wedge\mathbf{G}) = 0 + 0 + 0 = 0$$

$$\nabla\wedge\mathbf{F} = \begin{vmatrix} \hat{\mathbf{i}} & \hat{\mathbf{j}} & \hat{\mathbf{k}} \\ \dfrac{\partial}{\partial x} & \dfrac{\partial}{\partial y} & \dfrac{\partial}{\partial z} \\ 2x & x^2y & z^2 \end{vmatrix} = \hat{\mathbf{i}}(0) - \hat{\mathbf{j}}(0) + \hat{\mathbf{k}}(2xy) = 2xy\hat{\mathbf{k}}$$

$$\nabla\wedge\mathbf{G} = \begin{vmatrix} \hat{\mathbf{i}} & \hat{\mathbf{j}} & \hat{\mathbf{k}} \\ \dfrac{\partial}{\partial x} & \dfrac{\partial}{\partial y} & \dfrac{\partial}{\partial z} \\ x & y & 0 \end{vmatrix} = \hat{\mathbf{i}}(0) - \hat{\mathbf{j}}(0) + \hat{\mathbf{k}}(0) = 0$$

$$\therefore \qquad \mathbf{G}\cdot(\nabla\wedge\mathbf{F}) = (x\hat{\mathbf{i}} + y\hat{\mathbf{j}})\cdot(2xy\hat{\mathbf{k}}) = 0 \quad \text{and} \quad \mathbf{F}\cdot(\nabla\wedge\mathbf{G}) = 0$$

which verifies the identity.

(b) Considering the four terms on the right-hand side in turn,

$$\mathbf{G}\wedge(\nabla\wedge\mathbf{F}) = \begin{vmatrix} \hat{\mathbf{i}} & \hat{\mathbf{j}} & \hat{\mathbf{k}} \\ x & y & 0 \\ 0 & 0 & 2xy \end{vmatrix} = \hat{\mathbf{i}}(2xy^2) - \hat{\mathbf{j}}(2x^2y)$$

$$\mathbf{F}\wedge(\nabla\wedge\mathbf{G}) = 0$$

$$(\mathbf{G}\cdot\nabla)\mathbf{F} = \left(x\frac{\partial}{\partial x} + y\frac{\partial}{\partial y}\right)(2x\hat{\mathbf{i}} + x^2y\hat{\mathbf{j}} + z^2\hat{\mathbf{k}}) = x(2\hat{\mathbf{i}} + 2xy\hat{\mathbf{j}}) + y(x^2\hat{\mathbf{j}})$$

$$= 2x\hat{\mathbf{i}} + 3x^2y\hat{\mathbf{j}}$$

$$(\mathbf{F}\cdot\nabla)\mathbf{G} = \left(2x\frac{\partial}{\partial x} + x^2y\frac{\partial}{\partial y} + z^2\frac{\partial}{\partial z}\right)(x\hat{\mathbf{i}} + y\hat{\mathbf{j}}) = 2x\hat{\mathbf{i}} + x^2y\hat{\mathbf{j}}$$

Adding, the right-hand side of the required identity is

$$\hat{\mathbf{i}}(2xy^2 + 4x) + \hat{\mathbf{j}}(2x^2y)$$

The left-hand side is

$$\nabla(\mathbf{F}\cdot\mathbf{G}) = \nabla(2x^2 + x^2y^2) = \hat{\mathbf{i}}(4x + 2xy^2) + \hat{\mathbf{j}}(2x^2y)$$

which verifies the identity.

(c) Again considering the terms on the right-hand side in turn,

$$(\mathbf{G}\cdot\nabla)\mathbf{F} = 2x\hat{\mathbf{i}} + 3x^2y\hat{\mathbf{j}}$$

$$(\mathbf{F}\cdot\nabla)\mathbf{G} = 2x\hat{\mathbf{i}} + x^2y\hat{\mathbf{j}}$$

$$\mathbf{F}(\nabla\cdot\mathbf{G}) = (2x\hat{\mathbf{i}} + x^2y\hat{\mathbf{j}} + z^2\hat{\mathbf{k}})(2)$$

$$\mathbf{G}(\nabla\cdot\mathbf{F}) = (x\hat{\mathbf{i}} + y\hat{\mathbf{j}})(2 + x^2 + 2z)$$

Therefore the right-hand side of the identity is

$$2x\hat{\mathbf{i}} + 3x^2y\hat{\mathbf{j}} - 2x\hat{\mathbf{i}} - x^2y\hat{\mathbf{j}} + 2(2x\hat{\mathbf{i}} + x^2y\hat{\mathbf{j}} + z^2\hat{\mathbf{k}}) - (x\hat{\mathbf{i}} + y\hat{\mathbf{j}})(2 + x^2 + 2z)$$

$$= \hat{\mathbf{i}}(2x - 2xz - x^3) + \hat{\mathbf{j}}(-2y - 2yz + 3x^2y) + \hat{\mathbf{k}}(2z^2)$$

The left-hand side is

$$\nabla \wedge (\mathbf{F} \wedge \mathbf{G}) = \begin{vmatrix} \hat{\mathbf{i}} & \hat{\mathbf{j}} & \hat{\mathbf{k}} \\ \dfrac{\partial}{\partial x} & \dfrac{\partial}{\partial y} & \dfrac{\partial}{\partial z} \\ -z^2y & xz^2 & 2xy - x^3y \end{vmatrix}$$

$$= \hat{\mathbf{i}}(2x - x^3 - 2xz) - \hat{\mathbf{j}}(2y - 3x^2y + 2zy) + \hat{\mathbf{k}}(z^2 + z^2)$$

which agrees with the right-hand side and so verifies the identity.

Identity (a) is *proved* by considering general expressions for **F** and **G**:

$$\mathbf{F} = F_x\hat{\mathbf{i}} + F_y\hat{\mathbf{j}} + F_z\hat{\mathbf{k}} \quad \text{and} \quad \mathbf{G} = G_x\hat{\mathbf{i}} + G_y\hat{\mathbf{j}} + G_z\hat{\mathbf{k}}$$

$$\nabla \wedge \mathbf{F} = \begin{vmatrix} \hat{\mathbf{i}} & \hat{\mathbf{j}} & \hat{\mathbf{k}} \\ \dfrac{\partial}{\partial x} & \dfrac{\partial}{\partial y} & \dfrac{\partial}{\partial z} \\ F_x & F_y & F_z \end{vmatrix}$$

$$= \hat{\mathbf{i}}\left(\frac{\partial F_z}{\partial y} - \frac{\partial F_y}{\partial z}\right) - \hat{\mathbf{j}}\left(\frac{\partial F_z}{\partial x} - \frac{\partial F_x}{\partial z}\right) + \hat{\mathbf{k}}\left(\frac{\partial F_y}{\partial x} - \frac{\partial F_x}{\partial y}\right)$$

$$\therefore \qquad \mathbf{G}\cdot(\nabla\wedge\mathbf{F}) = G_x\left(\frac{\partial F_z}{\partial y}-\frac{\partial F_y}{\partial z}\right) - G_y\left(\frac{\partial F_z}{\partial x}-\frac{\partial F_x}{\partial z}\right) + G_z\left(\frac{\partial F_y}{\partial x}-\frac{\partial F_x}{\partial y}\right)$$

The expression for $\mathbf{F}\cdot(\nabla\wedge\mathbf{G})$ is similar, except that the symbols F and G are interchanged. Now

$$\mathbf{F}\wedge\mathbf{G} = \begin{vmatrix} \hat{\mathbf{i}} & \hat{\mathbf{j}} & \hat{\mathbf{k}} \\ F_x & F_y & F_z \\ G_x & G_y & G_z \end{vmatrix}$$

$$= \hat{\mathbf{i}}(F_yG_z - F_zG_y) - \hat{\mathbf{j}}(F_xG_z - G_xF_z) + \hat{\mathbf{k}}(F_xG_y - F_yG_x)$$

$$\therefore \qquad \nabla\cdot(\mathbf{F}\wedge\mathbf{G}) = \frac{\partial}{\partial x}(F_yG_z - F_zG_y) - \frac{\partial}{\partial y}(F_xG_z - G_xF_z) + \frac{\partial}{\partial z}(F_xG_y - F_yG_x)$$

Taking the derivatives implied on the right-hand side and comparing the result with $\mathbf{G}\cdot(\nabla\wedge\mathbf{F}) - \mathbf{F}\cdot(\nabla\wedge\mathbf{G})$ then proves the identity in general.

5 At any point at position vector $\mathbf{r}$ on a rotating rigid body, the linear velocity is $\mathbf{v} = \boldsymbol{\omega}\wedge(\mathbf{r}-\overrightarrow{OP})$ where $\overrightarrow{OP}$ is a (constant) position vector of a point on the axis of rotation. Now

$$\nabla\wedge\mathbf{v} = \nabla\wedge\boldsymbol{\omega}\wedge(\mathbf{r}-\overrightarrow{OP}) = \boldsymbol{\omega}(\nabla\cdot\mathbf{r}) - (\boldsymbol{\omega}\cdot\nabla)\mathbf{r}$$

using the identity in additional exercise 4(c) and the fact that $\boldsymbol{\omega}$ and $\overrightarrow{OP}$ are constant vectors. But since

$$\nabla\cdot\mathbf{r} = 3 \quad \text{and} \quad (\boldsymbol{\omega}\cdot\nabla)\mathbf{r} = \boldsymbol{\omega}$$

we have $\nabla\wedge\mathbf{v} = 2\boldsymbol{\omega}$, as required.

6 We have

$$\nabla\wedge(\nabla\wedge\mathbf{F}) = \begin{vmatrix} \hat{\mathbf{i}} & \hat{\mathbf{j}} & \hat{\mathbf{k}} \\ \dfrac{\partial}{\partial x} & \dfrac{\partial}{\partial y} & \dfrac{\partial}{\partial z} \\ \dfrac{\partial F_z}{\partial y}-\dfrac{\partial F_y}{\partial z} & \dfrac{\partial F_x}{\partial z}-\dfrac{\partial F_z}{\partial x} & \dfrac{\partial F_y}{\partial x}-\dfrac{\partial F_x}{\partial y} \end{vmatrix}$$

$$= \hat{\mathbf{i}}\left(\frac{\partial^2 F_y}{\partial y\,\partial x} - \frac{\partial^2 F_x}{\partial y^2} - \frac{\partial^2 F_x}{\partial z^2} + \frac{\partial^2 F_z}{\partial z\,\partial x}\right)$$

$$-\hat{\mathbf{j}}\left(\frac{\partial^2 F_y}{\partial x^2} - \frac{\partial^2 F_x}{\partial x\,\partial y} - \frac{\partial^2 F_z}{\partial z\,\partial y} + \frac{\partial^2 F_y}{\partial z^2}\right)$$

$$+\hat{\mathbf{k}}\left(\frac{\partial^2 F_x}{\partial x\,\partial z} - \frac{\partial^2 F_z}{\partial x^2} - \frac{\partial^2 F_z}{\partial y^2} + \frac{\partial^2 F_y}{\partial y\,\mathrm{d}z}\right)$$

Adding and subtracting $\partial^2 F_x/\partial x^2$ to the first term, $\partial^2 F_y/\partial y^2$ to the second term and $\partial^2 F_z/\partial z^2$ to the third term gives

$$\nabla \wedge (\nabla \wedge \mathbf{F}) = \left(\hat{\mathbf{i}}\frac{\partial}{\partial x} + \hat{\mathbf{j}}\frac{\partial}{\partial y} + \hat{\mathbf{k}}\frac{\partial}{\partial z}\right)\left(\frac{\partial F_x}{\partial x} + \frac{\partial F_y}{\partial y} + \frac{\partial F_z}{\partial z}\right) - \nabla^2 \mathbf{F}$$

which is the required identity.

7 (a) The normal component (to the surface $r = R$) of each of the vectors $\mathbf{v}_i$ and $\mathbf{v}_e$ is simply the scalar product of these vectors with the unit vector in the radial direction $\hat{\mathbf{r}}$, namely $\mathbf{v}_i \cdot \hat{\mathbf{r}}$ and $\mathbf{v}_e \cdot \hat{\mathbf{r}}$ respectively. Now

$$\mathbf{v}_i \cdot \hat{\mathbf{r}} = f(\hat{\mathbf{a}} \cdot \hat{\mathbf{r}})$$

and $$\mathbf{v}_e \cdot \hat{\mathbf{r}} = kf\frac{R^3}{r^5}[3\mathbf{r}\cdot\hat{\mathbf{r}}(\hat{\mathbf{a}}\cdot\mathbf{r}) - \hat{\mathbf{a}}\cdot\hat{\mathbf{r}}r^2] = kf\frac{R^3}{r^5}[2r^2(\hat{\mathbf{a}}\cdot\hat{\mathbf{r}})]$$

since $\hat{\mathbf{a}}\cdot\mathbf{r} = \hat{\mathbf{a}}\cdot\hat{\mathbf{r}}r$. If these terms are to coincide when $r = R$ (so that the normal components of the vector field are continuous across the surface of the sphere), then we require

$$f(\hat{\mathbf{a}}\cdot\hat{\mathbf{r}}) = 2kf(\hat{\mathbf{a}}\cdot\hat{\mathbf{r}}) \quad \text{or} \quad k = \tfrac{1}{2}$$

The tangential components of $\mathbf{v}_i$ and $\mathbf{v}_e$ can be obtained by taking the vector product of each of them with a unit vector in the radial direction because the two vectors $\mathbf{v}_i \wedge \hat{\mathbf{r}}$ and $\mathbf{v}_e \wedge \hat{\mathbf{r}}$ are perpendicular to $\hat{\mathbf{r}}$ and so lie in the tangent plane to the surface of the sphere $r =$ constant. Now

$$\mathbf{v}_i \wedge \hat{\mathbf{r}} = f(\hat{\mathbf{a}} \wedge \hat{\mathbf{r}}) \quad \text{and} \quad \mathbf{v}_e \wedge \hat{\mathbf{r}} = kf\frac{R^3}{r^5}[-r^2\hat{\mathbf{a}} \wedge \hat{\mathbf{r}}] \quad \text{since } \hat{\mathbf{r}} \wedge \hat{\mathbf{r}} = 0$$

If these components are to be equal at $r = R$, the requirement is clearly that $k = -1$.

(b) First consider $\Phi_i = -f(\hat{\mathbf{a}}\cdot\mathbf{r})$. Then from the identity of additional exercise 4(b), and noting that $\hat{\mathbf{a}}$ is a constant vector and that $\nabla \wedge \mathbf{r} = 0$, we find

$$\begin{aligned} -\nabla\Phi_i = f\nabla(\hat{\mathbf{a}}\cdot\mathbf{r}) &= f(\hat{\mathbf{a}}\cdot\nabla)\mathbf{r} \\ &= f\left[\left(a_x\frac{\partial}{\partial x} + a_y\frac{\partial}{\partial y} + a_z\frac{\partial}{\partial z}\right)\mathbf{r}\right] \\ &= f[a_x\hat{\mathbf{i}} + a_y\hat{\mathbf{j}} + a_z\hat{\mathbf{k}}] \\ &= f\hat{\mathbf{a}} \end{aligned}$$

This verifies that $\mathbf{v}_i$ can be represented by $-\nabla\Phi_i$.

Using similar results and the identity of additional exercise 4(c), we find

$$\begin{aligned} \nabla \wedge \mathbf{A}_i = \tfrac{1}{2}f\nabla \wedge (\hat{\mathbf{a}} \wedge \mathbf{r}) &= \tfrac{1}{2}f[-(\hat{\mathbf{a}}\cdot\nabla)\mathbf{r} + \hat{\mathbf{a}}(\nabla\cdot\mathbf{r})] \\ &= \tfrac{1}{2}f(-\hat{\mathbf{a}} + 3\hat{\mathbf{a}}) \\ &= f\hat{\mathbf{a}} \end{aligned}$$

verifying that $\mathbf{v}_i$ can also be represented by $\nabla \wedge \mathbf{A}_i$.

For the final part of the question, consider

$$\Phi_e = kfR^3\frac{(\hat{\mathbf{a}}\cdot\mathbf{r})}{r^3}$$

then $$-\nabla\Phi_e = -kfR^3\left[\frac{1}{r^3}\nabla(\hat{\mathbf{a}}\cdot\mathbf{r}) + (\hat{\mathbf{a}}\cdot\mathbf{r})\nabla\left(\frac{1}{r^3}\right)\right]$$

Now $\nabla(\hat{\mathbf{a}}\cdot\mathbf{r}) = \hat{\mathbf{a}}$ and $\nabla\left(\frac{1}{r^3}\right) = -3\left(\frac{\hat{\mathbf{r}}}{r^4}\right)$

$$\therefore \quad -\nabla\Phi_e = -kfR^3\left[\frac{\hat{\mathbf{a}}}{r^3} - \frac{3}{r^4}(\hat{\mathbf{a}}\cdot\mathbf{r})\hat{\mathbf{r}}\right]$$

$$= -kf\frac{R^3}{r^5}[\hat{\mathbf{a}}r^2 - 3\mathbf{r}(\hat{\mathbf{a}}\cdot\mathbf{r})]$$

thus verifying that $\mathbf{v}_e$ may be represented by $-\nabla\Phi_e$.

To evaluate curl $\mathbf{A}_e$, we need to use property (2b) of Section 3.6:

$$\nabla\wedge\mathbf{A}_e = kfR^3\left[\frac{\nabla\wedge(\hat{\mathbf{a}}\wedge\mathbf{r})}{r^3} + \nabla\left(\frac{1}{r^3}\right)\wedge(\hat{\mathbf{a}}\wedge\mathbf{r})\right]$$

Now from the calculation of $\nabla\wedge\mathbf{A}_i$ above, $\nabla\wedge(\hat{\mathbf{a}}\wedge\mathbf{r}) = 2\hat{\mathbf{a}}$, and so

$$\nabla\wedge\mathbf{A}_e = kfR^3\left[\frac{2\hat{\mathbf{a}}}{r^3} - \frac{3}{r^4}(\hat{\mathbf{r}}\wedge(\hat{\mathbf{a}}\wedge\mathbf{r}))\right]$$

But using equation (1.61) for the vector triple product,

$$\hat{\mathbf{r}}\wedge(\hat{\mathbf{a}}\wedge\mathbf{r}) = \hat{\mathbf{a}}(\hat{\mathbf{r}}\cdot\mathbf{r}) - \mathbf{r}(\hat{\mathbf{a}}\cdot\hat{\mathbf{r}}) = \hat{\mathbf{a}}r - \hat{\mathbf{r}}(\hat{\mathbf{a}}\cdot\mathbf{r})$$

and so finally

$$\nabla\wedge\mathbf{A}_e = kfR^3\left[\frac{2\hat{\mathbf{a}}}{r^3} - \frac{3}{r^4}(\hat{\mathbf{a}}r - \hat{\mathbf{r}}(\hat{\mathbf{a}}\cdot\mathbf{r}))\right]$$

$$= kf\frac{R^3}{r^5}[-\hat{\mathbf{a}}r^2 + 3\mathbf{r}(\hat{\mathbf{a}}\cdot\mathbf{r})]$$

showing that $\mathbf{v}_e$ can also be represented by $\nabla\wedge\mathbf{A}_e$.

Chapter 4

1 The work dW done in moving a short distance $d\rho$ is $\mathbf{F}\cdot d\mathbf{l}$, where $d\mathbf{l} = d\rho\hat{\boldsymbol{\rho}}$; therefore

$$dW = -\frac{\kappa}{\rho}\hat{\boldsymbol{\rho}}\cdot d\rho\hat{\boldsymbol{\rho}} = -\kappa\frac{d\rho}{\rho}$$

The total work done in moving from a distance ρ_1, say, to $2\rho_1$ is

$$W = -\kappa\int_{\rho_1}^{2\rho_1}\frac{d\rho}{\rho} = -\kappa(\ln 2\rho_1 - \ln\rho_1) = -\kappa\ln 2$$

which is independent of ρ_1.

2 The vector field is

$$\mathbf{F} = r^2\mathbf{r} = (x^2 + y^2 + z^2)(x\hat{\mathbf{i}} + y\hat{\mathbf{j}} + z\hat{\mathbf{k}})$$

Now, using vector identity 2(b) in Section 3.6,

$$\nabla \wedge \mathbf{F} = r^2\, \nabla \wedge \mathbf{r} + \nabla r^2 \wedge \mathbf{r}$$

But $\quad \nabla r^2 = \dfrac{\partial r^2}{\partial x}\hat{\mathbf{i}} + \dfrac{\partial r^2}{\partial y}\hat{\mathbf{j}} + \dfrac{\partial r^2}{\partial z}\hat{\mathbf{k}} = 2x\hat{\mathbf{i}} + 2y\hat{\mathbf{j}} + 2z\hat{\mathbf{k}}$

so $\quad \nabla r^2 \wedge \mathbf{r} = \begin{vmatrix} \hat{\mathbf{i}} & \hat{\mathbf{j}} & \hat{\mathbf{k}} \\ 2x & 2y & 2z \\ x & y & z \end{vmatrix} = 0$

Also $\quad \nabla \wedge \mathbf{r} = \begin{vmatrix} \hat{\mathbf{i}} & \hat{\mathbf{j}} & \hat{\mathbf{k}} \\ \dfrac{\partial}{\partial x} & \dfrac{\partial}{\partial y} & \dfrac{\partial}{\partial z} \\ x & y & z \end{vmatrix} = 0$

Hence $\nabla \wedge \mathbf{F} \equiv 0$, and $\mathbf{F}$ is conservative. We now require a potential Φ such that $\mathbf{F} = \nabla\Phi$. Equating corresponding components,

$$\frac{\partial \Phi}{\partial x} = x^3 + y^2x + z^2x \quad \text{or} \quad \Phi = \frac{x^4}{4} + \frac{y^2x^2}{2} + \frac{z^2x^2}{2} + f(y, z)$$

$$\frac{\partial \Phi}{\partial y} = x^2y + y^3 + z^2y \quad \text{or} \quad \Phi = \frac{x^2y^2}{2} + \frac{y^4}{4} + \frac{z^2y^2}{2} + g(x, z)$$

$$\frac{\partial \Phi}{\partial z} = x^2z + y^2z + z^3 \quad \text{or} \quad \Phi = \frac{x^2z^2}{2} + \frac{y^2z^2}{2} + \frac{z^4}{4} + h(x, y)$$

The deduction from these expressions is that

$$\Phi = \frac{x^4 + y^4 + z^4}{4} + \frac{x^2y^2 + x^2z^2 + y^2z^2}{2} = \frac{r^4}{4} + \text{an arbitrary constant}$$

3 We are integrating along the half-ellipse shown in Figure 8.

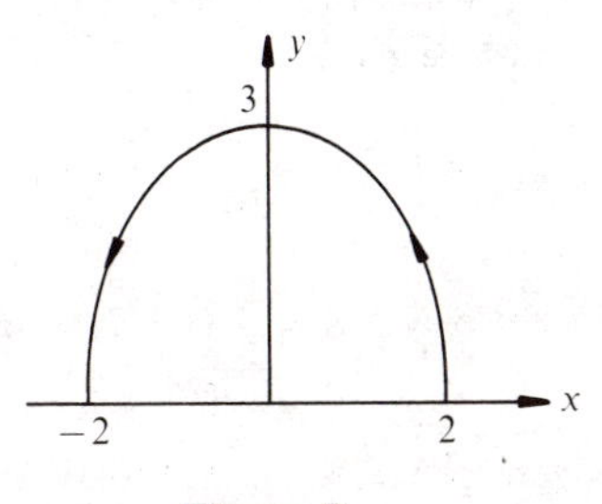

Figure 8

$$\int_C \mathbf{F}\cdot\mathrm{d}\mathbf{l} = \int_C \{3x^2\,\mathrm{d}x + 4xy\,\mathrm{d}y\}$$

$$= \int_0^{\pi} \{(12\cos^2 t)(-2\sin t)\,\mathrm{d}t + (24\cos t\sin t)(3\cos t)\,\mathrm{d}t\}$$

$$= \int_0^{\pi} \{-24\cos^2 t\sin t + 72\cos^2 t\sin t\}\,\mathrm{d}t$$

$$= 48\int_0^{\pi} \cos^2 t\sin t\,\mathrm{d}t$$

$$= -\frac{48}{3}\Big[\cos^3 t\Big]_0^{\pi}$$

$$= 32$$

4 The work done is $-5\int_C \mathbf{E}\cdot\mathrm{d}\mathbf{l}$, that is

$$W = -5\int_C \{2y\,\mathrm{d}x + 2x\,\mathrm{d}y\}$$

Along path (a), with $y = \frac{1}{3}(x+2)$,

$$W = -5\int_4^1 \{\tfrac{2}{3}(x+2)\,\mathrm{d}x + \tfrac{2}{3}x\,\mathrm{d}x\} = -5\int_4^1 (\tfrac{4}{3}x + \tfrac{4}{3})\,\mathrm{d}x = 70\ \mathrm{J}$$

Along path (b), with $x = y^2$,

$$W = -5\int_2^1 \{(2y)(2y)\,\mathrm{d}y + 2y^2\,\mathrm{d}y\} = -5\int_2^1 6y^2\,\mathrm{d}y = 70\ \mathrm{J}$$

Along path (c), with $x = \dfrac{4}{(7-3y)}$ and $\mathrm{d}x = \dfrac{12}{(7-3y)^2}\,\mathrm{d}y$,

$$W = -5\int_2^1 \left\{(2y)\frac{12}{(7-3y)^2}\,\mathrm{d}y + \frac{8}{(7-3y)}\,\mathrm{d}y\right\}$$

$$= -5\int_2^1 \frac{56}{(7-3y)^2}\,\mathrm{d}y = 70\ \mathrm{J}$$

5 We require curl **A** to be identically zero, which implies that each component of it must be zero.

$$x\text{-component of } \nabla\wedge\mathbf{A} = \frac{\partial}{\partial y}(4x+\gamma y+2z) - \frac{\partial}{\partial z}(\beta x - 3y - z) = \gamma + 1$$

$$y\text{-component of } \nabla\wedge\mathbf{A} = \frac{\partial}{\partial z}(x+2y+\alpha z) - \frac{\partial}{\partial x}(4x+\gamma y+2z) = \alpha - 4$$

$$z\text{-component of } \nabla\wedge\mathbf{A} = \frac{\partial}{\partial x}(\beta x - 3y - z) - \frac{\partial}{\partial y}(x+2y+\alpha z) = \beta - 2$$

Hence $\alpha = 4$, $\beta = 2$ and $\gamma = -1$, and we obtain

$$\mathbf{A} = (x + 2y + 4z)\hat{\mathbf{i}} + (2x - 3y - z)\hat{\mathbf{j}} + (4x - y + 2z)\hat{\mathbf{k}}$$

The corresponding scalar potential Φ is such that

$$\frac{\partial \Phi}{\partial x} = x + 2y + 4z \quad \text{or} \quad \Phi = \tfrac{1}{2}x^2 + 2xy + 4xz + f(y, z)$$

$$\frac{\partial \Phi}{\partial y} = 2x - 3y - z \quad \text{or} \quad \Phi = 2xy - \tfrac{3}{2}y^2 - yz + g(x, z)$$

$$\frac{\partial \Phi}{\partial z} = 4x - y + 2z \quad \text{or} \quad \Phi = 4xz - yz + z^2 + h(x, y)$$

Hence $\Phi = \frac{1}{2}(x^2 - 3y^2 + 2z^2) + 2xy - yz + 4xz + \text{a constant}$

The required line integral is given by $\Phi(1, 2, 1) - \Phi(1, 0, 0) = \frac{3}{2} - \frac{1}{2} = 1$. Evaluating the line integral along the specific path $x = 1$, $z = \frac{1}{2}y$ gives

$$\begin{aligned}
&\int_C \{(2 - 3y - z)\,\mathrm{d}y + (4 - y + 2z)\,\mathrm{d}z\} \\
&\quad = \int_0^2 \{(2 - 3y - \tfrac{1}{2}y)\,\mathrm{d}y + \tfrac{1}{2}(4 - y + y)\,\mathrm{d}y\} \\
&\quad = \int_0^2 (4 - \tfrac{7}{2}y)\,\mathrm{d}y = 1
\end{aligned}$$

6 We solve this problem (a) using cylindrical coordinates, and (b) by the method of projections.

(a) The unit outward normal to the given surface is

$$\hat{\mathbf{n}} = \frac{\nabla(x^2 + y^2)}{|\nabla(x^2 + y^2)|} = \frac{2x\hat{\mathbf{i}} + 2y\hat{\mathbf{j}}}{2\sqrt{(x^2 + y^2)}} = \frac{x\hat{\mathbf{i}} + y\hat{\mathbf{j}}}{4}$$

so the flux integral is

$$I = \iint_S \mathbf{F} \cdot \hat{\mathbf{n}}\,\mathrm{d}S = \tfrac{1}{4} \iint_S (zx + xy)\,\mathrm{d}S$$

or, in cylindrical coordinates with $\rho = 4$,

$$I = \int_0^5 \int_0^{\pi/2} (4z \cos\phi + 16 \cos\phi \sin\phi)\,\mathrm{d}\phi\,\mathrm{d}z = 90$$

(b) Using the method of projections, we project S onto the xz-plane (see Figure 9) so that the flux integral becomes a double integral over the rectangle D shown, that is,

$$\begin{aligned}
\iint_S \mathbf{F} \cdot \hat{\mathbf{n}}\,\mathrm{d}S &= \iint_D \mathbf{F} \cdot \hat{\mathbf{n}} \frac{\mathrm{d}x\,\mathrm{d}z}{|\hat{\mathbf{n}} \cdot \hat{\mathbf{j}}|} = \tfrac{1}{4} \iint_D (zx + xy) \frac{\mathrm{d}x\,\mathrm{d}z}{|\frac{1}{4}y|} \\
&= \iint_D \left(\frac{zx}{y} + x\right) \mathrm{d}x\,\mathrm{d}z
\end{aligned}$$

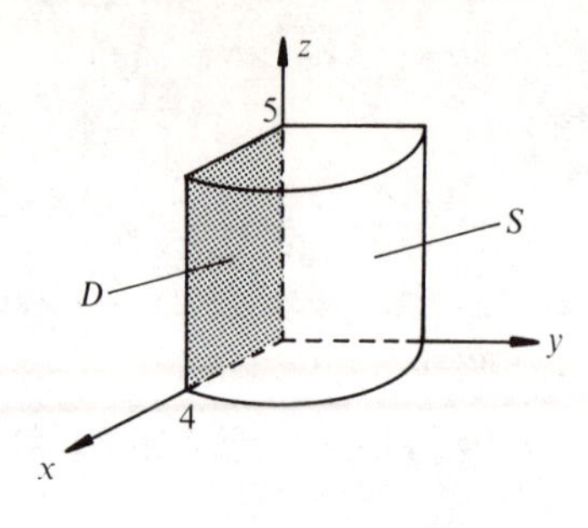

Figure 9

(replacing $|\frac{1}{4}y|$ by $\frac{1}{4}y$ since y is positive at all points on S). Hence

$$\iint_S \mathbf{F}\cdot\hat{\mathbf{n}}\,dS = \int_0^5\int_0^4 \left\{\frac{zx}{\sqrt{(16-x^2)}}+x\right\} dx\,dz$$

$$= \int_0^5 \left[-z\sqrt{(16-x^2)}+\tfrac{1}{2}x^2\right]_0^4 dz = \int_0^5 (4z+8)\,dz = 90$$

7 (a) For the given spherical surface, the unit outward normal is $\hat{\mathbf{n}} = \mathbf{r}/|\mathbf{r}| = \mathbf{r}/a$. Hence the flux integral is

$$\oiint_S \mathbf{r}\cdot\hat{\mathbf{n}}\,dS = \oiint_S \mathbf{r}\cdot\frac{\mathbf{r}}{a}\,dS = \oiint_S \frac{|\mathbf{r}|^2}{a}\,dS = \oiint_S a\,dS = 4\pi a^3$$

For the corresponding volume integral, $\nabla\cdot\mathbf{r} = \nabla\cdot(x\hat{\mathbf{i}}+y\hat{\mathbf{j}}+z\hat{\mathbf{k}}) = 3$, so we have

$$\iiint_V \nabla\cdot\mathbf{r}\,dV = \iiint_V 3\,dV = 3(\tfrac{4}{3}\pi a^3) = 4\pi a^3$$

(b) We have six surfaces (see Figure 10):

on $ABGF$, $\hat{\mathbf{n}} = +\hat{\mathbf{i}}$ and $x = 2$, so $\mathbf{r}\cdot\hat{\mathbf{n}} = +x = 2$

on $OCDE$, $\hat{\mathbf{n}} = -\hat{\mathbf{i}}$ and $x = 0$, so $\mathbf{r}\cdot\hat{\mathbf{n}} = -x = 0$

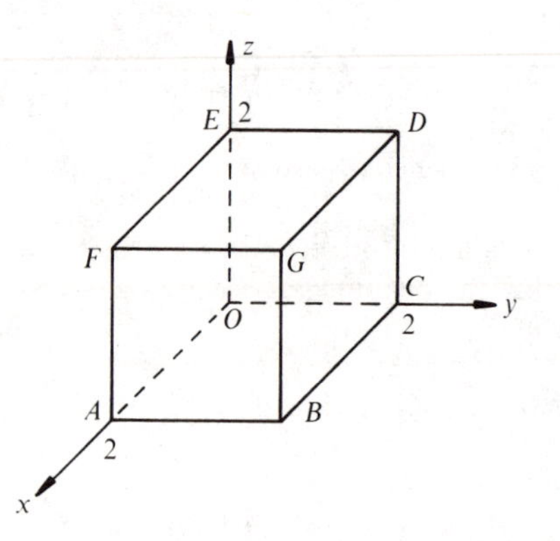

Figure 10

$$\text{on } FGDE, \quad \hat{\mathbf{n}} = +\hat{\mathbf{k}} \text{ and } z = 2, \quad \text{so} \quad \mathbf{r}\cdot\hat{\mathbf{n}} = +z = 2$$

$$\text{on } ABCO, \quad \hat{\mathbf{n}} = -\hat{\mathbf{k}} \text{ and } z = 0, \quad \text{so} \quad \mathbf{r}\cdot\hat{\mathbf{n}} = -z = 0$$

$$\text{on } AOEF, \quad \hat{\mathbf{n}} = -\hat{\mathbf{j}} \text{ and } y = 0, \quad \text{so} \quad \mathbf{r}\cdot\hat{\mathbf{n}} = -y = 0$$

$$\text{on } BCDG, \quad \hat{\mathbf{n}} = +\hat{\mathbf{j}} \text{ and } y = 2, \quad \text{so} \quad \mathbf{r}\cdot\hat{\mathbf{n}} = +y = 2$$

Hence, adding the three non-zero contributions,

$$\oiint_S \mathbf{r}\cdot\hat{\mathbf{n}}\, \mathrm{d}S = 3\iint_{ABGF} 2\, \mathrm{d}S = 6(\text{area of one face}) = 24$$

For the volume integral,

$$\iiint_V \nabla\cdot\mathbf{r}\, \mathrm{d}V = 3\iiint_V \mathrm{d}V = 3(\text{volume of the cube}) = 24$$

8 The unit outward normal to the hemisphere is

$$\hat{\mathbf{n}} = \tfrac{1}{5}(x\hat{\mathbf{i}} + y\hat{\mathbf{j}} + z\hat{\mathbf{k}})$$

so that $\mathbf{F}\cdot\hat{\mathbf{n}} = \frac{1}{5}[x(x+y) - 2zy + yz] = \frac{1}{5}(x^2 + xy - zy)$

Transforming to spherical polar coordinates,

$$\mathbf{F}\cdot\hat{\mathbf{n}} = \tfrac{1}{5}(r^2 \sin^2\theta \cos^2\phi + r^2 \sin^2\theta \cos\phi \sin\phi - r^2 \sin\theta \cos\theta \sin\phi)$$

For integrating over the hemisphere $(0 \leqslant \theta \leqslant \pi/2,\ 0 \leqslant \phi \leqslant 2\pi)$, the second and third terms will give zero when we carry out the integration with respect to ϕ. Therefore

$$\oiint_S \mathbf{F}\cdot\hat{\mathbf{n}}\, \mathrm{d}S = \tfrac{25}{5}\int_0^{2\pi}\int_0^{\pi/2} (\sin^2\theta \cos^2\phi)(25 \sin\theta)\, \mathrm{d}\theta\, \mathrm{d}\phi$$

$$= 125\int_0^{2\pi} \cos^2\phi\, \mathrm{d}\phi \int_0^{\pi/2} \sin^3\theta\, \mathrm{d}\theta = (125\pi)(\tfrac{2}{3}) = \tfrac{250}{3}\pi$$

9 We have

$$\nabla \wedge \mathbf{F} \equiv \operatorname{curl} \mathbf{F} = \begin{vmatrix} \hat{\mathbf{i}} & \hat{\mathbf{j}} & \hat{\mathbf{k}} \\ \dfrac{\partial}{\partial x} & \dfrac{\partial}{\partial y} & \dfrac{\partial}{\partial z} \\ y & -x & 0 \end{vmatrix} = -2\hat{\mathbf{k}}$$

Hence $\oiint_S \nabla \wedge \mathbf{F}\cdot\hat{\mathbf{n}}\, \mathrm{d}S = -2\oiint_S \hat{\mathbf{k}}\cdot\hat{\mathbf{n}}\, \mathrm{d}S$. Since $\hat{\mathbf{n}} = \frac{1}{2}(x\hat{\mathbf{i}} + y\hat{\mathbf{j}} + z\hat{\mathbf{k}})$, $\hat{\mathbf{k}}\cdot\hat{\mathbf{n}} = \frac{1}{2}z$ and the flux integral is

$$I = -\oiint_S z\, \mathrm{d}S$$

In spherical coordinates,

$$I = -\int_0^{2\pi}\int_0^{\pi/2} r\cos\theta\, r^2 \sin\theta\, \mathrm{d}\theta\, \mathrm{d}\phi$$

or, since $r=2$ on the spherical surface,

$$I=-8\int_0^{2\pi}\mathrm{d}\phi\int_0^{\pi/2}\tfrac{1}{2}\sin 2\theta\,\mathrm{d}\theta=-8(2\pi)(\tfrac{1}{2})=-8\pi$$

The line integral is

$$\oint_C \mathbf{F}\cdot\mathrm{d}\mathbf{l}=\oint_C (y\,\mathrm{d}x-x\,\mathrm{d}y)$$

where C is the circle $x^2+y^2=4$. Putting $\theta=\pi/2$ and $r=2$, we have $x=2\cos\phi$ and $y=2\sin\phi$, so the line integral becomes

$$\int_0^{2\pi}\{(2\sin\phi)(-2\sin\phi)\,\mathrm{d}\phi-2\cos\phi\;2\cos\phi\,\mathrm{d}\phi\}$$

$$=-4\int_0^{2\pi}(\sin^2\phi+\cos^2\phi)\,\mathrm{d}\phi=-8\pi$$

10 The surface is shown in Figure 11. The unit normal to this surface is

$$\hat{\mathbf{n}}=\frac{\nabla(x+2y+3z)}{|\nabla(x+2y+3z)|}=\frac{\hat{\mathbf{i}}+2\hat{\mathbf{j}}+3\hat{\mathbf{k}}}{\sqrt{14}}$$

Hence $\displaystyle\oiint_S \mathbf{F}\cdot\hat{\mathbf{n}}\,\mathrm{d}S=\frac{1}{\sqrt{14}}\oiint_S (x-4y-3z)\,\mathrm{d}S$

Projecting S onto the xy-plane gives the triangle D shown in Figure 11, and

$$\oiint_S \mathbf{F}\cdot\hat{\mathbf{n}}\,\mathrm{d}S=\frac{1}{\sqrt{14}}\iint_D (x-4y-3z)\frac{\mathrm{d}x\,\mathrm{d}y}{|\hat{\mathbf{n}}\cdot\hat{\mathbf{k}}|}$$

$$=\tfrac{1}{3}\iint_D \{x-4y-(6-x-2y)\}\,\mathrm{d}x\,\mathrm{d}y$$

$$=\tfrac{1}{3}\iint_D (2x-2y-6)\,\mathrm{d}x\,\mathrm{d}y$$

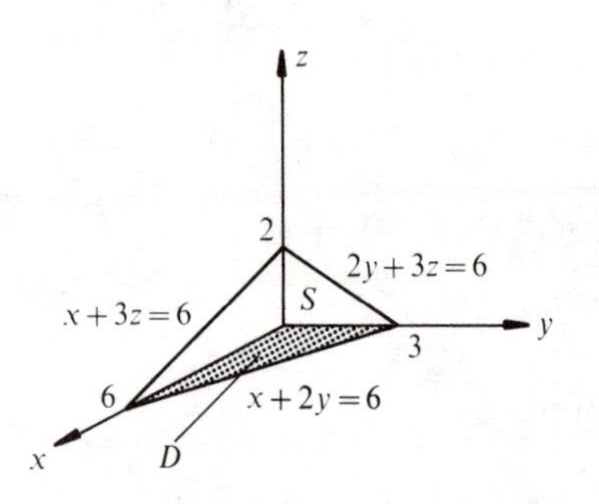

Figure 11

Performing the double integral over D by integrating first with respect to x, we obtain the iterated integral

$$\tfrac{1}{3}\int_0^3 \mathrm{d}y \int_0^{6-2y} (2x-2y-6)\,\mathrm{d}x = \tfrac{1}{3}\int_0^3 \Big[x^2-2xy-6x\Big]_0^{6-2y} \mathrm{d}y$$

$$= \tfrac{1}{3}\int_0^3 (8y^2-24y)\,\mathrm{d}y = -12$$

11 We have

$$\nabla\cdot\mathbf{F} \equiv \operatorname{div}\mathbf{F} = 4x-2x = 2x$$

so the required volume integral is

$$I = \iiint_V 2x\,\mathrm{d}V$$

where V is the region bounded by the slanting plane shown in Figure 12

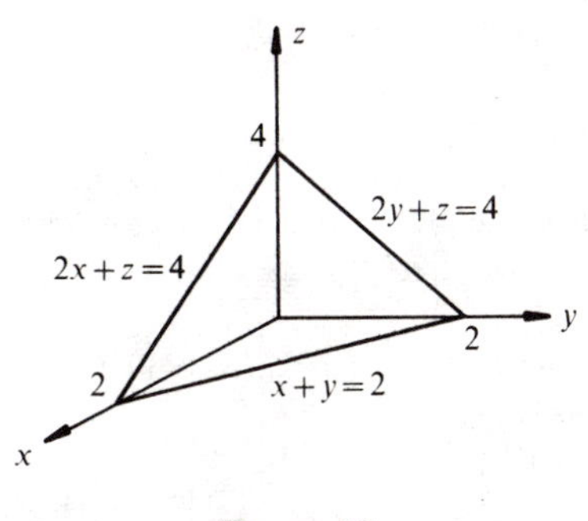

Figure 12

and the three coordinate planes. Hence, carrying out the z integration first,

$$I = 2\int_0^2 x\,\mathrm{d}x \int_0^{2-x} \mathrm{d}y \int_0^{4-2x-2y} \mathrm{d}z$$

$$= 2\int_0^2 x\,\mathrm{d}x \int_0^{2-x} (4-2x-2y)\,\mathrm{d}y$$

$$= 2\int_0^2 x\Big[4y-2xy-y^2\Big]_0^{2-x} \mathrm{d}x$$

$$= 2\int_0^2 (4x-4x^2+x^3)\,\mathrm{d}x = \tfrac{8}{3}$$

12 The volume V required is given by the integral $\iiint_V \mathrm{d}V$ over the required volume. We use cylindrical coordinates (ρ, ϕ, z), so that $\mathrm{e}^{-(x^2+y^2)}$ becomes

$e^{-\rho^2}$. Hence we obtain

$$V = \int_0^1 \rho \, d\rho \int_0^{2\pi} d\phi \int_0^{e^{-\rho^2}} dz = 2\pi \int_0^1 \rho \, e^{-\rho^2} \, d\rho$$

The final integration is carried out with the aid of the substitution $u = \rho^2$ to give $V = \pi(1 - e^{-1})$.

13 The total mass M of gas is given by the volume integral

$$M = \iiint_V \sigma \, dV$$

where the volume is conveniently divided into two parts – over the cylinder V_1 and the hemisphere V_2. The mean density is simply the total mass of the gas divided by the total volume of the container. For the gas in the cylindrical portion,

$$M_1 = C \iiint_{V_1} e^{-z} \rho \, d\rho \, d\phi \, dz = C \int_0^h e^{-z} \, dz \int_0^{2\pi} d\phi \int_0^a \rho \, d\rho$$

$$= \pi a^2 C \int_0^h e^{-z} \, dz = \pi a^2 C(1 - e^{-h})$$

For the gas in the hemispherical portion, we use spherical coordinates. Note carefully, however, that if the origin of spherical coordinates is at the centre of the hemisphere and z is measured from the base of the holder, we must replace $(z - h)$ by $r \cos \theta$. Hence

$$M_2 = C \iiint_{V_2} e^{-(h + r\cos\theta)} r^2 \sin\theta \, dr \, d\theta \, d\phi$$

$$= C \, e^{-h} \int_0^a r^2 \, dr \int_0^{\pi/2} e^{-r\cos\theta} \sin\theta \, d\theta \int_0^{2\pi} d\phi$$

$$= 2\pi C \, e^{-h} \int_0^a r^2 \, dr \int_0^{\pi/2} e^{-r\cos\theta} \sin\theta \, d\theta$$

(note the limits on the θ integral – we are integrating over a hemisphere). Using the substitution $u = \cos\theta$, for example, the θ integral gives $(1 - e^{-r})/r$. Therefore

$$M_2 = 2\pi C \, e^{-h} \int_0^a r(1 - e^{-r}) \, dr = 2\pi C \, e^{-h}(\tfrac{1}{2}a^2 + a \, e^{-a} + e^{-a} - 1)$$

Dividing this expression by the volume of the gas holder $(\pi a^2 h + \frac{2}{3}\pi a^3)$ then gives the required mean density.

14 The required integral is obtained using the volume integral $\iiint_V dV$ over the shaded region in Figure 13, where the angle θ_1 is $\pi/4$. Hence, using spherical coordinates,

$$V = \int_0^{2\pi} d\phi \int_0^{\pi/4} \sin\theta \, d\theta \int_0^4 r^2 \, dr = 64\pi(2 - \sqrt{2})/3$$

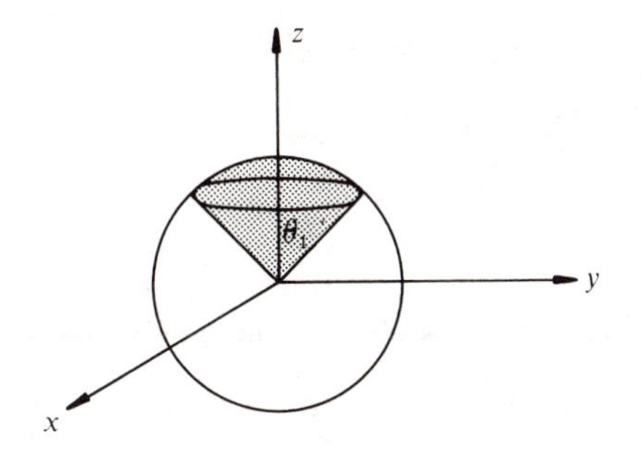

Figure 13

15 The total mass is

$$M = \iiint_V \sigma \, dV$$

taken over the volume of the sphere. Using spherical coordinates,

$$M = k \iiint_V (a - r\cos\theta)\, r^2 \sin\theta \, dr \, d\theta \, d\phi$$

$$= 2\pi k \int_0^a r^2 \, dr \int_0^\pi (a - r\cos\theta)\sin\theta \, d\theta$$

$$= 2\pi k \int_0^a r^2 \Big[-a\cos\theta + \tfrac{1}{4} r\cos 2\theta \Big]_0^\pi \, dr$$

$$= 4\pi a k \int_0^a r^2 \, dr = 4\pi k \tfrac{1}{3} a^4$$

Since the mass density depends only on z and not on x or y, the centre of gravity must lie on the z-axis, with a coordinate $\bar{z}$ such that

$$M\bar{z} = \iiint_V \sigma z \, dV$$

The integral on the right-hand side is evaluated in spherical coordinates:

$$\iiint_V \sigma z \, dV = k \iiint_V (a - r\cos\theta)(r\cos\theta)\, r^2 \sin\theta \, dr \, d\theta \, d\phi$$

$$= 2\pi k \int_0^a r^3 \, dr \int_0^\pi (\tfrac{1}{2} a \sin 2\theta - r\cos^2\theta \sin\theta) \, d\theta$$

$$= 2\pi k \int_0^a r^3 \Big[-\tfrac{1}{4} a \cos 2\theta + \tfrac{1}{3} r \cos^3\theta \Big]_0^\pi \, dr$$

$$= -\tfrac{4}{3}\pi k \int_0^a r^4 \, dr = -\tfrac{4}{15}\pi k a^5$$

Hence $\bar{z} = -\tfrac{1}{5} a$.

Chapter 5

1 (a) By Stokes' theorem,

$$\oint_C \mathbf{r}\cdot d\mathbf{l} = \iint_S \nabla \wedge \mathbf{r}\cdot\hat{\mathbf{n}}\, dS$$

where S is an arbitrary surface bounded by C. But $\nabla \wedge \mathbf{r} \equiv 0$ (see additional exercise 2 in Chapter 4), and the identity is proved.

(b) By the divergence theorem,

$$\oiint_S \mathbf{r}\cdot\hat{\mathbf{n}}\, dS = \iiint_V \nabla\cdot\mathbf{r}\, dV$$

But $\nabla\cdot\mathbf{r} = \nabla\cdot(x\hat{\mathbf{i}} + y\hat{\mathbf{j}} + z\hat{\mathbf{k}}) = 3$

Hence $\displaystyle\oiint_S \mathbf{r}\cdot\hat{\mathbf{n}}\, dS = 3\iiint_V dV = 3$(volume of region enclosed by S)

(c) By the result just proved, we require 3(volume under the surface S), where the volume must be calculated from the triple integral

$$I = 4\int_0^2 dx \int_0^{\sqrt{(4-x^2)}} dy \int_0^{4-(x^2+y^2)} dz$$

$$= 4\int_0^2 dx \int_0^{\sqrt{(4-x^2)}} \{4-(x^2+y^2)\}\, dy$$

This iterated integral is derived from a double integral over a circle and is best evaluated using polar coordinates. Hence

$$I = 4\int_0^{\pi/2} d\phi \int_0^2 (4-\rho^2)\rho\, d\rho = 2\pi\left[2\rho^2 - \tfrac{1}{4}\rho^4\right]_0^2 = 8\pi$$

so the required volume is $3(8\pi) = 24\pi$.

2 All these questions require the use of the divergence theorem.

(a) $$\oiint_S \nabla r^2\cdot\hat{\mathbf{n}}\, dS = \iiint_V \nabla\cdot(\nabla r^2)\, dV = \iiint_V \nabla^2 r^2\, dV$$

The Laplacian,

$$\nabla^2 r^2 = \frac{\partial^2}{\partial x^2}(x^2+y^2+z^2) + \text{two similar terms} = 2+2+2 = 6$$

Hence $\displaystyle\oiint_S \nabla r^2\cdot\hat{\mathbf{n}}\, dS = 6\iiint_V dV$

from which the result follows.

(b) Using Green's second identity (see Section 5.3),

$$\oiint_S (\Phi\,\nabla\vartheta - \vartheta\,\nabla\Phi)\cdot\hat{\mathbf{n}}\, dS = \iiint_R (\Phi\,\nabla^2\vartheta - \vartheta\,\nabla^2\Phi)\, dV$$

$$= \iiint_R (\Phi\, h\vartheta - \vartheta\, h\Phi)\, dV = 0$$

(c) Using Green's first identity,

$$\oiint_S \Phi\, \nabla\Phi \cdot \hat{\mathbf{n}}\, \mathrm{d}S = \iiint_R (\Phi\, \nabla^2\Phi + \nabla\Phi \cdot \nabla\Phi)\, \mathrm{d}V$$

But if Φ is harmonic, then it satisfies Laplace's equation, that is, $\nabla^2\Phi = 0$. Also, $\nabla\Phi \cdot \nabla\Phi = |\nabla\Phi|^2$, and the result is proved.

(d) This result follows by simply replacing $\nabla\Phi$ by $\mathbf{A}$ in the result just proved.

3 We must evaluate the volume integral $\iiint_V \nabla \cdot \mathbf{F}\, \mathrm{d}V$ over the volume enclosed by the hemisphere, and also the flux integral $\oiint_S \mathbf{F} \cdot \hat{\mathbf{n}}\, \mathrm{d}S$ over the hemispherical surface and its base.

We have $\nabla \cdot \mathbf{F} = y - z$, so the volume integral is

$$I = \iiint_V (y - z)\, \mathrm{d}V$$

which becomes, in spherical coordinates,

$$I = \int_0^{2\pi} \int_0^{\pi/2} \int_0^1 (r \sin\theta \sin\phi - r \cos\theta)\, r^2 \sin\theta\, \mathrm{d}r\, \mathrm{d}\theta\, \mathrm{d}\phi$$

Because of the $\sin\phi$ term, the first term will integrate to zero. Hence

$$I = -2\pi \int_0^{\pi/2} \int_0^1 r^3 \cos\theta \sin\theta\, \mathrm{d}r\, \mathrm{d}\theta = -2\pi \int_0^{\pi/2} \tfrac{1}{2} \sin 2\theta\, \mathrm{d}\theta \int_0^1 r^3\, \mathrm{d}r$$

$$= -\tfrac{1}{4}\pi$$

The surface integral over the base S_1 of the hemisphere is readily done:

$$\hat{\mathbf{n}} = -\hat{\mathbf{k}} \quad \text{so} \quad \mathbf{F} \cdot \hat{\mathbf{n}} = -3$$

therefore

$$\iint_{S_1} \mathbf{F} \cdot \hat{\mathbf{n}}\, \mathrm{d}S = -3 \iint_{S_1} \mathrm{d}S = -3\pi$$

For the hemispherical surface S_2,

$$\hat{\mathbf{n}} = \frac{\mathbf{r}}{|\mathbf{r}|} = \frac{x\hat{\mathbf{i}} + y\hat{\mathbf{j}} + z\hat{\mathbf{k}}}{1}$$

therefore

$$\mathbf{F} \cdot \hat{\mathbf{n}} = x^2 y - y^2 z + 3z$$

Converting to spherical coordinates (with $r = 1$ because we are integrating over the surface of the sphere),

$$\mathbf{F} \cdot \hat{\mathbf{n}} = \sin^3\theta \cos^2\phi \sin\phi - \sin^2\theta \sin^2\phi \cos\theta + 3 \cos\theta$$

Hence, putting $\mathrm{d}S = \sin\theta\,\mathrm{d}\theta\,\mathrm{d}\phi$,

$$\iint_{S_2} \mathbf{F}\cdot\hat{\mathbf{n}}\,\mathrm{d}S = -\int_0^{2\pi}\int_0^{\pi/2} \sin^3\theta\,\sin^2\phi\,\cos\theta\,\mathrm{d}\theta\,\mathrm{d}\phi$$

$$+3\int_0^{2\pi}\int_0^{\pi/2} \cos\theta\,\sin\theta\,\mathrm{d}\theta\,\mathrm{d}\phi$$

(where the first term in $\mathbf{F}\cdot\hat{\mathbf{n}}$ has been omitted because it integrates to zero).

Carrying out two straightforward integrations gives

$$\iint_{S_2} \mathbf{F}\cdot\hat{\mathbf{n}}\,\mathrm{d}S = -\tfrac{1}{4}\pi + 3\pi$$

and hence the complete flux integral has value $-\frac{1}{4}\pi$, so verifying the divergence theorem.

4 We first evaluate the volume integral of $\nabla\cdot\mathbf{F}$ over the volume V shown in Figure 14, where $\nabla\cdot\mathbf{F} = 4xy - 2y + 8xz$. Carrying out the x integration (from $x = 0$ to $x = 2$) first reduces the volume integral to the double integral

$$\iint_R (8y - 4y + 16z)\,\mathrm{d}y\,\mathrm{d}z$$

where R is the quarter-circle $y^2 + z^2 = 9$, $y \geqslant 0$, $z \geqslant 0$. Converting to polar coordinates, with

$$y = \rho\cos\phi, \quad z = \rho\sin\phi, \qquad 0 \leqslant \rho \leqslant 3, \quad 0 \leqslant \phi \leqslant \pi/2$$

gives $\displaystyle\int_0^3 \mathrm{d}\rho \int_0^{\pi/2} (4\rho^2\cos\phi + 16\rho^2\sin\phi)\,\mathrm{d}\phi$

from which two straightforward integrations give the value of the volume integral as 180.

For the flux integral, we must evaluate the normal flux $\mathbf{F}\cdot\hat{\mathbf{n}}$ over the five surfaces enclosing the volume V, namely the coordinate surfaces $x = 0$, $y = 0$ and $z = 0$, the flat surface $x = 2$ and the curved surface $y^2 + z^2 = 9$.

For the surface $y = 0$, $\hat{\mathbf{n}} = -\hat{\mathbf{j}}$ so $\mathbf{F}\cdot\hat{\mathbf{n}} = -y^2 = 0$, so the flux is zero.

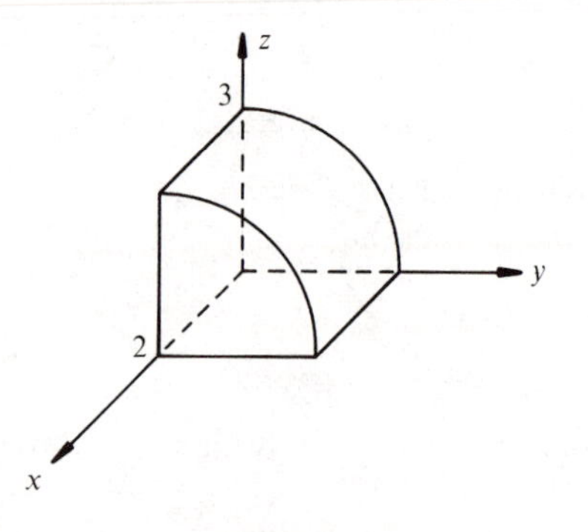

Figure 14

For the surface $z=0$, $\hat{\mathbf{n}}=-\hat{\mathbf{k}}$ so $\mathbf{F}\cdot\hat{\mathbf{n}}=-4xz^2=0$, and again the flux is zero.

For the surface $x=0$, $\hat{\mathbf{n}}=-\hat{\mathbf{i}}$ so $\mathbf{F}\cdot\hat{\mathbf{n}}=-2x^2y=0$, giving zero flux.

For the surface $x=2$, $\hat{\mathbf{n}}=+\hat{\mathbf{i}}$ so $\mathbf{F}\cdot\hat{\mathbf{n}}=2x^2y=8y$, so the flux over this surface is given by the double integral

$$\iint_R 8y\,\mathrm{d}S=\int_0^{\pi/2}\int_0^3 8\rho\cos\phi\,\rho\,\mathrm{d}\rho\,\mathrm{d}\phi=72$$

Finally, for the curved surface, $\hat{\mathbf{n}}=\frac{1}{3}(y\hat{\mathbf{j}}+z\hat{\mathbf{k}})$ so $\mathbf{F}\cdot\hat{\mathbf{n}}=-\frac{1}{3}y^3+\frac{4}{3}xz^3$, which has to be integrated over the curved surface $0\leqslant x\leqslant 2$, $0\leqslant\phi\leqslant\pi/2$. Putting

$$y=3\cos\phi,\qquad z=3\sin\phi,\qquad \mathrm{d}S=3\,\mathrm{d}\phi\,\mathrm{d}x$$

we obtain the flux integral

$$\begin{aligned}I&=\int_0^{\pi/2}\int_0^2(-27\cos^3\phi+108x\sin^3\phi)\,\mathrm{d}x\,\mathrm{d}\phi\\&=\int_0^{\pi/2}(-54\cos^3\phi+216\sin^3\phi)\,\mathrm{d}\phi=108\end{aligned}$$

so that the total flux is $72+108=180$, and the divergence theorem has been verified.

5 The given equation links a line integral with a surface integral. This suggests that we use Stokes' theorem to obtain a surface integral on each side. Using Stokes' theorem, the line integral transforms to

$$\oiint_S(\nabla\wedge\mathbf{E})\cdot\hat{\mathbf{n}}\,\mathrm{d}S$$

Also, $$\frac{\partial}{\partial t}\iint_S\mathbf{B}\cdot\hat{\mathbf{n}}\,\mathrm{d}S=\iint_S\frac{\partial\mathbf{B}}{\partial t}\cdot\hat{\mathbf{n}}\,\mathrm{d}S$$

because we can interchange the order of the t differentiation with the integration over a surface S. Hence, finally,

$$\oiint_S(\nabla\wedge\mathbf{E})\cdot\hat{\mathbf{n}}\,\mathrm{d}S=-\oiint_S\frac{\partial\mathbf{B}}{\partial t}\cdot\hat{\mathbf{n}}\,\mathrm{d}S$$

But since S is an arbitrary surface, this result implies that $\nabla\wedge\mathbf{E}=-\partial\mathbf{B}/\partial t$, as required.

6 The line integral is

$$\oint_C\mathbf{F}\cdot\mathrm{d}\mathbf{l}=\oint_C\{x\,\mathrm{d}x+(2-x)\,\mathrm{d}y\}$$

where C is the circle $x^2+y^2=4$, $z=1$. Putting $x=2\cos t$ and $y=2\sin t$,

$$\oint_C\mathbf{F}\cdot\mathrm{d}\mathbf{l}=\int_0^{2\pi}\{(2\cos t)(-2\sin t)+(2-2\cos t)2\cos t\}\,\mathrm{d}t=-4\pi$$

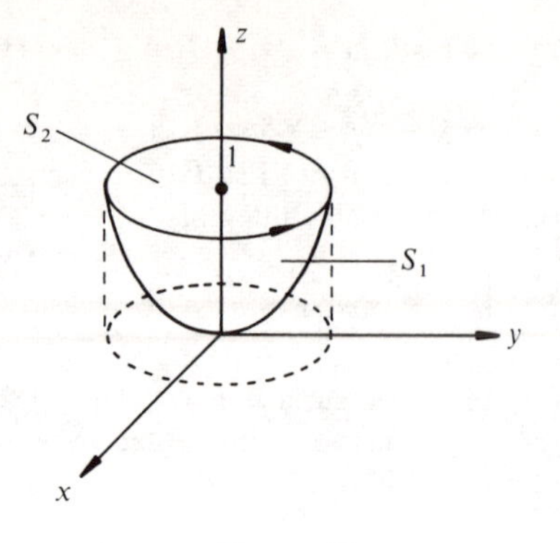

Figure 15

The curve C bounds an infinite number of surfaces, out of which we choose two simple ones S_1 and S_2 (see Figure 15).

(a) S_1 is the portion of the paraboloid $4z = x^2 + y^2$ beneath the plane $z = 1$. The normal to this surface is

$$\hat{\mathbf{n}} = \frac{\nabla f}{|\nabla f|}, \quad \text{where } f = x^2 + y^2 - 4z$$

Hence $\hat{\mathbf{n}} = \dfrac{2x\hat{\mathbf{i}} + 2y\hat{\mathbf{j}} - 4\hat{\mathbf{k}}}{\sqrt{(4x^2 + 4y^2 + 16)}}$

Since $\nabla \wedge \mathbf{F} = (2y - 2)\hat{\mathbf{i}} - \hat{\mathbf{k}}$

then $(\nabla \wedge \mathbf{F}) \cdot \hat{\mathbf{n}} = \dfrac{2x(2y - 2) + 4}{\sqrt{(4x^2 + 4y^2 + 16)}}$

Projecting S_1 onto the xy-plane (giving the interior of the circle $x^2 + y^2 = 4$), we obtain

$$\oiint_{S_1} (\nabla \wedge \mathbf{F}) \cdot \hat{\mathbf{n}} \, dS = \iint_R \{(\nabla \wedge \mathbf{F}) \cdot \hat{\mathbf{n}}\} \frac{dx\,dy}{|\hat{\mathbf{n}} \cdot \hat{\mathbf{k}}|}$$

$$= \iint_R \{2x(2y - 2) + 4\} \frac{dx\,dy}{(-4)}$$

$$= \iint_R (-xy + x - 1) \, dx\,dy$$

$$= 0 + 0 - (\text{area of circle}) = -4\pi$$

so verifying Stokes' theorem.

(b) S_2 is the portion of the plane $z = 1$ inside the circle $x^2 + y^2 = 4$. For this surface, $\hat{\mathbf{n}} = \hat{\mathbf{k}}$, therefore

$$(\nabla \wedge \mathbf{F}) \cdot \hat{\mathbf{n}} = -1$$

and $\displaystyle\oiint_{S_2} (\nabla \wedge \mathbf{F}) \cdot \hat{\mathbf{n}} \, dS = -\iint_{S_2} dS = -(\text{area of circle}) = -4\pi$

which again verifies Stokes' theorem.

Chapter 6

1 In cylindrical polars, the given vector field $\mathbf{A}$ is

$$\mathbf{A} = 0\hat{\boldsymbol{\rho}} + 0\hat{\boldsymbol{\phi}} + f(\rho)\hat{\mathbf{k}}$$

$$\therefore \quad \nabla \wedge \mathbf{A} = \frac{1}{\rho}\begin{vmatrix} \hat{\boldsymbol{\rho}} & \rho\hat{\boldsymbol{\phi}} & \hat{\mathbf{k}} \\ \dfrac{\partial}{\partial \rho} & \dfrac{\partial}{\partial \phi} & \dfrac{\partial}{\partial z} \\ 0 & 0 & f(\rho) \end{vmatrix} = \frac{1}{\rho}\hat{\boldsymbol{\rho}}(0) - \frac{1}{\rho}\rho\hat{\boldsymbol{\phi}}\left(\frac{\mathrm{d}f}{\mathrm{d}\rho}\right) + \frac{1}{\rho}\hat{\mathbf{k}}(0) = -\hat{\boldsymbol{\phi}}\frac{\mathrm{d}f}{\mathrm{d}\rho}$$

The direction of $\nabla \wedge \mathbf{A}$ is parallel or antiparallel to the unit vector $\hat{\boldsymbol{\phi}}$.

2 (a) $$\nabla f(r) = \hat{\mathbf{r}}\frac{\partial}{\partial r}(f(r)) + \frac{1}{r}\hat{\boldsymbol{\theta}}\frac{\partial}{\partial \theta}(f(r)) + \frac{1}{r\sin\theta}\hat{\boldsymbol{\phi}}\frac{\partial}{\partial \phi}(f(r)) = \hat{\mathbf{r}}\frac{\mathrm{d}f}{\mathrm{d}r}$$

since f is a function only of the single variable r.

(b) $$\nabla^2 f(r) = \frac{1}{r^2}\frac{\mathrm{d}}{\mathrm{d}r}\left(r^2\frac{\mathrm{d}f}{\mathrm{d}r}\right) + 0 + 0 = \frac{1}{r^2}\left(2r\frac{\mathrm{d}f}{\mathrm{d}r} + r^2\frac{\mathrm{d}^2 f}{\mathrm{d}r^2}\right) = \frac{\mathrm{d}^2 f}{\mathrm{d}r^2} + \frac{2}{r}\frac{\mathrm{d}f}{\mathrm{d}r}$$

(c) If $f(r)$ satisfies Laplace's equation, then

$$\frac{\mathrm{d}^2 f}{\mathrm{d}r^2} + \frac{2}{r}\frac{\mathrm{d}f}{\mathrm{d}r} = 0$$

that is, $\dfrac{\mathrm{d}}{\mathrm{d}r}\left(r^2\dfrac{\mathrm{d}f}{\mathrm{d}r}\right) = 0$ or $r^2\dfrac{\mathrm{d}f}{\mathrm{d}r} = c$ (a constant)

$$\therefore \quad \frac{\mathrm{d}f}{\mathrm{d}r} = \frac{c}{r^2}$$

leading to

$$f(r) = -\frac{c}{r} + d, \qquad r \neq 0$$

where c and d are constants. This scalar field corresponds to a point source at the origin.

3 We write the position vector $\mathbf{r}$ in terms of the new coordinates:

$$\mathbf{r} = x\hat{\mathbf{i}} + y\hat{\mathbf{j}} + z\hat{\mathbf{k}} = (u^2 - v^2)\hat{\mathbf{i}} + 2uv\hat{\mathbf{j}} + w\hat{\mathbf{k}}$$

Now $\dfrac{\partial \mathbf{r}}{\partial u} = 2u\hat{\mathbf{i}} + 2v\hat{\mathbf{j}}, \quad \dfrac{\partial \mathbf{r}}{\partial v} = -2v\hat{\mathbf{i}} + 2u\hat{\mathbf{j}}, \quad \dfrac{\partial \mathbf{r}}{\partial w} = \hat{\mathbf{k}}$

$$\therefore \quad \hat{\mathbf{u}} = \frac{\partial \mathbf{r}}{\partial u}\bigg/\left|\frac{\partial \mathbf{r}}{\partial u}\right| = \frac{2u\hat{\mathbf{i}} + 2v\hat{\mathbf{j}}}{\sqrt{(4u^2 + 4v^2)}}, \qquad \hat{\mathbf{v}} = \frac{\partial \mathbf{r}}{\partial v}\bigg/\left|\frac{\partial \mathbf{r}}{\partial v}\right| = \frac{-2v\hat{\mathbf{i}} + 2u\hat{\mathbf{j}}}{\sqrt{(4v^2 + 4u^2)}},$$

$$\hat{\mathbf{w}} = \frac{\partial \mathbf{r}}{\partial w}\bigg/\left|\frac{\partial \mathbf{r}}{\partial w}\right| = \hat{\mathbf{k}}$$

$$\therefore \quad \hat{\mathbf{u}}\cdot\hat{\mathbf{v}} = 0, \quad \hat{\mathbf{u}}\cdot\hat{\mathbf{w}} = 0 \quad \text{and} \quad \hat{\mathbf{v}}\cdot\hat{\mathbf{w}} = 0$$

verifying that (u, v, w) form an orthogonal system of coordinates. The scale factors are $2\sqrt{(u^2 + v^2)}$, $2\sqrt{(u^2 + v^2)}$ and 1 respectively.

4 These results follow immediately from formulae (6.25), (6.36) and (6.44) and the scale factors h_u, h_v and h_z developed explicitly in Example 6.5.

5 Using (6.37),

$$\nabla\cdot\mathbf{A} = \frac{1}{\rho}\frac{\partial}{\partial\rho}(\rho\tfrac{10}{3}\rho) + 0 + 0 = \tfrac{20}{3}$$

Hence $\displaystyle\iiint_V \nabla\cdot\mathbf{A}\,\mathrm{d}V = \frac{20}{3}\iiint_V \mathrm{d}V = \tfrac{20}{3}\pi(4-1)10 = 200\pi$

The annulus is enclosed by four surfaces (see Figure 16).

For S_3 and S_4, $\hat{\mathbf{n}} = +\hat{\mathbf{k}}$ and $-\hat{\mathbf{k}}$ respectively, so $\mathbf{A}\cdot\hat{\mathbf{n}} = 0$ in both cases and these surfaces make no contribution to the surface integrals.

For S_1, $\hat{\mathbf{n}} = -\hat{\boldsymbol{\rho}}$ so $\mathbf{A}\cdot\hat{\mathbf{n}} = -\tfrac{10}{3}\rho$ and

$$\begin{aligned}\iint_{S_1} \mathbf{A}\cdot\hat{\mathbf{n}}\,\mathrm{d}S &= -\tfrac{10}{3}\iint_{S_1} \rho(\rho\,\mathrm{d}\phi\,\mathrm{d}z)\\ &= -\tfrac{10}{3}\iint \mathrm{d}\phi\,\mathrm{d}z = -\tfrac{10}{3}(\text{area of } S_1)\\ &= -\tfrac{10}{3}(2\pi\times 1)(10) = \tfrac{200}{3}\pi\end{aligned}$$

For S_2, $\hat{\mathbf{n}} = +\hat{\boldsymbol{\rho}}$ so $\mathbf{A}\cdot\hat{\mathbf{n}} = +\tfrac{10}{3}\rho$ and

$$\iint_{S_2} \mathbf{A}\cdot\hat{\mathbf{n}}\,\mathrm{d}S = +\tfrac{10}{3}\iint_{S_2} \rho(\rho\,\mathrm{d}\phi\,\mathrm{d}z) = \tfrac{40}{3}(\text{area of } S_2) = \tfrac{800}{3}\pi$$

Therefore

$$\oiint_S \mathbf{A}\cdot\hat{\mathbf{n}}\,\mathrm{d}S = \tfrac{800}{3}\pi - \tfrac{200}{3}\pi + 0 + 0 = 200\pi$$

and the divergence theorem is verified.

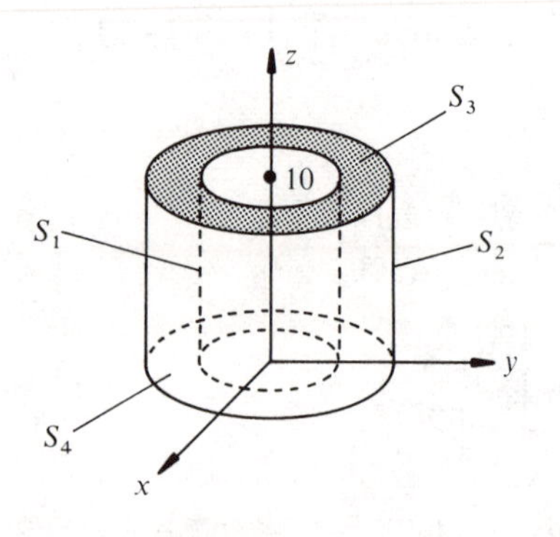

Figure 16

6 Using (6.46),

$$\nabla \wedge \mathbf{F} = \frac{1}{\rho}\begin{vmatrix} \hat{\boldsymbol{\rho}} & \rho\hat{\boldsymbol{\phi}} & \hat{\mathbf{k}} \\ \dfrac{\partial}{\partial \rho} & \dfrac{\partial}{\partial \phi} & \dfrac{\partial}{\partial z} \\ 0 & \rho F_\phi & 0 \end{vmatrix}$$

where $F_\phi = 2\rho^2(z+1)\sin^2\phi$. Therefore

$$\nabla \wedge \mathbf{F} = -\frac{1}{\rho}\frac{\partial}{\partial z}(\rho F_\phi)\hat{\boldsymbol{\rho}} + \frac{1}{\rho}\frac{\partial}{\partial \rho}(\rho F_\phi)\hat{\mathbf{k}}$$

For the surface of integration S, we choose $\hat{\mathbf{n}} = +\hat{\boldsymbol{\rho}}$, so

$$(\nabla \wedge \mathbf{F})\cdot\hat{\mathbf{n}} = -\frac{1}{\rho}\frac{\partial}{\partial z}(\rho F_\phi) = -2\rho^2\sin^2\phi = -8\sin^2\phi \quad \text{on } S$$

Therefore

$$\iint_S (\nabla \wedge \mathbf{F})\cdot\hat{\mathbf{n}}\,\mathrm{d}S = -8\int_1^{1.5}\mathrm{d}z\int_{\pi/4}^{\pi/2}\sin^2\phi\,2\,\mathrm{d}\phi = -8\int_{\pi/4}^{\pi/2}\sin^2\phi\,\mathrm{d}\phi$$

$$= -4(\tfrac{1}{4}\pi + \tfrac{1}{2}) = -5.14$$

We now consider the line integral $\oint_C \mathbf{F}\cdot\mathrm{d}\mathbf{l}$ around the four curves forming the boundary of S (see Figure 17). Since $\mathbf{F}$ is normal to the z-direction there is no contribution to the line integral along the sides parallel to the z-axis. Note the sense of travel around C to correspond with our choice of $\hat{\mathbf{n}} = +\hat{\boldsymbol{\rho}}$ for the surface integral.

For side C_1, $z = 1$, $\rho = 2$ and $\mathrm{d}\mathbf{l} = 2\,\mathrm{d}\phi\,\hat{\boldsymbol{\phi}}$, so

$$\int_{C_1}\mathbf{F}\cdot\mathrm{d}\mathbf{l} = 32\int_{\pi/4}^{\pi/2}\sin^2\phi\,\mathrm{d}\phi = 16(\tfrac{1}{4}\pi + \tfrac{1}{2})$$

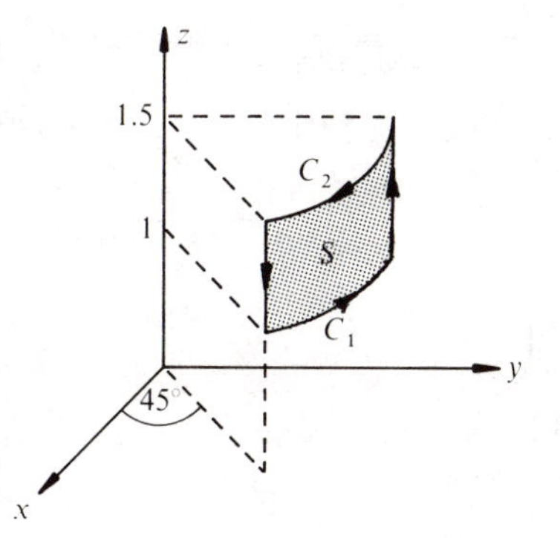

Figure 17

Similarly, for C_2, $z = 1.5$, $\rho = 2$ and $\mathbf{dl} = -2\,\mathrm{d}\phi\,\hat{\boldsymbol{\phi}}$, so

$$\int_{C_2} \mathbf{F}\cdot\mathbf{dl} = -20\int_{\pi/4}^{\pi/2} \sin^2\phi\,\mathrm{d}\phi = -20(\tfrac{1}{4}\pi + \tfrac{1}{2})$$

Adding all the contributions,

$$\oint_C \mathbf{F}\cdot\mathbf{dl} = -4(\tfrac{1}{4}\pi + \tfrac{1}{2})$$

and Stokes' theorem is verified.

Chapter 7

1 Using (6.38) for the divergence in spherical coordinates and realizing that $\mathbf{D}$ has only one non-zero component,

$$\nabla\cdot\mathbf{D} = 0 + \frac{1}{r\sin\theta}\frac{\partial}{\partial\theta}\left(\frac{0.1}{r}\sin\theta\cos\theta\right) + 0 = \frac{0.1}{r^2\sin\theta}\cos 2\theta$$

By the Maxwell equation (7.5a), this equals the density ρ_V of electric charge in the region. The total charge is then

$$Q = \iiint_V \rho_V\,\mathrm{d}V = 0.1\int_0^{2\pi}\mathrm{d}\phi\int_0^{\pi/4}\cos 2\theta\,\mathrm{d}\theta\int_2^5 \mathrm{d}r$$

where we have used the volume element $\mathrm{d}V = r^2\sin\theta\;\mathrm{d}r\,\mathrm{d}\theta\,\mathrm{d}\phi$. The volume integral is thus simply a product of three separate integrals and is readily evaluated to give 0.3π C.

2 (a) The required differential form is (7.7b), namely

$$\nabla\wedge\mathbf{H} = \mathbf{J} \quad \text{or} \quad \nabla\wedge\mathbf{B} = \mu_0\mathbf{J}$$

Using (6.46) for the curl in cylindrical coordinates,

$$\nabla\wedge\mathbf{B} = \frac{1}{\rho}\begin{vmatrix} \hat{\boldsymbol{\rho}} & \rho\hat{\boldsymbol{\phi}} & \hat{\mathbf{k}} \\ \dfrac{\partial}{\partial\rho} & \dfrac{\partial}{\partial\phi} & \dfrac{\partial}{\partial z} \\ 0 & \rho B_\phi & 0 \end{vmatrix}, \quad \text{where } B_\phi = 0.06\left(\frac{\rho}{2} - \frac{\rho^2}{3\times 10^{-2}}\right)$$

$$\text{Hence } \mathbf{J} = \frac{1}{\mu_0}\nabla\wedge\mathbf{B} = \frac{0.06}{\mu_0\rho}\frac{\partial}{\partial\rho}\left(\frac{\rho^2}{2} - \frac{\rho^3}{3\times 10^{-2}}\right)\hat{\mathbf{k}}$$

$$= \frac{0.06}{\mu_0}\left(1 - \frac{\rho}{10^{-2}}\right)\hat{\mathbf{k}}\ \ \mathrm{A\,m^{-2}}$$

that is, the current is flowing in the z-direction. The total current in the conductor is, using (7.1),

$$I = \iint_S \mathbf{J}\cdot\hat{\mathbf{n}}\,\mathrm{d}S$$

where S is a cross-section of the conductor. Because $\mathbf{J}$ depends only on ρ, this surface integral can be reduced to a single integral by putting

$$\hat{\mathbf{n}}\,dS = 2\pi\rho\,d\rho\,\hat{\mathbf{k}} \quad \text{(an annulus)}$$

Therefore

$$I = \frac{0.06}{\mu_0} 2\pi \int_0^{0.01} (1 - 100\rho)\rho\,d\rho = 5\text{ A}$$

(b) Rather more easily, we can use the integral form (7.14), namely

$$I = \oint_C \mathbf{H}\cdot d\mathbf{l} \quad \text{or} \quad I = \frac{1}{\mu_0}\oint_C \mathbf{B}\cdot d\mathbf{l}$$

where C is the circumference of the conductor:

$$I = \frac{1}{\mu_0}\int_C B_\phi \hat{\boldsymbol{\phi}}\cdot 0.01\,d\phi\,\hat{\boldsymbol{\phi}} = \frac{0.01}{\mu_0}\int_0^{2\pi} B_\phi\,d\phi = \frac{0.01}{\mu_0} B_\phi(0.01)2\pi$$

(because B_ϕ depends only on ρ and not on ϕ). Therefore

$$I = \frac{0.01}{2\times 10^{-7}}(0.06)\left(\frac{0.01}{2} - \frac{(0.01)^2}{3\times 10^{-2}}\right) = 5\text{ A}$$

3 Using (7.14) with $\mathbf{B} = \mu_0\mathbf{H}$,

$$\oint_C \mathbf{B}\cdot d\mathbf{l} = \mu_0 I$$

where, if C is a circle of radius ρ_1 inside the conductor, I will be the current enclosed by this circle. By symmetry,

$$\oint_C \mathbf{B}\cdot d\mathbf{l} = \oint_C |\mathbf{B}|\hat{\boldsymbol{\phi}}\cdot\rho_1\,d\phi\,\hat{\boldsymbol{\phi}} = |\mathbf{B}(\rho_1)|2\pi\rho_1$$

Also,
$$I = \iint_S \mathbf{J}(\rho)\cdot\hat{\mathbf{n}}\,dS = 2\pi\int_0^{\rho_1} J(\rho)\,\rho\,d\rho = 2\pi\frac{J_0}{a^n}\int_0^{\rho_1}\rho^{n+1}\,d\rho$$

$$= \frac{2\pi J_0}{(n+2)a^n}\rho_1^{n+2}$$

So
$$\mathbf{B} = \frac{\mu_0 J_0}{(n+2)a^n}\rho^{n+1}\hat{\boldsymbol{\phi}} \quad \text{at any radius } \rho$$

Hence,
$$\nabla\wedge\mathbf{B} = \frac{1}{\rho}\begin{vmatrix} \hat{\boldsymbol{\rho}} & \rho\hat{\boldsymbol{\phi}} & \hat{\mathbf{k}} \\ \dfrac{\partial}{\partial\rho} & \dfrac{\partial}{\partial\phi} & \dfrac{\partial}{\partial z} \\ 0 & K\rho^{n+2} & 0 \end{vmatrix} \quad \text{where } K = \frac{\mu_0 J_0}{(n+2)a^n}$$

Evaluating the determinant gives

$$\nabla\wedge\mathbf{B} = \mu_0 J_0\left(\frac{\rho}{a}\right)^n = \mu_0\mathbf{J}$$

as required.

Given that $\mathbf{A} = A_z(\rho)\hat{\mathbf{k}}$ where $A_z(\rho)$ is to be determined,

$$\nabla \wedge \mathbf{A} = \frac{1}{\rho}\begin{vmatrix} \hat{\boldsymbol{\rho}} & \rho\hat{\boldsymbol{\phi}} & \hat{\mathbf{k}} \\ \dfrac{\partial}{\partial \rho} & \dfrac{\partial}{\partial \phi} & \dfrac{\partial}{\partial z} \\ 0 & 0 & A_z(\rho) \end{vmatrix} = -\frac{\partial A_z}{\partial \rho}\hat{\boldsymbol{\phi}}$$

Hence, using $\mathbf{B} = \nabla \wedge \mathbf{A}$,

$$-\frac{\partial A_z}{\partial \rho} = K\rho^{n+1} \quad \text{so} \quad A_z = -\frac{K\rho^{n+2}}{(n+2)} + \text{a constant}$$

Hence $\nabla^2\mathbf{A} \equiv \nabla^2 A_z\hat{\mathbf{k}} = -\dfrac{K}{(n+2)}\dfrac{1}{\rho}\dfrac{\partial}{\partial \rho}(\rho(n+2)\rho^{n+1})\hat{\mathbf{k}}$ (using (6.40b))

which gives

$$-\mu_0 J_0\left(\frac{\rho}{a}\right)^n\hat{\mathbf{k}} = -\mu_0\mathbf{J}$$

as required.

4 Integrating $\partial\mathbf{D}/\partial t = 2\times10^{-6}\cos(\omega t - 5z)\hat{\mathbf{i}}$ and ignoring constants gives

$$\mathbf{D} = \frac{2\times10^{-6}}{\omega}\sin(\omega t - 5z)\,\hat{\mathbf{i}} \quad \mathrm{C\,m^{-2}}$$

Hence $\mathbf{E} = \dfrac{\mathbf{D}}{4\varepsilon_0} = \dfrac{10^{-6}}{2\omega\varepsilon_0}\sin(\omega t - 5z)\,\hat{\mathbf{i}} \quad \mathrm{V\,m^{-1}}$

B is calculated using

$$\frac{\partial\mathbf{B}}{\partial t} = -\nabla\wedge\mathbf{E} = \frac{5\times10^{-6}}{2\omega\varepsilon_0}\cos(\omega t - 5z)\,\hat{\mathbf{j}}$$

Hence $\mathbf{B} = \dfrac{5\times10^{-6}}{2\omega^2\varepsilon_0}\sin(\omega t - 5z)\,\hat{\mathbf{j}} \quad \mathrm{Wb\,m^{-2}}$

and $\mathbf{H} = \dfrac{\mathbf{B}}{5\mu_0} = \dfrac{10^{-6}}{2\omega^2\varepsilon_0\mu_0}\sin(\omega t - 5z)\,\hat{\mathbf{j}} \quad \mathrm{A\,m^{-1}}$

We recalculate $\partial\mathbf{D}/\partial t$ using (7.6b) (with $\mathbf{J} = 0$ for a perfect dielectric):

$$\frac{\partial\mathbf{D}}{\partial t} = \nabla\wedge\mathbf{H} = -\frac{\partial H_y}{\partial z}\hat{\mathbf{i}}$$

$$\therefore \quad \frac{\partial\mathbf{D}}{\partial t} = \frac{5\times10^{-6}}{2\omega^2\varepsilon_0\mu_0}\cos(\omega t - 5z)\,\hat{\mathbf{i}}$$

Comparing this with the given expression for $\partial\mathbf{D}/\partial t$,

$$\frac{5\times10^{-6}}{2\omega^2\varepsilon_0\mu_0} = 2\times10^{-6} \quad \text{so} \quad \omega^2 = \frac{5}{4\mu_0\varepsilon_0}$$

Numerically, it can be shown that $1/\sqrt{(\mu_0\varepsilon_0)} = 3\times10^8$, which gives $\omega = 3.35\times10^8\ \mathrm{rad\,s^{-1}}$.

Chapter 8

1 The equation of motion of the particle is

$$m\frac{d^2\mathbf{r}}{dt^2} + 2m\boldsymbol{\omega} \wedge \frac{d\mathbf{r}}{dr} = \mathbf{R} - m\mathbf{g}$$

where, because the table is smooth, the reaction force **R** is vertical – as is the force due to gravity. Because the motion of the particle is restricted to lie in a horizontal plane, both these forces are perpendicular to the velocity of the particle. Hence

$$\frac{d\mathbf{r}}{dt} \wedge \frac{d^2\mathbf{r}}{dt^2} + 2\frac{d\mathbf{r}}{dt} \wedge \left(\boldsymbol{\omega} \wedge \frac{d\mathbf{r}}{dt}\right) = 0$$

Using (1.61) for the triple vector product and also noting that $v^2 = (d\mathbf{r}/dt)\cdot(d\mathbf{r}/dt)$, we may write the second term on the left-hand side as

$$2\frac{d\mathbf{r}}{dt} \wedge \left(\boldsymbol{\omega} \wedge \frac{d\mathbf{r}}{dt}\right) = 2\boldsymbol{\omega}v^2 - 2\left(\boldsymbol{\omega}\cdot\frac{d\mathbf{r}}{dt}\right)\frac{d\mathbf{r}}{dt}$$

The square of the modulus of this vector is

$$\begin{aligned}4\left[\omega^2v^4 - 2\left(\boldsymbol{\omega}\cdot\frac{d\mathbf{r}}{dt}\right)^2 v^2 + \left(\boldsymbol{\omega}\cdot\frac{d\mathbf{r}}{dt}\right)^2 v^2\right] &= 4\left[\omega^2v^4 - \left(\boldsymbol{\omega}\cdot\frac{d\mathbf{r}}{dt}\right)^2 v^2\right] \\ &= 4[\omega^2v^4 - \omega^2v^4\sin^2\theta] \\ &= 4\omega^2v^4\cos^2\theta\end{aligned}$$

since the angle between the rotation vector and the velocity vector is $(\pi/2 - \theta)$.

From Section 2.7, we know that the particle needs a force of magnitude $m\kappa v^2$ if it is to continue to move along an arc of curvature κ. Now

$$mv^2\kappa = mv^2\frac{\left|\frac{d\mathbf{r}}{dt} \wedge \frac{d^2\mathbf{r}}{dt^2}\right|}{\left|\frac{d\mathbf{r}}{dt}\right|^3} \quad \text{(from (2.49))}$$

$$= mv^2\frac{\left|2\frac{d\mathbf{r}}{dt} \wedge \left(\boldsymbol{\omega} \wedge \frac{d\mathbf{r}}{dt}\right)\right|}{\left|\frac{d\mathbf{r}}{dt}\right|^3} = mv^2\left(\frac{2\omega v^2\cos\theta}{v^3}\right) = 2m\omega v\cos\theta$$

that is,

$$\kappa = \frac{2\omega\cos\theta}{v} = \text{constant}$$

Hence the motion of the particle is circular with radius of curvature $1/\kappa = v/(2\omega\cos\theta)$.

2 Let $\boldsymbol{\Omega}$ be the angular velocity of the rolling sphere. Consider a coordinate system fixed at the common centre of A and B (see Figure 18).

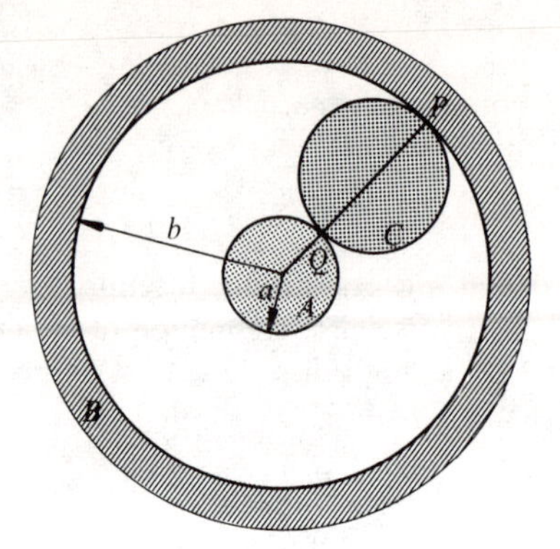

Figure 18

The velocity of Q (due to the motion of A) is, using (8.40a),

$$\mathbf{v}_Q = 0 + \boldsymbol{\omega}_A \wedge a\hat{\mathbf{r}}$$

whilst from C,

$$\mathbf{v}_Q = \mathbf{v}_C + \boldsymbol{\Omega} \wedge [-\tfrac{1}{2}(b-a)\hat{\mathbf{r}}]$$

where $\mathbf{v}_C$ is the velocity of the centre of C. If the motion is to be rolling, then

$$\mathbf{v}_C + \boldsymbol{\Omega} \wedge [-\tfrac{1}{2}(b-a)\hat{\mathbf{r}}] = \boldsymbol{\omega}_A \wedge a\hat{\mathbf{r}} \tag{4}$$

The velocity of P (due to the motion of B) is

$$\mathbf{v}_P = 0 + \boldsymbol{\omega}_B \wedge b\hat{\mathbf{r}}$$

whilst from C,

$$\mathbf{v}_P = \mathbf{v}_C + \boldsymbol{\Omega} \wedge [\tfrac{1}{2}(b-a)\hat{\mathbf{r}}]$$

Therefore, for rolling motion at P we require

$$\mathbf{v}_C + \boldsymbol{\Omega} \wedge [\tfrac{1}{2}(b-a)\hat{\mathbf{r}}] = \boldsymbol{\omega}_B \wedge b\hat{\mathbf{r}} \tag{5}$$

Adding (4) and (5),

$$2\mathbf{v}_C = \boldsymbol{\omega}_A \wedge a\hat{\mathbf{r}} + \boldsymbol{\omega}_B \wedge b\hat{\mathbf{r}}$$

which may be rewritten as

$$\mathbf{v}_C = \left(\frac{a\boldsymbol{\omega}_A + b\boldsymbol{\omega}_B}{a+b}\right) \wedge \tfrac{1}{2}(a+b)\hat{\mathbf{r}}$$

Since the position vector of the centre of C is $\tfrac{1}{2}(a+b)\hat{\mathbf{r}}$, this last equation implies that the motion of C may be regarded as resulting from an angular velocity

$$\boldsymbol{\Omega} = \frac{a\boldsymbol{\omega}_A + b\boldsymbol{\omega}_B}{a+b}$$

Equations (4) and (5) contain a further piece of information obtained by subtraction:

$$\mathbf{\Omega} \wedge (b-a)\hat{\mathbf{r}} = \boldsymbol{\omega}_B \wedge b\hat{\mathbf{r}} - \boldsymbol{\omega}_A \wedge a\hat{\mathbf{r}}$$

that is, $[\mathbf{\Omega}(b-a) - \boldsymbol{\omega}_B b + \boldsymbol{\omega}_A a] \wedge \hat{\mathbf{r}} = 0$

After substituting our expression for $\mathbf{\Omega}$, this becomes

$$(\boldsymbol{\omega}_A - \boldsymbol{\omega}_B) \wedge \hat{\mathbf{r}} = 0$$

But since $\boldsymbol{\omega}_A$ and $\boldsymbol{\omega}_B$ are fixed vectors and $\hat{\mathbf{r}}$ is variable in direction (depending on the position of C), this will only be satisfied for all $\hat{\mathbf{r}}$ if

$$\boldsymbol{\omega}_A = \boldsymbol{\omega}_B$$

and hence

$$\mathbf{v}_C = \boldsymbol{\omega}_A \wedge \tfrac{1}{2}(a+b)\hat{\mathbf{r}}$$

We conclude that rolling motion is only possible in the special case when the outer and inner spheres have the same angular velocities, so that all three spheres rotate as a rigid body about the common centre of A and B.

3 Now $\mathbf{v}_A = v_A\hat{\mathbf{j}}$, $\mathbf{v}_B = -V\hat{\mathbf{k}}$ and $\mathbf{v}_C = \alpha\hat{\mathbf{i}} + \beta\hat{\mathbf{j}}$, since C does not move off the xy-plane. If $\boldsymbol{\omega}$ is the angular velocity of the frame, then

$$\mathbf{v}_C = \mathbf{v}_B + \boldsymbol{\omega} \wedge \overrightarrow{BC} \qquad \textbf{(6)}$$

$$\mathbf{v}_A = \mathbf{v}_B + \boldsymbol{\omega} \wedge \overrightarrow{BA} \qquad \textbf{(7)}$$

or $$v_A\hat{\mathbf{j}} = -V\hat{\mathbf{k}} + \boldsymbol{\omega} \wedge (\hat{\mathbf{j}} - \hat{\mathbf{k}})\frac{l}{\sqrt{2}}$$

Taking the scalar product of both sides with $(\hat{\mathbf{j}} - \hat{\mathbf{k}})$, thus eliminating the second term on the right-hand side, we find

$$v_A = V$$

Hence (7) becomes

$$V\hat{\mathbf{j}} = -V\hat{\mathbf{k}} + \boldsymbol{\omega} \wedge (\hat{\mathbf{j}} - \hat{\mathbf{k}})\frac{l}{\sqrt{2}}$$

If we write $\boldsymbol{\omega} = a\hat{\mathbf{i}} + b\hat{\mathbf{j}} + c\hat{\mathbf{k}}$, where a, b and c are to be determined, then

$$V(\hat{\mathbf{j}} + \hat{\mathbf{k}}) = \frac{l}{\sqrt{2}}\begin{vmatrix} \hat{\mathbf{i}} & \hat{\mathbf{j}} & \hat{\mathbf{k}} \\ a & b & c \\ 0 & 1 & -1 \end{vmatrix} = \frac{l}{\sqrt{2}}[\hat{\mathbf{i}}(-b-c) - \hat{\mathbf{j}}(-a) + \hat{\mathbf{k}}(a)]$$

leading to

$$b = -c \quad \text{and} \quad a = \frac{V\sqrt{2}}{l}$$

Also, from (6),

$$\mathbf{v}_C = -V\hat{\mathbf{k}} + \boldsymbol{\omega} \wedge (\hat{\mathbf{i}} - \hat{\mathbf{k}})\frac{l}{\sqrt{2}}$$

$$\text{that is, } \alpha\hat{\mathbf{i}} + \beta\hat{\mathbf{j}} + V\hat{\mathbf{k}} = \frac{l}{\sqrt{2}}\begin{vmatrix} \hat{\mathbf{i}} & \hat{\mathbf{j}} & \hat{\mathbf{k}} \\ \dfrac{V\sqrt{2}}{l} & b & -b \\ 1 & 0 & -1 \end{vmatrix}$$

$$= \frac{l}{\sqrt{2}}\left[\hat{\mathbf{i}}(-b) - \hat{\mathbf{j}}\left(-\frac{V\sqrt{2}}{l} + b\right) + \hat{\mathbf{k}}(-b)\right]$$

Equating coefficients and then solving for α, β and b, we find

$$\alpha = V, \qquad \beta = 2V, \qquad b = -\frac{V\sqrt{2}}{l}$$

leading to

$$\mathbf{v}_C = V(\hat{\mathbf{i}} + 2\hat{\mathbf{j}}) \quad \text{and} \quad \boldsymbol{\omega} = \frac{V\sqrt{2}}{l}(\hat{\mathbf{i}} - \hat{\mathbf{j}} + \hat{\mathbf{k}})$$

4 Let the indexed versions of $\mathbf{r}$ and $\hat{\mathbf{n}}$ be x_i and n_i respectively and the indexed version of $\mathbf{r}'$ be x_i'. Now since

$$(\hat{\mathbf{n}} \wedge \mathbf{r})_i = \varepsilon_{ijk}n_j x_k \quad \text{and} \quad \mathbf{r}\cdot\hat{\mathbf{n}} = x_k n_k$$

the ith component of Rodrigue's formula is

$$x_i' = x_i \cos\theta + (x_k n_k)n_i(1 - \cos\theta) + \varepsilon_{ijk}n_j x_k \sin\theta$$

For small rotations,

$$\cos\theta \approx 1 - \tfrac{1}{2}\theta^2 \quad \text{and} \quad \sin\theta \approx \theta$$

$$\therefore \qquad x_i' \approx x_i(1 - \tfrac{1}{2}\theta^2) + (x_k n_k)n_i(\tfrac{1}{2}\theta^2) + \varepsilon_{ijk}n_j x_k\theta$$

If terms up to first order in θ are kept, this becomes

$$x_i' \approx x_i + \varepsilon_{ijk}n_j x_k\theta$$

or, in terms of standard vector notation,

$$\mathbf{r}' \approx \mathbf{r} + \boldsymbol{\omega} \wedge \mathbf{r}, \quad \text{where } \boldsymbol{\omega} = \hat{\mathbf{n}}\theta$$

If we perform a second 'small' rotation by an angle ϕ about an axis $\hat{\mathbf{m}}$ (or m_j), then x_i' goes to x_i'', where

$$x_i'' \approx x_i' + \varepsilon_{ijk}m_j x_k'\phi \approx x_i + \varepsilon_{ijk}n_j x_k\theta + \varepsilon_{ijk}m_j x_k\phi$$

where, once again, we have only retained terms of first order in θ and ϕ. This last equation can be expressed in standard vector form (with $\boldsymbol{\omega}' = \hat{\mathbf{m}}\phi$) as

$$\mathbf{r}'' \approx \mathbf{r} + \boldsymbol{\omega} \wedge \mathbf{r} + \boldsymbol{\omega}' \wedge \mathbf{r} = \mathbf{r} + (\boldsymbol{\omega} + \boldsymbol{\omega}') \wedge \mathbf{r}$$

This implies that the result of two successive rotations is fully defined by the single vector $\boldsymbol{\omega} + \boldsymbol{\omega}'$. Note also that order is not important, since $\boldsymbol{\omega} + \boldsymbol{\omega}' = \boldsymbol{\omega}' + \boldsymbol{\omega}$.

Index